Green Materials and Manufacturing Processes

Green Materials and Manufacturing Processes

Editors

Stefano Guarino
Flaviana Tagliaferri

Basel • Beijing • Wuhan • Barcelona • Belgrade • Novi Sad • Cluj • Manchester

Editors

Stefano Guarino
Faculty of Engineering
University Niccolò Cusano
Rome
Italy

Flaviana Tagliaferri
Faculty Engineering Sciences
Hochschule Mittweida -
University of Applied Sciences
Mittweida
Germany

Editorial Office
MDPI
St. Alban-Anlage 66
4052 Basel, Switzerland

This is a reprint of articles from the Special Issue published online in the open access journal *Materials* (ISSN 1996-1944) (available at: www.mdpi.com/journal/materials/special_issues/GMMP).

For citation purposes, cite each article independently as indicated on the article page online and as indicated below:

Lastname, A.A.; Lastname, B.B. Article Title. *Journal Name* **Year**, *Volume Number*, Page Range.

ISBN 978-3-0365-9051-6 (Hbk)
ISBN 978-3-0365-9050-9 (PDF)
doi.org/10.3390/books978-3-0365-9050-9

Contents

About the Editors

Stefano Guarino

Prof. Stefano Guarino is a full professor at the University Niccolò Cusano (Rome, Italy). Graduated in mechanical engineering, in 2004 he achieved his doctorate in Engineering of Energy–Environment at the Department of Mechanical Engineering of the University of Rome, 'Tor Vergata' (Italy). He is currently vice-rector for internationalization, and he holds the courses titled "Manufacturing Processes" and "Technology and Systems in Industry 4.0". The research activities are focused on the field of manufacturing processes and materials. Since 2005, he has been the coordinator of numerous national and international research projects.

Flaviana Tagliaferri

Dr. Flaviana Tagliaferri is a researcher at the Mittweida University of Applied Sciences (Germany), under the chair for System Electronics. She holds a course titled "Industry 4.0: Responsible Consumption and Production" and is involved in several national and European research projects.

She is also a lecturer at Niccolò Cusano University, Rome (Italy), teaching "Manufacturing Materials and Technologies for Food Industry". She graduated in 2008 with the best score in Mechanical Engineering at 'Tor Vergata' University, Rome (Italy) and achieved her doctorate in 2012 at the Department of Aerospace Engineering of the University of Naples, 'Federico II' (Italy).

Preface

The green approach is no longer (and can no longer be) a goal of the future, but it is a concrete and practical necessity of the present. Recently, we have finally seen an acceleration in the development and application of green materials and manufacturing processes, and the importance of environmentally responsible materials or sustainable manufacturing processes has never been higher. This Special Issue aims to line up some recently developed green materials and remarkable progress and developments in manufacturing processes to take stock of new trends.

Stefano Guarino and Flaviana Tagliaferri
Editors

Article

Environmental and Economic Assessment of Repairable Carbon-Fiber-Reinforced Polymers in Circular Economy Perspective

Elisabetta Abbate *, **Maryam Mirpourian**, **Carlo Brondi**, **Andrea Ballarino** and **Giacomo Copani**

STIIMA-CNR—Institute of Intelligent Industrial Technologies and Systems for Advanced Manufacturing, National Research Council, Via Alfonso Corti 12, 20133 Milan, Italy; maryam.mirpourian@stiima.cnr.it (M.M.); carlo.brondi@stiima.cnr.it (C.B.); andrea.ballarino@stiima.cnr.it (A.B.); giacomo.copani@stiima.cnr.it (G.C.)
* Correspondence: elisabetta.abbate@stiima.cnr.it

Abstract: The explosive growth of the global market for Carbon-Fiber-Reinforced Polymers (CFRP) and the lack of a closing loop strategy of composite waste have raised environmental concerns. Circular economy studies, including Life Cycle Assessment (LCA) and Life Cycle Costing (LCC), have investigated composite recycling and new bio-based materials to substitute both carbon fibers and matrices. However, few studies have addressed composite repair. Studies focused on bio-based composites coupled with recycling and repairing are also lacking. Within this framework, the paper aims at presenting opportunities and challenges of the new thermosetting composite developed at the laboratory including the criteria of repairing, recycling, and use of bio-based materials in industrial applications through an ex ante LCA coupled with LCC. Implementing the three criteria mentioned above would reduce the environmental impact from 50% to 86% compared to the baseline scenario with the highest benefits obtained by implementing the only repairing. LCC results indicate that manufacturing and repairing parts built from bio-based CFRP is economically sustainable. However, recycling can only be economically sustainable under a specific condition. Managerial strategies are proposed to mitigate the uncertainties of the recycling business. The findings of this study can provide valuable guidance on supporting decisions for companies making strategic plans.

Keywords: carbon fiber; composites; automotive; ex ante LCA; economic assessment; LCC; recycling; repair; circular economy; sustainable manufacturing

Citation: Abbate, E.; Mirpourian, M.; Brondi, C.; Ballarino, A.; Copani, G. Environmental and Economic Assessment of Repairable Carbon-Fiber-Reinforced Polymers in Circular Economy Perspective. *Materials* **2022**, *15*, 2986. https://doi.org/10.3390/ma15092986

Academic Editor: Alain Celzard

Received: 16 February 2022
Accepted: 14 April 2022
Published: 20 April 2022

Publisher's Note: MDPI stays neutral with regard to jurisdictional claims in published maps and institutional affiliations.

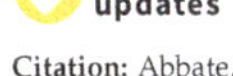

1. Introduction

1.1. General Framework

Thermosetting Carbon-Fiber-Reinforced Polymers (CFRPs) are engineered composites made out of carbon fibers (CFs) as the reinforcement and epoxy resin as the polymer matrix [1], which acts as load transfer elements across fibers [2]. Among different types of resin (epoxy, phenolic, polyester, urethane and vinyl ester) [2], epoxy is the chosen resin when mechanical and resistance performance is required [3]. Indeed, CFRP composites are characterized by outstanding mechanical properties such as high stiffness, long life span, non-corrosive and high fatigue resistance [1,2,4]. Lightweight is an additional property of CFRP that can reduce, for instance, the overall weight of a vehicle up to 10% compared to steel and aluminum [5] with a reduction in fuel consumption or an increase in batteries duration for electric vehicles [6], resulting in a lower environmental impact during the use phase [7]. Indeed, CFRP composites are increasingly replacing other materials such as steel [8] and aluminum [9] in a wide range of sectors such as sports equipment, wind energy, aircraft, construction and automotive [4,10,11]. In recent decades, the global demand for CF increased from 16 kt to 72 kt [12], reaching 100 kt in 2019 [6]. Furthermore, the global demand for CF and CFRP is projected to reach, respectively, 117 kt and 194 kt in 2022, which portrays an annual compound growth rate, respectively, of 11.50% and 11.98% [13,14].

Although the undeniable potentialities of CFRPs, environmental and economic concerns on the spreading of CFRPs have been raised by many studies, including Life Cycle Assessment (LCA) and Life Cycle Costing (LCC) studies. LCA and LCC are the main adopted tools, respectively, for the environmental and cost burdens assessment in the entire life cycle of a product or a service [15,16]. Both tools allow not only the calculation of the environmental and cost impacts but also the identification of the life cycle stages with the most relevant impact and the potential drivers to reduce the impact [17]. Based on those studies, current challenges on CFRP composites are mainly associated with its waste production and management [8,18]. Firstly, albeit mechanical, thermal and chemical recycling are potential technologies for CF recovery [10] currently in the market [19], landfill and incineration are still the main routes of CFRP waste management. However, they are both in contrast with the Circular Economy concept [14,20,21]. Secondly, recycling of thermosetting CFRP is difficult because of its intrinsic properties, which do not allow it to be remolded and reshaped once it is cured [3,4]. Moreover, even when CFs are recovered, the recycled CFs could have a lower quality than virgin CFs [22]. Thirdly, 40% of CFRP scraps are generated during the manufacturing of the product [14]. Finally, several types of damages may occur both during product manufacturing, for instance, porosity or undesired bodies in the matrix, and during the use phase, for instance, delamination, matrix crack and fiber–matrix debonding [23]. However, in the case of a damaged CFRP product, repairing techniques are limited. Hence, its entire substitution is the most widespread approach, resulting in a further increase in costs and resources [24]. Weak CFRP waste management combined with the surging demand for CFRP has been leading to a tremendous amount of waste. It is estimated that the global CFRP waste will reach 20 kt per year by 2025. Furthermore, taking into consideration that the monetary value of CF is around 26.5 euro/kg [25], from an economic and environmental perspective, a considerable amount of composite waste accumulated every year results in a loss of valuable and energy intensive materials [12,14].

Among possible options for reducing the environmental and cost impact of CFRP composites, repairing, recycling and using alternative bio-based materials have been identified on legislative, industrial and research levels. Firstly, at the legislative level, Council Directive on End-of-Life (EoL) Vehicles defined the target on vehicles to reuse and recycle up to 85% by 2015, which is currently under review by the European commission [26]. At the industrial level, repairing technologies could extend the product lifetime, minimizing resource depletion [27] and costs [28]. If repair is not possible, a potential reduction in the environmental impact still occurs through used CF recovery [14]. Finally, at the research level, extensive R&D activities have been conducted in the material field to develop a new type of composite that can be recovered and reused [29] or to substitute CFs and traditional matrices with bio-based materials [7,30].

Within this framework, the environmental and economic assessment of composites in a circular economy perspective is urgent for the growth of CFRP, guiding potential future development and strategies to minimize its environmental burden. However, studies that focus simultaneously on the implementation of recycling, repairing and the use of bio-based materials are still missing. Section 1.2 provides details about the goal of the study. Sections 1.3 and 1.4 provide the current status, issues and challenges, respectively, of LCAs and LCCs for the implementation of the above-mentioned actions on CFRP composites.

1.2. Goal of the Study

The goal of this paper is to perform an ex ante LCA coupled with LCC of new types of composites developed in a laboratory by Politecnico di Milano that can be repaired and recycled. In particular, this paper aims to:

1. Highlight the importance of simultaneously evaluating repairing, recycling and using bio-based materials in CFRP sectors from a Circular Economy perspective to define the main drivers both on the environment and the economy;
2. Couple LCA with LCC to obtain a broad assessment of the composites;

3. Emphasize the importance of providing a preliminary environmental and economical assessment of emerging technologies;
4. Outline the importance of needed efforts on repairing composite products to reduce their environmental impact, resource depletion and cost.

Section 2 describes the methodology adopted to perform both the LCA and LCC, including the scenarios developed, the data used and the assumptions made. Section 3 shows the results obtained regarding the environmental and economic aspects with a discussion of the main outcomes and limitations.

1.3. Life Cycle Assessment on Composites

Nowadays, the LCA tool is adopted by companies in order to estimate the current environmental footprint among several environmental categories and as a decision support tool to reduce their environmental impact [31]. As a matter of fact, the International Organization of Standardization (ISO) developed standards on LCA in order to consolidate the methodology [32] and provide guidelines to LCA practitioners and companies willing to apply this technique [33]. The tool estimates the current environmental impact of a product through its entire life cycle, from the raw material extraction to its End-of-Life (EoL), analyzing the entire supply chain. Moreover, it provides possible indications on reducing the environmental impact and a range of results according to the developed scenarios. LCA is divided into four iterative steps, which are *Goal and scope definition*, *Life Cycle Inventory (LCI)*, *Life Cycle Impact Assessment (LCIA)* and *Interpretation* [33]. Regarding the LCI step, primary data are directly provided by companies; secondary data are available from datasets and literature [34].

Although the LCA tool is mainly applied to existing technologies [31,35,36], a preliminary environmental assessment of emerging technology further enhances the knowledge of this technology from an environmental point of view [37], improving its design and further developing the product on an industrial scale [38]. As a consequence, this could attempt at obtaining a product in the market with as little environmental impact as possible [35]. This is the case of the ex ante LCA. It simulates the environmental impact of an emerging technology on an industrial level, for instance, a technology developed at laboratory, in order to (1) estimate its environmental impact among possible scenarios in case it will be developed on an industrial scale [31,39] and (2) compare it with existing technologies that provide the same or similar service [40]. Difficulties and challenges have emerged in developing an ex ante LCA mainly regarding the data collection [41] and scenario selection due to the absence of data, differences between industrial and laboratory scale and uncertainties on future scenarios of development [42] and large-scale technology diffusion on the market [43]. Indeed, the ex ante methodology is applied in order to solve the following two main issues related to primary data from a laboratory: (1) scale-up from consumption to industrial scale [42,44] and (2) unavailable primary data [42,45]. On one hand, technologies used as well as consumption of energy and materials on a laboratory scale are different from the ones at the industrial level. As a matter of fact, the consumption of raw materials and energy can be much higher on a laboratory scale than the one on the industrial scale [42,46]. On the other hand, data on energy and auxiliary materials consumption are usually not available at the laboratory scale [42], and hence, calculations are needed in order to compute the LCI. Efforts on establishing a methodology through which an ex ante LCA can be performed has been made by different authors [40]. Cucurachi et al. [31] identified five types of ex ante LCAs in order to uniform the present literature on this topic, provided a definition of 'emerging technology' and outlined the main challenges of performing ex ante LCAs. Moni et al. [42] focused on the challenges to perform LCA of emerging technologies, identifying possible pathways and recommendations in order to overcome those limits. Piccino et al. [46] provided an operational methodology of an ex ante LCA through which data on a laboratory scale can be transformed to industrial-scale data. However, few LCAs directly applied ex ante methodology in real case studies.

Concerning the application of LCA to composites, many studies can be found in the literature. LCAs on glass fiber polymers [47,48] and carbon fiber polymers [49–51] as well as thermoplastic [52,53] and thermosetting composites [5,52] among different applications, for instance, automotive [8,49,54], aviation [7,55] and construction [48,56,57] sectors. More recent comparative LCAs on CFRP waste management included mechanical, chemical and thermal recycling as alternatives to landfills and incineration [58]. Meng et al. [12] analyzed recycling technologies of CF, concluding that all recycling routes could achieve environmental benefits in terms of Global Warming Potential (GWP) and Primary Energy Demand.

As concerns composite repair, the use of composite patches is one the most suitable methods to repair damages to composite products [59]. In particular, Mohammadi et al. [59] discussed the mechanical and structural advantages of the use of composite patches. However, LCA studies on this topic are still missing. Tapper et al. [60] mentioned maintenance and repair of vehicles which were, however, excluded in the study. Raugei et al. [49] analyzed a thermoplastic composite outlining its properties that enable repairing and recycling. The LCA also analyzed the maintenance stage of a vehicle, which included a percentage of substitution due to damages in its entire lifetime. However, the repairing was not included in the LCA analysis. As concerns the ecodesign efforts for material circularity, efforts have been made both on substituting the CF and traditional matrices with bio-based materials. Hermansson et al. [30] outlined the environmental benefit of using lignin as feedstock to produce CF, collecting several LCA studies compared to the traditional CF. Cotton, flax, hemp, jute and bamboo are some of the bio-based fibers that can substitute the traditional PAN used as a precursor to the CF [7,48]. Moreover, *Mischantus* fiber was analyzed by Roy et al. [61] as potential promising feedstock in substitution of the traditional CF. However, incineration with energy recovery of this type of material was mentioned as a good alternative to chemical and mechanical recycling, which are instead less feasible and efficient [7]. As concerns bio-based resin, the use of wood pulp or biofuels could contribute to reducing the environmental impacts of composites [7]. At present, LCA studies on composites that simultaneously focused on repairing, recycling and using bio-based materials are missing. Furthermore, ex ante LCAs that preliminarily evaluate the potential of new composites are lacking. Within this framework, this LCA attempts to provide (i) preliminary potentialities of the new composites for further development on an industrial scale and (ii) an environmental assessment of the implementation of recycling, repairing and using bio-based materials from a circular economy perspective. The preliminary literature review on composites performed in this study is reported in Table S1 of Section S1 of the Supplementary Materials.

1.4. Economic Sustainability of Composites EoL

The economic assessment of a new project or a technology under different system configurations is of paramount importance to attain competitive edge since it helps business management to compare the profitability of the new business under a wide range of scenarios. To increase the competitive advantage of a product, the Life cycle cost (LCC) analysis is a proper tool that is used to compare the profitability of the new business under a broad range of hypotheses [16,62,63]. The U.S. military first introduced the LCC concept in the 1960s [64]. The idea was then used by different industries, such as energy, transportation, healthcare, etc., to assess a project's economic sustainability and helped investors compare the cost-effectiveness of alternative business decisions [16]. LCC analysis starts with the cost breakdown structure (CBS) and revenue breakdown structure (RBS) of a product or a new project. This approach measures and calculates the costs and revenues of the product at its different stages of life. Then costs and revenues are discounted using the discount rate. The discount rate represents the rate of inflation at the time when the financial model is performed [64]. From a higher-level standpoint, the LCC procedure aggregates all the discounted cash flows incurred over a project's entire life span. These discounted cash flows can be used to identify the profitability of the new project based on financial indicators such as net present value (NPV) and discounted payback period [65].

Although economic assessment and LCC are widely applied in different sectors, a very limited number of studies have investigated the economic sustainability of composites EoL solutions in the literature. Many studies on composites have evaluated the economic sustainability of products during the manufacturing and use phase. For example, Witik et al. [58] investigated the financial sustainability of CF composite products using diverse manufacturing processes (autoclave vs. out-of-autoclave). Delogu et al. [25] carried out a comparative analysis to evaluate the economic performance of CFRP during the manufacturing process compared to the use phase. Strogonov [66] analyzed the cost reduction of products made out of composites compared to steel products during the manufacturing process. However, very few studies have focused on evaluating the economic performance for the EoL of composite waste. In this regard, for example, Castella et al. [67] assessed the economic sustainability of a composite product in the automotive industry during different value chain stages: raw material fabrication, manufacturing, use phase and the end of life. It was found that using hybrid composites instead of steel can lower the lifecycle cost by 16%. La Rosa et al. [62] also focused on the economic performance of carbon fiber (CF)-thermoset composite during manufacturing and solvolysis recycling. The finding of this study proved that the chemical recycling is not economically sustainable. However, research on this topic is still lacking. In a broader sense, it can be argued that most studies on composite EoL were focused on the technical performance of recycling technologies [1,4,10,68]. Among these studies concentrating on the closing loop techniques, a few have briefly evaluated the economic impacts of recycling technologies. For example, Karuppannan Gopalraj and Kärki [10] carried out an extensive literature review on the technical aspects of the mechanical recycling, thermal recycling and chemical recycling and briefly pointed out that thermal and chemical recycling can be economically sustainable under certain conditions. Finally, it can be stated that although within the circular economy paradigm, repair is defined as one of the most effective approaches to address waste [69] and composite repair has been proven technically feasible [59], the economic assessment on this topic is still missing. Therefore, this paper aims to address this knowledge gap by evaluating the economic sustainability of EoL composites through repair and thermal recycling (CO_2 assisted pyrolysis).

2. Materials and Methods

This section describes the methods adopted to perform the LCA and LCC of a couple paddle of shifters used in the automotive sector. In particular, Section 2.1 describes the methodology adopted for the LCA, which includes the goal of the study, the functional unit, the system boundaries, the scenarios developed, the data collection of the LCA study, the characterization method used and impact categories. Finally, Section 2.2 focuses on the collection of data used for the costing analysis.

2.1. Environmental Assessment

2.1.1. Goal and Scope Definition

This ex ante LCA is used as an exploratory methodology for the introduction of new technologies on the industrial level. The scope of this LCA is to provide a comparative analysis of CFRPs, and scenarios were built according to the following criteria:

1. Type of materials used for the production of the CFRP product;
2. Possibility of repairing the damaged part of the CFRP product;
3. Different End-of-Life (EoL) of the CFRP.

As concerns the first criteria, two types of composites were developed. Generally, the composite is made of CFs, an epoxy resin [70] and a curing agent that define the properties of the composite [71]. In this study, both composites were developed using the same type of epoxy resin and two different types of dynamic curing agents. The first type of composite was produced using a fossil-fuel-based dynamic curing agent (i.e., CFRP-ff); the second type of composite was produced using a bio-based dynamic curing agent (i.e., CFRP-bio).

As concerns the second criteria, this LCA compares two scenarios of each aforementioned composite in order to evaluate the potential environmental benefits of repair during the use phase. On one hand, the impossibility of repairing the composite requires the substitution of the product. This means that the waste management of the damaged CFRP product and the production of the second CFRP product are considered. On the other hand, the possibility of repairing implies that no additional CFRP product is required to be manufactured. In this case, only the repairing process was included. In this study, one damage in the entire life of the car was assumed. The third criteria aimed at comparing different EoLs. In accordance with other studies on composites [8,12,49,52], as the base scenario of LCA, 50% of the CFRP waste was assumed to be sent to a landfill, and the remaining 50% was assumed to be sent to incineration. Two additional scenarios, respectively, mechanical and thermal recycling, were analyzed to further evaluate the environmental benefits of both recycling compared to the current waste management procedure. Further details on scenarios are reported in Section 'Scenario Description'.

Based on the properties of the developed CFRP, it can be used in automotive and aeronautic sectors. The functional unit (FU) is one couple of paddle shifters used in the automotive sector. The system boundaries are shown in Figure 1. The chosen LCA database is ecoinvent attributional, cut-off v3.6. Additional data used for the inventory were taken from the literature. The chosen LCIA method is ReCiPe in accordance with other LCA studies on composites [53,72–74].

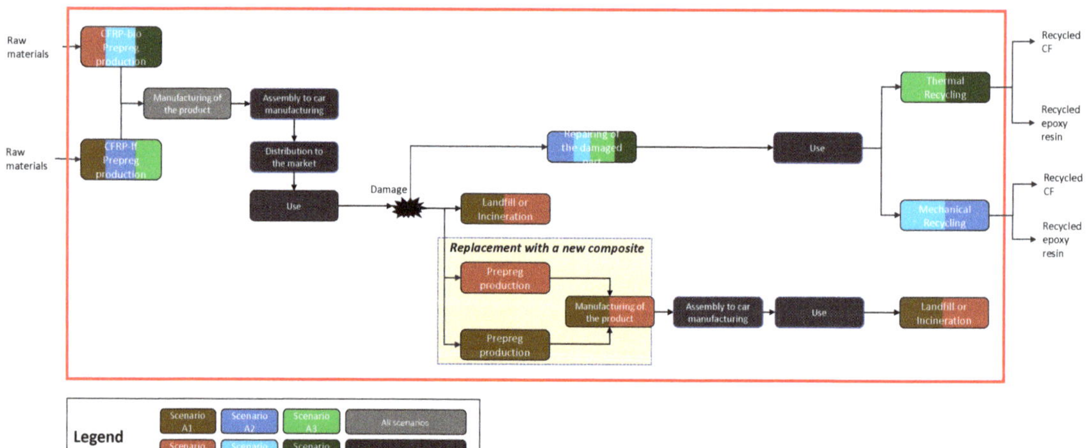

Figure 1. System boundaries of each scenario of the comparative LCA.

Scenario Description

Six scenarios, listed in Table 1, were defined according to the criteria described in Section 2.1.1. Figure 1 describes the steps included in the study for each scenario. Groups of scenario A and B represent the life cycle, respectively, of CFRP-ff and CFRP-bio. Scenario A1 and B1 are representative of the most likely situation of a composite EoL: landfill and/or incineration [19]. Moreover, in the case of damage during the use phase, the composite is substituted with a new composite, which results in another production and manufacturing process of the composite.

Scenario A2 and B2 include the repairing of the composite after the damage instead of its substitution. At the end of the entire life of the car, the composite is sent to mechanical recycling.

Scenario A3 and B3 include the repairing of the composite after the damage instead of its substitution. At the end of the entire life of the car, the composite is sent to thermal recycling. Both thermal and mechanical recycling allow the recovery of epoxy resin and CF.

Table 1. Description of the scenarios of the comparative LCA.

Number of Scenario	Name of Composite	Composition	Repairing	End-of-Life
A1			No	50% Landfill + 50% Incineration
A2	CFRP-ff	curing agent 1+epoxy resin+CF	Yes	Mechanical recycling
A3			Yes	Thermal recycling
B1			No	50% Landfill + 50% Incineration
B2	CFRP-bio	curing agent 2+epoxy resin+CF	Yes	Mechanical recycling
B3			Yes	Thermal recycling

2.1.2. Life Cycle Inventory
Production of the Prepreg

The production of prepreg, which is the pre-impregrated CF with the epoxy resin, is the most widespread manufacturing route at the industrial level [14]. Firstly, the epoxy resin is mixed with a curing agent, and secondly, the CFs are pre-impregnated with the cured epoxy resin in order to obtain the prepreg [55]. The production of CFRP-bio requires a further transesterification catalyst to be added to the mixing and additional pretreatment for the activation of the curing agent prior to the mixing of the epoxy resin with the curing agent. Figure 2 summarizes the production processes of prepreg at the laboratory scale for both composites.

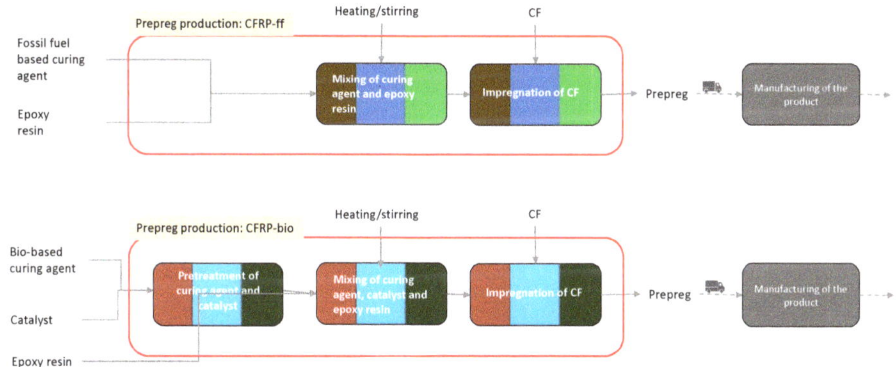

Figure 2. Prepreg production process on a laboratory scale. CFRP-ff (i.e., group of scenario A) includes two main steps: firstly, mixing the dynamic curing agent with the epoxy resin, and secondly, the impregnation of the CF. CFRP-bio (i.e., group of scenario B) includes the same steps of scenario A plus two additional pretreatment steps needed for the dynamic curing agent and catalyst. Colors represent the different scenarios.

An ex ante evaluation was needed in order to estimate the energy and materials consumption at the industrial level starting from the data at laboratory scale. Table 2 shows the main differences between the primary data provided from laboratory tests and the primary data needed for the LCA modeling usually provided at the industrial scale. The transformation of primary data from laboratory to industrial scale was performed for the following prepreg production steps shown in Figure 2: *Mixing of curing agent and epoxy resin, mixing of curing agent, catalyst and epoxy resin* and *pretreatment of curing agent and catalyst.* The scale-up of those steps was conducted according to the methodology introduced by Piccinno et al. [46]. In particular, for each process listed in the first column of

Table 2, the equations provided by Piccinno et al. [46] were implemented, and assumptions were introduced in order to obtain the inputs to the LCA model starting the primary data provided from laboratory tests (i.e., second column of Table 2). The overall electricity consumption of the first step of prepreg production at the industrial scale was estimated with the equations shown in Table 3 taken from Piccinno et al. [46]. The ratio between the epoxy resin and the curing agent was taken from De Luzuriaga et al. [3]. Adjustments were made for CFRP-bio to include the catalyst in the prepreg composition.

Tables 4 and 5 summarize, respectively, the two steps of production, outlining the main information on materials and energy consumption both for CFRP-ff and CFRP-bio. The electricity consumption for the impregnation of the CF in Table 5 was taken from Song et al. [54]. Background data such as epoxy resin, CF, the fossil-fuel-based curing agent, the bio-based curing agent and the catalyst were taken from the literature and ecoinvent database. The epoxy resin was modeled according to ecoinvent database *market for epoxy resin, liquid | epoxy resin, liquid | Cutoff, U-RER*. CF production was modeled according to literature data. Data from Khalil [55] were used to model CF production from PAN. Data from Duflou et al. [8] were used to model Polyacrylonitrile (PAN) starting from Acrylonitrile, which is modeled in the used database. The fossil-fuel-based curing agent was modeled according to the ecoinvent database with an aromatic amine *market for meta-phenylene diamine | meta-phenylene diamine | Cutoff*. The material was chosen in accordance with studies on epoxy resins [75,76]. The bio-based curing agent derives from different routes, e.g., from fructose, polysaccharides [77], starch, glucose [78] and lignin [79]. Lignin biomass was used as reference production routes starting from tetrahydrofuran (THF), and data from Kim et al. [79] were used. The catalyst was modeled according to the ecoinvent database with the *market for ammonium thiocyanate | ammonium thiocyanate | Cutoff* process.

Table 2. Differences in primary data available between laboratory and industrial scales. For each production process, specific data are needed as input to build the LCA model. On the laboratory scale, data on production processes are, for instance, temperature, duration of the process, pH, etc. However, those inputs are not directly usable in an LCA study. On the industrial level, the available data are directly usable in an LCA, such as energy and auxiliary material consumption.

Production Process	Laboratory Scale (Unit = [g])	Industrial Scale (Unit = [kg or t])
Heating	Temperature, Duration	Energy and auxiliary material consumption
Stirring	Temperature, Rotation velocity, Duration	Energy and auxiliary material consumption
Grinding	Duration, Dimension of particles	Energy and auxiliary material consumption
Filtration	Auxiliary materials, Duration	energy and auxiliary material consumption

Table 3. Calculation methodology adopted from Piccinno et al. [46] to perform the scale up of the energy consumption from laboratory scale to industrial scale. C_p = Specific heat capacity [J/(kg*K)]; m_{mix} = mass of the reaction mixture (kg); T_r = temperature of reaction (K); t = time of reaction (s); ρ_{mix} = density of the mixture (kg/m).

Production Process	Calculation
Heating	$Q_{react} = [C_p * m_{mix} * (T_r - 298.15K) + 3.303 * (T_r - 298.15K) * t]/0.75$
Stirring	$E_{stir} = 0.018\rho_{mix} * t$
Grinding	8–16 kWh/ton
Filtration	1–10 kWh/ton

Table 4. Mixing of the epoxy resin and the dynamic curing agent according to the two scenarios. Data are referred to 1 kg of uncured vitrimer. Epoxy resin/dynamic curing agent ratio from de Luzuriaga et al. [3].

Input	CFRP-ff	CFRP-bio
Epoxy resin (kg)	0.7	0.70
Fossil fuel based curing agent (kg)	0.3	-
Bio-based curing agent (kg)	-	0.27
Catalyst (kg)	-	0.03
Electricity (kWh)	5.9×10^{-2}	1.3×10^{-1}
Steam (kg)	-	33.4
Cooling water (kg)	-	0.59
Other auxiliary materials (kg)	-	21.8

Table 5. Impregration of the CF with the epoxy resin. Data refer to 1 kg of prepreg. Data on energy consumption were taken from Song et al. [54].

Input	CFRP-ff	CFRP-bio
Epoxy resin (kg)	0.51	0.52
Carbon Fiber (CF) (kg)	0.49	0.48
Electricity (MJ)	40	40

Manufacturing the Product

Manufacturing the CFRP product mainly includes the cutting phase of the prepreg and the lay-up phase. The latter phase is the core manufacturing process through which the prepreg is modeled in order to obtain the final shape of the desired product, which, in our case, is on a couple of paddle shifters. The lay-up phase can be performed with different technologies [72]. Although primary data from a laboratory were available, the manufacturing of the paddle shifters was modeled according to literature data for two main reasons: i. experiments in a laboratory are performed with small quantities of materials and a higher level of auxiliary materials consumption and scraps and waste generation compared to the expected quantities consumed and generated at the industrial level [42,46]; ii. the autoclave is one well-established technology at the industrial level for CFRP products. Hence, the auxiliary material and energy consumption were reasonably assumed not to be linked to the prepreg composition.

The manufacturing phase was modeled according to Forcellese et al. [72], including the mold and the master production. The data inventory is reported in Table 6. The auxiliary materials used during the lay-up phase, such as the vacuum bag in Polyamide 66, the breather (Polyethylene terephthalate), the release film (Polytetrafluoroethylene) and the release agent, were assumed to be sent to a landfill at the end of the production since they were not reusable. According to the scope of the study, the FU is the production of a CFRP product in the automotive sector. As a consequence, the manufacturing of the product is followed by the final assembly to the car. However, according to Raugei et al. [49], the environmental impact due to vehicle manufacturing is low compared to the other life cycle stage. Hence, assembly was assumed negligible compared to the entire life cycle, and it was not considered in the study. Further details about distances and transportation are described in Section 'Transportation'.

Table 6. Inventory data for the manufacturing of the paddle shifters taken from Forcellese et al. [72].

Material	Quantity (kg)
Input	
Prepreg (CFRP-ff or CFRP-bio)	15.7
Electric energy (kWh)	17.074
Master mold in polyurethane foam	30.5
Mold (in CFRP)	0.095
Polyamide 66 (PA66)	0.5
Polyethylene terephthalate (PET)	0.375
Polytetrafluoroethylene (PTFE)	0.055
Organic solvent	0.03
output	
Paddle shifters	11
scraps of prepreg	4.7
scraps of master mold	9.5

Use and Repairing

The use phase of the paddle shifters is directly linked to the use phase of the car, which is mainly related to fuel consumption determined by, for instance, the shape and weight of the car [5]. The latter depends on the materials used. As a matter of fact, the use phase of a car component is usually considered in LCA studies in order to evaluate its indirect environmental impact [73]. However, the weights of the two CFRPs analyzed in this study are almost the same. Moreover, the weight of the FU is 76 g, which is reasonably low compared to the weight of the whole car, which is usually in the range of 500–1000 kg [80]. Hence, the comparison of their environmental impact during the use phase was reasonably assumed not to be relevant to the comparative LCA.

Assuming the overall life cycle of the paddle shifters equals the life cycle of the car, damage to the paddle shifters in the entire life cycle of the car was considered in the study. Two main options were analyzed in this analysis. The first option, represented by scenarios A1 and B1, describes the substitution of the product with a new one. As shown in Figure 1, the damaged product is sent to a landfill or for incineration, and a new product is manufactured with the same technologies of the first composite. It was assumed that both paddle shifters were damaged and substituted. The second option represented by scenarios A2, A3, B2 and B3 describes repairing of the damaged part and avoiding the substitution of a new paddle shifter. The repairing of the damaged part is performed using laser technology under development at the laboratory scale. Consequently, the operation of substitution with patches is not required. Based on the development of this technology, it was assumed in the present analysis that 1 cm^2 of resin was damaged. Although the electricity is the main contribution to this phase, the value is omitted for confidentiality reasons. No waste is generated during the process.

End of Life

Scenario A1 and B1 represent the most likely situation and the worst scenarios from a Circular Economy prospective [10,19]. As a matter of fact, the paddle shifters are sent to a landfill and for incineration at the end of their life. Moreover, as described in Section 'Use and Repairing', after one paddle shifter is damaged, it is sent to a landfill and/or for incineration, and a new one is manufactured. The process of landfill and incineration were modeled using the ecoinvent database.

Scenario A2 and B2 describe the situation in which the paddle shifters are mechanically recycled. The SELFRAG is a technology for mechanical recycling that aims at disintegrating the composite through the principle of electrodynamic fragmentation: the composite is subjected to electrical discharges that break the composite. The process occurs using water [81,82]. SELFRAG technology has been used in laboratories in order to understand the potentiality of recycling both the composite and the resin. Currently, 60% of the CF can be recycled, while the possibility to recycle the resin is still uncertain and under study. Hence, an overall 60% of recycling was considered in this study without specifying the composition of the recyclates. Data were collected from the official report of SELFRAG [81]. Inventory data are reported in Table 7.

Scenario A3 and B3 analyze the thermal recycling through CO_2-assisted pyrolysis. Thermal recycling of composite operates at high temperatures in the range of 350–750 °C [10,18,19,83], and at the end of the thermal process, the CFs are recovered. Korec is CO_2-assisted pyrolisis technology commercially available to recover glass fiber composites; it is currently being tested in laboratories in order to analyze the potentiality of CFRP recycling. Inventory data are reported in Table 8. According to KOREC data, the glass fiber is recovered with an efficiency of 99% [84]. The same recovery efficiency was assumed for the CF. Eighty-five percent of the resin can be recovered, and up to sixty percent of recycled resin can be mixed with virgin resin [85]. CO_2 is recovered and reused internally in the process, and consequently, its consumption was assumed to be negligible. The electricity consumption of the process was taken from Nunes et al. [52], with slight differences in the process. During the process, solid residue is generally generated in the case of inert mineral fillers in the composite. Non-condensable gas is produced during the process and is used for energy purposes, fulfilling the energy requirement. However, it was not possible to characterize and quantify the solid residue for our specific composite due to its small quantities.

Table 7. Inventory data for the mechanical recycling of paddle shifters. The consumption of the electrodes were allocated to 1 kg of CFRP waste on the mass basis. The data are valid both for CFRP-ff and CFRP-bio.

Material	Quantity
Input	
Paddle shifters waste (kg)	1
Electric consumption (kWh)	1
electrodes (item)	0.005
Water (L)	5
output	
recycled CFRP (kg)	0.6
waste CFRP (kg)	0.4
Water losses (L)	0.25

Table 8. Inventory data for CO_2-assisted pyrolysis of the paddle shifters. The CO_2 was not included in the mass balance since it is recirculated in the process. CO_2 losses were not considered. The data are valid both for CFRP-ff and CFRP-bio. Data of electricity consumption were taken from Nunes et al. [52].

Material	Quantity
Input	
Paddle shifters waste (kg)	1
Electric consumption (MJ)	54
output	
CFs recycling efficiency (%)	0.99
Epoxy resin recycling efficiency (%)	0.85
solid residue (kg)	0.25

Transportation

Table 9 shows the transportation included in the study from the raw material production until the composite's EoL. Since the analysis was preliminary, and it was not focused on evaluating the environmental impact of a product that is available on the market, the supply chain was unknown. Hence, the following assumptions were introduced. The production of the prepreg and the manufacturing of the paddle shifters were assumed to be located in Italy in different cities. The production of the CF, the curing agents and the epoxy resin was assumed to be located in Europe by estimating an average distance to the production site of the prepreg. Distance from the manufacturing site to the car assembly site was not considered since the assembly phase was excluded from the system boundaries. The repairing site was assumed to be the same as the manufacturing site. Distance from the user to the repairing site was conservatively assumed to be 500 km due to the high variability of the car user's location. Information on mechanical and thermal recycling locations were estimated from information on companies provided by Giorgini et al. [19]. Landfill and incineration were assumed to be located 50 km from the collection point.

Table 9. Inventory data of transportation.

Transportation	Distance (km)
CF to prepreg production site	20,145
Fossil-fuel-based epoxy resin to prepreg production site	10,130
Bio-based epoxy resin to prepreg production site	10,130
Prepreg production to paddle shifter manufacturing site	1600
CFRP waste to landfill/incineration	50
CFRP waste to repairing site	150
CFRP waste to mechanical recycling	1000
CFRP waste to thermal recycling	1000

2.1.3. Life Cycle Impact Assessment Method

The LCIA was performed using OpenLCA software, version 1.11.0 (GreenDelta, Berlin, Germany) with the Ecoinvent 3.6 database for the LCI phase and ReCiPe mid-point (hierarchist) as LCIA methodology. The chosen LCIA methodoloy allows the assignment of the LCI results to indicator results according to the chosen environmental impact categories, which represent the product profile [33]. The following eighteen mid-point metrics were adopted to perform the LCA. Fine particular matter formation ($kgPM_{2.5}eq.$), Fossil resource scarcity (kgoileq.), freshwater ecotoxicity (kg1,4-DBeq.), freshwater eutrophication (kgPeq.), Global warming ($kgCO_2eq$), human carcinogenic toxicity (kg1,4-DCBeq.), human non-carcinogenic toxicity (kg1,4-DCBeq.), ionizing radiation (kBqCo-60 eq), land use (m^2a crop eq.), terrestrial acidification ($kgSO_2eq.$), marine ecotoxicity (kg1,4-DCBeq.), marine eutrophication (kgNeq.), Mineral resource scarcity (kgCueq.), Ozone formation Human health ($kgNO_x$ eq), Ozone formation terrestrial ecosystems ($kgNO_x$ eq), terrestrial acidification ($kgSO_2eq$), terrestrial ecotoxicity (kg1,4-DCBeq.), Stratospheric Ozone depletion (kgCFC11eq.) and water consumption (m^3) were quantified for each stage of the paddle shifters' life cycle.

2.2. Economic Sustainability Assessment

To assess the economic sustainability of the new EoL composite solutions, which are repair and thermal recycling, multiple supply chain perspectives have been taken into account: the one of the composite parts producers and the one of composite materials recyclers. The economic assessment of producer is performed for the two new composites developed in a laboratory, i.e., CFRP-ff and CFRP-bio, as described in Section 2.1. Regarding the repair, producers of composite parts are supposed to integrate the new CF composites into their production system. Due to the properties of the new composites, they can be repaired [86]. From a business standpoint, the producer would use the new material only if it could take advantage of the repair of the material. Since the new CF composite price is higher than the conventional ones, for the manufacturer, the main incentive of using the new materials is to exploit the repair opportunity to address the defective products coming from the production process. Therefore, in this new paradigm, a producer would also become a remanufacturer by integrating the repair facilities into its existing production plant. Here, it is important to highlight that, based on the interviews with the production manager of the case study company, it becomes evident that one of the main issues of CF composite products is their high defect rate. Thus, in our hypothesis, the potential increase in the cost of the raw materials could be compensated by the potential savings due to the introduction of in-house repair practices enabled by the new composites. In addition, another aspect of this business is the fact that product repair helps the producer abolish its landfilling cost. In the current context, when products made from conventional CFRP have a defect during the production process, they cannot be repaired and are sent to a landfill. Thus, the manufacturer bears their landfilling cost, while in this new setting, producer landfilling costs associated with defective parts are eliminated. Regarding the system boundaries, the economic assessment consists of material production, manufacturing, and repair of defective products (products that are damaged during the production process) for a producer of the paddle shifter.

On the contrary, the economic assessment of thermal recycling represents an entirely new business for a composites recycler. In this framework, the recycler receives CFRP waste that is recyclable. The new thermal recycling technology enables both carbon fibers (CF) and resin to be recovered and sold for secondary applications. In this study, we restricted the economic simulation to the sales of CF, neglecting the possible revenue stream from recycled resin. This was justified by the relatively higher added-value of the fibers compared to resin and to the multiple criteria that have to be assessed for recycled resin to be used in secondary applications, which cannot be forecast at the moment due to the novelty of the developed material. The system boundaries consist of a recycler's economic sustainability receiving recyclable composite waste, i.e., CFRP-ff and CFRP-bio wastes, and selling the recovered CF. The economic sustainability of composite parts manufacturers and composites recyclers has been assessed through LCC techniques. In this study, parametric modeling is used to perform the LCC analysis. Parametric models provide a high level of flexibility in changing the parameters. These models are appropriate for identifying cost drivers and carrying out the sensitivity analysis. They allow the user to define different scenarios under a wide range of assumptions.

2.2.1. Case Study Definition

The case studies considered in the economic assessment of a producer of CFRP parts and the recycler are presented in this section.

1. Two main industrial cases are taken into account for the repair producer. One case investigates the economic sustainability of the repair business when the paddle shifter is made from CFRP-ff, and the other assesses the profitability when the product is made from CFRP-bio.

2. One main industrial case is taken into account for the thermal recyclers receiving either CFRP-ff or CFRP-bio waste. From an economic standpoint, the recycling processes of CFRP-ff waste or CFRP-bio waste are very much alike for the recycler for the following three reasons. First, the recycling process of both materials is precisely the same. Second, the acquisition cost of CFRP-ff and CFRP-bio wastes is the same for a recycler. Based on the interview with the operation manager of a recycling company, it is hypothesized that the recycler would pay no waste acquisition cost since companies in charge of dismantling would be relieved from sustaining recycling costs. In this framework, the recycler only bears the transportation cost of waste to its plant. Finally, since the percentage of CF in new CFRPs materials is almost the same (49% for CFRP-ff and 48% for CFRP-bio), it becomes evident that the financial results of a recycler receiving CFRP-ff waste or CFRP-bio are almost the same. In addition, in this study, the economic modeling of the recycler is not only restricted to the paddle shifter. Since recycling is a high-volume business in which its sustainability is based on the economy of scales, and the baseline product considered in our case study is a low-volume niche component produced with a limited quantity of raw materials. Thus, we hypothesized

that the recycler receives various products from multiple customers built with the new recyclable material in order to aggregate massive volumes to be treated.

2.2.2. Cost and Revenue Breakdown Structure of Manufacturer

Based on the various steps of the production and repair processes discussed in Section 1.4, specific CBS and RBS have been defined for the producer. The input data related to the manufacturing costs of the paddle shifter were collected during two rounds of interviews in 2021, with the production manager, technical manager and marketing manager of the case study company, as shown in Tables 10 and 11. In addition, data related to laser technology were provided by Politecnico di Milano in 2021. The costs of the business model of a producer that integrates new CFRP composites into its production process and repairs its defective parts are as follows:

- The manufacturing cost includes the machinery, depreciation, labor, energy and overhead cost. Due to data confidentiality, the manufacturing cost of product is shown as a percentage of the whole product cost;
- The material cost of a product represents the cost of new repairable/recyclable composites: CFRP-ff or CFRP-bio. Due to data confidentiality, the material cost of product is shown as a percentage of the whole product cost;
- The investment cost of repair facilities includes purchasing the laser technology and its accessories;
- The maintenance cost includes one operator cleaning and taking care of the optics of laser technology;
- The repair cost consists of the energy consumption and labor needed to repair the damaged area. As described in Section 'Use and Repairing', the dimension of the damaged area is considered to be 1 cm^2. Given that the energy consumption of 1 cm^2 is 0.111 kWh, and as the time needed to heal the damaged part is 5 s, the total repair cost of 1 cm^2 is EUR 0.045 (as shown in Equations (1)–(3)).

$$Total\ cost_{repair} = Energy\ Cost_{repair} + Labor\ Cost_{repair} \tag{1}$$

$$Energy\ Cost_{repair} = Electricity\ consumption * Cost_{Electricity\ power} \tag{2}$$

$$Labor\ Cost_{repair} = Time\ needed\ to\ repair\ the\ damaged\ area * Cost_{Personnel\ wages} \tag{3}$$

The revenues stream of manufacturer includes:

- Selling of the product made either from CFRP-ff or CFRP-bio. Here, it is important to mention that the selling price of the paddle shifter made from the new repairable/recyclable CFRP is assumed to be the same as the product made from conventional CFRP. Since, based on the interviews with the case study company, the market might not accept an increase in the product price, and the best strategy is to keep the selling price at the same level of the conventional product. Here, it is also important to mention that the profit margin of this product is relatively high (50%) due to the high value-added of composite in the high-end automotive market.

Table 10. Input data of manufacturing cost of the paddle shifter.

Data	Unit of Measure	Value
Yearly product volume	piece/year	800
Cost of conventional CFRP (prepreg)	EUR/kg	50
Profit margin	%	50
Defect rate	%	20
Material cost compared to the total product cost	%	2.2
Manufacturing cost compared to the total product cost	%	97.8

Table 11. Characteristics and cost of the repair technology. Electricity power cost were taken from [87]. Personnel wages data were taken from [88].

Data	Unit of Measure	Value
Investment cost	EUR	200k
Yearly maintenance cost of laser facility	EUR	10k
Dimension of damage area	cm^2	1
Electricity consumption to repair the damage area	kWh	0.111
Electrical power cost	EUR/kWh	0.1254
Personnel wages	EUR/h	28
Time needed to repair the damage area	seconds	5

2.2.3. Cost and Revenue Breakdown Structure of Recycler

Based on the various steps of the recycling processes discussed in Section 2.2, specific CBS and RBS have been defined for the recycler. The input data of the thermal recycling business model, as shown in Table 12, were provided by Politecnico di Milano, an operation manager of a thermal recycling company located in Northern Italy and KOREC technology in 2021.

In the following, all the recycler's costs and then the recycler's revenue stream is presented:

- The investment cost includes purchasing the CO_2-assisted pyrolysis facilities. The investment cost of thermal recycling is EUR 550,000 for a plant of 250 ton/year;
- The maintenance cost includes the price of 1 h of work of an operator who cleans, lubricates and adjusts the thermal recycling machine after every 8 h cycle;
- The recycling cost includes the processing cost in terms of the depreciation, labor and energy cost;
- The overhead cost includes all administrative and accounting fees;
- The transportation cost includes delivering the EoL of CFRP waste to thermal recycler facilities. As mentioned in Table 9, the distance between the waste management company and the recycler is assumed to be 1000 km. Within these assumptions, the operation manager of the recycling company provided us the transportation cost of waste to be 0.20 EUR/kg, with a coefficient of +0.05 EUR/kg and −0.05 EUR/kg.

The revenue stream of the recycler is:

- Selling of the recycled CF to secondary applications. As mentioned in Table 8, the recovery rate of CF within the CO_2-assisted pyrolysis is 99%. Here, it is important to mention that the market price for selling the recycled CF is 5 EUR/kg, with a coefficient of +1 EUR/kg and −1 EUR/kg.

Table 12. Characteristics and costs related to thermal recycling. Electricity power costs were taken from [87]. Personnel wages data were taken from [88].

Data	Unit of Measure	Value
Investment cost of thermal recycling facilities	EUR	550k
Capacity of plants	ton/year	250
Working days per year	day	240
Working hours per day	hour	8
Electrical power cost	EUR/kWh	0.1254
Personnel wages	EUR/h	28
Overhead and administration cost	%	20
Maintenance time (after every cycle of 8 h)	hour	1
Gas cost (CO_2)	EUR/ton	150
Transport cost	EUR/kg	0.2
Recycled CF price	EUR	5

2.2.4. Assumptions

The assumption that were made to carry out the economic assessment are as following:

1. The cost estimation of new repairable/recyclable composites is an important piece of information that is needed to carry out the LCC of the repair business model of a manufacturer using CFRP-ff or CFRP-bio in its production system. From the laboratory data, it is evident that the costs of the two new CF composites are higher than the conventional ones. Thus, proper cost estimation of the new material is critical. In this study, the top-down approach is applied to calculate the relative costs of CFRP-ff and CFRP-bio. The rationale behind using this method is the fact that there are uncertainties for cost analysis of the new CF composites that do not allow a detailed level cost analysis of the product. These uncertainties arise from the fact that these materials are developed in the laboratory, where the material costs are much higher and not comparable to the industry costs. For the top-down method, the expert judgement is used to calculate the relative costs of CFRP-ff and CFRP-bio. Three experts were selected and interviewed to evaluate the relative cost of the new CFRPs. A production manager, technical manager and the senior R&D manager of the case study company are all experienced in composite materials' formulation and production and composite production technologies and processes. To estimate the industrial cost of the CFRP-ff and CFRP-bio (prepreg), the first step was to add up the costs of the production material and manufacturing as the initial building blocks of the total cost. Given that the production procedure of the CFRP-ff and CFRP-bio is the same as the conventional procedure, the manufacturing cost of the new materials was considered similar to that of the conventional materials. In other words, the main difference between newly developed CFRP and the conventional one lies in the material cost. Using the top-down method first, the relative material cost of the CFRP-ff and CFRP-bio is compared to the price of conventional CFRP. According to composite experts, the relative material cost of the CFRP-ff at the industrial scale can be 30% higher than the price of a conventional one with the coefficient of -20% to $+20\%$. In other words, a 30% cost increase in the CFRP cost is considered as the base case, while the best case is where material increases by 10%, and the worst case is when a material cost increases by 50%. On the other hand, as the material cost of CFRP-bio is more expensive than the CFRP-ff, experts' judgment indicated that the relative material cost for CFRP-bio can be 50% higher than the conventional CFRP with the coefficient of -20% and $+20\%$. To estimate the final production cost of the new CFRP in addition to the material cost, it also important to know the cost breakdown structure of the CFRP. According to Graf et al. [89], in CFRP components, the material cost accounts for 36%, followed by 64% for the manufacturing cost. Based on this cost structure, when the material cost of CFRP-ff increases by 10%, 30% or 50%, the total cost of CFRP increases by 3.6%, 10.8% and 18%. Concerning CFRP-bio, when the material cost increases by 30%, 50% and 70%, the total cost increases by 10.8%, 18% and 25.2%.
2. In this study, a net present value (NPV) and discounted payback period of the investment are used to evaluate the economic performance of a system. The discount rate is assumed to be 10%, and the NPV is calculated over a period of 10 years.

2.2.5. Sensitivity Analysis

Sensitivity analysis is used to investigate how the output is affected by the system's uncertainty. As a result, a sensitivity analysis is performed with assumptions on the parameters with higher uncertainty [62,90]. In this study, the following steps were adopted to carry out the sensitivity analysis:

1. Identification of the uncertain parameters;
2. Definition of the potential ranges for the parameters resulting in various scenarios;
3. The economic assessment of producer and thermal recycler carried out through feeding different value parameters into the economic model.

Based on the steps mentioned above, in the following, the most uncertain parameters that were used to assess the profitability of producer and recycler business models, and their ranges are presented. Concerning the repair business model of the producer, among various input data affecting the LCC, the relative cost of new materials was identified as the most uncertain parameter as it has the highest impact on the NPV and payback period. Tables 13 and 14 present the variability of the parameter, resulting in three scenarios for each type of composite. The investment cost is not considered as a variable parameter since it is not affected by uncertainty, and we could rely on the result of the economic system. In addition, the repair cost is not also considered as an uncertain parameter since this cost is relatively low and has no impact on the financial output of the economic indicator. In this regard, if the repair cost increased by 1000%, the NPV would increase by 0.24%, representing that the impact of this parameter on the final results is almost negligible.

Table 13. Parameters used for sensitivity analysis of the repair business of paddle shifters made from CFRP-ff. The conventional CFRP cost is 50 EUR/kg.

Range	Relative Cost Increase Compared to the Conventional CFRP(%)	Cost of CFRP-ff (EUR/kg)
R1	3.6%	51.8
R2	10.8%	55.4
R3	18%	59

Table 14. Parameters used for sensitivity analysis of the repair business of paddle shifter made from CFRP-bio. The conventional CFRP cost is 50 EUR/kg.

Range	Relative Cost Increase Compared to the Conventional CFRP(%)	Cost of CFRP-bio (EUR/kg)
R1	10.8%	55.4
R2	18%	59
R3	25.2%	62.6

Regarding the thermal recycler business model, among various input data affecting the LCC, the following parameters were used to conduct the sensitivity analysis: the annual amount of CFRP waste, the selling price of recycled CF (rCF) and the transportation cost to deliver the waste to the recycler's plant. These parameters were affected by a considerable level of uncertainty. To capture the uncertainties surrounding the recycled CF selling price and the transportation cost, experts provided us with ranges for these parameters as it is presented in Section 2.2.3. The amount of waste that the recycler can receive each year is considered a variable since this parameter can have a high impact on the NPV and payback period. The amount of waste that the recycler can receive each year was hypothesized as a percentage of the saturation rate of the plant: 75%, 85%, 90%, 95% and 100%. Given that the capacity of the plant is 250 ton/year, the potential ranges of the waste are presented in Table 15. The combination of these three parameters with their ranges led to 45 scenarios.

Table 15. Parameters used for sensitivity analysis of the thermal recycling business of a recycler.

Ranges	Price of Recycled CF (EUR/kg)	Transportation Cost (EUR/kg)	Annual Amount of Waste (ton/Year)
R1	4	0.15	187.5
R2	5	0.20	212.5
R3	6	0.25	225
R4	-	-	237.5
R5	-	-	250

3. Results and Discussion

3.1. LCA Results

This section provides preliminary results of a new technology still under development at the laboratory scale in order to identify potential opportunities to be further developed at the industrial level [31]. This section shows the variability across scenarios of the following three challenges of composites in the circular economy perspective: use of bio-based materials, repairing and recycling. Table 16 shows the LCIA of the FU of the study for the six scenarios performed across the eighteen environmental categories of the ReCiPe methodology listed in Section 2.1.3. Scenario A1 shows the highest results in almost all the impact categories, except for *Fine particulate matter formation—PM, Freshwater ecotoxicity—FEX, Freshwater eutrophication—FE, Land use—LU* and *Stratospheric ozone depletion—SOD*. The highest value for *Land use—LU* is shown in scenario B1, meaning that the variability of this category is more related to differences in the curing agent of the two composites. The lowest value is shown for scenario A3. Scenario B2 and B3 show the lowest values for several impact categories, meaning that a combination of bio-based materials, repairing and recycling could result in a reduction in the environmental impacts. The highest values for *Fine particulate matter formation—PM, Freshwater ecotoxicity—FEX, Freshwater eutrophication—FE* and *Stratospheric ozone depletion—SOD* are shown in scenario A3, with slightly lower values for scenario B3. Those results are mainly related to the thermal process of recycling and in particular to the high energy intensity of the process. Normalization of the LCA results is provided in Figures S1–S9 of Section S2.1 of the Supplementary Materials.

Table 16. LCA results of the six scenarios performed across the impact categories. symbol (+) shows the highest value compared to the other scenarios; symbol (−) shows the lowest value compared to other scenarios. The visualization shows A1 has the highest results for almost all categories, and B2 and B3 have the lowest values in several impact categories.

Impact Category	Unit	A1	A2	A3	B1	B2	B3
PM	$kgPM_{2.5}eq$	1.20×10^{-5}	6.66×10^{-6}	5.43×10^{-5} (+)	1.15×10^{-5}	6.39×10^{-6} (−)	5.41×10^{-5}
FRS	$kgoileq$	2.67×10^{-3} (+)	1.38×10^{-3}	1.89×10^{-3}	2.50×10^{-3}	1.29×10^{-3} (−)	1.80×10^{-3}
FEX	$kg1,4\text{-}DCB$	3.89×10^{-4}	3.16×10^{-4}	4.60×10^{-4} (+)	3.84×10^{-4}	3.14×10^{-4} (−)	4.57×10^{-4}
FE	$kgPeq$	2.89×10^{-6}	1.68×10^{-6}	1.17×10^{-5} (+)	2.83×10^{-6}	1.65×10^{-6} (−)	1.17×10^{-5}
GW	$kgCO_2eq$	8.13×10^{-3} (+)	4.21×10^{-3}	4.14×10^{-3}	7.60×10^{-3}	3.94×10^{-3}	3.87×10^{-3} (−)
HCT	$kg1,4\text{-}DCB$	3.51×10^{-4} (+)	2.19×10^{-4}	3.34×10^{-4}	3.46×10^{-4}	2.17×10^{-4} (−)	3.32×10^{-4}
HT	$kg1,4\text{-}DCB$	6.54×10^{-3} (+)	4.37×10^{-3}	4.19×10^{-3}	6.47×10^{-3}	4.33×10^{-3}	4.15×10^{-3} (−)
IR	$kBqCo\text{-}60eq$	1.02×10^{-3} (+)	5.37×10^{-4}	5.29×10^{-4}	1.02×10^{-3}	5.35×10^{-4}	5.26×10^{-4} (−)
LU	$m^2a\ crop\ eq$	6.99×10^{-6}	4.76×10^{-6}	4.71×10^{-6} (−)	7.70×10^{-6} (+)	5.12×10^{-6}	5.06×10^{-6}
MEX	$kg1,4\text{-}DCB$	5.09×10^{-4} (+)	4.08×10^{-4}	3.89×10^{-4}	4.99×10^{-4}	4.03×10^{-4}	3.84×10^{-4} (−)
ME	$kgNeq$	1.13×10^{-5} (+)	5.65×10^{-6}	5.69×10^{-6}	2.23×10^{-6}	1.12×10^{-6} (−)	1.16×10^{-6}
MRS	$kgCueq$	2.04×10^{-6} (+)	1.30×10^{-6}	1.16×10^{-6}	2.02×10^{-6}	1.29×10^{-6}	1.15×10^{-6} (−)
OHH	$kgNO_xeq$	1.73×10^{-5} (+)	9.16×10^{-6}	9.67×10^{-6}	1.65×10^{-5}	8.76×10^{-6} (−)	9.27×10^{-6}
OTE	$kgNO_xeq$	1.83×10^{-5} (+)	9.69×10^{-6}	1.07×10^{-5}	1.74×10^{-5}	9.26×10^{-6} (−)	1.03×10^{-5}
SOD	$kgCFC11eq$	5.64×10^{-8}	2.83×10^{-8}	2.14×10^{-6} (+)	5.17×10^{-8}	2.59×10^{-8} (−)	2.14×10^{-6}
TA	$kgSO_2eq$	3.32×10^{-5} (+)	1.81×10^{-5}	1.78×10^{-5}	3.18×10^{-5}	1.74×10^{-5}	1.72×10^{-5} (−)
TE	$kg1,4\text{-}DCB$	1.86×10^{-2} (+)	1.68×10^{-2}	1.67×10^{-2}	1.69×10^{-2}	1.60×10^{-2}	1.59×10^{-2} (−)
W	m^3	3.11×10^{-2} (+)	1.73×10^{-2}	1.69×10^{-2}	3.08×10^{-2}	1.72×10^{-2}	1.67×10^{-2} (−)

The contribution of the environmental impacts across the life stages of the product is shown in Figure 3. *CF production* and *Manufacturing* contribute to the entire life cycle for scenarios A1 and B1 in the range of 29% and 49% for all the impact categories, except for *Marine Eutrophication* of scenario A1, which is equal to 9%. This result is mainly due to the several contributions of *substitution* of the composite. The contribution of those two processes further increases considering scenarios with repairing (i.e., A2, A3, B2 and B3) being in the range of 42% and 98% for almost all the impact categories, except for *Marine Eutrophication* of scenarios A2 and A3, which is equal to 19%, and for *Fine particulate matter formation—PM, Freshwater ecotoxicity—FEX, Freshwater eutrophication—FE* and *Stratospheric ozone depletion—SOD* of scenario A3 and B3, mainly due to the energy intensive process of thermal recycling. For those categories, the thermal process covers 86%–99% of the impacts. The other EoL treatments have low contributions compared to the overall life cycle stages. However, further clarification on different EoLs is shown below in this section. *Fossil-fuel-*

based curing agent synthesis and *bio-based curing agent synthesis* show similar contributions across scenarios for almost all impact categories of between 1.8% and 13%. The exception is shown for *Marine Eutrophication* of groups in scenario A due to the higher contribution to the overall life cycle stages being in the range of 40% and 80%.

Focusing on *Global warming* category, Figure 4 shows the contribution of each life cycle stage across the six scenarios. Excluding the EoL phase of composites, *Manufacturing* and *CF production* phases cover the majority of the CO_2eq emissions (from 42% up to 86% of the total). However, their contribution is the same across the scenarios since the process is the same. The synthesis of both curing agents have different contributions (from 4% to 13%), mainly due to the substitution. The substitution of the composite results to double the environmental impact in the overall life cycle of the product. Indeed, the substitution covers the majority of the impact since a second product with the same properties and manufacturing pathway is required. Focusing on the curing agents, the substitution of the fossil-fuel-based curing agent with the bio-based curing agent reduces the impact by up to 40%. Finally, the recycling of the product avoids the production of CFs and epoxy resin from virgin material, reducing the potential environmental impacts by 18–37%. The negative bar represents the avoided primary production of CFs and epoxy resins due to the recycling of CFRP waste. Differences in the negative impacts of recycling is due to different efficiencies of recycling both materials.

As concerns the use of alternative bio-based materials, the blue bars in Figure 5, which refer to scenarios A1 and B1, show that CFRP-ff and CFRP-bio have similar impacts across almost all the environmental categories. The substitution of the dynamic fossil-fuel-based curing agent with the bio-based material slightly reduces the impact for almost all the categories. A more relevant reduction is shown for the *Marine eutrophication* impact category. Indeed, the synthesis of the fossil-fuel-based cured epoxy resin covers 40% of the overall impact of this category. Furthermore, 99% of the impact of the curing agent synthesis is represented by the *meta-phenylene diamine* production—the proxy material used to model the dynamic curing agent. Sensitivity analysis on curing agents modeling are described in Section 3.1.1. CFRP-bio shows a slightly higher impact in the *Land use* impact category, mainly due to the pre-treatment of the epoxy resin mixing with the catalyst and the curing agent.

As concerns the repairing of damages in the composite, Figure 5 shows a clear reduction in the impacts in almost all the environmental categories in the case of repairing the composite (scenario A2 and B2) instead of substituting the composite (scenario A1 and B1). Repairing the composite results in a reduction in the impact of between 30% and 50% for fifteen impact categories. *Freshwater, Marine and Terrestrial ecotoxicity* are the environmental categories with a minor reduction in the impacts due to reparation compared to substitution (in the range of 4% and 18%) because of the laser machine contribution. Figure 5 shows little difference in results between the scenario A group and the scenario B group. This outlines that differences in the curing agents slightly affect the final results. It is noted that the comparison between scenarios with substitution and scenarios with repair excluded the EoL evaluation in order to compare only the impact variation due to the repair.

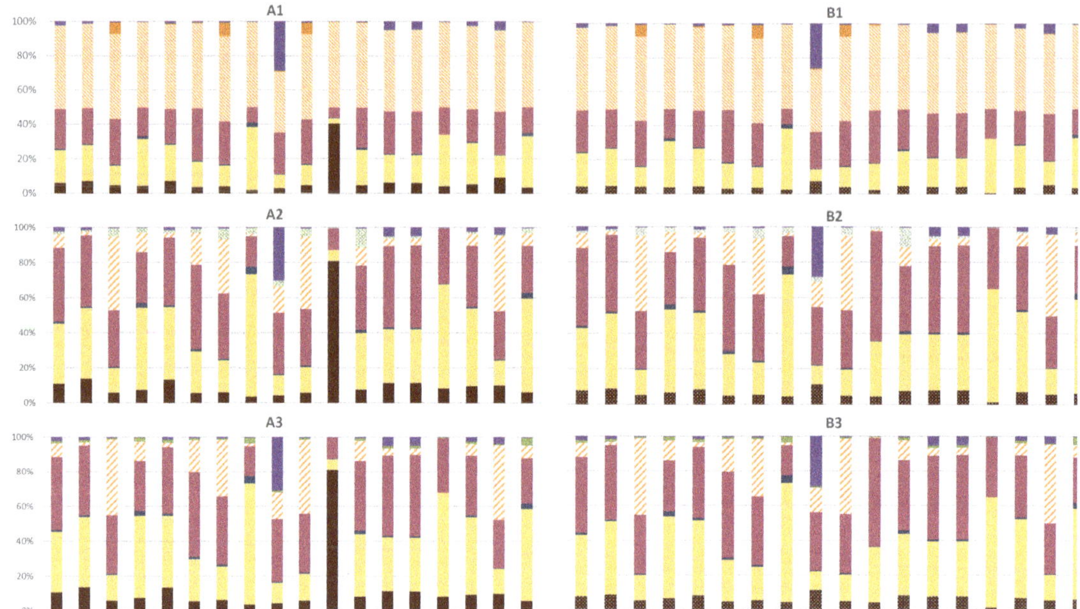

Figure 3. Impactcontribution of the life cycle stages of the product across the impact categories analyzed for the six scenarios. *CF production* and *Manufacturing* cover the majority of the impact. This contribution decreases for scenario A1 and B1 because of the *substitution*. Similar contributions are shown for the synthesis of both curing agents for almost all categories. Limited contributions are observed for *Prepreg*. Thermal recycling is the only EoL option that covers the majority of the impact for three categories (i.e., A3 and B3).

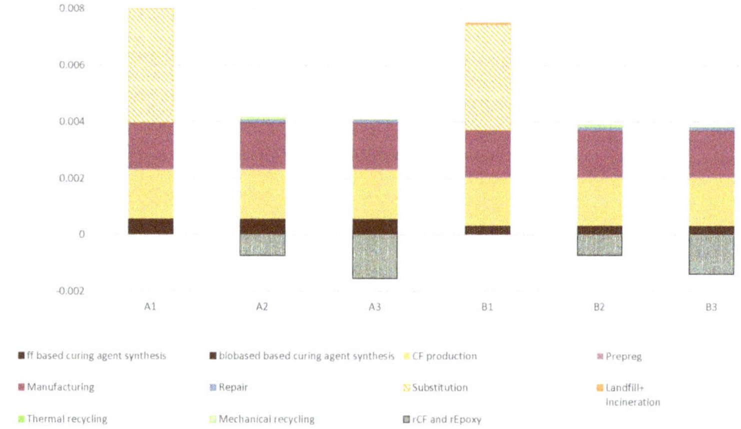

Figure 4. Impact contribution for the 6 scenarios across each life cycle of the FU for the *Global Warming* impact category. Negative impact values of *rCF and rEpoxy* are due to the avoided impact of virgin CF and epoxy resin because of their recycling. The quantitative avoided impacts are related to the chosen recycling option and its efficiency of recovery.

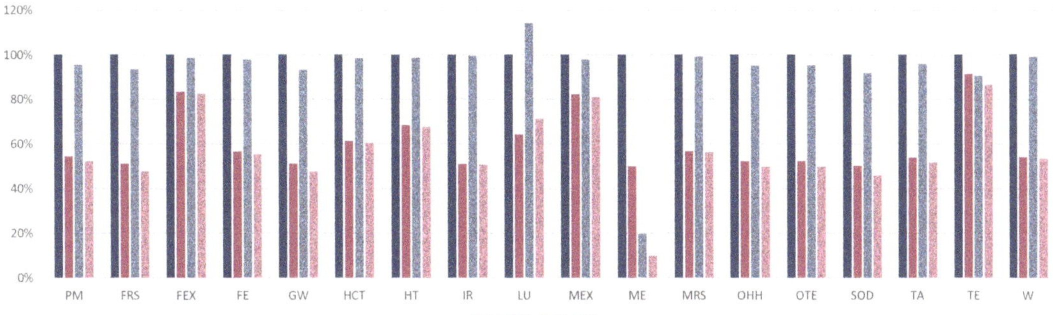

Figure 5. Comparative LCA between CFRP-ff with substitution (scenario A1), CFRP-ff with repair and mechanical recycling (scenario A2), CFRP-ff with repair and thermal recycling (scenario A3), CFRP-bio with substitution (scenario B1), CFRP-bio with repair and mechanical recycling (scenario B2) and CFRP-bio with repair and thermal recycling (scenario B3). Repairing the composite reduces the impact by up to 50% compared to scenarios with substitution. Less reduction in the impact when repairing are shown for the *Freshwater, Marine and Terrestrial Ecotoxicity* impact categories. The substitution of the curing agent slightly contributed to reducing the impacts among all the categories. The same occurs comparing mechanical and thermal recycling.

As concerns the recycling, Figure 6 shows the results regarding the three scenarios considered for both composites: landfill and incineration, mechanical recycling and thermal recycling. As a reference, the scenarios with the highest value were set as 100%, and the other scenarios were scaled accordingly. Comparing landfill/incineration with mechanical recycling, across the eighteen environmental categories, the impact decreases from 21% in the *Mineral resource scarcity* category up to 99% in the *Stratospheric ozone depletion* category. Comparing landfill/incineration with thermal recycling, the results are less stable. Indeed, thermal recycling shows the highest values for nine impact categories because of the high intensity of the thermal process. However, the thermal recycling can recover CFs and epoxy resin. This comparative assessment between different EoL treatments was not performed considering the avoided production of new materials due to, respectively, mechanical and thermal recycling. Indeed, from a circular economy perspective, both recycling treatments reduce the request of new raw materials, which does not occur in the case of landfills and incineration. Further assessments should include the additional production of CF and epoxy resin that is instead recovered during mechanical and thermal recycling. The negative impact values of *rCF and rEpoxy* shown in Figure 4 are due to the avoided impact of virgin CF and epoxy resin because of their recycling, which are related to the chosen recycling option and the efficiency of recovery. As a matter of fact, the pyrolysis treatment recovers 99% of CF and 85% of epoxy resin, while mechanical recycling recovers 60% of the CFs.

Summarizing the obtained results, LCIA shows that CF production and manufacturing of the product through autoclave technology are the most relevant processes in the entire life cycle of the product. Indeed, as shown in Figure 3, in scenarios with substitution (i.e., A1 and B1), both processes contributes to the entire life cycle in the range of 29% and 49% for all the impact categories, except for *Marine Eutrophication* of scenario A1, which is equal to 9%. The contribution of those two processes further increases considering scenarios with reparation (i.e., A2, A3, B2 and B3) to the range of 42% and 98%, except for *Marine Eutrophication* of scenarios A2 and A3, which is equal to 19%, and for *Fine particulate matter formation—PM, Freshwater ecotoxicity—FEX, Freshwater eutrophication—FE* and *Stratospheric ozone depletion—SOD* of scenario A3 and B3, mainly due to the energy intensive process of thermal recycling. Although comparison with other LCA studies on composites is not always possible due to differences in the FU, these results are aligned with other studies [49,58,61,72]. According to these results, efforts to identify alternative fibers to substitute CF have also been made by other studies [30,61,74]. However, the present case study did not include alternative bio-based fibers since laboratory tests on the possibility of repairing and recycling were not performed. Further research could be conducted in studying the possibility of repairing and recycling these bio-based fibers. Limitations were also due to *Production of the prepreg*, which was based on laboratory data that were scaled up to industrial scale. Those data should be further updated with primary data on pilot tests in order to verify the robustness of this process. It is noted that the analysis did not include the use phase of the product because it was assumed that the impact to be allocated to a small component compared to the entire car was limited. However, contributions of the use phase could change according to the product considered in the study and its relevance in the use phase. Indeed, Roy et al. [61] outlined that the use phase of a car component covers the majority of the impacts of the overall life cycle.

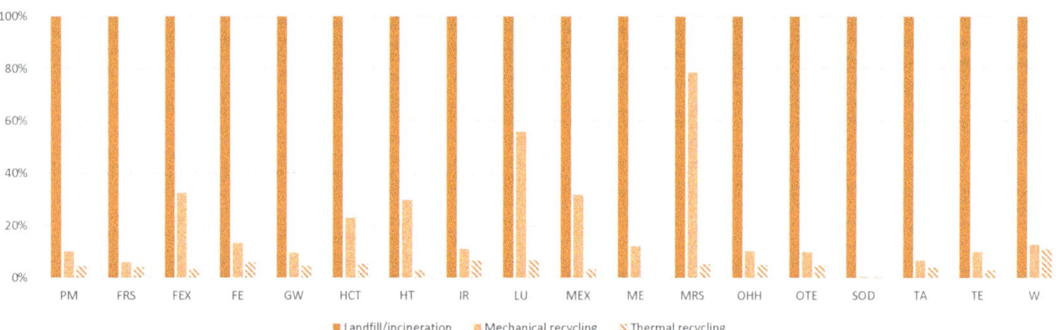

Figure 6. Comparative LCA between EoLs of composites. Incineration and landfill are representative of the most common composite waste management and hence are the baseline of the analysis. Relevant reduction in all the impact categories both for the thermal and mechanical recycling pathways. Limited environmental benefits compared to the baseline are shown for mechanical recycling in *Mineral resources scarcity*.

Across the scenarios developed, B3 included the substitution of the fossil-fuel-based curing agent with the bio-based curing agent, the reparation of the composite and the pyrolysis through thermal recycling of the composite at the EoL. As shown in Figure 5, B3 is the scenario with the highest environmental benefit among the impact categories compared to the baseline scenario (i.e., A1). This result outlines that the three above-mentioned drivers of the circular economy perspective could provide a reduction in the environmental impact from 10% up to 86% depending on the category considered. However, the highest benefits in terms of environmental burdens could be obtained with the implementation of repairing procedures on an industrial scale with a reduction up to 44–50% for eleven impact categories. Indeed, the analysis of environmental impacts due to the substitution of a damaged product to guarantee the entire life cycle of a service, for instance the car life cycle, is not usually included in LCA studies. Furthermore, substitution could occur also for aesthetic damages that do not compromise the structural properties of a product.

Finally, the lowest environmental benefits are observed substituting the traditional material with the bio-based material, which results in a variation in the range of −9% and +10%. According to the results obtained in this LCA, repairing is the driver with the highest environmental benefits compared to recycling and using alternative bio-based materials.

3.1.1. Sensitivity Analysis of LCA

Uncertainties on LCA results are mainly due to (1) primary data from the laboratory scale that were scaled-up to the industrial scale with assumptions and (2) the absence of literature data of some materials modeled and further modeling according to proxy data provided by the chosen database. Thus, sensitivity analysis was performed within this context to evaluate the robustness of the chosen modeling and how the results are affected by modeling assumptions. Furthermore, a sensitivity analysis was performed on the baseline scenario (i.e., A1), assuming that the damaged product is sent to mechanical or thermal recycling. As a matter of fact, the following sensitivity analysis were performed:

1. Fossil-fuel-based dynamic curing agent modeling;
2. Electrodes modeling for mechanical recycling;
3. Different EoLs of the damaged product in scenario A1.

The fossil-fuel-based dynamic curing agent was modeled with *market for meta-phenylene diamine | meta-phenylene diamine | Cutoff*. As explained in Section 2.1.2, meta-phenylene diamine was chosen as the aromatic amine in accordance with the literature [75,76]. As shown in Figure 5, the two chosen types of curing agents (i.e., fossil fuel and bio-based) showed a similar impact for all the impact categories, except for *Marine eutrophication*. The following two proxy materials were modeled in order to verify the trend: *market for diethanolamine | diethanolamine | Cutoff* and *market for diethanolamine | diethanolamine | Cutoff*. Variations in the results were observed to be in the range of -10% and +3% with respect of the baseline scenario (i.e., A1). Furthermore, in both cases, *Marine eutrophication* decreased by up to 20% compared to scenario A1, resulting in a similar value as CFRP-bio. Sensitivity analysis on the fossil-fuel-based curing agent modeling showed that the chosen model did not significantly vary with respect to other available options of modeling. However, this evaluation is limited to available datasets, and further research on curing agents modeling could change the results.

The use of electrodes was modeled with *market for electrode, negative, LiC6 | electrode, negative, LiC6 | Cutoff*. Due to the absence of information on the electrode properties, uncertainties of the modeling were analyzed. Two more scenarios of mechanical recycling were developed, substituting the modeling of electrode production with *market for electrode, negative, Ni | electrode, negative, Ni | Cutoff* and *market for electrode, positive, LaNi5 | electrode, positive, LaNi5 | Cutoff* in order to verify the robustness of the modeling. Figure 7 shows high variability in the results for *Fine particulate matter formation*, *Stratospheric ozone depletion*, *Terrestrial acidification* and *Terrestrial ecotoxicity*. However, *Freshwater eutrophication*, *Global Warming*, *Ionizing radiation* and *Marine eutrophication* are the impact categories with the lowest variability across the three possible models. Sensitivity analysis on electrodes used for mechanical recycling showed significant variation among the possible alternative datasets for some impact categories (Figure 7). Further research on the used electrodes could enhance the robustness of the data.

An alternative case was analyzed in order to evaluate the weight of the recycling in the case of substitution. This sensitivity analysis aimed to evaluate how the results might vary according to the selected EoL in case substitution is still necessary. Instead of sending the damaged product to a landfill or for incineration, this sensitivity analysis assumes that the damaged product is sent to mechanical recycling. The analysis comprises the production of CF, epoxy resin and prepreg, the manufacturing of the paddle shifters and the substitution. However, the contribution is almost the same as the base case (i.e., substitution without recycling). This means that the EoL treatment has a minimum weight compared to the rest of the life cycle stages. However, a reduction in the impact can be observed due to the mechanical recycling of the composite if further analysis of the avoided production of new CFs is considered.

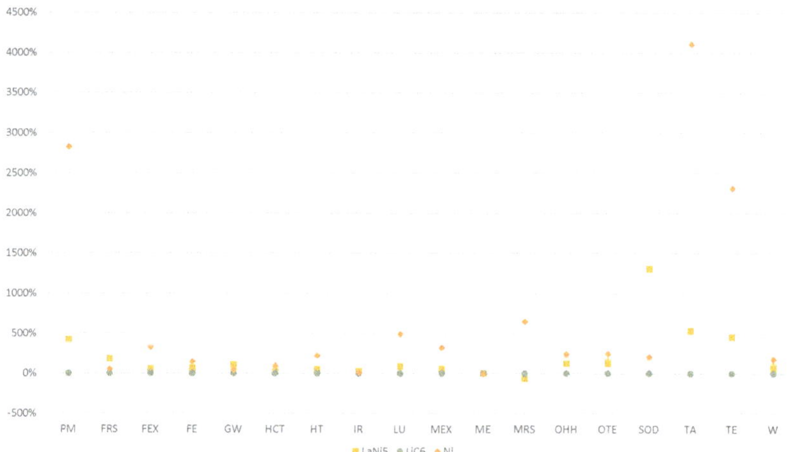

Figure 7. Sensitivity analysis on mechanical recycling. According to different providers of electrode production, some of the impact categories are affected by the results. Compared to LiC6, which is the base case, the highest and the lowest variability is observed, respectively, for *Terrestrial acidification* and *Marine eutrophication* categories.

3.2. LCC Results

In this section, the LCC results of the repair producer and thermal recycling businesses of a recycler are reported.

3.2.1. Repair Business Model of a Producer

The results of the economic assessment of the repair business model of the producer when the product is made from CFRP-ff are demonstrated in Table 17. As can be observed, the financial modeling indicates that for all three scenarios, with different ranges of CFRP-ff costs, the manufacturing and repair business model of a producer is always profitable as NPV is positive. Moreover, there is limited variability (0.8%) across the scenarios concerning NPV. The discounted payback period for all scenarios is achieved within the fifth year. Moving next to Table 18, the result of the economic assessment of a producer using CFRP-bio is demonstrated. The financial results show that this business is also profitable with limited variability (0.8%) across the scenarios with respect to NPV, and the discounted payback period is also achieved in the fifth year. Here, it is also essential to emphasize that the investment cost for the laser is high, whereas the sale volume of this product is relatively low, particularly in the context of automotive, where sale figures are high. Therefore, although the payback period is achieved within the fifth year, the payback period decreases if the sale volume increases.

Table 17. Result of the economic assessment of the repair business of producer/remanufacturer for product made from CFRP-ff based on various ranges of material costs.

Scenarios	Cost of CFRP-ff (EUR/kg)	NPV (EUR)	Discounted Payback Period (Years)
S1	51.8	+168,302	5
S2	55.4	+167,629	5
S3	59	+166,957	5

Table 18. Result of the economic assessment of the repair business of producer/remanufacturer for product made from CFRP-bio based on various ranges of material costs.

Scenarios	Cost of CFRP-bio (EUR/kg)	NPV (EUR)	Discounted Payback Period (Years)
S1	55.4	+167,629	5
S2	59	+166,957	5
S3	62.6	+166,284	5

To clearly understand why NPVs of all scenarios are positive, it is important to highlight that by manufacturing products from CFRP-ff, the total cost of the product increases by 0.08% to 0.39% compared to the total cost of the product built from the conventional CFRP. Concerning the product made from CFRP-bio, the total cost of the product increases by 0.23% to 0.55% compared to the total cost of the conventional CFRP product. Moreover, the repair cost is relatively low, 0.05%, compared to the total production cost. Therefore, it is clear that product repair can provide save a lot of costs as the producer can recapture the monetary value of the manufacturing process. The economic assessment of a product built from the new CF composites (CFRP-ff and CFRP-bio) and taking advantage of the reparability of the material to address the defective products shows that, overall, the manufacturing coupled with the repair business is very promising. In the current context, when the CF composite product is damaged, the manufacturer loss includes the entire production cost (both material and manufacturing cost will be discarded). By integrating the repair facilities into the manufacturing plant and bearing a relatively low repair cost, the manufacturer would avoid their loss of the whole production process. Therefore, the reparability of the new material is a significant opportunity for manufacturers to save money.

3.2.2. Thermal Recycling for a Recycler

The economic assessment results of a thermal recycler are summarized in Figure 8. The results display a positive NPV (10 years) for 4% of scenarios (2/45 scenarios) and show significant variability across the 45 scenarios. The positive NPV is achieved when the recycled CF price is at the maximum value (6 EUR/kg), and the transportation cost is at the minimum value (0.15 EUR/kg), and the plant is at least 95% saturated. The average discounted payback period for the two scenarios showcasing a positive NPV is obtained within 9.5 years. The sensitivity analysis shows that recycled CF prices are the most significant parameter impacting the NPV and discounted payback period. Moreover, the transportation cost is also the second most impactful parameter on both the NPV and the discounted payback period. Finally, a secondary element affecting the profitability of results is the annual amount of CFRP waste.

The economic sustainability of the thermal recycler that receives CFRP-ff/bio waste indicates that the business can be profitable only under a specific condition. Therefore, the economic results are affected by uncertainties that arise from the market price of recycled CF, the transportation cost, and the amount of waste. In this regard, a recycler cannot control the price of the recycled CF since the variabilities associated with this parameter are caused by the adverse economic cycle, uncertain financial stability and low demand for products at the global level [91,92]. However, a recycler can reduce the variabilities associated with the transportation cost and amount of CFRP waste. From a business point of view, strategic management of the reverse supply chain is critical for the profitability of the business of a thermal recycler. Strategic partnerships with organizations operating in the reverse supply chain, such as compliance schemes or waste collectors, are potential pathways to keep transportation costs minimal. In addition, such partnership can also guarantee a steady supply of a certain volume of waste, which is crucial for the successful implementation of the business.

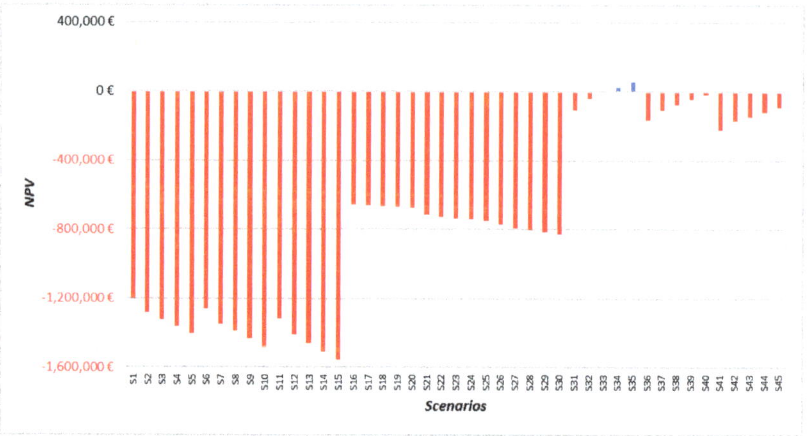

Figure 8. NPV of the thermal recycler based on various scenarios. The three parameters that affect the NPVs are the selling price of the recycled CF, the transportation and the volume of waste. The NPV is positive when the CF selling price is at a maximum, the transportation is at a minimum and the plant is saturated at least by 95%.

4. Conclusions

The present study attempted to provide an LCA coupled with an LCC of a new CFRP developed in a laboratory that could be repaired and recycled. The scope of this LCA was to analyze in a single study the following three relevant drivers that could enhance a circular economy: repairing, recycling and using bio-based materials. Firstly, according to the LCA results, the substitution of a CFRP product can double the overall environmental impact in the entire life cycle of a service. The introduction of repairing the product could reduce the environmental impacts from 17% to 50% according to each category (Figure 5). Furthermore, the substitution of the composite product may also occur for aesthetic damages, for instance, in a luxury car, which means that the damaged product becomes waste, even though its mechanical properties are not compromised. Nevertheless, the substitution of a CFRP product is not usually included in LCA studies, and hence, the environmental and cost effects of manufacturing an additional product are not quantified. Indeed, damages to a product and its subsequent substitution is unexpected and not desired considering the entire life cycle of a product. Secondly, the recovery of rCf and epoxy resin through mechanical and thermal recycling allowed reducing the environmental impact related to the waste management up to 99% for some categories (Figure 6). Finally, only the curing agent was analyzed as a bio-based material in the composite (group of scenarios B) in order to guarantee the properties of the developed composite to be repairable. Hence, the curing agent is a small percentage if compared to the overall weight of the composite. As a result, the two materials (i.e., CFRP-ff and CFRP-bio) had similar environmental impacts across the eighteen impact categories (Figure 5). The results may change increasing the percentage of bio-based materials to produce a CFRP composite.

From an economic point of view, the LCC assessment showed that repair technologies and processes are highly convenient for composite parts' manufacturers due to the typically high defect rate of composites and the high reproduction costs that have to be sustained. Under these conditions, investing in laser-based technology and the higher cost of the new repairable composite materials that should be sustained are largely covered by cost savings in short time periods. Regarding the thermal recycling of the newly assessed material and technologies, it appeared that the business can be economically sustainable only if the recovered CF is sold at its maximum price, the transportation cost is at its lowest and the recycler receives high volumes of waste. In this regard, this business is characterized by significant uncertainties, which can be reduced through a vertical integration or strategic partnerships with actors involved in the reverse supply chain in order to guarantee big treatment volumes and a stable transportation cost.

The LCA and LCC results obtained are affected by uncertainties due to the absence of data or scaling up of laboratory data to the industrial scale, as analyzed in Sections 3.1.1 and 2.2.5. The LCA and LCC allowed collecting preliminary environmental and costing information that is crucial for further developing the product at pilot and industrial scales. Further research on those two types of composites and repairing technologies with a pilot test are fundamental to obtaining a new type of product that can be repaired, which could push on the LCA and LCC's robustness. Moreover, performing LCA and LCC during all those steps could be crucial to both understanding how the results change according to data availability and production scale (i.e., from laboratory to industrial scale) and monitoring the environmental and costing efficiency as decision support tools for its development.

Supplementary Materials: The following supporting information can be downloaded at: https://www.mdpi.com/article/10.3390/ma15092986/s1, Table S1: LCA studies on composites; Figure S1: Normalization of LCA results of the six scenarios | human health; Figure S2: Normalization of LCA results of the six scenarios - terrestrial ecosystem; Figure S3: Normalization of LCA results of the six scenarios - freshwater ecosystem; Figure S4: Normalization of LCA results of scenario A1. Water consumption and Toxicity are the two categories with higher environmental impact. CF production, manufacturing and substitution are the steps with highest contribution; Figure S5: Normalization of LCA results of scenario A2. Water consumption and Toxicity are the two categories with higher environmental impact. CF production and manufacturing are the steps with highest contribution; Figure S6: Normalization of LCA results of scenario A3. Stratospheric ozone depletion, Water consumption and Toxicity are the two categories with higher environmental impact. CF production, manufacturing and thermal recycling are the steps with highest contribution; Figure S7: Normalization of LCA results of scenario B1. Water consumption and Toxicity are the two categories with higher environmental impact. CF production, manufacturing and substitution are the steps with highest contribution; Figure S8: Normalization of LCA results of scenario A2. Water consumption and Toxicity are the two categories with higher environmental impact. CF production and manufacturing are the steps with highest contribution; Figure S9: Normalization of LCA results of scenario A3. Stratospheric ozone depletion, Water consumption and Toxicity are the two categories with higher environmental impact. CF production, manufacturing and thermal recycling are the steps with the highest contribution. References [93–95] are cited in the supplementary materials.

Author Contributions: Conceptualization, E.A., M.M., C.B. and G.C.; methodology, E.A. and M.M.; software, E.A.; validation, E.A. and M.M.; formal analysis, E.A., M.M., C.B. and G.C.; investigation, E.A. and M.M.; resources, E.A., M.M., C.B. and G.C.; data curation, E.A. and M.M.; writing—original draft preparation, E.A. and M.M.; writing—review and editing, E.A., M.M., C.B. and G.C.; visualization, E.A., M.M., C.B. and G.C.; supervision, E.A., C.B., G.C. and A.B.; project administration, E.A., C.B., G.C. and A.B.; funding acquisition, G.C. and A.B. All authors have read and agreed to the published version of the manuscript.

Funding: This research was funded by Fondazione Cariplo, grant number 2018–1004, as part of the project 'COMPOSER—Carbon-fiber reinforced cOMPOsites for Sustainable circular Economy models based on Repair and Remanufacturing for Reuse'.

Institutional Review Board Statement: Not applicable.

Informed Consent Statement: Not applicable.

Data Availability Statement: The used primary data directly represent laboratory test performed by Politecnico di Milano. Due to confidentiality reasons, these data are and will not be publicly available.

Acknowledgments: This work was supported through the 'COMPOSER' project. For technical aspects, we acknowledge the Department of Chemistry, Materials 'Giulio Natta' of Politecnico di Milano, that performed laboratory tests for the development of the composite and Marcello Colledani, Ali Gökhan Demir and Marco Diani from the Chemical Engineering and Mechanical Engineering Department of Politecnico di Milano who performed experiments for the repair and recycling technologies.

Conflicts of Interest: The authors declare no conflict of interest.

References

1. Asmatulu, E.; Twomey, J.; Overcash, M. Recycling of fiber-reinforced composites and direct structural composite recycling concept. *J. Compos. Mater.* **2014**, *48*, 593–608. [CrossRef]
2. Altin Karataş, M.; Gökkaya, H. A review on machinability of carbon fiber reinforced polymer (CFRP) and glass fiber reinforced polymer (GFRP) composite materials. *Def. Technol.* **2018**, *14*, 318–326. [CrossRef]
3. Luzuriaga, A.R.D.; Martin, R.; Markaide, N.; Rekondo, A.; Cabañero, G.; Rodriguez, J.; Odriozola, I. Epoxy resin with exchangeable disulfide crosslinks to obtain reprocessable, repairable and recyclable fiber-reinforced thermoset composites. *Mater. Horizons* **2016**, *3*, 241–247. [CrossRef]
4. Yang, Y.; Boom, R.; Irion, B.; van Heerden, D.J.; Kuiper, P.; de Wit, H. Recycling of composite materials. *Chem. Eng. Process. Process Intensif.* **2012**, *51*, 53–68. [CrossRef]
5. Das, S. Life cycle assessment of carbon fiber-reinforced polymer composites. *Int. J. Life Cycle Assess.* **2011**, *16*, 268–282. [CrossRef]
6. Assocompositi. *2020 State of the Industry Report*; Technical Report; Assocompositi: Milano, Italy, 2020.
7. Bachmann, J.; Hidalgo, C.; Bricout, S. Environmental analysis of innovative sustainable composites with potential use in aviation sector—A life cycle assessment review. *Sci. China Technol. Sci.* **2017**, *60*, 1301–1317. [CrossRef]
8. Duflou, J.R.; De Moor, J.; Verpoest, I.; Dewulf, W. Environmental impact analysis of composite use in car manufacturing. *Cirp Ann.—Manuf. Technol.* **2009**, *58*, 9–12. [CrossRef]
9. Scelsi, L.; Bonner, M.; Hodzic, A.; Soutis, C.; Wilson, C.; Scaife, R.; Ridgway, K. Potential emissions savings of lightweight composite aircraft components evaluated through life cycle assessment. *Express Polym. Lett.* **2011**, *5*, 209–217. [CrossRef]
10. Karuppannan Gopalraj, S.; Kärki, T. A review on the recycling of waste carbon fibre/glass fibre-reinforced composites: Fibre recovery, properties and life-cycle analysis. *SN Appl. Sci.* **2020**, *2*, 433. [CrossRef]
11. Xian, G.; Guo, R.; Li, C. Combined effects of sustained bending loading, water immersion and fiber hybrid mode on the mechanical properties of carbon/glass fiber reinforced polymer composite. *Compos. Struct.* **2022**, *281*, 115060. [CrossRef]
12. Meng, F.; Olivetti, E.A.; Zhao, Y.; Chang, J.C.; Pickering, S.J.; McKechnie, J. Comparing Life Cycle Energy and Global Warming Potential of Carbon Fiber Composite Recycling Technologies and Waste Management Options. *ACS Sustain. Chem. Eng.* **2018**, *6*, 9854–9865. [CrossRef]
13. Witten, E.; Sauer, M.; Kuhnel, M. *Composites Market Report 2017*; AVK (Federation of Reinforced Plastics): Frankfurt, Germany, 2017; pp. 1–44.
14. Zhang, J.; Chevali, V.S.; Wang, H.; Wang, C.H. Current status of carbon fibre and carbon fibre composites recycling. *Compos. Part B Eng.* **2020**, *193*, 108053. [CrossRef]
15. Heijungs, R.; Suh, S. *The Computational Structure of Life Cycle Assessment*; Eco-Efficiency in Industry and Science: Delft, The Netherlands, 2002. [CrossRef]
16. Shafiee, M.; Alghamdi, A.; Sansom, C.; Hart, P.; Encinas-Oropesa, A. A through-life cost analysis model to support investment decision-making in concentrated solar power projects. *Energies* **2020**, *13*, 1553. [CrossRef]
17. Jacquemin, L.; Pontalier, P.y.; Sablayrolles, C. Life cycle assessment (LCA) applied to the process industry: A review. *Int. J. Life Cycle Assess.* **2012**, *17*, 1028–1041. [CrossRef]
18. Fortunato, G.; Anghileri, L.; Griffini, G.; Turri, S. Simultaneous recovery of matrix and fiber in carbon reinforced composites through a diels-alder solvolysis process. *Polymers* **2019**, *11*, 1007. [CrossRef]
19. Giorgini, L.; Benelli, T.; Brancolini, G.; Mazzocchetti, L. Recycling of carbon fiber reinforced composite waste to close their life cycle in a cradle-to-cradle approach. *Curr. Opin. Green Sustain. Chem.* **2020**, *26*, 100368. [CrossRef]
20. Liu, P.; Meng, F.; Barlow, C.Y. Wind turbine blade end-of-life options: An eco-audit comparison. *J. Clean. Prod.* **2019**, *212*, 1268–1281. [CrossRef]
21. Hadi, P.; Ning, C.; Ouyang, W.; Xu, M.; Lin, C.S.; McKay, G. Toward environmentally-benign utilization of nonmetallic fraction of waste printed circuit boards as modifier and precursor. *Waste Manag.* **2015**, *35*, 236–246. [CrossRef]
22. Vo Dong, P.A.; Azzaro-Pantel, C.; Cadene, A.L. Economic and environmental assessment of recovery and disposal pathways for CFRP waste management. *Resour. Conserv. Recycl.* **2018**, *133*, 63–75. [CrossRef]
23. Ghobadi, A. Common Type of Damages in Composites and Their Inspections. *World J. Mech.* **2017**, *7*, 24–33. [CrossRef]
24. Jacques, K.; Bax, L.; Vasiliadis, H.; Magallon, I.; Ong, K. *Polymer Composites for Automotive Sustainability*; Technical Report; SUSCHEM: Bruxelles, Belgium, 2016.
25. Delogu, M.; Zanchi, L.; Dattilo, C.A.; Ierides, M. Parameters affecting the sustainability trade-off between production and use stages in the automotive lightweight design. *Procedia CIRP* **2018**, *69*, 534–539. [CrossRef]
26. European Parliament and Council. Directive 2000/53/EC on end-of-life vehicles. *Off. J. Eur. Union* **2000**. Available online: https://eur-lex.europa.eu/legal-content/EN/TXT/PDF/?uri=CELEX:32000L0053&from=EN (accessed on 15 February 2022).
27. Marcelo, G.; Cherubini, E.; Orsi, P.; Roberto, S. Waste management Life Cycle Assessment: The case of a reciprocating air compressor in Brazil. *J. Clean. Prod.* **2014**, *70*, 164–174. [CrossRef]
28. Yin, Z.; Li, C.; Tie, Y.; Duan, Y. Impact damage detection in patch-repaired CFRP laminates using nonlinear lamb waves. *Sensors* **2021**, *21*, 219. [CrossRef] [PubMed]

29. Zaman, A.U.; Gutub, S.A.; Soliman, M.F.; Wafa, M.A. Sustainability and human health issues pertinent to fibre reinforced polymer composites usage: A review. *J. Reinf. Plast. Compos.* **2014**, *33*, 1069–1084. [CrossRef]
30. Hermansson, F.; Janssen, M.; Svanström, M. Prospective study of lignin-based and recycled carbon fibers in composites through meta-analysis of life cycle assessments. *J. Clean. Prod.* **2019**, *223*, 946–956. [CrossRef]
31. Cucurachi, S.; Van Der Giesen, C.; Guinée, J. Ex-ante LCA of Emerging Technologies. *Procedia CIRP* **2018**, *69*, 463–468. [CrossRef]
32. Finkbeiner, M.; Inaba, A.; Tan, R.B.; Christiansen, K.; Klüppel, H.J. The New International Standards for Life Cycle Assessment: ISO 14040 and ISO 14044. *Int. J. Life Cycle Assess.* **2006**, *11*, 80–85. [CrossRef]
33. ISO. *BS EN ISO 14044: Environmental Management—Life Cycle Assessment—Requirements and Guidelines*; Technical Report; International Organization for Standardization: Geneva, Switzerland, 2020.
34. Rovelli, D.; Brondi, C.; Andreotti, M.; Abbate, E.; Zanforlin, M.; Ballarino, A. A Modular Tool to Support Data Management for LCA in Industry: Methodology, Application and Potentialities. *Sustainability* **2022**, *14*, 3746. [CrossRef]
35. Fazeni, K.; Lindorfer, J.; Prammer, H. Methodological advancements in Life Cycle Process Design: A preliminary outlook. *Resour. Conserv. Recycl.* **2014**, *92*, 66–77. [CrossRef]
36. van der Giesen, C.; Cucurachi, S.; Guinée, J.; Kramer, G.J.; Tukker, A. A critical view on the current application of LCA for new technologies and recommendations for improved practice. *J. Clean. Prod.* **2020**, *259*, 120904. [CrossRef]
37. Rebitzer, G.; Ekvall, T.; Frischknecht, R.; Hunkeler, D.; Norris, G.; Rydberg, T.; Schmidt, W.P.; Suh, S.; Weidema, B.P.; Pennington, D.W. Life cycle assessment Part 1: Framework, goal and scope definition, inventory analysis, and applications. *Environ. Int.* **2004**, *30*, 701–720. [CrossRef] [PubMed]
38. Walczak, K.A.; Hutchins, M.J.; Dornfeld, D. Energy system design to maximize net energy production considering uncertainty in scale-up: A case study in artificial photosynthesis. *Procedia CIRP* **2014**, *15*, 306–312. [CrossRef]
39. Villares, M.; Işıldar, A.; Mendoza Beltran, A.; Guinee, J. Applying an ex-ante life cycle perspective to metal recovery from e-waste using bioleaching. *J. Clean. Prod.* **2016**, *129*, 315–328. [CrossRef]
40. Thonemann, N.; Schulte, A.; Maga, D. How to conduct prospective life cycle assessment for emerging technologies? A systematic review and methodological guidance. *Sustainability* **2020**, *12*, 1192. [CrossRef]
41. Hesser, F. Environmental advantage by choice: Ex-ante LCA for a new Kraft pulp fibre reinforced polypropylene composite in comparison to reference materials. *Compos. Part B Eng.* **2015**, *79*, 197–203. [CrossRef]
42. Moni, S.M.; Mahmud, R.; High, K.; Carbajales-Dale, M. Life cycle assessment of emerging technologies: A review. *J. Ind. Ecol.* **2020**, *24*, 52–63. [CrossRef]
43. Cornago, S.; Rovelli, D.; Brondi, C.; Crippa, M.; Morico, B.; Ballarino, A.; Dotelli, G. Stochastic consequential Life Cycle Assessment of technology substitution in the case of a novel PET chemical recycling technology. *J. Clean. Prod.* **2021**, *311*, 127406. [CrossRef]
44. Gavankar, S.; Suh, S.; Keller, A.A. The Role of Scale and Technology Maturity in Life Cycle Assessment of Emerging Technologies: A Case Study on Carbon Nanotubes. *J. Ind. Ecol.* **2015**, *19*, 51–60. [CrossRef]
45. Hung, C.R.; Ellingsen, L.A.W.; Majeau-Bettez, G. LiSET: A Framework for Early-Stage Life Cycle Screening of Emerging Technologies. *J. Ind. Ecol.* **2020**, *24*, 26–37. [CrossRef]
46. Piccinno, F.; Hischier, R.; Seeger, S.; Som, C. From laboratory to industrial scale: A scale-up framework for chemical processes in life cycle assessment studies. *J. Clean. Prod.* **2016**, *135*, 1085–1097. [CrossRef]
47. Akhshik, M.; Panthapulakkal, S.; Tjong, J.; Sain, M. Life cycle assessment and cost analysis of hybrid fiber-reinforced engine beauty cover in comparison with glass fiber-reinforced counterpart. *Environ. Impact Assess. Rev.* **2017**, *65*, 111–117. [CrossRef]
48. Zhou, C.; Shi, S.Q.; Chen, Z.; Cai, L.; Smith, L. Comparative environmental life cycle assessment of fiber reinforced cement panel between kenaf and glass fibers. *J. Clean. Prod.* **2018**, *200*, 196–204. [CrossRef]
49. Raugei, M.; Morrey, D.; Hutchinson, A.; Winfield, P. A coherent life cycle assessment of a range of lightweighting strategies for compact vehicles. *J. Clean. Prod.* **2015**, *108*, 1168–1176. [CrossRef]
50. La Rosa, A.D.; Banatao, D.R.; Pastine, S.J.; Latteri, A.; Cicala, G. Recycling treatment of carbon fibre/epoxy composites: Materials recovery and characterization and environmental impacts through life cycle assessment. *Compos. Part B Eng.* **2016**, *104*, 17–25. [CrossRef]
51. Rankine, R.K.; Chick, J.P.; Harrison, G.P. Energy and carbon audit of a rooftop wind turbine. *Proc. Inst. Mech. Eng. Part A J. Power Energy* **2006**, *220*, 643–654. [CrossRef]
52. Nunes, A.O.; Viana, L.R.; Guineheuc, P.M.; da Silva Moris, V.A.; de Paiva, J.M.F.; Barna, R.; Soudais, Y. Life cycle assessment of a steam thermolysis process to recover carbon fibers from carbon fiber-reinforced polymer waste. *Int. J. Life Cycle Assess.* **2018**, *23*, 1825–1838. [CrossRef]
53. Duflou, J.R.; Yelin, D.; Van Acker, K.; Dewulf, W. Comparative impact assessment for flax fibre versus conventional glass fibre reinforced composites: Are bio-based reinforcement materials the way to go? *Cirp Ann.—Manuf. Technol.* **2014**, *63*, 45–48. [CrossRef]
54. Song, Y.S.; Youn, J.R.; Gutowski, T.G. Life cycle energy analysis of fiber-reinforced composites. *Compos. Part A Appl. Sci. Manuf.* **2009**, *40*, 1257–1265. [CrossRef]
55. Khalil, Y.F. Eco-efficient lightweight carbon-fiber reinforced polymer for environmentally greener commercial aviation industry. *Sustain. Prod. Consum.* **2017**, *12*, 16–26. [CrossRef]

56. Ahmed, I.M.; Tsavdaridis, K.D. Life cycle assessment (LCA) and cost (LCC) studies of lightweight composite flooring systems. *J. Build. Eng.* **2018**, *20*, 624–633. [CrossRef]

57. Yılmaz, E.; Arslan, H.; Bideci, A. Environmental performance analysis of insulated composite facade panels using life cycle assessment (LCA). *Constr. Build. Mater.* **2019**, *202*, 806–813. [CrossRef]

58. Witik, R.A.; Teuscher, R.; Michaud, V.; Ludwig, C.; Månson, J.A.E. Carbon fibre reinforced composite waste: An environmental assessment of recycling, energy recovery and landfilling. *Compos. Part A Appl. Sci. Manuf.* **2013**, *49*, 89–99. [CrossRef]

59. Mohammadi, S.; Yousefi, M.; Khazaei, M. A review on composite patch repairs and the most important parameters affecting its efficiency and durability. *J. Reinf. Plast. Compos.* **2021**, *40*, 3–15. [CrossRef]

60. Tapper, R.J.; Longana, M.L.; Norton, A.; Potter, K.D.; Hamerton, I. An evaluation of life cycle assessment and its application to the closed-loop recycling of carbon fibre reinforced polymers. *Compos. Part B Eng.* **2020**, *184*, 107665. [CrossRef]

61. Roy, P.; Defersha, F.; Rodriguez-Uribe, A.; Misra, M.; Mohanty, A.K. Evaluation of the life cycle of an automotive component produced from biocomposite. *J. Clean. Prod.* **2020**, *273*, 123051. [CrossRef]

62. La Rosa, A.D.; Greco, S.; Tosto, C.; Cicala, G. LCA and LCC of a chemical recycling process of waste CF-thermoset composites for the production of novel CF-thermoplastic composites. Open loop and closed loop scenarios. *J. Clean. Prod.* **2021**, *304*, 127158. [CrossRef]

63. Norris, G.A. Integrating life cycle cost analysis and LCA. *Int. J. Life Cycle Assess.* **2001**, *6*, 118–120. [CrossRef]

64. Ghosh, C.; Maiti, J.; Shafiee, M.; Kumaraswamy, K.G. Reduction of life cycle costs for a contemporary helicopter through improvement of reliability and maintainability parameters. *Int. J. Qual. Reliab. Manag.* **2018**, *35*, 545–567. [CrossRef]

65. Bernhard, R.H. *Engineering Economics for Capital Investment Analysis*; Rentice Hall: Englewood Cliffs, NJ, USA, 1992; Volume 38, pp. 72–73. [CrossRef]

66. Strogonov, K.V. Energy and economic assessment of composite products production on the example of pipe products. In Proceedings of the 2021 3rd International Youth Conference on Radio Electronics, Electrical and Power Engineering (REEPE), Moscow, Russia, 11–13 March 2021; pp. 5–8.

67. Castella, P.S.; Blanc, I.; Ferrer, M.G.; Ecabert, B.; Wakeman, M.; Manson, J.A.; Emery, D.; Han, S.H. Integrating life cycle costs and environmental impacts of composite rail car-bodies for a Korean train. *Int. J. Life Cycle Assess.* **2009**, *14*, 429–442. [CrossRef]

68. Conroy, A.; Halliwell, S.; Reynolds, T. Composite recycling in the construction industry. *Compos. Part A Appl. Sci. Manuf.* **2006**, *37*, 1216–1222. [CrossRef]

69. Gharfalkar, M.; Court, R.; Campbell, C.; Ali, Z.; Hillier, G. Analysis of waste hierarchy in the European waste directive 2008/98/EC. *Waste Manag.* **2015**, *39*, 305–313. [CrossRef] [PubMed]

70. Henriksen, M.L.; Ravnsbæk, J.B.; Bjerring, M.; Vosegaard, T.; Daasbjerg, K.; Hinge, M. Epoxy Matrices Modified by Green Additives for Recyclable Materials. *ChemSusChem* **2017**, *10*, 2936–2944. [CrossRef] [PubMed]

71. Jin, F.L.; Li, X.; Park, S.J. Synthesis and application of epoxy resins: A review. *J. Ind. Eng. Chem.* **2015**, *29*, 1–11. [CrossRef]

72. Forcellese, A.; Marconi, M.; Simoncini, M.; Vita, A. Life cycle impact assessment of different manufacturing technologies for automotive CFRP components. *J. Clean. Prod.* **2020**, *271*, 122677. [CrossRef]

73. Duflou, J.R.; Deng, Y.; Van Acker, K.; Dewulf, W. Do fiber-reinforced polymer composites provide environmentally benign alternatives? A life-cycle-assessment-based study. *MRS Bull.* **2012**, *37*, 374–382. [CrossRef]

74. Isola, C.; Sieverding, H.L.; Raghunathan, R.; Sibi, M.P.; Webster, D.C.; Sivaguru, J.; Stone, J.J. Life cycle assessment of photodegradable polymeric material derived from renewable bioresources. *J. Clean. Prod.* **2017**, *142*, 2935–2944. [CrossRef]

75. Ashcroft, W.R. Curing agents for epoxy resins. In *Chemistry and Technology of Epoxy Resins*; Springer: Dordrecht, The Netherlands, 1993; pp. 37–71. [CrossRef]

76. White, H.J. Amine curing agents for epoxy resins. *J. Prot. Coat. Linings* **1989**, *6*, 47–56.

77. Davidson, M.G.; Elgie, S.; Parsons, S.; Young, T.J. Production of HMF, FDCA and their derived products: A review of life cycle assessment (LCA) and techno-economic analysis (TEA) studies. *Green Chem.* **2021**, *23*, 3154–3171. [CrossRef]

78. Dessbesell, L.; Souzanchi, S.; Venkateswara Rao, K.T.; Carrillo, A.A.; Bekker, D.; Hall, K.A.; Lawrence, K.M.; Tait, C.L.J.; Xu, C. Production of 2,5-furandicarboxylic acid (FDCA) from starch, glucose, or high-fructose corn syrup: Techno-economic analysis. *Biofuels Bioprod. Biorefining* **2019**, *13*, 1234–1245. [CrossRef]

79. Kim, H.; Lee, S.; Ahn, Y.; Lee, J.; Won, W. Sustainable Production of Bioplastics from Lignocellulosic Biomass: Technoeconomic Analysis and Life-Cycle Assessment. *ACS Sustain. Chem. Eng.* **2020**, *8*, 12419–12429. [CrossRef]

80. Zervas, E.; Lazarou, C. Influence of European passenger cars weight to exhaust CO2 emissions. *Energy Policy* **2008**, *36*, 248–257. [CrossRef]

81. Weh, A. *High Voltage Pulse Fragmentation Technology to Recycle Fibre-Reinforced Composites*; European Commission: Geneva, Switzerland, 2015; p. 12.

82. Shuaib, N.A.; Mativenga, P.T. Carbon Footprint Analysis of Fibre Reinforced Composite Recycling Processes. *Procedia Manuf.* **2017**, *7*, 183–190. [CrossRef]

83. Naqvi, S.R.; Prabhakara, H.M.; Bramer, E.A.; Dierkes, W.; Akkerman, R.; Brem, G. A critical review on recycling of end-of-life carbon fibre/glass fibre reinforced composites waste using pyrolysis towards a circular economy. *Resour. Conserv. Recycl.* **2018**, *136*, 118–129. [CrossRef]

84. Saviano, L. Recycling of Fibreglass Waste. (Original Version in Italian: "Riciclo dei Rifiuti in Vetroresina". Korec. 2020. Available online: http://www.ko-rec.com/wp-content/uploads/2020/01/impaginato-pubblicato_bassa-risoluzione.pdf (accessed on 15 February 2022).
85. Korec. Innovative Technology for the Fibreglass Waste Recycling. (Original version in Italian: "Processo Innovativo per il Riciclo dei Rifiuti in Vetroresina". Available online: http://www.ko-rec.com/ (accessed on 15 February 2022).
86. Da Via, F.; Suriano, R.; Boumezgane, O.; Grande, A.M.; Tonelli, C.; Turri, S. Self-healing behavior in blends of PDMS-based polyurethane ionomers. *Polym. Adv. Technol.* **2022**, *33*, 556–565. [CrossRef]
87. European Commission. Electricity Price Statistics, 2021; Statistics Explained. Available online: https://ec.europa.eu/eurostat/statistics-explained/index.php?title=Electricity_price_statistics (accessed on 15 February 2022).
88. European Commission. Wages and Labour Cost, 2020; Statistics Explained. Available online: https://ec.europa.eu/eurostat/statistics-explained/index.php/Wages_and_labour_cost (accessed on 15 February 2022).
89. Graf, J.; Gruber, K.; Shen, Y.; Reinhart, G. An Approach for the Sensory Integration into the Automated Production of Carbon Fiber Reinforced Plastics. *Procedia CIRP* **2016**, *52*, 280–285. [CrossRef]
90. Alam, S.; Hedayati Dezfuli, F. Experiment-Based Sensitivity Analysis of Scaled Carbon-Fiber-Reinforced Elastomeric Isolators in Bonded Applications. *Fibers* **2016**, *4*, 4. [CrossRef]
91. Yazdani, M.; Gonzalez, E.D.; Chatterjee, P. A multi-criteria decision-making framework for agriculture supply chain risk management under a circular economy context. *Manag. Decis.* **2019**, *59*, 1801–1826. [CrossRef]
92. Brillinger, A.S.; Els, C.; Schäfer, B.; Bender, B. Business model risk and uncertainty factors: Toward building and maintaining profitable and sustainable business models. *Bus. Horizons* **2019**, *63*, 121–130. [CrossRef]
93. Pegoretti, T.D.S.; Mathieux, F.; Evrard, D.; Brissaud, D.; Arruda, J.R.D.F. Use of recycled natural fibres in industrial products: A comparative LCA case study on acoustic components in the Brazilian automotive sector. *Resour. Conserv. Recycl.* **2014**, *84*, 1–14. [CrossRef]
94. Li, X.; Bai, R.; McKechnie, J. Environmental and financial performance of mechanical recycling of carbon fibre reinforced polymers and comparison with conventional disposal routes. *J. Clean. Prod.* **2016**, *127*, 451–460. [CrossRef]
95. National Institute for Public Health and Environment; Ministry of Health Welfare and Sport. Normalization scores ReCiPe 2016. 2020. Available online: https://www.rivm.nl/en/documenten/normalization-scores-recipe-2016 (accessed on 15 February 2022).

Review

Literature Review on the Utilization of Rice Husks: Focus on Application of Materials for Digital Fabrication

Kohei Morimoto [1,*], **Kazutoshi Tsuda** [2] **and Daijiro Mizuno** [2]

1   Graduate School of Design, Nagaoka Institute of Design, Niigata 9402088, Japan
2   Center for the Possible Futures, Kyoto Institute of Technology, Kyoto 6060951, Japan; tsudakazutoshi@kit.ac.jp (K.T.); daijirom@kit.ac.jp (D.M.)
*   Correspondence: kmorimoto@nagaoka-id.ac.jp; Tel.: +81-258-21-3557

Abstract: To achieve a sustainable society, it is important to use biological resources effectively to the extent that they are renewable. Rice husk, which is abundantly produced in various regions, is a useful biomass resource. To promote their use further, it is important to expand the fields in which they are used. Therefore, this study reviews the research on rice-husk-based materials that can be used in digital fabrication, such as those used with 3D printers and Computer Numerical Control (CNC) machines, which have become increasingly popular in recent years. After outlining the characteristics of each machining method, the authors surveyed and analyzed the original research on rice-husk-based materials for 3D printers and particleboard available in digital fabrication machines for 2D machining. This review identifies issues and proposes solutions for expanding the use of rice-husk-based materials. It also indicates the need for further research on various aspects, such as the workability and maintainability of the equipment.

Keywords: green materials; rice husk; agricultural byproduct; 3D printing; digital fabrication; particleboard

Citation: Morimoto, K.; Tsuda, K.; Mizuno, D. Literature Review on the Utilization of Rice Husks: Focus on Application of Materials for Digital Fabrication. *Materials* **2023**, *16*, 5597. https://doi.org/10.3390/ma16165597

Academic Editors: Stefano Guarino and Flaviana Tagliaferri

Received: 19 June 2023
Revised: 14 July 2023
Accepted: 10 August 2023
Published: 12 August 2023

1. Introduction

In recent years, a growing body of research has focused on design strategies aimed at promoting sustainability. One prominent model proposed as a strategy for achieving circular design, as discussed in the study conducted by Moreno et al. [1], is "Design for resource conservation." Furthermore, the butterfly system diagram put forth by the Ellen MacArthur Foundation serves as a comprehensive overview of the circular economy and delineates the following principles: "The second principle of the circular economy is to circulate products and materials at their highest value. This means keeping materials in use, either as a product or, when that can no longer be used, as components or raw materials" [2]. Hence, effective utilization of biological resources, ensuring that they are renewable, has emerged as a crucial approach for attaining sustainability.

Agricultural residues, such as rice husks from rice cultivation, can be considered major biological resources. Rice husks are an important by-product of the rice milling process and are a major agricultural waste product. Table 1 shows the statistical data on global rice paddy production released by Food and Agriculture Organization of the United Nations (FAO) [3]. Global rice production in 2020 was about 756 million tons, and rice production has increasing compared to 2010, especially in Asia and Africa. In addition to their mechanical properties, rice husks are useful materials from which its main components, cellulose, lignin, and silica, can be extracted and used as functional materials. Therefore, there have been many studies on rice husk reuse for a long time. Table 2 summarizes the number of submissions of review articles published since 2013. These results show that the number of submissions is increasing, indicating a high level of interest in rice husk utilization.

Table 1. Annual production quantity of rice paddies in 2010 and 2020 (FAOSTAT).

Region	Production Quantity [$\times 10^6$ ton]	
	2010	2020
World	694.5	756.7
Africa	26.0	37.9
Americas	36.5	38.1
Asia	627.5	676.6
Europe	4.3	4.1
Oceania	0.2	0.1

Table 2. Number of review articles published on rice husks (ScienceDirect). A keyword search for "rice husk" was performed in ScienceDirect, and the results were sorted by checking the "Review Article" checkbox. The results for 2013–2022 are listed and categorized into those where the research content and application methods were analyzed broadly and those where a more detailed review was conducted by narrowing down the application cases.

	Research Target	2013	2014	2015	2016	2017	2018	2019	2020	2021	2022	Total
General	Rice husk	0	0	1	0	0	0	0	1	0	1	3
	Rice husk ash	0	1	0	1	0	0	1	0	0	0	3
	Rice husk derived silica	0	0	0	0	1	0	0	0	0	0	1
Application	Cement/concrete	1	0	0	0	2	2	1	3	4	2	15
	Adsorbents, soil conditioners, fertilizers	0	0	0	0	0	1	2	0	2	0	5
	Energy	0	0	1	2	1	0	0	0	0	2	6
	Fiber-reinforced polymer composites	0	0	0	0	0	0	0	0	1	0	1
	Application to metal matrix	0	0	0	0	0	0	0	0	0	1	1
	Application to alkali activated materials	0	0	0	0	0	0	0	0	0	1	1
	Total	1	1	2	3	4	3	4	4	7	7	36

In order to promote further utilization in the future, it is important not only to develop technologies in existing areas of utilization, but also to increase utilization in new areas. One of these new areas involves the use of materials for digital fabrication. One reason is the recent development of digital fabrication technologies, such as 3D printers and laser cutters, and the other is that rice husk utilization materials have traditionally been applied in the fields of architecture and product design, where this technology is particularly actively used. The most common application in this field is to mix it with concrete and cement materials, but other research papers have been published on other materials that can be applied to construction and products, as shown in Table 3.

Based on the above, it is expected that further research on rice husk materials that can be used in digital fabrication will promote their use in the architectural and product design fields and make even more effective use of unused resources. In fact, research papers have already been published on the use of rice husks as a material for 3D printers in 2020 and beyond. If other digital fabrication equipment, such as laser cutters and CNC machines, can also make use of this material, it will see even greater use.

Hence, this study provides a comprehensive survey of original papers focusing on the use of rice-husk-based materials for 3D printers and particleboard as a viable material for digital fabrication. By identifying recent research trends, summarizing existing research gaps, and providing recommendations for addressing these challenges, this study aims to contribute to the development of this field.

Table 3. Number of research articles published on the application of rice husks that can be used in the construction and product fields (ScienceDirect and Springer). Keyword searches for "rice husk" were conducted on Elsevier's ScienceDirect and Springer's article search sites, and the results from 2013 to 2022 that included the keyword in the title were listed. A total of 1243 search results were found in ScienceDirect and 535 in Springer. The titles of each article were then checked, and the articles that corresponded to the application items in Table 3 were extracted and classified.

Application	2013	2014	2015	2016	2017	2018	2019	2020	2021	2022	Total
Polymer composite	2	5	3	5	8	5	10	16	21	14	89
Metal matrix composite	3	1	2	0	1	3	2	2	1	4	19
Rubber composite	1	0	0	1	0	1	1	2	4	2	12
Glass	1	0	2	0	1	0	3	1	0	2	10
Particleboard	1	0	0	0	1	2	0	2	0	3	9
Building materials (board/panel)	0	0	0	1	0	1	1	3	1	0	7
3D printing materials	0	0	0	0	0	0	0	1	1	1	3
Others	1	0	1	1	0	0	0	1	1	6	11
	9	6	8	8	11	12	17	28	29	32	160

2. The Physical Properties of Rice Husks

Rice husk constitutes around 20% of the weight of paddy rice before milling [4]. It consists of cellulose (25–35%), hemicellulose (18–21%), lignin (26–31%), silica (15–17%), soluble matter (2–5%), and moisture (about 7.5%) [5]. In contrast to other biomass varieties, rice husk is inherently high in silica and ash content. [6]. Furthermore, due to their higher lignin content compared to wood, rice-husk-based products exhibit greater hydrophobicity compared to wood-based products [7]. The combustion of rice husks as fuel generates rice husk ash as a by-product, constituting about 20–25% of the weight of rice husks [8]. Generally, rice husk ash contains approximately 85–95% amorphous silica, but the physical properties and characteristics of rice husk ash depend on the processing parameters, including combustion methods, separation processes, and milling [9]. Materials derived from rice husk ash are characterized by their porous nature and are used as adsorbents and fillers.

3. Targeted Digital Fabrication Machines

Digital fabrication machine is a generic term for machine tools that process materials based on computer-designed data; a typical example is the 3D printer.

Three-dimensional (3D) printing technology, also called additive manufacturing, is a method of layering materials based on data created by 3D computer-aided design (3D CAD) or 3D data generated by a 3D scanner. Various layering methods exist, including applying thermally melted resin or paste materials layer-by-layer, curing liquid resins with ultraviolet light, and sintering powdered materials. Two typical types of machine are shown in Figure 1.

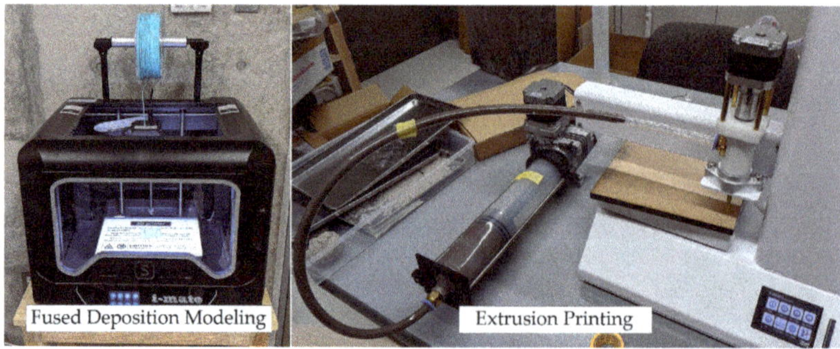

Figure 1. 3D printer (fused deposition modeling type and extrusion printing type).

In addition to 3D printers, digital fabrication also uses equipment such as computer numerical control (CNC) machines and laser cutters that can cut board materials and other flat materials based on 2D data created with CAD or Graphic Soft. A CNC milling machine is a tool that cuts materials according to 2D data using an end mill. The equipment is shown in Figure 2.

Figure 2. Computer numerical control (CNC) machine.

This tool can also utilize 3D data to perform 3D cutting. It is also known as "subtractive manufacturing" because it removes unnecessary parts from the materials. A laser cutter is a machine tool that cuts or engraves objects using a laser beam instead of an end mill. Both technologies are expected to be utilized in the fields of architecture and product design. Therefore, this section introduces research on materials utilizing rice husks for 3D printers and board materials that can be used in 2D processing. The target machines and materials are shown in Figure 3.

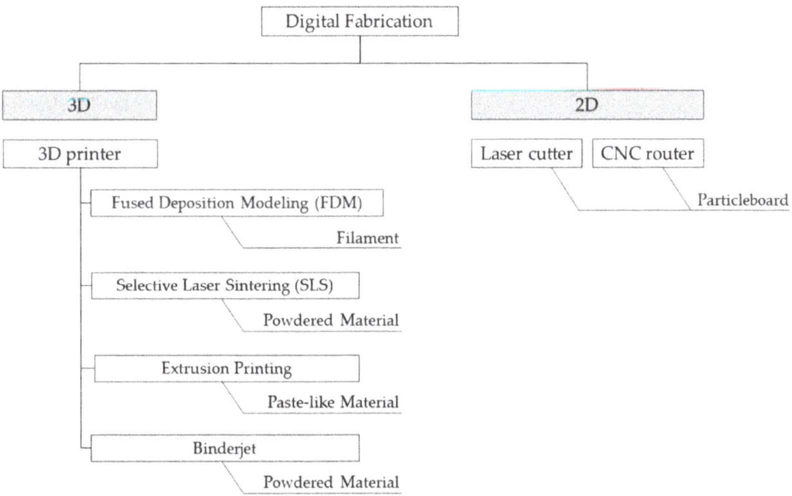

Figure 3. The target machines and materials in this article.

4. Articles on Materials for 3D Printer Utilizing Agricultural Wastes Including Rice Husks

4.1. Method

Searches on ScienceDirect and Springer only produced one review article and three research articles on the utilization of rice husk as a material for 3D printers. Therefore, in this section, Google Scholar was additionally used to search for the keyword "rice husk 3D printer". We extract relevant articles from 2013 to 2022 that contained the keyword in the title and extracted relevant articles after reviewing the abstracts. Among them, we select only those articles that mentioned studies utilizing rice husks as a raw material. After categorizing them into review articles and original articles, we will provide an overview of each and mention some important issues.

4.2. Review Articles on Materials for 3D Printer Utilizing Agricultural Wastes Including Rice Husks

Among the various 3D printing methods, fused deposition modeling (FDM) is the most popular method, in which a string of thermoplastic polymer filaments is melted and layered with heat. Polylactic acid (PLA) and acrylonitrile butadiene styrene (ABS) are typical materials used in FDM. Numerous studies have been reported on blending natural fibers with these materials to produce composite filaments, and several review articles have been published. Five review articles were extracted as a result of the survey using the above methods.

Most recently, Rajendran et al. summarized an approach using natural-fiber-reinforced polymer composites as filaments for 3D printers. They analyzed the effects of fiber treatment, material preparation methods, and compatibilizer additions and reviewed methods for producing filaments from composites [10]. Ahmed et al. also summarized a study on the properties of polymers in 3D printing filaments with added natural fibers [11]. They noted that in many studies, the average particle size of the fiber material used as filler was described, but detailed fiber structure information was lacking. They suggested that it is important to carefully analyze the quality and composition of the filaments before printing. Mazzanti et al. noted that further investigations on other semicrystalline polymers would be interesting, noting that materials less common in FDM, such as polyethylene (PE) and polypropylene (PP), can benefit from composites with natural fibers and particles [12]. The review also specifically focused on studies using PLA as a matrix, discussing the properties of PLA-based bio-composites and printing requirements, as well as future opportunities for applications, upcycling and recycling, and biorefineries [13]; Seker et al. proposed a method for manufacturing acoustic panels using additive manufacturing and presented the acoustic properties of composites reinforced with natural fibers, noting that small voids form between the deposition lines in the 3D-printed composites. The authors concluded that these voids provided an advantage for the acoustic panels to effectively absorb sound [14].

As mentioned above, there are several review articles on 3D printing filaments with added natural fibers. However, these articles mainly focus on FDM 3D printers, and there are no papers that organize material studies for other 3D printers. Therefore, this paper will research papers on the application of rice husks as a material for various 3D printing methods, including FDM.

4.3. Original Articles on 3D Printer Materials Using Rice Husks

Examples of research articles on 3D printer materials using rice husks are listed in Table 4. For each article, the method, matrix, and filler or binder materials are provided.

Table 4. Original articles on materials for 3D printers using rice husks.

3D Printing Method	Matrix	Filler or Binder	Authors	Year
FDM	PLA	Rice husk	Tsou et al. [15]	2019
FDM	PLA	Rice husk	Wu and Tsou [16]	2019
FDM	PLA	Rice husk/wood	le Guen et al. [17]	2019
FDM	Recycled polypropylene	Rice husk	Morales et al. [18]	2021
FDM	ABS	Rice husk ash	Peng et al. [19]	2021
SLS	Co-polyamide	Rice husk	Li et al. [20]	2022
Material Extrusion	Cement	Rice husk ash	Muthukrishnan et al. [21]	2022

Table 4. *Cont.*

3D Printing Method	Matrix	Filler or Binder	Authors	Year
Material Extrusion	Cement	Rice husk ash	Palaniappan [22]	2020
Material Extrusion	Soil	Rice husk and lime	Ferretti et al. [23]	2022
Material Extrusion	Rice husk	Guar gum (as a binder)	Nida et al. [24]	2021
Material Extrusion	Al_2O_3-SiO_2 ink (includes rice-husk-derived silica)	PVA solution, glycerol, and dispersant (as a binder)	Hossain et al. [25]	2022
Binder jetting	Miscanthus particles, wood flour, seashell powder, fruit stone flour, rice husk (only mentioned)	Lignin sulfonate/sodium silicate/polyvinyl alcohol (as a binder)	Zeidler et al. [26]	2018

4.3.1. Fused Deposition Modeling

The most common way to utilize natural resources as a material for 3D printers is to mix them with a polymer as a matrix to generate a composite filament. As an example, filaments mixed with wood and bamboo powders are shown in Figure 4.

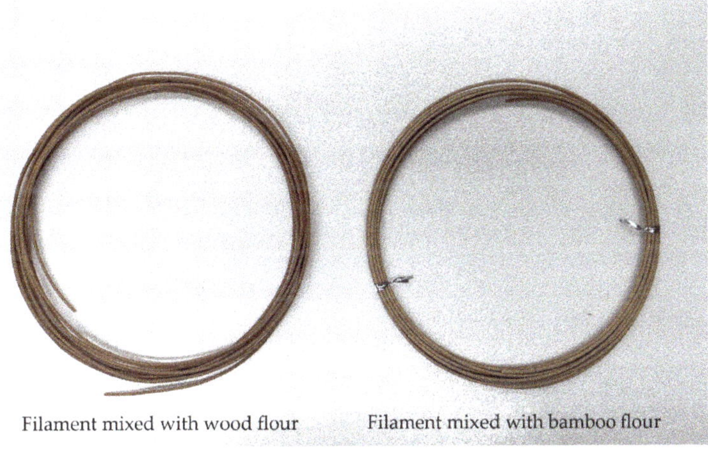

Figure 4. Filaments mixed with natural fiber materials.

This method is also the most common way to apply rice husks as a 3D printer material, and various studies have been conducted to improve material performance and modeling quality. For example, Tsou et al. added methylene diphenyl diisocyanate as an interfacial compatibilizer to improve the tensile strength and impact resistance (Izod impact) of composite materials. They also observed that a porous morphology was generated in the composite material and that the pore size increased considerably with the rice husk content [15] Contrastingly, Wu et al. fabricated filaments with acrylic acid-grafted PLA (PLA-g-AA) and coupling-agent-treated rice husks to improve the properties of bio-composites. They reported that the developed material achieved improved tensile properties and water resistance [16]. le Guen et al. produced 3D-printed filaments by blending wood and rice husk powders with PLA and analyzed their properties. They observed differences in the visual evaluation and process stability, attributed it to the silica content and particle size in the powders, and considered that the particle size of the materials may have a considerable influence on their rheological behavior [17].

There have also been studies that used materials other than PLA as the base material. Morales et al. studied rice husks and filaments using recycled polypropylene (rPP). rPP with rice husk fibers has some problems, such as increased water absorption and swelling diameter and faster degradation than rPP alone; however, it can reduce the warpage caused by the printing process and can be used in applications where its mechanical properties are acceptable [18]. Research on blends of rice husk ash and ABS was conducted, noting that a powder mixture of ABS and rice husk ash is highly useful in the manufacture of optical components by 3D printing because of its ability to increase the refractive index and decrease the absorption coefficient [19].

4.3.2. Selective Laser Sintering (SLS)

The SLS method is a 3D printing technology in which a powder bed is selectively scanned by a laser and processed during layer-by-layer sintering. It is characterized by low material loss and the ability to create highly accurate models. Polycarbonate and polystyrene powders are the commonly used materials. Although there are not many examples of the utilization of rice husks using this method, a method of fabricating composites for SLS by adding rice husk powder to CO-polyamide powder, which is utilized as a hot melt adhesive, has been proposed [20]. This study confirmed the effect of rice husk content on the mechanical properties, dimensional accuracy, and residual ash content of parts, as well as the sintering properties of the material, and clarified the requirements for optimizing the SLS 3D printer method for investment casting parts, which are frequently applied in many cases.

4.3.3. Extrusion Printing

Extrusion printing is a 3D printing method in which a paste-like material is injected into a cylinder and extruded through a nozzle using a screw or air pressure. The most well-known example of this technique is the use of a large 3D printer to produce cement to form architectural structures. Food 3D printers, which have been actively developed in recent years, are also based on this method. In this method, the rheological properties of the paste material to be extruded, as well as the means and time required for curing after printing, are major issues.

The building industry has seen an increase in the research and use of 3D printing, and research has been conducted on the 3D printing of cement using rice husk ash. Muthukrishnan fabricated a rectangular structure with rice husk ash added, utilizing a screw-type extruder, the same system used for 3D printing in architecture, and evaluated the properties of the structure against bare mortar [21]. He then considered that the structure to which rice husk ash was added showed less shape change over time, which was attributed to its faster curing in the mortar and suppressed shape deformation owing to self-loading. Therefore, he concluded that the addition of rice husk ash improved the model. Palaniappan et al. detailed the potential of 3D printing in the construction industry in their study utilizing waste materials, such as cow dung ash, rice husk ash, sugarcane bagasse ash, ground granulated blast-furnace slag (GGBS), and marble dust as concrete materials [22]. Research is also being conducted on the application of rice husks as a material for the 3D printer architecture of soil structures. This is based on the traditional building technique called cob, which is a method for building structures by layering moist clayey soil. The Italian 3D printer company WASP produces a 3D printer for soil structures. The material used was a composite of soil mixed with rice husks and lime. The contact surface between the rice husks and the soil matrix increases, promoting the formation of Si-O-Si bonds and improving the stiffness of the material [23]. Nida et al. investigated the effect of the addition of guar gum (GG) to ground rice hulls of different particle sizes, with and without the addition of GG, on printability. The ratio of powdered rice hulls to guar gum was determined, and its application in food packaging was proposed [24]. A case study of in situ mullite ($3Al_2O_3$-$2SiO_2$) foam development using silica extracted from rice husk ash was investigated. Al_2O_3-SiO_2 ink was prepared using an aqueous binder containing α-alumina and two types of silica (rice-husk-ash-extracted biogenic nano-silica and commercial silica). The material was extruded by air pressure through a conical nozzle, dried in an air furnace, and fired in a muffle furnace. Its application to high-performance products (e.g., insulators, filters, and high-temperature operating devices such as burners) was proposed [25].

4.3.4. Binder Jetting

The binder jetting method produces a three-dimensional model object by repeatedly placing the powdered material on a bed and injecting a liquid binder in layers. As with SLS, there is little loss of material; however, there is a problem in that the strength after modeling is low, and post-processing is sometimes required. For example, Zeidler et al. applied lignin sulfonate, sodium silicate (water glass), and polyvinyl alcohol as binders to Miscanthus sinensis particles, wood flour, shell powder, and fruit stone powder; evaluated the properties of the moldings; and discussed the optimal combinations. Miscanthus/PVA was proposed as an appropriate combination, and it was also confirmed that spray coating with a sodium silicate solution can improve the mechanical strength and water resistance. In contrast, they mentioned that the mechanical strength was weaker than that of polymer-based components, making them suitable for applications where they are not subjected to loads, and suggested their application in food packaging. The report mentioned that rice husks were not included because no considerable results have been achieved; however, they are interesting to investigate because they are abundantly available in many regions and are expected to be developed in the future [26].

5. Articles on Materials for Particleboard Using Rice Husks

Particleboard is a material typically produced by hot-pressing wood chips sprayed with an adhesive that serves as a binder. An intermediate layer composed of coarse particles and fibers is sandwiched between layers of fine particles to achieve strength and a light weight. However, obtaining the wood chips used as raw materials is fraught with deforestation and raw material shortage problems. Urea-formaldehyde, which is mainly used as an adhesive, has been pointed out to be harmful to the human body. Particleboards can also be processed by CNC or laser cutters and are widely used as building and furniture materials, along with veneer boards. Agricultural waste would be an effective means of resource recycling and would help stabilize the supply of board materials if it could be utilized as raw materials for these boards.

This study surveys research articles on the development of particleboards utilizing rice husks, as there are currently no studies on the use of rice husks in digital fabrication.

Original articles were searched using the keywords "rice husk" and "particleboard" on the ScienceDirect and Springer websites, and all keywords must be included in the title. Table 5 shows the extracted articles organized by material/binder/characteristic treatment.

Table 5. Original article on particleboard utilizing rice husks.

Matrix	Binder	Treatment	Authors	Year
Rice husks/sawdust	Synthetic adhesives Fevicol/Ponal/Woodfix		Olupot et al. [27]	2022
Mixture of groundnut shell and rice husk particles	Cassava starch blended with urea-formaldehyde		Akinyemi et al. [28]	2022
Mixture of insect rearing residue and rice particles	Citric acid/tapioca starch		Huang et al. [29]	2022
Rice husk particles/hemp fibers	Cornstarch		Battegazzore et al. [30]	2018
Rice husk particles/hemp fibers	Cornstarch	Add ammonium dihydrogen phosphate to the binder	Battegazzore et al. [31]	2018

Table 5. *Cont.*

Matrix	Binder	Treatment	Authors	Year
Rice husk particles	Soy protein concentrate	Impregnate the board with tung oil	Nicolao et al. [32]	2020
Rice husk particles	Soy protein concentrate	Add two jute fabric layers to the board	Chalapud et al. [33]	2020
Rice husk particles	Phenol-formaldehyde resin	Sandwiched between two wood strand layers	Kwon et al. [34]	2013
Rice husk particles	Branched polyethylene imine and poly acrylic acid	Layer by Layer-coating on rice husks	Battegazzore et al. [35]	2017

There are many examples of research on particleboards made from rice hulls, in which different combinations of materials and binders were used to evaluate mechanical strength, water resistance, and heat resistance. Olupot et al. found that the addition of sawdust improved the properties of rice-husk-based, low-density particleboard. They utilized commercially available synthetic adhesives as binders instead of the commonly used urea-formaldehyde and phenol-formaldehyde for particleboard manufacturing, with promising results [27]. Akinyemi et al. processed agricultural waste, peanut shells, and rice husks into particles and used them to produce composite panelboards. In addition, cassava starch blended with urea-formaldehyde was used as an adhesive to reduce the environmental impact, presenting its potential as an eco-friendly interior panelboard material [28]. Huang et al. evaluated the feasibility of manufacturing particleboard by combining insect breeding residues and rice husks in different ratios and using citric acid/tapioca starch as a natural binder. Although a decrease in the rupture factor was observed as the percentage of insect breeding residue increased, they identified a combination ratio of materials that met the Japanese Industrial Standard (JIS) specifications, showing its potential as a board material [29].

Some studies have fabricated and characterized fiberboard and particleboard using hemp staple fibers and rice husk particles, respectively. Cornstarch was employed as the binder, and the boards were prepared by hot-pressing the binder-impregnated material. Fiberboard is stiffer than particleboard and has been proven to be able to withstand greater loads, meeting standards that allow it to be recommended as an interior material [30]. The same material and binder combination is also used as a fire retardant, which is a challenge when used as a building material. Ammonium dihydrogen phosphate was added as a flame retardant, and boards were manufactured to find the optimal combination and meet the high flame-retardant standards [31].

As confirmed from the above studies, it is difficult to guarantee the strength particleboards containing rice hulls, which have a smaller aspect ratio compared common particleboards; and, when naturally derived adhesives are used, there is a high possibility that various types of performance will be degraded. Therefore, some attempts have been made to improve performance with additional treatments. For example, Nicolao et al. studied how to improve the function of manufactured rice husk boards by utilizing soy protein as a binder. They achieved functional improvements, including water resistance, by impregnating the manufactured boards with tung oil [32]. Chalapud et al. also fabricated boards using the same binder. They achieved improved mechanical strength by sandwiching the surface of the board between jute cloth [33]. Another method to increase the mechanical strength through sandwich construction is to build wood strand layers on the surface. The binder was a common formaldehyde adhesive; however, it was a good substitute for wood strand boards [34].

Finally, a special case is the manufacturing method of layer-by-layer treatment [35]. In this method, a binder is coated on the surface of the rice husks by vapor deposition and then hot-pressed to produce the board. Strong mechanical properties were obtained due to structures bonded via strong electrostatic interactions that occurred at the molecular scale. Furthermore, the polymer system used branched polyethyleneimine and polyacrylic acid, and layer-by-layer approaches have environmentally friendly features, such as the use of water as a solvent and optimal performance at room temperature and low solution concentrations. Such an efficient surface modification method has shown promising sustainable potential.

6. Discussion

This study reviewed the literature on the effective use of rice husks, focusing on their utilization in the fields of architecture and product design. The results of the survey of review articles on the use of rice husks showed a high number of review articles on the use of incinerated ash as a concrete filler, indicating a high level of interest. They have been published continuously since 2017 and the number is still increasing, suggesting that this method is a promising means of incinerated ash utilization. In addition, it was confirmed that rice husks have the potential for a wide variety of applications. In many cases, husks are utilized as a base material or filler or the silica and lignin components of the husks are extracted and utilized. In deciding which method to choose, the performance of the rice-husk-derived materials and their economics and the balance between the yield of each region and the amount of treatment with each utilization method are important guidelines. Therefore, it will be necessary in the future to analyze the amount of rice husks processed and the cost aspects of each utilization method.

The most common method when using rice husks as a material for 3D printers is to mix them as a filler in filaments used for FDM 3D printers, which are the most widespread type of 3D printer. PLA was the most common polymer used in these studies as the base polymer for mixing rice husks. In general, adding natural fiber fillers reduces the mechanical strength and durability; hence, efforts have been made to compensate for these problems. However, additives are often included to overcome these problems, causing costs to increase. In the future, it will be necessary to consider both the purpose and the cost of the material. While it was noted that there was insufficient information on particle size in each of the studies, other studies indicated that the particle size of the material could have a significant impact on its rheological behavior. Therefore, it is important in the future to establish a method for measuring filler particle size and to unify this information.

Rice husks also have the potential to be used as an additive to materials, especially in cement extrusion printing. Because the addition of rice husks leads to retention of the shape of the modeled object until hardening, it is very effective as a material for 3D printing without using a mold. Although it is still a developing technology, rice husks have been used as materials for concrete in the past and their usage is expected to expand along with the spread of 3D printing in construction.

The possibility of using rice husks as materials for SLS has been confirmed. However, the powder materials used in this method tends to scatter easily, affecting the installation environment of the equipment. If rice husks are used, their particle size and specific gravity should be compared with those of ordinary nylon materials to evaluate their ease of handling, equipment cleanliness, and work environment.

Extrusion printing is a useful method for a wide range of applications from architecture to the manufacture of functional products. However, the material in this method is wet, and the state of the material changes with time until modeling. Material drying always occurs using this method. In addition, when adding particles, such as rice husks, material separation may occur depending on the rheological properties of the binder. Furthermore, when mixing organic materials, excess material and residue in the cylinder can cause mold and corrosion. In the case of rice husk ash, the risk is low. However, if the rice husk powder is used directly, this issue should be considered. Based on the above, to expand the use of rice husks in this method, it is necessary to obtain knowledge on the speed of aging and deterioration of the physical properties of the material before firing or hardening. It is also important to develop equipment that is easy to clean.

The use of rice husks in particleboards has been studied, regardless of the digital fabrication trend. All of these particleboards tend to have reduced mechanical properties compared to common wood particleboards, and the focus is on how to reduce performance loss and maintain this within the standard. These studies have focused on improving properties such as mechanical strength and have not mentioned the workability of the material. In particular, there is no research on the machinability of CNC or laser cutters for these materials. Therefore, to promote the use of rice-husk-derived particleboard as a material for digital fabrication, it is necessary to evaluate its workability.

7. Conclusions

The following is a summary of the issues and solutions associated with the use of rice husks as materials for digital fabrication.

- Applications suited to material properties

Mixing rice husk fibers with neat materials, such as polymer composites, tends to lead to performance degradation. However, the addition of other materials or treatments to improve functionality can increase the costs. Therefore, it is important not only to aim for improved functionality, but to also consider applications that match the characteristics of the material.

- Evaluation of actual operation (long-term quality of materials/maintainability of equipment)

The paste used in the material extrusion method tends to change over time, and the material remaining in the syringe or at the nozzle tip can affect the printing characteristics. However, previous studies have not considered the aging of these materials or operational concerns. Therefore, future studies should investigate the changes in the material properties over time and consider the maintainability of the equipment.

- Processability of board materials

When rice-husk-derived particleboard is cut by CNC, the silica component in the material may shorten the life of the end mills. Depending on the internal density and interparticle adhesion strength, the material on the cut surface may peel off when milled. If the material peels off during processing by CNC, it may be improved by reducing the grain size or increasing the adhesive strength. However, the amount of silica cannot be changed significantly; therefore, when machining with CNC, it is effective to deal with this problem by slowing down the machining speed or using end mills for metal rather than wood. When processing with a laser cutter, it is also necessary to verify the reaction of the rice husk components with the laser beam. Therefore, to utilize a rice-husk-based particleboards for digital fabrication, research on their processability and optimal processing conditions would help expand their use.

Despite the abovementioned issues, the use of rice husks as a structural material for buildings and industrial products is an effective method of material recycling because it leads to mass disposal of agricultural waste and long-term carbon fixation.

This study describes the research trend on the use of rice husks for digital fabrication and discussed the possibilities of rice husk use, as well as issues and solutions. The promotion of material research for digital fabrication, which is expected to be further utilized in the future, will lead to further effective utilization of agricultural waste. This review will help to expand the use of rice husks in the architectural and product design fields.

Author Contributions: Conceptualization, K.M., K.T. and D.M.; methodology, K.M. and K.T.; formal analysis, K.M.; investigation, K.M.; resources, K.M.; data curation, K.M.; writing—original draft preparation, K.M.; writing—review and editing, K.M.; visualization, K.M.; supervision, K.T. and D.M.; project administration, K.M., K.T. and D.M.; funding acquisition, K.M. All authors have read and agreed to the published version of the manuscript.

Funding: This research received no external funding.

Institutional Review Board Statement: Not applicable.

Informed Consent Statement: Not applicable.

Data Availability Statement: All the data are available in the manuscript.

Conflicts of Interest: The authors declare no conflict of interest.

References

1. Moreno, M.; de los Rios, C.; Rowe, Z.; Charnley, F. A conceptual framework for circular design. *Sustainability* **2016**, *8*, 937. [CrossRef]
2. Circulate Products and Materials, the Ellen MacArthur Foundation. Available online: https://ellenmacarthurfoundation.org/circulate-products-and-materials (accessed on 29 May 2023).
3. Food and Agriculture Organization of the United Nations. Crops and Livestock Products, "FAOSTAT". Available online: https://www.fao.org/faostat/en/#data/QCL (accessed on 30 May 2023).
4. Norhasnan, N.H.A.; Hassan, M.Z.; Nor, A.F.M.; Zaki, S.A.; Dolah, R.; Jamaludin, K.R.; Aziz, S.A. Physicomechanical Properties of Rice Husk/Coco Peat Reinforced Acrylonitrile Butadiene Styrene Blend Composites. *Polymers* **2021**, *13*, 1171. [CrossRef]
5. Luduena, L.; Fasce, D.; Alvarez, V.A.; Stefani, P.M. Nanocellulose from rice husk following alkaline treatment to remove silica. *BioResources* **2011**, *6*, 2. [CrossRef]
6. Shukla, S.S.; Chava, R.; Appari, S.; Bahurudeen, A.; Kuncharam, B.V.R. Sustainable use of rice husk for the cleaner production of value-added products. *J. Environ. Chem. Eng.* **2022**, *10*, 106899. [CrossRef]
7. Kariuki, S.W.; Wachira, J.; Kawira, M.; Murithi, G. Crop residues used as lignocellulose materials for particleboards formulation. *Heliyon* **2020**, *6*, e05025. [CrossRef] [PubMed]
8. Das, S.K.; Adediran, A.; Kaze, C.R.; Mustakim, S.M.; Leklou, N. Production, characteristics, and utilization of rice husk ash in alkali activated materials: An overview of fresh and hardened state properties. *Constr. Build. Mater.* **2022**, *345*, 128341.
9. Siddika, A.; Mamun, M.A.A.; Alyousef, R.; Mohammadhosseini, H. State-of-the-art-review on rice husk ash: A supplementary cementitious material in concrete. *J. King Saud Univ. Eng. Sci.* **2021**, *33*, 294–307. [CrossRef]
10. Rajendran, R.N.R.; Leong, J.S.; Chan, W.N.; Tan, J.R.; Shamsuddin, Z.S.B. Current state and challenges of natural fibre-reinforced polymer composites as feeder in FDM-based 3D printing. *Polymers* **2021**, *13*, 2289. [CrossRef] [PubMed]
11. Ahmed, W.; Alnajjar, F.; Zaneldin, E.; Al-Marzouqi, A.H.; Gochoo, M.; Khalid, S. Implementing FDM 3D printing strategies using natural fibers to produce biomass composite. *Materials* **2020**, *13*, 4065. [CrossRef]
12. Mazzanti, V.; Malagutti, L.; Mollica, F. FDM 3D printing of polymers containing natural fillers: A review of their mechanical properties. *Polymers* **2019**, *11*, 1094. [CrossRef]
13. Bhagia, S.; Bornani, K.; Agrawal, R.; Satlewal, A.; Ďurkovič, J.; Lagaňa, R.; Bhagia, M.; Yoo, C.G.; Zhao, X.; Kunc, V.; et al. Critical review of FDM 3D printing of PLA biocomposites filled with biomass resources, characterization, biodegradability, upcycling and opportunities for biorefineries. *Appl. Mater. Today* **2021**, *24*, 101078. [CrossRef]
14. Sekar, V.; Fouladim, M.H.; Namasivayam, S.N.; Sivanesan, S. Additive manufacturing: A novel method for developing an acoustic panel made of natural fiber-reinforced composites with enhanced mechanical and acoustical properties. *J. Eng.* **2019**, *2019*, 4546863. [CrossRef]
15. Tsou, C.H.; Yao, W.H.; Wu, C.S.; Wu, C.S.; Tsou, C.Y.; Hung, W.S.; Chen, J.C.; Guo, J.; Yuan, S.; Wen, E.; et al. Preparation and characterization of renewable composites from Polylactide and Rice husk for 3D printing applications. *J. Polym. Res.* **2019**, *26*, 227. [CrossRef]
16. Wu, C.S.; Tsou, C.H. Fabrication, characterization, and application of biocomposites from poly(lactic acid) with renewable rice husk as reinforcement. *J. Polym. Res.* **2019**, *26*, 44. [CrossRef]
17. le Guen, M.J.; Hill, S.; Smith, D.; Theobald, B.; Gaugler, E.; Barakat, A.; Mayer-Laigle, C. Influence of rice husk and wood biomass properties on the manufacture of filaments for fused deposition modeling. *Front. Chem.* **2019**, *7*, 735. [CrossRef] [PubMed]
18. Morales, M.A.; Atencio, M.C.L.; Maranon, A.; Hernandez, C.; Michaud, V.; Porras, A. Development and characterization of rice husk and recycled polypropylene composite filaments for 3D printing. *Polymers* **2021**, *13*, 1067. [CrossRef] [PubMed]
19. Peng, H.Y.; Yang, C.; Wei, Y.; Ruan, Y.; Hsu, Y.; Hsieh, C.; Cheng, C.P. Terahertz complex refractive index properties of acrylonitrile butadiene styrene with rice husk ash and its possible applications in 3D printing techniques. *Opt. Mater. Express* **2021**, *11*, 2777. [CrossRef]
20. Li, H.; Guo, Y.; Shuhui, T.; Jian, L.; Idriss, A.I. Study on selective laser sintering process and investment casting of rice husk/Co-polyamide (Co-PA hotmelt adhesive) composite. *J. Thermoplast. Compos. Mater.* **2022**, *36*, 2311–2331. [CrossRef]
21. Muthukrishnan, S.; Kua, H.W.; Yu, L.N.; Chung, J.K.H. Fresh properties of cementitious materials containing rice husk ash for construction 3D printing. *J. Mater. Civ. Eng.* **2020**, *32*, 04020195. [CrossRef]
22. Palaniappan, M. An optimal utilization of waste materials in concrete to enhance the strength property: An experimental approach and possibility of 3D printing technology. In *Sustainability for 3D Printing*; Sandhu, K., Singh, S., Prakash, C., Subburaj, K., Ramakrishna, S., Eds.; Springer: Cham, Switzerland, 2022; pp. 149–158.
23. Ferretti, E.; Moretti, M.; Chiusoli, A.; Naldoni, L.; de Fabritiis, F.; Visonà, M. Rice-husk shredding as a means of increasing the long-term mechanical properties of earthen mixtures for 3D printing. *Materials* **2022**, *15*, 743. [CrossRef] [PubMed]
24. Nida, S.; Anukiruthika, T.; Moses, J.A.; Anandharamakrishnan, C. 3D printing of grinding and milling fractions of rice husk. *Waste Biomass Valorization* **2021**, *12*, 81–90. [CrossRef]

25. Hossain, S.S.; Baek, I.W.; Son, H.J.; Park, S.; Bae, C.J. 3D printing of porous low-temperature in-situ mullite ceramic using waste rice husk ash-derived silica. *J. Eur. Ceram. Soc.* **2022**, *42*, 2408–2419. [CrossRef]

26. Zeidler, H.; Klemm, D.; Böttger-Hiller, F.; Fritsch, S.; le Guen, M.J.; Singamneni, S. 3D printing of biodegradable parts using renewable biobased materials. *Procedia Manuf.* **2018**, *21*, 117–124. [CrossRef]

27. Olupot, P.W.; Menya, E.; Lubwama, F.; Ssekaluvu, L.; Nabuuma, B.; Wakatuntu, J. Effects of sawdust and adhesive type on the properties of rice husk particleboards. *Results Eng.* **2022**, *16*, 100775. [CrossRef]

28. Akinyemi, B.A.; Kolajo, T.E.; Adedolu, O. Blended formaldehyde adhesive bonded particleboards made from groundnut shell and rice husk wastes. *Clean Technol. Environ. Policy* **2022**, *24*, 1653–1662. [CrossRef]

29. Battegazzore, D.; Alongi, J.; Duraccio, D.; Frache, A. All natural high-density fiber- and particleboards from hemp fibers or rice husk particles. *J. Polym. Environ.* **2018**, *26*, 1652–1660. [CrossRef]

30. Huang, H.K.; Hsu, C.H.; Hsu, P.K.; Cho, Y.M.; Chou, T.H.; Cheng, Y.S. Preparation and evaluation of particleboard from insect rearing residue and rice husks using starch/citric acid mixture as a natural binder. *Biomass Convers. Biorefin.* **2022**, *12*, 633–641. [CrossRef]

31. Battegazzore, D.; Alongi, J.; Duraccio, D.; Frache, A. Reuse and valorisation of hemp fibres and rice husk particles for fire resistant fibreboards and particleboards. *J. Polym. Environ.* **2018**, *26*, 3731–3744. [CrossRef]

32. Nicolao, E.S.; Leiva, P.; Chalapud, M.C.; Ruseckaite, R.A.; Ciannamea, E.M.; Stefani, P.M. Flexural and tensile properties of biobased rice husk-jute-soybean protein particleboards. *J. Build. Eng.* **2020**, *30*, 101261. [CrossRef]

33. Chalapud, M.C.; Herdt, M.; Nicolao, E.S.; Ruseckaite, R.A.; Ciannamea, E.M.; Stefani, P.M. Biobased particleboards based on rice husk and soy proteins: Effect of the impregnation with tung oil on the physical and mechanical behavior. *Constr. Build. Mater.* **2020**, *230*, 116996. [CrossRef]

34. Kwon, J.H.; Ayrilmis, N.; Han, T.H. Enhancement of flexural properties and dimensional stability of rice husk particleboard using wood strands in face layers. *Compos. B Eng.* **2013**, *44*, 728–732. [CrossRef]

35. Battegazzore, D.; Alongi, J.; Frache, A.; Wågberg, L.; Carosio, F. Layer by Layer-functionalized rice husk particles: A novel and sustainable solution for particleboard production. *Mater. Today Commun.* **2017**, *13*, 92–101. [CrossRef]

materials

Article

Impact of Diverse Parameters on the Physicochemical Characteristics of Green-Synthesized Zinc Oxide–Copper Oxide Nanocomposites Derived from an Aqueous Extract of *Garcinia mangostana* L. Leaf

Yu Bin Chan [1], Mohammod Aminuzzaman [1,2,*], Lai-Hock Tey [1,*], Yip Foo Win [1], Akira Watanabe [3], Sinouvassane Djearamame [4] and Md. Akhtaruzzaman [5]

[1] Department of Chemical Science, Faculty of Science, Universiti Tunku Abdul Rahman (UTAR), Kampar Campus, Jalan Universiti, Bandar Barat, Kampar 31900, Malaysia; yubinchan1221@gmail.com (Y.B.C.); yipfw@utar.edu.my (Y.F.W.)

[2] Centre for Photonics and Advanced Materials Research (CPAMR), Universiti Tunku Abdul Rahman (UTAR), Sungai Long Campus, Jalan Sungai Long, Bandar Sungai Long, Kajang 43000, Malaysia

[3] Institute of Multidisciplinary Research for Advanced Materials (IMRAM), Tohoku University, Sendai 980-8577, Japan; akira.watanabe.c6@tohoku.ac.jp

[4] Department of Biomedical Science, Faculty of Science, Universiti Tunku Abdul Rahman (UTAR), Kampar Campus, Jalan Universiti, Bandar Barat, Kampar 31900, Malaysia; sinouvassane@utar.edu.my

[5] Solar Energy Research Institute (SERI), Universiti Kebangsanan Malaysia (UKM), Bangi 43600, Malaysia; akhtar@ukm.edu.my

* Correspondence: mohammoda@utar.edu.my (M.A.); teylh@utar.edu.my (L.-H.T.)

check for
updates

Citation: Chan, Y.B.; Aminuzzaman, M.; Tey, L.-H.; Win, Y.F.; Watanabe, A.; Djearamame, S.; Akhtaruzzaman, M. Impact of Diverse Parameters on the Physicochemical Characteristics of Green-Synthesized Zinc Oxide–Copper Oxide Nanocomposites Derived from an Aqueous Extract of *Garcinia mangostana* L. Leaf. *Materials* **2023**, *16*, 5421. https://doi.org/10.3390/ma16155421

Academic Editor: Slavko Bernik

Received: 27 May 2023
Revised: 6 July 2023
Accepted: 21 July 2023
Published: 2 August 2023

Abstract: Compared to conventional metal oxide nanoparticles, metal oxide nanocomposites have demonstrated significantly enhanced efficiency in various applications. In this study, we aimed to synthesize zinc oxide–copper oxide nanocomposites (ZnO-CuO NCs) using a green synthesis approach. The synthesis involved mixing 4 g of $Zn(NO_3)_2 \cdot 6H_2O$ with different concentrations of mangosteen (*G. mangostana*) leaf extract (0.02, 0.03, 0.04 and 0.05 g/mL) and 2 or 4 g of $Cu(NO_3)_2 \cdot 3H_2O$, followed by calcination at temperatures of 300, 400 and 500 °C. The synthesized ZnO-CuO NCs were characterized using various techniques, including a UV-Visible spectrometer (UV-Vis), photoluminescence (PL) spectroscopy, Fourier Transform Infrared (FTIR) spectroscopy, X-ray powder diffraction (XRD) analysis and Field Emission Scanning Electron Microscope (FE-SEM) with an Energy Dispersive X-ray (EDX) analyzer. Based on the results of this study, the optical, structural and morphological properties of ZnO-CuO NCs were found to be influenced by the concentration of the mangosteen leaf extract, the calcination temperature and the amount of $Cu(NO_3)_2 \cdot 3H_2O$ used. Among the tested conditions, ZnO-CuO NCs derived from 0.05 g/mL of mangosteen leaf extract, 4 g of $Zn(NO_3)_2 \cdot 6H_2O$ and 2 g of $Cu(NO_3)_2 \cdot 3H_2O$, calcinated at 500 °C exhibited the following characteristics: the lowest energy bandgap (2.57 eV), well-defined Zn-O and Cu-O bands, the smallest particle size of 39.10 nm with highest surface area-to-volume ratio and crystalline size of 18.17 nm. In conclusion, we successfully synthesized ZnO-CuO NCs using a green synthesis approach with mangosteen leaf extract. The properties of the nanocomposites were significantly influenced by the concentration of the plant extract, the calcination temperature and the amount of precursor used. These findings provide valuable insights for researchers seeking innovative methods for the production and utilization of nanocomposite materials.

Keywords: *Garcinia mangostana* L.; green product; green synthesis; nanocomposites; zinc oxide-copper oxide

1. Introduction

Compared to individual semiconductor metal oxide nanoparticles (NPs), such as zinc oxide (ZnO), copper oxide (CuO), nickel oxide (NiO), etc., the mixing of these NPs has gained significant attention due to their excellent application in sensor, electrical and electronic products. Mixing semiconductor metal oxides allows for control over their structural, morphological and surface properties, making them important in various practical applications [1]. Among the *p-n* type mixed semiconductors, ZnO-CuO nanocomposites (NCs) garnered considerable interest from researchers. Copper is preferred to combine with ZnO due to its ability to easily overlap *d*-electrons with a valence bond of ZnO [2]. This results in enhanced surface area, smaller particle size and the formation of ZnO-CuO heterojunctions, which strengthen the optical and electronic properties [2,3]. Consequently, ZnO-CuO NCs find application in environmental remediation, photo-catalysis, fuel cell, solar cell, antibacterial, UV protection and optoelectronics devices [1,4–6]. For example, the effectiveness in degrading methylene blue was higher by using ZnO-CuO NCs (98%) compared to ZnO (81%) [5].

Green synthesis of nanomaterial offers a simpler, more cost-effective, eco-friendly alternative with lower energy consumption compared to conventional methods [7–12]. Generally, various parts of plants, including flowers, leaves, stems, roots and seeds, are utilized in the green synthesizing of nanomaterials [13–15]. During the green synthesis process, phytochemicals present in plants, such as phenols, aldehydes, ketones, carboxylic acids, nitrogenous compounds, flavonoids, alkaloids, terpenoids, tannins and pigments, accumulate and later interact with metals to cap, stabilize and reduce to NPs [9,16,17]. However, achieving the desired morphology and shape remains a challenge in the green synthesis of NPs and NCs. As a result, extensive research has been conducted to optimize the synthesis conditions, including plant extract concentration, temperature and precursor concentration, to synthesize NPs and NCs with desired structural, morphological and optical properties [18,19].

While aqueous extract from *Aloe barbadansis* leaf [3], *Calotropis gigantea* leaf [4], *Theobroma cacao* seed bark [6], *Dovyalis caffra* leaf [20], *Verbascum sinaiticum* Benth [21], *Sambucus nigra* L. shoot [22], *Alchornea cordifolio* leaf [23] and *Calotropis gigamtae* leaf [24] has been utilized for synthesizing ZnO-CuO NCs. The use of *Garcinia mangostana* L., commonly known as mangosteen, in synthesizing ZnO-CuO NCs has not been explored. Mangosteen is a seasonal fruit in the *Clusiacae* family and is commonly found in tropical countries [25–29]. It contains numerous phytochemicals, such as xanthones, flavonoids and terpene [30–33], which have the potential to form stable colloidal nanomaterials.

In this study, we synthesized ZnO-CuO NCs using a mangosteen leaf aqueous extract in a green, fast and simple manner. The mangosteen leaf aqueous extract-mediated ZnO-CuO NCs were optimized by varying the concentration of the mangosteen leaf aqueous extract (0.02, 0.03, 0.04 and 0.05 g/mL), calcination temperatures (300, 400 and 500 °C) and the amount of $Cu(NO_3)_2 \cdot 3H_2O$ (2.0 and 4.0 g). In this paper, we investigated the effects of these parameters (plant concentration, calcination temperature and precursor weight) on the optical, structural and morphological properties of the mangosteen leaf aqueous extract-mediated ZnO-CuO NCs.

2. Materials and Methods

2.1. Materials

The mangosteen leaves were collected from a neighborhood in Kampar, Malaysia. Zinc nitrate hexahydrate, $Zn(NO_3)_2 \cdot 6H_2O$, was purchased from HiMedia Laboratories Pvt. Ltd. (Nashik, India), and copper nitrate trihydrate, $Cu(NO_3)_2 \cdot 3H_2O$ was purchased from HmbG (Hamburg, Germany). Both chemicals were used without further purification. All glassware was washed with deionized water and dried in an oven before use.

2.2. Characterization

The selection of optimized parameters in green synthesizing ZnO-CuO NCs was based on their structural, morphological and optical properties. The absorption spectra were recorded by a UV-Visible (UV-Vis) spectrophotometer (Thermo Scientific GENESYS 10S, Waltham, MA, USA). The recombination of electron-hole pairs (e^-/h^+) of the synthesized samples was investigated using photo luminance (PL) spectroscopy (Perkin Elmer LS 55 Fluorescence Spectrometer, Waltham, MA, USA) with an excitation wavelength of 350 nm in the range of 350 to 600 nm. The Fourier Transform Infrared (FTIR) spectroscopy study was carried out at room temperature in the range of 4000 to 400 cm^{-1} with a resolution of 4 cm^{-1} by using KBr pellets in a Perkin Elmer RX1 spectrophotometer. X-ray powder diffraction (XRD) patterns were taken in the reflection mode with Cu Kα (λ = 1.5406 Å) radiation in the 2θ range of 10° to 80° by using a Shimadzu XRD 6000 X-ray diffractometer with continuous scanning which was operated at 40 kV/30 mA and 0.02 min^{-1}. The morphological, microstructural and elemental compositional of all synthesized samples was determined using a Field Emission Scanning Electron Microscope (FE-SEM) (JEOL JSM-6710F, Tokyo, Japan) with Energy Dispersive X-ray (EDX) analyzer (X-max, 150 Oxford Instruments, Abingdon-*on*-Thames, UK).

2.3. Preparation of Mangosteen Leaf Aqueous Extract

The freshly plucked mangosteen leaves were washed with tap water to remove dust and dried in an oven at 50 °C for 48 h and further dried in a vacuum oven at 60 °C for 8 h. Then, the leaves were ground into a fine powder by using a grinder. Then, 5 g of leaf powder was added to 100 mL of deionized water and heated with stirring at 70–80 °C for 20 min to obtain 0.05 g/mL of leaf aqueous extract. Upon cooling, the leaf aqueous extract was vacuum filtrated, and a reddish-brown filtrate was collected and immediately used for ZnO-CuO NCs synthesis.

2.4. Synthesis of ZnO-CuO NCs

With minor modification from Chan et al. [34], the synthesis of ZnO-CuO NCs using mangosteen leaf aqueous extract was performed. The reaction parameters, which included mangosteen leaf aqueous extract concentration, calcination temperature and weight of Cu(NO$_3$)$_2$·3H$_2$O added, were optimized.

2.4.1. Leaf Aqueous Extract Optimization

The 50 mL of mangosteen leaf aqueous extract (0.02, 0.03, 0.04 and 0.05 g/mL) was mixed separately with 4.0 g of Zn(NO$_3$)$_2$·6H$_2$O and 2 g of Cu(NO$_3$)$_2$·3H$_2$O. Immediately, a greenish-brown solution formed. The solution was heated at 70–80 °C with constant stirring until the formation of a brown paste. The paste was then cooled to room temperature and calcinated at 500 °C for 2 h using the Muffle furnace to obtain a fine black-blue ZnO-CuO powder.

2.4.2. Calcination Temperature Optimization

After the selection of the optimized mangosteen leaf aqueous extract concentration at 0.05 g/mL, the synthesis of ZnO-CuO NCs was repeated using 4 g of Zn(NO$_3$)$_2$·6H$_2$O and 2 g of Cu(NO$_3$)$_2$·3H$_2$O. The cooled brown paste was calcinated at 300, 400 and 500 °C for 2 h to have more energy savings during the ZnO-CuO NCs synthesis.

2.4.3. Precursor Optimization

After the selection of the optimized mangosteen leaf aqueous extract concentration at 0.05 g/mL and calcination temperature at 500 °C, the synthesis steps were repeated using 4 g of Zn(NO$_3$)$_2$·6H$_2$O with different weights of Cu(NO$_3$)$_2$·3H$_2$O (2 and 4 g). Until the formation of brown paste. It was then calcinated at 500 °C for 2 h.

3. Results

3.1. UV-Vis Spectroscopy Analysis

Figure 1 shows the UV-Vis spectra of the mangosteen leaf aqueous extract, Cu(NO$_3$)$_2$·3H$_2$O, Zn(NO$_3$)$_2$·6H$_2$O and mangosteen leaf aqueous extract-mediated ZnO-CuO NCs with their energy bandgap. The absorption peak position had no significant changes in ZnO-CuO NCs synthesized at different controlled parameters. The mangosteen leaf aqueous extract absorption peak was located at 479 cm^{-1}, while for Cu(NO$_3$)$_2$·3H$_2$O and Zn(NO$_3$)$_2$·6H$_2$O, it was located at 295 and 305 cm^{-1}, respectively. On the other hand, the ZnO-CuO NCs absorption peak was located at 369–375 cm^{-1}.

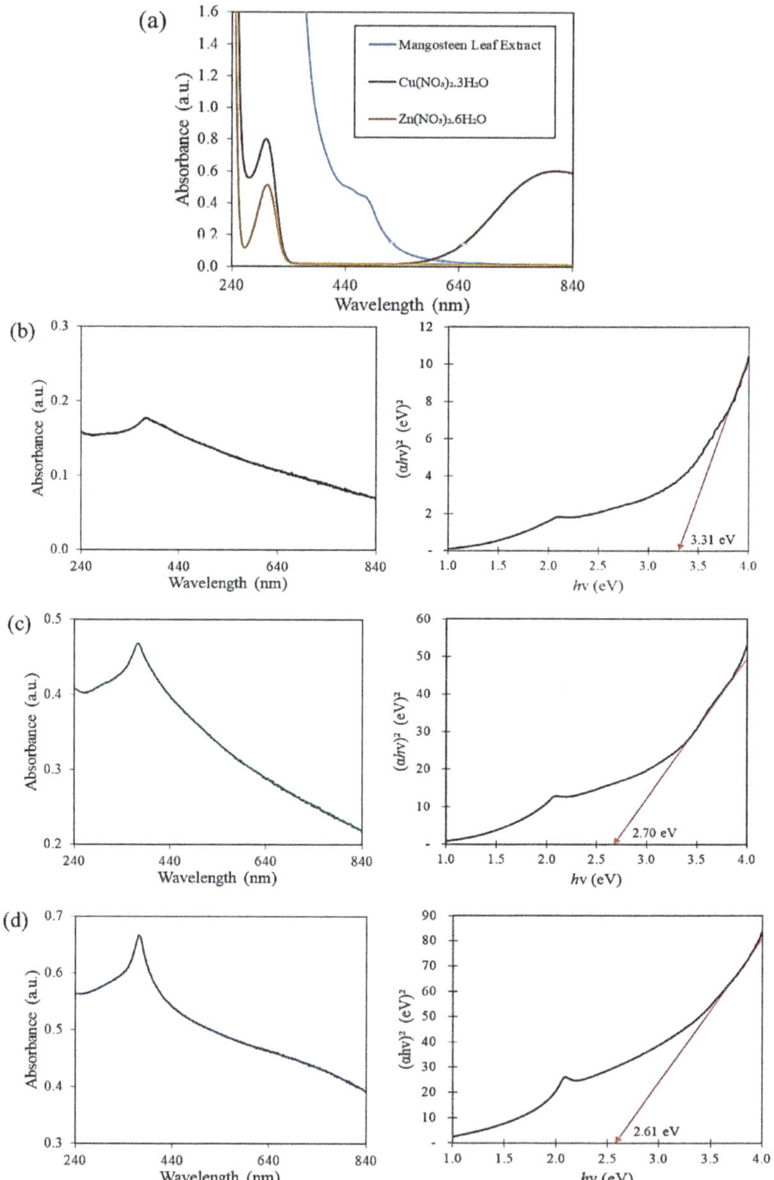

Figure 1. *Cont.*

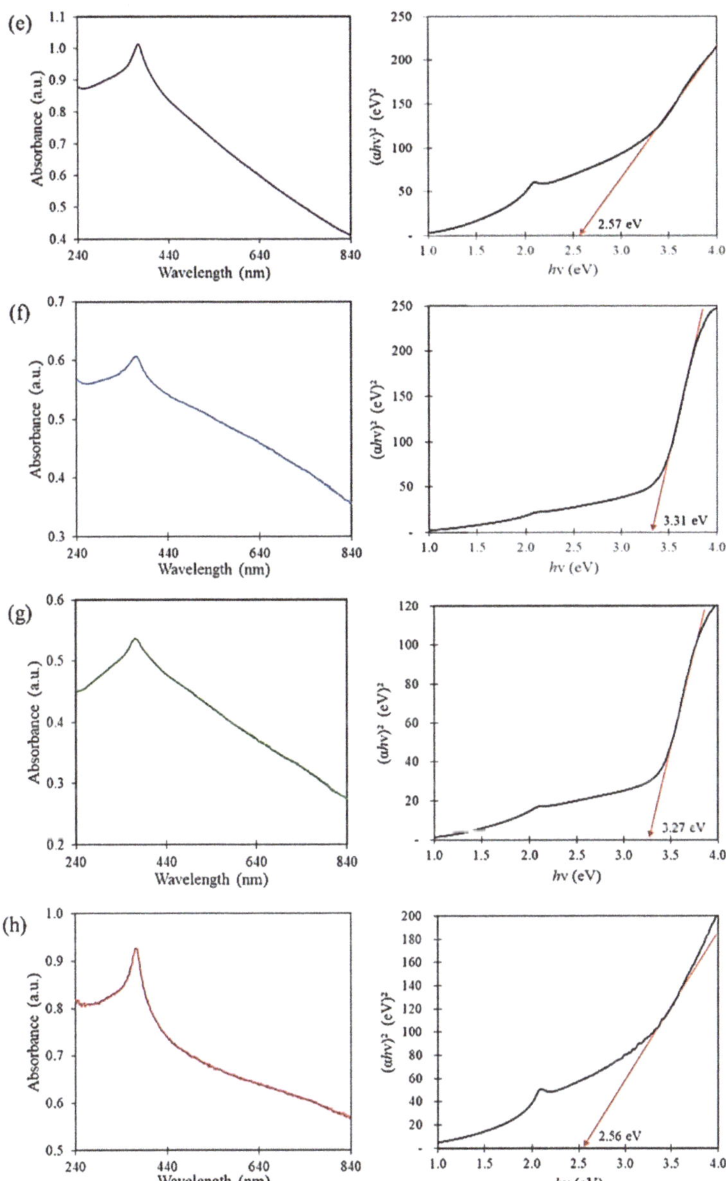

Figure 1. UV-Vis spectra (**left**) and energy bandgap (**right**) of (**a**) mangosteen leaf aqueous extract and green-synthesized ZnO-CuO NCs using (**b**) 0.02 g/mL, (**c**) 0.03 g/mL, (**d**) 0.04 g/mL and (**e**) 0.05 g/mL of mangosteen leaf extract; calcinated at (**f**) 300 °C and (**g**) 400 °C, respectively, by using 2 g of $Cu(NO_3)_2 \cdot 3H_2O$. Meanwhile, (**h**) is the UV-Vis spectrum (left) and energy bandgap (right) of 4 g of $Cu(NO_3)_2 \cdot 3H_2O$ calcinated at 500 °C by using 0.05 g/mL of mangosteen leaf aqueous extract.

The energy bandgap of the mangosteen leaf aqueous extract-mediated ZnO-CuO NCs at different synthesizing conditions is tabulated in Table 1. The energy band gap of the ZnO-CuO NPs was expressed in eV and calculated using a Tauc-plot approach using Equation (1).

$$\alpha h v = A\left(h v - E_g\right)^n \tag{1}$$

where h is Plank's constant (6.626×10^{-34} Js), n is the exponential factor for electronic transition ($n = \frac{1}{2}$ for the indirect band, $n = 2$ for the direct band) and α is the absorption coefficient. The energy bandgap showed no significant difference when using 2 g (2.57 eV) and 4 g (2.56 eV) of $Cu(NO_3)_2 \cdot 3H_2O$ in synthesizing ZnO-CuO NCs. In contrast, a plant aqueous extract concentration-dependent and calcination temperature-dependent shifts were observed as the energy bandgap decreased from 3.31 eV to 2.57 eV and higher leaf aqueous extract concentrations and calcination temperatures were applied.

Table 1. Energy bandgap of green-synthesized ZnO-CuO NCs at different synthesizing condition parameters.

Mangosteen Leaf Aqueous Extract Concentration (g/mL)	Calcination Temperature (°C)	Weight of $Zn(NO_3)_2 \cdot 6H_2O$ (g)	Weight of $Cu(NO_3)_2 \cdot 3H_2O$ (g)	Energy Bandgap (eV)
0.02	500	4.0	2.0	3.31
0.03	500	4.0	2.0	2.70
0.04	500	4.0	2.0	2.61
0.05	500	4.0	2.0	2.57
0.05	300	4.0	2.0	3.31
0.05	400	4.0	2.0	3.27
0.05	500	4.0	4.0	2.56

3.2. FTIR Spectroscopy Analysis

The FTIR spectra interpretation of mangosteen leaf aqueous extract-mediated ZnO-CuO NCs at different controlled parameters is shown in Table 2, and their FTIR spectra are shown in Figure 2. The 3401–3436 cm^{-1} and 1629–1636 cm^{-1} bands corresponded to v(O-H) and v(C=O) or v(C=C). Moreover, 1384 cm^{-1} and 1099–1114 cm^{-1} bands were assigned to v(C-C aromatic) and v(C-O). The bond vibration of CuO and ZnO was indicated by the bands at 649–674 cm^{-1} and 447–524 cm^{-1}, respectively.

Table 2. Interpretation of FTIR spectra of green-synthesized ZnO-CuO NCs at different synthesizing condition parameters with the data presented in cm^{-1}.

Functional Groups	Parameters								
	Mangosteen Leaf Aqueous Extract Concentration (g/mL)				Calcination Temperature (°C)			Weight of $Cu(NO_3)_2 \cdot 3H_2O$ (g)	
	0.02	0.03	0.04	0.05 *	300	400	500 *	2 *	4
v(O-H)	3429	3435	3435	3435	3436	3435	3435	3435	3401
v(C=O), v(C=C)	1629	1636	1636	1636	1633	1635	1636	1636	1635
v(C-C aromatic)	1384	1384	1384	1384	1384	1384	1384	1384	1384
v(C-O)	1114	1108	1108	1108	1102	1099	1108	1108	1106
v(Cu-O)	674	660	649	651	657	657	651	651	672
v(Zn-O)	447	455	485	486	502	498	486	486	524

* The ZnO-CuO NCs were green-synthesized in the same conditions.

The v(C-C aromatic) and v(C=O) or v(C=C) intensities increased when higher concentrations of mangosteen leaf aqueous extract were used. On the other hand, v(C-C aromatic) and v(C=O) or v(C=C) intensities decreased, while v(Cu-O) intensity increased at elevated calcination temperatures. Additionally, the bands, which included v(C-C aromatic), v(C=O) or v(C=C), v(C-O) and v(Cu-O) intensities improved when more $Cu(NO_3)_2 \cdot 3H_2O$ was added.

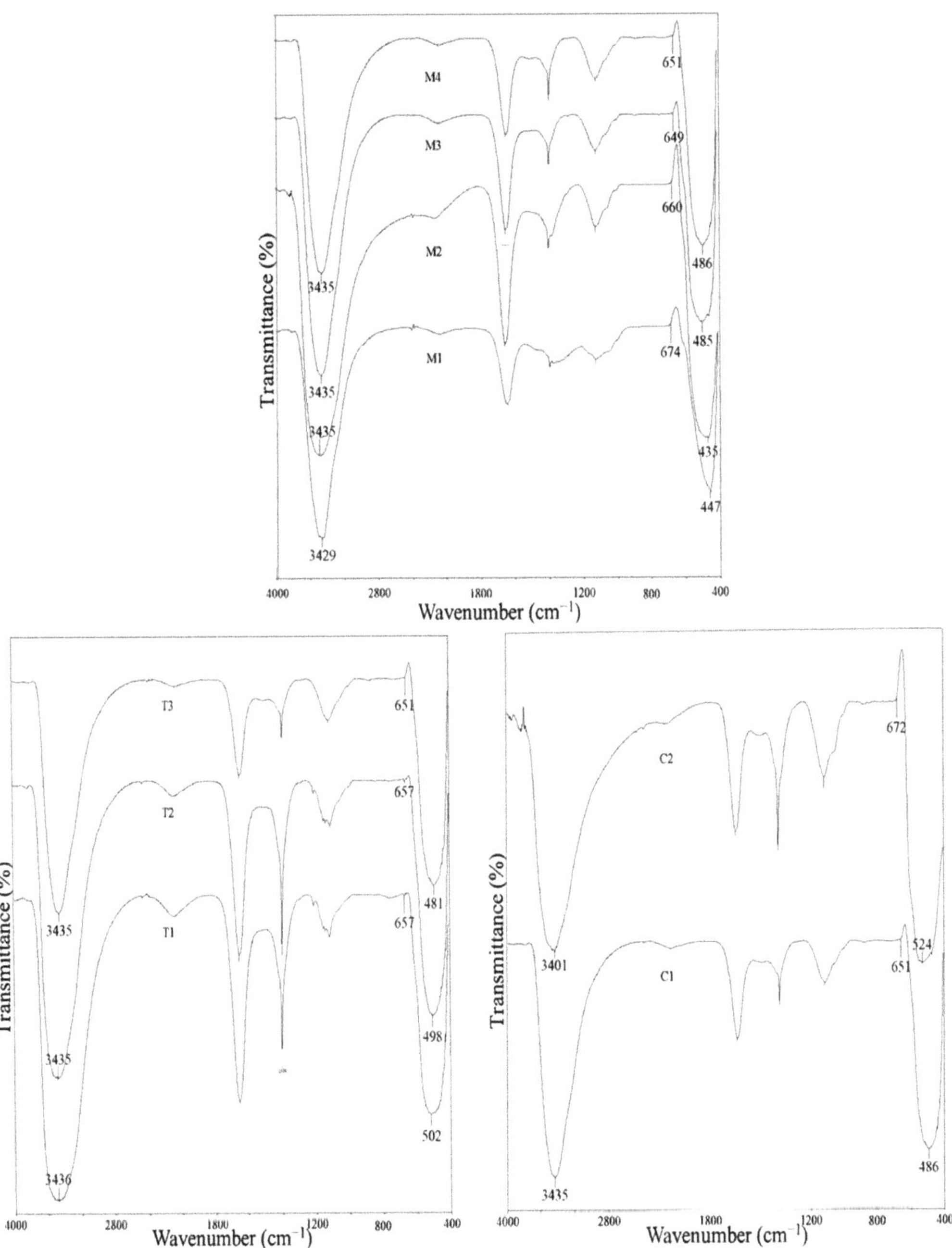

Figure 2. FTIR spectra of green-synthesized ZnO-CuO NCs at different controlled parameters: mangosteen leaf aqueous extract concentration (M1 = 0.02 g/mL, M2 = 0.03 g/mL, M3 = 0.04 g/mL and M4 = 0.05 g/mL), calcination temperature (T1 = 300 °C, T2 = 400 °C and T3 = 500 °C) and weight of $Cu(NO_3)_2 \cdot 3H_2O$ added (C1 = 2 g and C2 = 4 g), respectively.

3.3. PL Spectroscopy Analysis

The potential recombination of the photo-generated electron-hole (e^-/h^+) pairs of the mangosteen leaf aqueous extract-mediated ZnO-CuO NCs and the occurrence of their electronic transfer in NCs were determined by using PL spectroscopy (Figure 3). Overall, the ZnO-CuO NCs emission peaked in the violet region (390–405 nm). From Figure 3, it can be observed that the PL intensity was more affected by calcination temperature as high temperature-calcinated ZnO-CuO NCs had lower charge carrier separation compared to lower temperature samples.

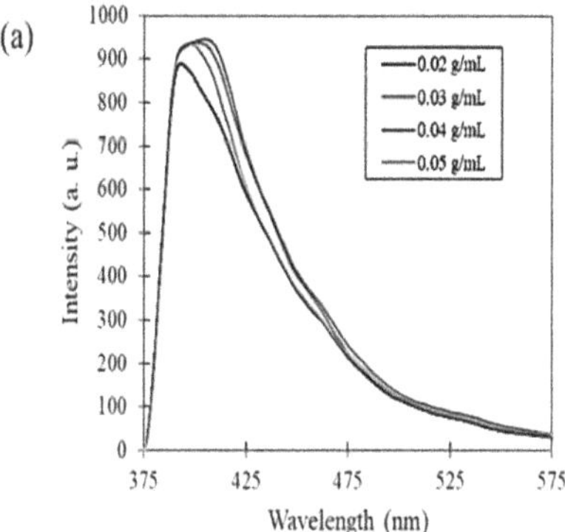

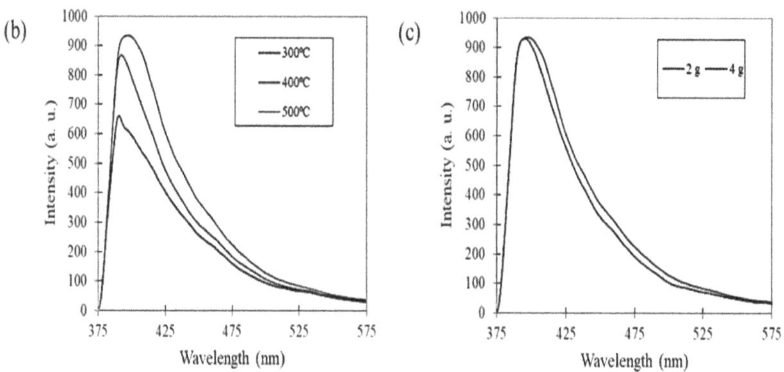

Figure 3. PL spectra of green-synthesized ZnO-CuO NCs under different parameters: (**a**) mangosteen leaf aqueous extract concentration, (**b**) calcination temperature and (**c**) weight of $Cu(NO_3)_2 \cdot 3H_2O$ added, respectively.

3.4. XRD Spectroscopy Analysis

The mangosteen leaf aqueous extract-mediated ZnO-CuO NCs with the reference card number ICDD 01-081-9217 were in a hexagonal-wurtzite phase, $a = 3.2459$ Å and $c = 5.1975$ Å, with space group *P63mc*. All peaks were very sharp and intense, indicating the samples were of a crystalline nature. ZnO-CuO NCs had diffraction peaks at 2θ values of 31.74, 34.41, 36.22, 47.57, 56.58, 62.90 and 69.03°, matched with the ZnO phase, indexed as (1 0 0), (0 0 2), (1 0 1), (1 0 2), (1 1 0), (1 0 3) and (2 0 1), respectively. Meanwhile, those 2θ values of 32.60, 35.58, 38.80, 48.59, 61.57, 66.30 and 68.01° matched with CuO phase were indexed as (−1 1 0), (0 0 2), (1 1 1), (−2 0 2), (−1 1 3), (0 2 2) and (−2 2 0), respectively. On the other hand, the CuO-indexed peaks' intensity magnitude was highest when using 4 g of $Cu(NO_3)_2 \cdot 3H_2O$, especially (1 1 1). The ZnO-CuO NCs spectra are shown in Figure 4.

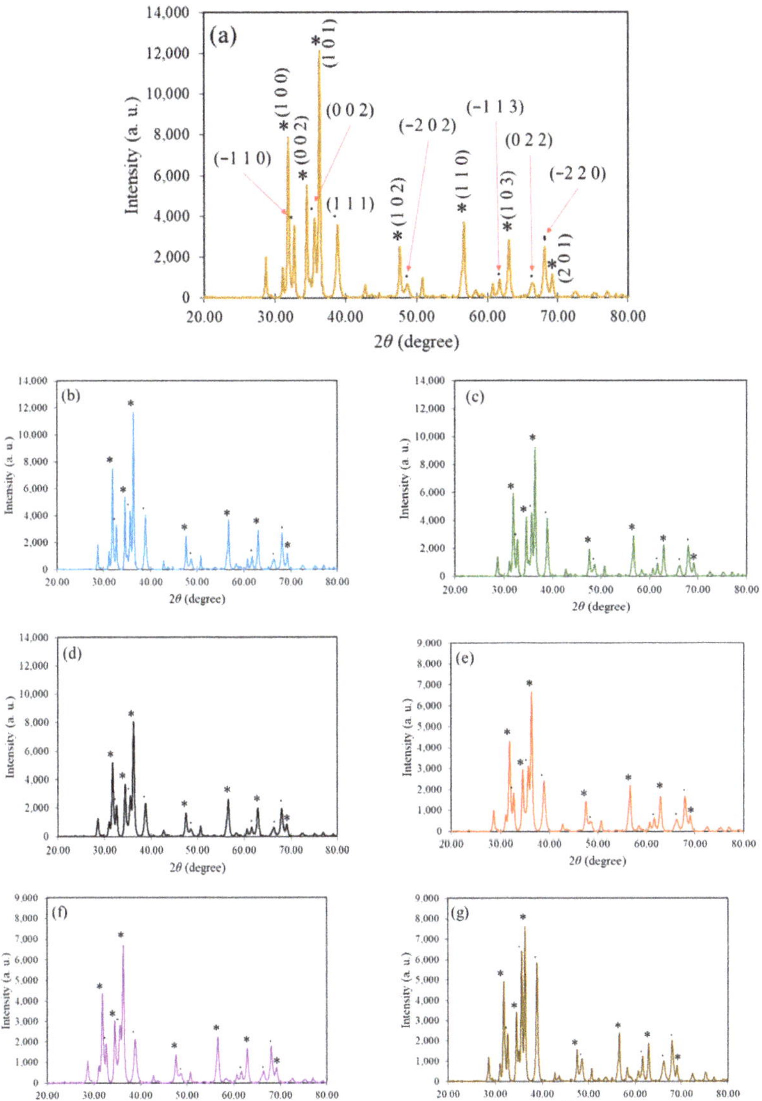

Figure 4. XRD spectra green-synthesized ZnO-CuO NCs using (**a**) 0.02 g/mL, (**b**) 0.03 g/mL, (**c**) 0.04 g/mL and (**d**) 0.05 g/mL of mangosteen leaf extract; calcinated at (**e**) 300 °C and (**f**) 400 °C, respectively, using 2 g of $Cu(NO_3)_2 \cdot 3H_2O$. Meanwhile, (**g**) is the XRD spectrum of 4 g of $Cu(NO_3)_2 \cdot 3H_2O$ calcinated at 500 °C using 0.05 g/mL of mangosteen leaf aqueous extract. All the lattice faces are represented in (**a**). Note that "*" represents ZnO, while "·" represents CuO.

The crystallinity of green-synthesized ZnO-CuO NCs was significantly affected by the mangosteen leaf aqueous concentration compared to calcination temperature and added $Cu(NO_3)_2 \cdot 3H_2O$ weight (Table 3). As shown in Equation (2), Debye–Scherrer's formula was used to calculate the crystalline size of ZnO-CuO NCs [35].

$$D = \frac{0.94\lambda}{\beta cos\theta} \tag{2}$$

where D is the crystalline size of NPs, λ is the X-ray wavelength, β is the full-width half-maximum (FWHM) of the peak and θ is the Bragg angle. In general, the crystalline size of the ZnO-CuO NCs was in the range of 18.17 to 28.51 nm. The decrement in crystalline size of ZnO-CuO NCs at elevated mangosteen leaf aqueous extract concentrations (from 28.51 nm to 18.17 nm) and calcination temperature (from 22.25 nm to 18.17 nm) was obtained. In contrast, a slight increment in the crystalline size of ZnO-CuO NCs, from 18.17 nm to 22.29 nm, when the weight of the added $Cu(NO_3)_2 \cdot 3H_2O$ increased from 2 g to 4 g.

Table 3. Crystalline sizes, dislocation density and micro strain of green-synthesized ZnO-CuO NCs at different parameters of synthesizing conditions.

Mangosteen Leaf Aqueous Extract Concentration (g/mL)	Calcination Temperature (°C)	Weight of $Zn(NO_3)_2 \cdot 6H_2O$ (g)	Weight of $Cu(NO_3)_2 \cdot 3H_2O$ (g)	ZnO Crystalline Size (nm)	CuO Crystalline Size (nm)	Crystalline Size (nm)	Dislocation Density (10^{14} cm^{-1})	Micro Strain (10^{-4})
0.02	500	4.0	2.0	31.98	25.03	28.51	12.31	1.35
0.03	500	4.0	2.0	28.67	25.23	26.95	13.77	1.32
0.04	500	4.0	2.0	27.42	23.23	25.32	15.59	1.40
0.05	500	4.0	2.0	21.10	15.23	18.17	30.30	2.77
0.05	300	4.0	2.0	23.13	21.37	22.25	20.19	1.69
0.05	400	4.0	2.0	21.94	18.52	20.23	24.42	1.86
0.05	500	4.0	4.0	20.47	20.31	22.29	20.13	1.63

The ZnO-CuO NCs dislocation density was estimated using Williamson and Smallman's formula [35] in Equation (3).

$$\delta = \frac{1}{D^2} \tag{3}$$

where δ is the dislocation density of NPs and D is the NPs' crystalline size. The ZnO-CuO NCs' dislocation density was in the range of 12.31×10^{14} to 30.30×10^{14} cm^{-1}. An increment in dislocation density was obtained when higher mangosteen leaf aqueous extract concentrations (from 12.31×10^{14} cm^{-1} to 30.30×10^{14} cm^{-1}) and calcination temperatures (from 20.19×10^{14} cm^{-1} to 30.30×10^{14} cm^{-1}) were applied. However, their dislocation density decreased from 30.30×10^{14} cm^{-1} to 20.13×10^{14} when more $Cu(NO_3)_2 \cdot 3H_2O$ was added during the green synthesis of ZnO-CuO NCs.

Equation (4) was used to calculate the micro strain of the ZnO-CuO NCs [35].

$$\varepsilon = \frac{\beta cos\theta}{4} \tag{4}$$

where ε is the micro strain of NPs, β is the FWMH of the peak and θ is the Bragg angle. Greater micro strain in ZnO-CuO NCs was found at higher concentrations of mangosteen leaf aqueous extract (from 1.35×10^{-4} to 2.77×10^{-4}) and calcination temperatures (from 1.69×10^{-4} to 2.77×10^{-4}), which was contradictory to when more $Cu(NO_3)_2 \cdot 3H_2O$ was added (from 2.77×10^{-4} to 1.63×10^{-4}).

3.5. FE-SEM Spectroscopy Analysis

The particle size of ZnO-CuO NCs was in the range of 39.10 to 74.53 nm, as tabulated in Table 4. The particle size decreased at elevated mangosteen leaf aqueous extract concentrations (61.46 nm decreased to 39.10 nm) and calcination temperatures (74.53 nm decreased to 39.10 nm). In contrast, a larger particle size was found when 4 g of $Cu(NO_3)_2 \cdot 3H_2O$ (65.18 nm) was used compared to 2 g of $Cu(NO_3)_2 \cdot 3H_2O$ (39.10 nm) in green synthesizing ZnO-CuO NCs. The trends of the particle size of the biogenic ZnO-CuO NCs were in accordance with the analyzed XRD results and tabulated in Table 4. The SEM micrographs are shown in Figure 5.

Table 4. Particle sizes and morphologies of green-synthesized ZnO-CuO NCs at different synthesizing condition parameters.

Mangosteen Leaf Aqueous Extract Concentration (g/mL)	Calcination Temperature (°C)	Weight of $Zn(NO_3)_2 \cdot 6H_2O$ (g)	Weight of $Cu(NO_3)_2 \cdot 3H_2O$ (g)	Particle Size (nm)	Morphology
0.02	500	4.0	2.0	61.46	Agglomerated with irregular nanostructure
0.03	500	4.0	2.0	56.27	Agglomerated with quasi-spherical nanostructure
0.04	500	4.0	2.0	55.23	Agglomerated with lobed nanostructure
0.05	500	4.0	2.0	39.10	Agglomerated and uniformly distributed with spherical nanostructure
0.05	300	4.0	2.0	74.53	Highly agglomerated with quasi-spherical nanostructure
0.05	400	4.0	2.0	53.71	Agglomerated with spherical nanostructure
0.05	500	4.0	4.0	65.18	Agglomerated with spherical nanostructure

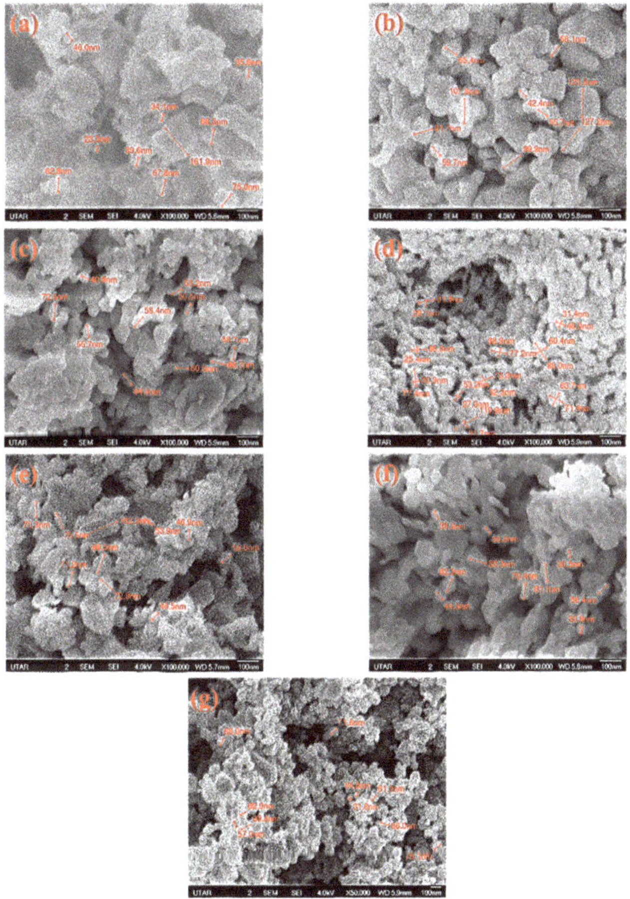

Figure 5. SEM micrographs of green-synthesized ZnO-CuO NCs using (**a**) 0.02 g/mL, (**b**) 0.03 g/mL, (**c**) 0.04 g/mL and (**d**) 0.05 g/mL of mangosteen leaf extract; calcinated at (**e**) 300 °C and (**f**) 400 °C, respectively, using 2 g of $Cu(NO_3)_2 \cdot 3H_2O$. Meanwhile, (**g**) is the SEM image of 4 g of $Cu(NO_3)_2 \cdot 3H_2O$ calcinated at 500 °C using 0.05 g/mL of mangosteen leaf aqueous extract.

3.6. EDX Spectroscopy Analysis

The copper-to-zinc atomic percentage ratio was similar to the copper precursor-to-zinc precursor weight ratio used in synthesizing ZnO-CuO NCs. Neither the mangosteen leaf aqueous extract concentration nor the calcination temperature applied significantly influenced the detected element atomic percentage, as stated in Table 5. On the other hand, compared to 2 g of $Cu(NO_3)_2\cdot3H_2O$, an obvious increment in copper atomic percentage (18.44%) and its intensity (around 8 keV) were observed by using 4 g of $Cu(NO_3)_2\cdot3H_2O$ in synthesizing ZnO-CuO NCs. The previous copper atomic percentage was only 12.55%. Overall, the synthesized ZnO-CuO NCs depicted the highest atomic percentage in oxygen (60.16–66.25%), followed by zinc atomic percentage (20.10–26.98%) and copper atomic percentage (11.64–18.44%). The presence of an oxygen peak indicated zinc and copper were in oxidized form, and no impurity was found in EDX spectra (Figure 6).

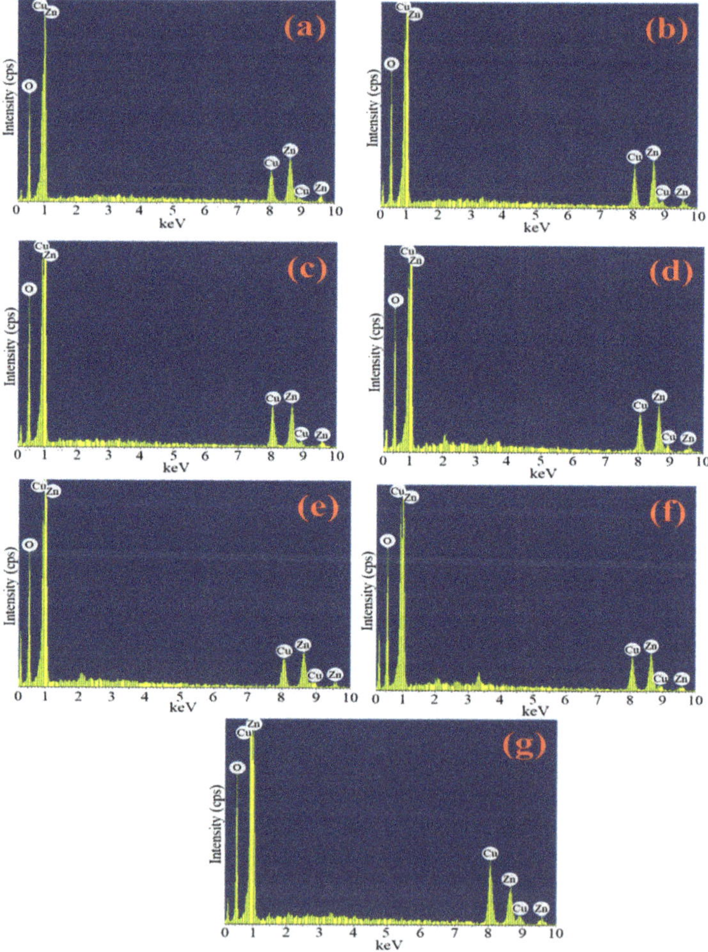

Figure 6. EDX spectra of green-synthesized ZnO-CuO NCs using (**a**) 0.02 g/mL, (**b**) 0.03 g/mL, (**c**) 0.04 g/mL and (**d**) 0.05 g/mL of mangosteen leaf extract; calcinated at (**e**) 300 °C and (**f**) 400 °C, respectively, using 2 g of $Cu(NO_3)_2\cdot3H_2O$ Meanwhile, (**g**) is the EDX spectrum of 4 g of $Cu(NO_3)_2\cdot3H_2O$ calcinated at 500 °C using 0.05 g/mL of mangosteen leaf aqueous extract.

Table 5. Atomic percentage of green-synthesized ZnO-CuO NCs at different parameters of synthesizing conditions.

Mangosteen Leaf Aqueous Extract- Concentration (g/mL)	Calcination Temperature (°C)	Weight of $Zn(NO_3)_2 \cdot 6H_2O$ (g)	Weight of $Cu(NO_3)_2 \cdot 3H_2O$ (g)	Oxygen Atomic Percentage	Copper Atomic Percentage	Zinc Atomic Percentage
0.02	500	4.0	2.0	62.28	13.03	24.68
0.03	500	4.0	2.0	60.16	12.86	26.98
0.04	500	4.0	2.0	63.25	13.10	23.65
0.05	500	4.0	2.0	62.41	12.55	25.03
0.05	300	4.0	2.0	66.25	12.28	21.47
0.05	400	4.0	2.0	66.20	11.64	22.16
0.05	500	4.0	4.0	61.45	18.44	20.10

3.7. Comparison with Other Studies

The lowest energy bandgap, and smallest crystalline and particle sizes of the mangosteen leaf aqueous extract-mediated ZnO-CuO NCs were selected to compare with other reports, as shown in Table 6. By using less $Cu(NO_3)_2 \cdot 3H_2O$, the selected ZnO-CuO NCs' energy bandgap, crystalline and particle sizes was comparable to other reports. This proved that ZnO-CuO NCs green synthesized in the current study were more cost-effective and eco-friendly when using a mangosteen leaf aqueous extract.

Table 6. Comparison of plant extract-mediated ZnO-CuO NCs with other studies.

Plant Extract	Plant Extract Concentration (g/mL)	Calcination Temperature (°C)	Zinc Salt Added (g)	Copper Salt Added (g)	Energy Bandgap (eV)	Crystalline Size (nm)	Particle Size (nm)	Reference
Theobroma cacao seed bark	0.20 [1]	400	0.03 M	0.01 M	-	10.00	20.0–50.0	[6]
D. caffra leaf	0.10	400	18.35 [2]	3.19 [3]	-	23.21	20.0–32.0	[20]
V. sinaiticum Benth	0.10	500	5.0	1.25	2.74	18.00	-	[21]
S. nigra L. shoot	0.14	400	1.6	0.8	-	-	20.0–130.0	[22]
G. mangostana L. leaf	0.05	500	4.0	2.0	2.57	18.17	25.4–60.4	Current study

[1] 0.1 mol, [2] 0.02 mol, [3] Extracted using methanol, fractional by distilled water and n-hexane.

4. Discussion

The appearance of an absorption peak at 479 cm^{-1} in mangosteen leaf aqueous extract can be attributed to the $\pi \rightarrow \pi^*$ transition [36]. On the other hand, the absorption peaks at 305 cm^{-1} and 308 cm^{-1} indicated the *d-d* transition of the $Cu(NO_3)_2 \cdot 3H_2O$ and $Zn(NO_3)_2 \cdot 6H_2O$, respectively. The presence of phytochemicals in the mangosteen leaf aqueous extract led to the occurrence of surface plasmon resonance (SPR) phenomena at a specific wavelength. The change in color of the leaf aqueous extract from light brown to brown upon the addition of the precursors revealed the reduction of zinc(II) ions to zinc(0) and copper(II) to copper(0), followed by oxidation into ZnO-CuO [17,37,38]. As a result, the absorption peaks of the ZnO-CuO NCs were red-shifted to a higher wavelength due to the formation of secondary electronic states, influenced by the metal oxide conjugation with electronic transitions between the valence band and conduction band and the exchange interaction of *s, p-d* spin within the atoms of metal and oxygen [23].

Regarding the energy bandgap, Ma et al. (2019) [39] reported that the copper precursor did not significantly affect it. However, Fouda et al. observed a significant decrease in the energy bandgap of ZnO-CuO NCs with an increase in the copper precursor amount [40]. Similarly, a decreasing trend in energy bandgap was observed when higher leaf aqueous extract concentrations and calcination temperatures were applied. According to the energy bandgap theory, the energy bandgap of NCs should increase or decrease due to the splitting of each level into a number of levels equal to the number of interacting atoms. In the case of hetero-structured NCs, the bands may overlap [41]. Moreover, the energy bandgap of ZnO-CuO NCs involved coupled transitions from the O_2 ($2p$) valance band to zinc(II) ($3d^1$–$4s$) and copper(II) ($3d^9$) ion conduction bands [42]. Additionally, the presence of CuO, acting as an impurity, reduces the energy bandgap in ZnO-CuO NCs [6,23], and this effect became more significant with higher concentrations of mangosteen leaf aqueous extract and higher calcination temperatures, suggesting the presence of a higher amount of CuO in ZnO-CuO NCs. Also, the redshift in the energy bandgap could be attributed to the interactions between electrons in the localized d-orbital of copper ions, which replaced zinc ions and the band electrons in the NCs [43]. This phenomenon makes the NCs efficient in light harvesting for photocatalytic applications [21].

The high PL indicates significant recombination of charge carriers, while low PL suggests maximum charge separation, which is beneficial for the photo-degradation of the processes [6,21,23,42]. The emission peaks of the ZnO-CuO NCs in the violet region (390–405 nm) were attributed to near-band-edge (NBE) emission caused by the defect states in ZnO and CuO [6,21,42]. Furthermore, the lower separation of charge carriers observed in ZnO-CuO NCs calcinated at high temperatures could be attributed to the reduced presence of oxygen vacancies, leading to the enhancement of NBE emission intensity [42].

The phytochemicals present in the mangosteen leaf aqueous extract, such as xanthones, flavonoids and terpene [30–33], were responsible for the observed functional groups. These compounds played a crucial role as capping, stabilizing and reducing agents during the green synthesis of ZnO-CuO NCs, primarily through electrostatic and steric stabilization mechanisms [32,44]. The vibration of the CuO and ZnO bonds was supported by previous studies [4–6,40,45,46]. The bands corresponding to metal oxides and hydroxides are typically located below 1000 cm^{-1} (fingerprint region) due to interatomic vibrations [47]. The sharp band observed in the Zn-O bond vibration confirmed the presence of a strong hexagonal-wurtzite single-phase of ZnO [18]. Additionally, the absence of Cu_2O could be inferred from the location of Cu-O bond vibration [42], as depicted in Figure 3. Furthermore, slight shifts in the bands indicated structural changes in ZnO-CuO NCs due to the incorporation of an additional element [43]. Changes in the intensity of the bands may be attributed to the variations in the interaction of functional groups from the plant extract under different controlled parameters.

The XRD patterns shown in Figure 4 confirmed the successful biosynthesis of ZnO-CuO NCs [35,40]. Previous literature reports have suggested that NCs with less than 15% of copper exhibited a one-phase wurtzite-like $Cu_xZn_{1-x}O$, while those with a higher copper content appeared as a tenorite-like oxide phase, $Zn_xCu_{1-x}O$ [20,23]. The higher peak intensity of ZnO peaks compared to CuO peaks indicate a higher percentage of ZnO in ZnO-CuO NCs [20,22]. Furthermore, the role of ZnO as a coating material led to lower peak crystallization of CuO [22]. The highest intensity at (1 0 1) corresponded to a ZnO crystal structure grown in the a-direction [39]. The intensity of indexed CuO peaks was highest when 4 g of $Cu(NO_3)_2 \cdot 3H_2O$ was used, indicating the contribution of copper to the formation of ZnO-CuO NCs [21,35,40], which also reflected its higher weight percentage [45,46]. The crystalline size of the ZnO-CuO NCs was similarly reported in Adeyemi et al.'s study [20]. The decrease in crystalline size of the ZnO-CuO NCs with increasing concentrations of mangosteen leaf aqueous extract and calcination temperatures demonstrated the effectiveness of phytochemicals in the plant extract for capping and stabilizing the ZnO-CuO NCs [48], particularly when a high concentration of mangosteen leaf aqueous extract and high calcination temperature were applied. In contrast, a slight increase in the crystalline size of the ZnO-CuO NCs was observed when more $Cu(NO_3)_2 \cdot 3H_2O$ was added, indicating that the crystallinity of the synthesized NCs was greatly influenced by the variations in the precursor added [18]. However, these results differed from those reported in Fouda et al.'s study, where the crystalline size of their ZnO-CuO NCs decreased with the addition of more copper precursors during synthesis [40]. The broadening of peaks in the XRD pattern of ZnO-CuO NCs was caused by the strain resulting from non-uniform lattice distortion and crystal phase dislocation due to the mismatch in the sizes of zinc and copper atoms [35,42]. Consequently, the presence of a greater number of interfaces in each volume led to a smaller crystalline size [34], and the level of micro strain in the synthesized material increased as the size decreased [42], which was consistent with the results obtained from the calculated crystalline size.

Agglomerated spherical nanostructures were observed in mangosteen leaf aqueous extract-mediated ZnO-CuO, as depicted in Figure 5. This can be attributed to several factors, including the high viscosity of the plant extract [49], the surface physicochemical characteristics [50–53], the strong forces of attraction between particles [44,54], and the oxidation of metal oxide NPs or NCs [55]. The agglomeration of ZnO-CuO NCs was also influenced by the reduction of salt precursors to zinc and copper ion nucleation mediated by the mangosteen leaf aqueous extract, indicating their role as capping and reducing agents during the formation of ZnO-CuO NCs [43]. The formation of spherical nanostructures (0.05 g/mL) progressively occurred with increasing concentrations of mangosteen leaf aqueous extract, transitioning from irregular nanostructures at low concentrations (0.02 g/mL) of leaf aqueous extract. This may be due to greater isotropic aggregation at the isoelectric point, resulting in strong particle cohesion and the formation of nearly spherical structures [53,56,57] accompanied by the coarsening and coalescence of the NCs [9,11].

Similar results have been reported in terms of EDX analysis by Elemike et al.'s study [23]. The presence of only zinc, copper and oxygen peaks in all ZnO-CuO NCs in the EDX spectra suggests the purity of the green-synthesized ZnO-CuO NCs [20].

Although Yulizar et al.'s study [6] suggested crosslinking between zinc hydroxide and copper hydroxide in the formation of ZnO-CuO NCs, the mechanism and bonding involved in the green synthesis of ZnO-CuO NCs were not clearly addressed by researchers. Phytochemicals present in plants with functional groups, such as -C-O-C-, -C-O-, -C=C- and –C=O- in flavonoids, alkaloids, phenols and anthracenes, have been hypothesized to play a significant role in reducing, capping and stabilizing green-synthesized nanomaterials [58,59]. Xanthones, such as 1, 5, 8-trihydroxy-3-methoxy-2-(3-methylbut-2-enyl) xanthone and 1, 6-dihydroxy-3-methoxy-2-(3-methyl-2-buthenyl)-xanthone, are the major compounds in mangosteen leaf [60]. During chelation, electrons from the precursors' zinc and copper atoms were donated to form positively charged zinc(II) and copper(II) ions, respectively, which then formed metal complexes with the phytochemicals. These metal complexes subsequently bonded with negatively charged oxygen(II) ions during calcination [61]. Another possible mechanism for the formation of ZnO-CuO NCs was bio reduction, where the divalent oxidation state of zinc and copper were reduced to a zero-valent state by the phytochemicals present in the mangosteen leaf aqueous extract, as indicated by the immediate color change during the green synthesis [62]. ZnO and CuO nuclei were formed after the metallic zinc and copper reacted with the dissolved oxygen in the precursor solution [61], and coordinate covalent bonds were subsequently formed between ZnO and CuO through the lone-pair electron from the oxygen atoms of the metal oxides. A strong framework of ZnO-CuO NCs was then produced during calcination [6]. The possible mechanism and bonding in green synthesizing ZnO-CuO NCs is represented in Scheme 1.

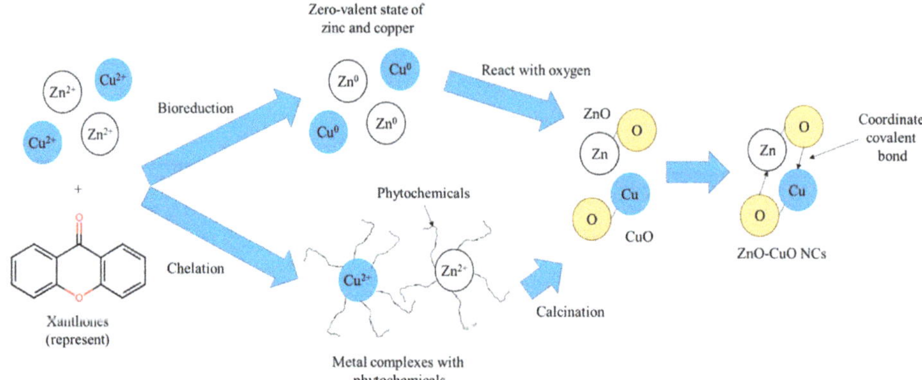

Scheme 1. Possible mechanism of green-synthesized ZnO-CuO NCs via bio reduction and chelation with the presence of phytochemicals (represented by xanthones). Coordinate covalent bond formation between ZnO and CuO by lone-pair electrons from the bonded oxygen atom was proposed to link both metal oxides in NCs.

5. Conclusions

ZnO-CuO NCs were successfully green synthesized using a mangosteen leaf aqueous extract at different concentrations (0.02, 0.03, 0.04 and 0.05 mg/mL), calcination temperatures (300, 400 and 500 °C) and weights of $Cu(NO_3)_2 \cdot 3H_2O$ (2 and 4 g). The properties of ZnO-CuO NCs were significantly influenced by the green synthesis parameters, including the concentration of the plant extract, calcination temperature and precursor weight. The energy bandgap and crystalline properties of the ZnO-CuO NCs were notably affected by the concentration of the mangosteen leaf aqueous extract and the calcination temperature. However, the intensity of the PL spectrum was solely dependent on the applied calcination temperature. Moreover, the atomic percentage of copper-to-zinc was primarily affected by the weight of the zinc and copper precursor used to synthesize ZnO-CuO NCs. The particle size and morphology were significantly influenced by varied parameters employed in the green synthesis of ZnO-CuO NCs. However, the locations of the FTIR bands in the ZnO-CuO NCs remained consistent throughout the study. The presence of coordinate covalent bonds between ZnO and CuO facilitated by the lone pair of electrons from the oxygen atoms was suggested. The study clearly illustrated the effects of plant extract concentrations, calcination temperatures and precursor amount on the optical, structural and morphological properties during the green synthesis of ZnO-CuO NCs. These findings provide valuable insights for researchers to synthesize ZnO-CuO NCs with specific properties for future applications.

Author Contributions: Conceptualization, M.A. (Mohammod Aminuzzaman), L.-H.T. and M.A. (Md. Akhtaruzzaman); methodology, M.A. (Mohammod Aminuzzaman); validation, Y.F.W.; formal analysis, Y.B.C., M.A. (Mohammod Aminuzzaman) and A.W.; investigation, Y.B.C. and S.D.; resources, Y.B.C.; data curation, Y.B.C.; writing—original draft preparation, Y.B.C.; writing—review and editing, M.A. (Mohammod Aminuzzaman), L.-H.T. and Y.F.W..; visualization, Y.B.C.; supervision, M.A. (Mohammod Aminuzzaman), L.-H.T. and M.A. (Md. Akhtaruzzaman); project administration, M.A. (Mohammod Aminuzzaman) and L.-H.T.; funding acquisition, M.A. (Mohammod Aminuzzaman) and L.-H.T. All authors have read and agreed to the published version of the manuscript.

Funding: This research was funded by the Universiti Tunku Abdul Rahman (UTAR) through UTARRF (IPSR/RMC/UTARRF202-C2/M01).

Institutional Review Board Statement: Not applicable.

Informed Consent Statement: Not applicable.

Data Availability Statement: The data presented in this study are available upon request from the corresponding author.

Acknowledgments: The authors extended our appreciation to the Universiti Tunku Abdul Rahman (UTAR) for providing financial support through UTARRF (IPSR/RMC/UTARRF/202-C2/M01) and research facilities to carry out the research work. Authors also extended their appreciation to Wong Ling Shing in giving advices and suggestions in improving the English fluency of the manuscript.

Conflicts of Interest: The authors declare no conflict of interest.

References

1. Das, S.; Srivastava, V.C. An overview of the synthesis of CuO-ZnO nanocomposite for environmental and other applications. *Nanotechnol. Rev.* **2018**, *7*, 267–282. [CrossRef]
2. Khan, S.A.; Noreen, F.; Kanwal, S.; Iqbal, A.; Hussain, G. Green synthesis of ZnO and Cu-doped ZnO nanoparticles from leaf extracts of *Abutilon indicum*, *Clerodendrum infortunatum*, *Clerodendrum inerme* and investigation of their biological and photocatalytic activities. *Mater. Sci. Eng. C* **2017**, *82*, 46–59. [CrossRef] [PubMed]
3. Vibitha, B.V.; Anitha, B.; Tharayil, N.J. Green synthesis of ZnO:CuO nanocomposites by *Aloe barbadansis* leaf extract: Structure and photo catalytic properties. In Proceedings of the AIP Conference Proceedings, International Conference on Energy and Environment 2019, Guimaraes, Portugal, 16–17 May 2019.
4. Rajith Kumar, C.R.; Betageri, V.S.; Nagaraju, G.; Pujar, G.H.; Onkarappa, H.S.; Latha, M.S. One-pot green synthesis of ZnO-CuO nanocomposite and their enhanced photocatalytic and antibacterial activity. *Adv. Nat. Sci. Nanosci. Nanotechnol.* **2020**, *11*, 015009. [CrossRef]
5. Sakib, A.A.M.; Masum, S.M.; Hoinkis, J.; Islam, R.; Molla, M.A.I. Synthesis of CuO/ZnO nanocomposites and their application in photodegradation of toxic textile dye. *J. Compos. Sci.* **2019**, *3*, 91–103. [CrossRef]
6. Yulizar, Y.; Bakri, R.; Apriandanu, D.O.B.; Hidayat, T. ZnO/CuO nanocomposite prepared in one-pot green synthesis using seed bark extract of *Theobroma cacao*. *Nano-Struct. Nano-Objects* **2018**, *16*, 300–305. [CrossRef]
7. Bano, S.; Pillai, S. Green synthesis of calcium oxide nanoparticles at different calcination temperatures. *World J. Sci. Technol. Sustain. Dev.* **2020**, *17*, 283–295. [CrossRef]
8. Hassan, S.E.-D.; Fouda, A.; Saied, E.; Farag, M.M.S.; Eid, A.M.; Barghoth, M.G.; Awad, M.A.; Hamza, M.F.; Awad, M.F. *Rhizopus oryzae*-mediated green synthesis of magnesium oxide nanoparticles (MgO-NPs): A promising tool for antimicrobial, mosquitocidal action, and tanning effluent treatment. *J. Fungi* **2021**, *7*, 372–396. [CrossRef]
9. Jameel, M.S.; Aziz, A.A.; Dheyab, M.A. Green synthesis: Proposed mechanism and factors influencing the synthesis of platinum nanoparticles. *Green. Process Synth.* **2020**, *9*, 386–398. [CrossRef]
10. Khan, A.; Shabir, D.; Ahmad, P.; Khandaker, M.U.; Faruque, M.R.; Din, I.U. Biosynthesis and antibacterial activity of MgO-NPs produced from *Camellia-sinensis* leaves extract. *Mater. Res. Express* **2021**, *8*, 015402. [CrossRef]
11. Mazli, S.R.A.; Yusoff, H.M.; Idris, N.H. Synthesis of zinc oxide nanoparticles by using *Aloe vera* leaf extract as pontential anode material in lithium ion battery. *Univ. Malaysia Teren. J. Undergrad. Res.* **2020**, *2*, 1–8.
12. Xu, J.; Huang, Y.; Zhu, S.; Abbes, N.; Jing, X.; Zhang, L. A review of the green synthesis of ZnO nanoparticles using plant extracts and their prospects for application in antibacterial textiles. *J. Eng. Fiber Fabr.* **2021**, *16*, 1–14. [CrossRef]
13. Efenberger-Szmechtyk, M.; Nowak, A.; Czyzowska, A. Plant extracts rich in polyphenols: Antibacterial agents and natural preservatives for meat and meat products. *Crit. Rev. Food Sci. Nutr.* **2020**, *61*, 149–178. [CrossRef]
14. Prasanth, R.; Dinesh Kumar, S.; Jayalakshmi, A.; Singaravelu, G.; Govindaraju, K.; Ganesh Kumar, V. Green synthesis of magnesium oxide nanoparticles and their antibacterial activity. *Indian. J. Geo Mar. Sci.* **2019**, *48*, 1210–1215.
15. Shammout, M.W.; Awwad, A.M. A novel route for the synthesis of copper oxide nanoparticles using *Bougainvillea* plant flowers extract and antifungal activity evaluation. *Int. Sci. Organ.* **2021**, *7*, 71–78.
16. Kumar, H.; Bhardwaj, K.; Dhanjal, D.S.; Nepovimova, E.; Şen, F.; Regassa, H.; Singh, R.; Verma, R.; Kumar, V.; Kumar, D.; et al. Fruit extract mediated green synthesis of metallic nanoparticles: A new avenue in pomology applications. *Int. J. Mol. Sci.* **2020**, *21*, 8458. [CrossRef] [PubMed]
17. Kureshi, A.A.; Vaghela, H.M.; Kumar, S.; Singh, R.; Kumari, P. Green synthesis of gold nanoparticles mediated by *Garcinia* fruits and their biological applications. *Pharm. Sci.* **2021**, *27*, 238–250. [CrossRef]
18. Kaningini, A.G.; Azizi, S.; Sintwa, N.; Mokalane, K.; Mohale, K.C.; Mudau, F.N.; Maaza, M. Effect of optimized precursor concentration, temperature, and doping on optical properties of ZnO nanoparticles synthesized via a green route using bush tea (*Athrixia phylicoides* DC.) leaf extracts. *ACS Omega* **2022**, *7*, 31658–31666. [CrossRef] [PubMed]
19. Kumar, I.; Mondal, M.; Sakthivel, N. Green synthesis of phytogenic nanoparticles. In *Green Synthesis, Characterization and Applications of Nanoparticles*; Shukla, A.K., Iravani, S., Eds.; Elsevier Inc.: Amsterdam, The Netherlands, 2019; pp. 37–73.
20. Adeyemi, J.O.; Onwudiwe, D.C.; Oyedeji, A.O. Biogenic synthesis of CuO, ZnO, and CuO–ZnO nanoparticles using leaf extracts of *Dovyalis caffra* and their biological properties. *Molecules* **2022**, *27*, 3206. [CrossRef]
21. Bekru, A.G.; Tufa, L.T.; Zelekew, O.A.; Goddati, M.; Lee, J.; Sabir, F.K. Green synthesis of a CuO ZnO nanocomposite for efficient photodegradation of methylene blue and reduction of 4-nitrophenol. *ACS Omega* **2022**, *7*, 30908–30919. [CrossRef]

22. Cao, Y.; Dhahad, H.A.; El-Shorbagy, M.A.; Alijani, H.Q.; Zakeri, M.; Heydari, A.; Bahonar, E.; Slouf, M.; Khatami, M.; Naderifar, M.; et al. Green synthesis of bimetallic ZnO–CuO nanoparticles and their cytotoxicity properties. *Sci. Rep.* **2021**, *11*, 23479. [CrossRef]
23. Elemike, E.E.; Onwudiwe, D.C.; Singh, M. Eco-friendly synthesis of copper oxide, zinc oxide and copper oxide–zinc oxide nanocomposites, and their anticancer applications. *J. Inorg. Organomet. Polym. Mater.* **2019**, *30*, 400–409. [CrossRef]
24. Govindasamy, G.A.; Mydin, R.B.S.M.N.; Harun, N.H.; Sreekantan, S. Calcination temperatures, compositions and antimicrobial properties of heterostructural ZnO–CuO nanocomposites from *Calotropis gigantea* targeted for skin ulcer pathogens. *Sci. Rep.* **2020**, *11*, 99. [CrossRef] [PubMed]
25. Hiew, C.W.; Lee, L.J.; Junus, S.; Tan, Y.N.; Chai, T.T.; Ee, K.Y. Optimization of microwave-assisted extraction and the effect of microencapsulation on mangosteen (*Garcinia mangostana* L.) rind extract. *Food Sci. Technol.* **2021**, *42*, e35521. [CrossRef]
26. Huang, X.; Zhou, X.; Dai, Q.; Qin, Z. Antibacterial, antioxidation, UV-blocking, and biodegradable soy protein isolate food packaging film with mangosteen peel extract and ZnO nanoparticles. *Nanomaterials* **2021**, *11*, 3337. [CrossRef]
27. Mohd Basri, M.S.; Ren, B.L.M.; Talib, R.A.; Zakaria, R.; Kamarudin, S.H. Novel mangosteen leaves-based marker ink color lightness, viscosity, optimized composition, and microstructural analysis. *Polymers* **2021**, *13*, 1581. [CrossRef]
28. Mulyono, D.; Irawati, Y.; Syah, M.J.A. Identification morphological variability of six mangosteen (*Garcinia mangostana* L.) as a conservation strategy for local varieties. *IOP Conf. Ser. Earth Environ. Sci.* **2021**, *739*, 012076. [CrossRef]
29. Syahputra, M.R.; Setiado, H.; Siregar, L.A.M.; Damanik, R.I. Morphological characteristics of mangosteen plants (*Garcinia mangostana* L.) in Langkat District, North Sumatera, Indonesia. *IOP Conf. Ser. Earth Environ. Sci.* **2021**, *782*, 042056. [CrossRef]
30. Andani, R.; Fajrina, A.; Asra, R.; Eriadi, A. Antibacterial activity test of mangosteen plants (*Garcinia mangostana* L.): A review. *Asian J. Pharm. Res. Dev.* **2021**, *9*, 164–171. [CrossRef]
31. Anggraeni, R.S. Antibacterial (*Staphylococcus aureus* and *Escherichia coli*) and Antifungal (*Saccharomyces cerevisiae*) activity assay on nanoemulsion formulation of ethanol extract of mangosteen leaves (*Garcinia mangostana* L). *J. Food Pharm. Sci.* **2021**, *9*, 351–365.
32. Jassim, A.M.N.; Shafy, G.M.; Mohammed, M.T.; Farhan, S.A.; Noori, O.M. Antioxidant, anti-inflammatory and wound healing of biosynthetic gold nanoparticles using mangosteen (*G. mangostana*). *Iraqi J. Ind. Res.* **2021**, *8*, 59–74. [CrossRef]
33. Tran, V.A.; Thi Vo, T.-T.; Thi Nguyen, M.-N.; Duy, N.D.; Doan, V.-D.; Nguyen, T.-Q.; Vu, Q.H.; Le, V.T.; Tong, T.D. Novel α-mangostin derivatives from mangosteen (*Garcinia mangostana* L.) peel extract with antioxidant and anticancer potential. *J. Chem.* **2021**, *2021*, 9985604. [CrossRef]
34. Chan, Y.B.; Selvanathan, V.; Tey, L.-H.; Akhtaruzzaman, M.; Anur, F.H.; Djearamane, S.; Watanabe, A.; Aminuzzaman, M. Effect of calcination temperature on structural, morphological and optical properties of copper oxide nanostructures derived from *Garcinia mangostana* L. leaf extract. *Nanomaterials* **2022**, *12*, 3589. [CrossRef] [PubMed]
35. Hitkari, G.; Chowdhary, P.; Kumar, V.; Singh, S.; Motghare, A. Potential of copper-zinc oxide nanocomposite for photocatalytic degradation of congo red dye. *Clean. Chem. Eng.* **2022**, *1*, 100003–100009. [CrossRef]
36. Rajendran, N.K.; George, B.P.; Houreld, N.N.; Abrahamse, H. Synthesis of zinc oxide nanoparticles using *Rubus fairholmianus* root extract and their activity against pathogenic bacteria. *Molecules* **2021**, *26*, 3029. [CrossRef]
37. Sivakavinesan, M.; Vanaja, M.; Annadurai, G. Dyeing of cotton fabric materials with biogenic gold nanoparticles. *Sci. Rep.* **2021**, *1*, 13249. [CrossRef] [PubMed]
38. Trang, N.L.N.; Hoang, V.T.; Dinh, N.X.; Tam, L.T.; Le, V.P.; Linh, D.T.; Cuong, D.M.; Khi, N.T.; Anh, N.H.; Nhung, P.T.; et al. Novel eco-friendly synthesis of biosilver nanoparticles as a colorimetric probe for highly selective detection of Fe (III) ions in aqueous solution. *J. Nanomater.* **2021**, *2021*, 5527519. [CrossRef]
39. Mansoor Al-Saeedi, A.M.; Mohamad, F.K.; Ridha, N.J. Synthesis and characterization CuO-ZnO binary nanoparticles. *J. Nanostructures* **2022**, *12*, 86–96.
40. Fouda, A.; Salem, S.S.; Wassel, A.R.; Hamza, M.F.; Shaheen, T.I. Optimization of green biosynthesized visible light active CuO/ZnO nano-photocatalysts for the degradation of organic methylene blue dye. *Heliyon* **2020**, *6*, e04896–e04908. [CrossRef]
41. Rao, G.T.; Ravikumar, R.V.S.S.N. Novel Fe-doped ZnO-CdS nanocomposite with enhanced visible light-driven photocatalytic performance. *Mater. Res. Innov.* **2020**, *25*, 215–220. [CrossRef]
42. Siddiqui, V.U.; Ansari, A.; Ansari, M.T.; Akram, M.K.; Siddiqi, W.A.; Alosaimi, A.M.; Hussein, M.A.; Rafatullah, M. Optimization of facile synthesized ZnO/CuO nanophotocatalyst for organic dye degradation by visible light irradiation using response surface methodology. *Catalysts* **2021**, *11*, 1509. [CrossRef]
43. Khan, M.I.; Fatima, N.; Shakil, M.; Tahir, M.B.; Riaz, K.N.; Rafique, M.; Iqbal, T.; Mahmood, K. Investigation of in-vitro antibacterial and seed germination properties of green synthesized pure and nickel doped ZnO nanoparticles. *Phys. B Phys. Condens. Matter.* **2021**, *601*, 412563. [CrossRef]
44. Yusefi, M.; Shameli, K.; Yee, O.S.; Teow, S.Y.; Hedayatnasab, Z.; Jahangirian, H.; Webster, T.J.; Kuca, K. Green synthesis of Fe_3O_4 nanoparticles stabilized by a garcinia mangostana fruit peel extract for hyperthermia and anticancer activities. *Int. J. Nanomed.* **2021**, *16*, 2515–2532. [CrossRef] [PubMed]
45. Mohammadi-Aloucheh, R.; Habibi-Yangjeh, A.; Bayrami, A.; Latifi-Navid, S.; Asadi, A. Enhanced anti-bacterial activities of ZnO nanoparticles and ZnO/CuO nanocomposites synthesized using *Vaccinium arctostaphylos* L. fruit extract. *Artif. Cells Nanomed. Biotechnol.* **2018**, *46*, 1200–1209. [CrossRef] [PubMed]

46. Mohammadi-Aloucheh, R.; Habibi-Yangjeh, A.; Bayrami, A.; Latifi-Navid, S.; Asadi, A. Green synthesis of ZnO and ZnO/CuO nanocomposites in *Mentha longifolia* leaf extract: Characterization and their application as anti-bacterial agents. *J. Mater. Sci. Mater. Electron.* **2018**, *29*, 13596–13605. [CrossRef]

47. Haneefa, M.M.; Jayandran, M.; Balasubramanian, V. Green synthesis characterization and antimicrobial activity evaluation of manganese oxide nanoparticles and comparative studies with salicylalchitosan functionalized nanoform. *Asian J. Pharm.* **2017**, *11*, 65–74.

48. Demissie, M.G.; Sabir, F.K.; Edossa, G.D.; Gonfa, B.A. Synthesis of zinc oxide nanoparticles using leaf extract of *Lippia adoensis* (Koseret) and evaluation of its antibacterial activity. *J. Chem.* **2020**, *2020*, 7459042. [CrossRef]

49. Siddiqui, V.U.; Ansari, A.; Chauhan, R.; Siddiqi, W.A. Green synthesis of copper oxide (CuO) nanoparticles by *Punica granatum* peel extract. *Mater. Today Proc.* **2021**, *36*, 751–755. [CrossRef]

50. Phang, Y.-K.; Aminuzzaman, M.; Akhtaruzzaman, M.; Muhammad, G.; Ogawa, S.; Watanabe, A.; Tey, L.-H. Green synthesis and characterization of CuO nanoparticles derived from papaya peel extract for the photocatalytic degradation of palm oil mill effluent (POME). *Sustainability* **2021**, *13*, 796. [CrossRef]

51. Sajjad, A.; Bhatti, S.H.; Ali, Z.; Jaffari, G.H.; Khan, N.A.; Rizvi, Z.F.; Zia, M. Photoinduced fabrication of zinc oxide nanoparticles: Transformation of morphological and biological response on light irradiance. *ACS Omega* **2021**, *6*, 11783–11793. [CrossRef] [PubMed]

52. Sharma, S.; Yadav, D.K.; Chawla, K.; Lai, N.; Lai, C. Synthesis and characterization of CuO nanoparticles by *Aloe barbadensis* leaves. *Quantum J. Eng. Sci. Technol.* **2021**, *2*, 1–9.

53. You, W.; Ahn, J.C.; Boopathi, V.; Arunkumar, L.; Rupa, E.J.; Akter, R.; Kong, B.M.; Lee, G.S.; Yang, D.C.; Kang, S.C.; et al. Enhanced antiobesity efficacy of tryptophan using the nanoformulation of *Dendropanax morbifera* extract mediated with ZnO nanoparticle. *Materials* **2021**, *14*, 824. [CrossRef] [PubMed]

54. Aminuzzaman, M.; Chong, C.-Y.; Goh, W.S.; Phang, Y.-K.; Tey, L.-H.; Chee, S.-Y.; Akhtaruzzaman, M.; Ogawa, S.; Watanabe, A. Biosynthesis of NiO nanoparticles using soursop (*Annona muricata* L.) fruit peel green waste and their photocatalytic performance on crystal violet dye. *J. Clust. Sci.* **2021**, *32*, 949–958. [CrossRef]

55. Ramzan, M.; Obodo, R.M.; Mukhtar, S.; Ilyas, S.Z.; Aziz, F.; Thovhogi, N. Green synthesis of copper oxide nanoparticles using *Cedrus deodara* aqueous extract for antibacterial activity. *Mater. Today Proc.* **2021**, *36*, 576–581. [CrossRef]

56. Baharudin, K.B.; Abdullah, N.; Derawi, D. Effect of calcination temperature on the physicochemical properties of zinc oxide nanoparticles synthesized by coprecipitation. *Mater. Res. Express* **2018**, *5*, 125018. [CrossRef]

57. Naseer, M.; Aslam, U.; Khalid, B.; Chen, B. Green route to synthesize zinc oxide nanoparticles using leaf extracts of *Cassia fistula* and *Melia azadarach* and their antibacterial potential. *Sci. Rep.* **2020**, *10*, 9055. [CrossRef]

58. Jeevanandam, J.; Chan, Y.S.; Danquah, M.K. Biosynthesis of metal and metal oxide nanoparticles. *ChemBioEng Rev.* **2016**, *3*, 55–67. [CrossRef]

59. Singh, J.; Dutta, T.; Kim, K.H.; Rawat, M.; Samddar, P.; Kumar, P. "Green" synthesis of metals and their oxide nanoparticles: Applications for environmental remediation. *J. Nanobiotechnol.* **2018**, *16*, 84. [CrossRef]

60. Obolskiy, D.; Pischel, I.; Siriwatanametanon, N.; Heinrich, M. *Garcinia mangostana* L.: A phytochemical and pharmacological review. *Phyther Res.* **2009**, *23*, 1047–1065. [CrossRef]

61. Selvanathan, V.; Aminuzzaman, M.; Tan, L.X.; Yip, F.W.; Eddy Cheah, S.G.; Heng, M.H.; Tey, L.-H.; Arullappan, S.; Algethami, N.; Alharthi, S.S.; et al. Synthesis, characterization, and preliminary in vitro antibacterial evaluation of ZnO nanoparticles derived from soursop (*Annona muricata* L.) leaf extract as a green reducing agent. *J. Mater. Res. Technol.* **2022**, *20*, 2931–2941. [CrossRef]

62. Fawcett, D.; Verduin, J.J.; Shah, M.; Sharma, S.B.; Poinern, G.E.J. A review of current research into the biogenic synthesis of metal and metal oxide nanoparticles via marine algae and seagrasses. *J. Nanosci.* **2017**, *2017*, 8013850. [CrossRef]

Article

Environmental and Economic Benefits of Using Pomegranate Peel Waste for Insulation Bricks

Ayman Ragab [1], Nasser Zouli [2,*], Ahmed Abutaleb [2], Ibrahim M. Maafa [2], M. M. Ahmed [3] and Ayman Yousef [2,3,*]

1 Department of Architecture, Faculty of Engineering, Aswan University, Aswan 81542, Egypt; ayman.ragab@aswu.edu.eg
2 Department of Chemical Engineering, College of Engineering, Jazan University, Jazan 45142, Saudi Arabia; azabutaleb@jazanu.cdu.oa (A.A.); imoaafa@jazanu.edu.sa (I.M.M.)
3 Department of Mathematics and Physics Engineering, College of Engineering at Mataria, Helwan University, Cairo 11718, Egypt; marwa_elnagar77@yahoo.com
* Correspondence: nizouli@jazanu.edu.sa (N.Z.); aymanhassan@jazanu.edu.sa (A.Y.)

Abstract: Rapid urbanization has negative effects on ecology, economics, and public health, primarily due to unchecked population growth. Sustainable building materials and methods are needed to mitigate these issues and reduce energy use, waste production, and environmental damage. This study highlights the potential of agricultural waste as a sustainable source of construction materials and provides valuable insights into the performance and benefits of using fired clay bricks made from pomegranate peel waste. In this study, fired clay bricks were produced using pomegranate peel waste as a sustainable building material. To optimize the firing temperature and percentage of pomegranate peel waste, a series of experiments was conducted to determine fundamental properties such as mechanical, physical, and thermal properties. Subsequently, the obtained thermal properties were utilized as input data in Design Builder software version (V.5.0.0.105) to assess the thermal and energy performance of the produced bricks. The results showed that the optimum firing temperature for the bricks was 900 °C with 10% pomegranate peel waste. The fabricated bricks reduced energy consumption by 6.97%, 8.54%, and 13.89% at firing temperatures of 700 °C, 800 °C, and 900 °C, respectively, due to their decreased thermal conductivity. CO_2 emissions also decreased by 4.85%, 6.07%, and 12% at the same firing temperatures. The payback time for the bricks was found to be 0.65 years at a firing temperature of 900 °C. These findings demonstrate the potential of fired clay bricks made from pomegranate peel waste as a promising construction material that limits heat gain, preserves energy, reduces CO_2 emissions, and provides a fast return on investment.

Keywords: biomass waste; thermal insulation; building performance; energy efficiency

Citation: Ragab, A.; Zouli, N.; Abutaleb, A.; Maafa, I.M.; Ahmed, M.M.; Yousef, A. Environmental and Economic Benefits of Using Pomegranate Peel Waste for Insulation Bricks. *Materials* **2023**, *16*, 5372. https://doi.org/10.3390/ma16155372

Academic Editors: Stefano Guarino and Flaviana Tagliaferri

Received: 13 June 2023
Revised: 20 July 2023
Accepted: 26 July 2023
Published: 31 July 2023

1. Introduction

There is a growing movement towards the use of renewable energy sources to address the world's expanding energy demands and mounting environmental concerns [1,2]. The Kingdom of Saudi Arabia's (KSA) strategy for growth in the year 2030 includes addressing the country's energy crisis. Saudi Arabia's fast economic growth over the last three decades may be attributed, in large part, to the country's abundant natural gas and oil resources. The KSA is home to over 31 million people, and its manufacturing sector is growing at a rate of 5% to meet the country's increasing need for energy [3]. However, the alarmingly rapid rise in national energy and oil consumption in the KSA stands out dramatically when compared to the rest of the world. Almost 70% of the projected population in 2030 will be under the age of 30, driving an annual need for new homes of 2.32 million. It is predicted that by 2040, if energy consumption drops by only 1% per year, savings of USD 35 billion will have been realized. Around 80% of the kingdom's reserves and industrial centers are supplied with electricity from the dominant energy grid network [3–5].

As the Kingdom of Saudi Arabia's population and industrial sector continue to grow, the country's energy consumption is rapidly increasing, highlighting the need for sustainable solutions. This poses a significant challenge for the construction industry, which is responsible for a significant portion of global energy consumption and carbon emissions. To address this challenge, there is an increasing focus on using sustainable materials in the building envelope, such as bio-based materials.

Bio-based architecture, or "green architecture", is a new approach to planning, designing, and constructing that promotes environmental sustainability by adhering to ethical, economic, and social norms at the forefront of technological development [6]. Bioarchitecture is a design approach that considers the usage of sustainable materials and renewable energy sources, such as those found in nature. Stones like lime; agricultural byproducts like potato peel, straw, and hemp; and farming byproducts like wool, etc., are all examples of naturally derived materials that are increasingly being used in contemporary architecture [7]. Paper-pulp, textile, and food-processing biowaste are only a few examples of frequently used industrial biowaste. They generate a lot of biomass waste, which is a source of interest in the quest to find innovative green materials with which to construct structures with a low impact on the environment.

Agricultural waste is a readily available residue from renewable resources and is abundant on the planet. However, the accumulation of agricultural waste in significant quantities every year poses both environmental hazards and economic challenges [8]. Therefore, the issue of waste management is a pressing concern globally, with large amounts of plant waste being dumped in landfills [9]. This waste not only takes up space but can also cause pollution of the air and water [10]. The sustainable management of agricultural waste is therefore important in terms of reducing its negative impacts on the environment and for exploring its potential for to generate economic value through the development of innovative practices and products [11]. However, there are innovative solutions to this problem, such as using plant waste as a substitute for clay in bricks [12]. This not only reduces waste but also creates a building material that is less dense and more porous, making it potentially advantageous during earthquakes. Moreover, the efficacy of using plant waste as a replacement for clay in bricks has been demonstrated in reducing the energy consumption required for climatic control in various regions [6,13–15]. Recent reports indicate that pomegranate production is on the rise in Saudi Arabia [16], particularly in regions such as Al-Qassim, Al-Madinah, and Al-Ahsa, which provide ideal environmental conditions for cultivation [17]. Despite the health benefits and natural color associated with pomegranate peels, they are often discarded in landfills. To address this issue, researchers have proposed using pomegranate peel waste as a substitute for clay in bricks. Pomegranate peels contain high levels of lignocellulose, a material that can enhance the properties of bricks and reduce the amount of clay required in their production [18]. By promoting the reuse of pomegranate peels and other plant waste, we can minimize waste and promote sustainable practices, thereby mitigating the harmful environmental impacts of waste disposal.

Agricultural waste is a significant environmental concern due to its potential contribution to greenhouse gas emissions and pollution if not appropriately managed. However, recent research has demonstrated that agricultural waste can be an affordable and sustainable source of raw materials for various industries, including brick manufacturing. In recent years, there has been a growing interest in the use of agricultural waste in the production of firing bricks, which are widely used in the construction industry. Various studies have examined the feasibility of using agricultural waste as a substitute for traditional raw materials in the manufacturing of firing bricks.

Huy et al. [19] conducted a study to evaluate the feasibility of using raw rice husk, bottom ash, and fly ash in the production of unburnt building bricks. This study involved the design of two mixtures with varying W/B ratios and the replacement of bottom ash content with rice husk at percentages of 0%, 3%, 6%, and 9% by mass. The results indicated that incorporating

9% rice husk led to a significant reduction in the unit weight and thermal conductivity of the bricks, with values in the range of 1.06 ÷ 1.08 T/m^3 0.201 ÷ 0.216 W/m.K, respectively.

Hassan et al. [20] examined the thermal characteristics of several brick samples made from clay, sludge, and sugarcane bagasse ash and assessed their energy-saving potential. The experimental findings revealed that the thermal conductivity of clay bricks was enhanced significantly by incorporating sludge and sugarcane bagasse ash as compared to traditional brick types used in Egypt. The mean thermal conductivity of the brick samples produced by mixing clay, sludge, and sugarcane bagasse ash (0.11 to 0.26 W/m. K) was found to be lower than that of traditional brick types (0.33–1.6 W/m. K). The results indicate that the predicted energy consumption for the traditional wall systems is 7525 kWh, whereas the proposed wall system, i.e., walls constructed by incorporating sludge and sugarcane bagasse ash into the clay brick, has an annual energy consumption of 6285 kWh, resulting in an overall reduction of 16.5% in energy consumption. Mehrzad et al. [15] utilized environmentally friendly sugarcane bagasse waste to create fiber samples of varying densities and thicknesses and tested them for thermal insulation and sound absorption. The thermal conductivity of the samples ranged from 0.034 to 0.042 W/mK.

Aravind et al. [14] reported the creation of a sustainable exterior wall panel with thermal insulation and its mechanical, thermal, and durability features. Foam concrete and rice husks were utilized to create the wall panel instead of fly ash, which is typically used to create cement. The created wall panels' thermal conductivity and durability were significantly affected by the percentage of rice husk and fly ash present in the material. Marques et al. [21] developed a new polymer-based composite material that utilized waste materials from rice husk and expanded cork industries. Boards made from a variety of composite mixtures were tested for their mechanical, thermal, and acoustic qualities. The results indicate that construction solutions based on these composite materials may be employed in buildings to minimize energy consumption during the lifecycle of a building.

In a study conducted by Ghorbani et al. [10], the effect of potato peel powder (PPP) on sound insulation and other physicochemical parameters was investigated. Results showed that the addition of 7% dried PPP resulted in improved compressive strength, apparent dry density, saturated density, and thermal conductivity. These findings suggest that the use of dried PPP in building materials can have several practical implications in construction. Incorporating 7% dried PPP into bricks can potentially reduce the weight of structures, making them more cost-effective and easier to transport. Moreover, the sound insulation properties of the bricks can be beneficial in partition walls, which can improve the acoustic performance of buildings. Additionally, the study provides a cost-effective solution for crop residue management. Ramos et al. [22] compared the thermal properties and environmental effects of two corncob particleboards made using different glue binders and found that both could be used as sustainable construction materials for wall thermal insulation. Chee-Ming [23] utilized oil palm fruit and pineapple leaves to create unfired and fired clay bricks, discovering that fibers significantly reduce porosity in burned bricks compared to unfired specimens while maintaining the same strength. Elinwa [24] used sawdust ash to create lightweight bricks, recommending burning the bricks at 600 °C for good porosity and reasonable strength. J. Vėjeliene et al. [25] investigated straw insulation and found that laboratory-produced samples with most straw perpendicular to heat flow had lower thermal conductivity than plant specimens obtained from straw bales and rolls with most straw parallel to heat flow.

Huixia et al. [26] conducted a study to investigate the micro–macro characteristics of sustainable mortar containing construction waste fines (0–0.15 mm) as a replacement for cement and sand. The construction waste fines utilized in the study included both waste concrete fines (WCF) and waste brick fines (WBF). The results revealed that the incorporation of WCF as cement replacement led to a reduction in the strength and permeability resistance of the blended mortar. However, the addition of WBF content initially improved the strength and permeability resistance of the mortar, although further increases in WBF content resulted in a decline in these properties. However, this study ignores all thermal properties of the produced mortar.

Ahmed et al. [6] conducted a study evaluating the effectiveness of incorporating different percentages of pomegranate peel waste (5%, 7.5%, 10%, and 15%) into clay bricks used for the external walls of social residential buildings in New Aswan City, Egypt. The study showed a gradual increase in energy savings with an increase in the percentage of pomegranate peel waste. However, the study did not investigate the effect of varying firing temperatures on the properties of the bricks.

Saudi Arabia experiences extreme temperatures during the summer months, with daytime temperatures often exceeding 40 °C and nighttime temperatures remaining above 30 °C. As a result, air conditioning is essential for maintaining comfortable indoor temperatures, which can lead to high energy consumption. Another factor contributing to the increased energy consumption for cooling in Saudi buildings is inefficient building design. Many buildings in Saudi Arabia are constructed with inadequate insulation, which can result in heat gain during the day and heat loss at night.

The purpose of this article is to investigate the impact of replacing traditional brick with a newly proposed brick containing 10% pomegranate peel waste fired at different temperatures (700 °C, 800 °C, and 900 °C) on the indoor thermal performance and energy consumption of a residential building located in Jazan City, Saudi Arabia. The study aims to assess the potential benefits of using this new brick in terms of improving indoor thermal comfort and reducing the energy required for cooling purposes, with the goal of promoting sustainable building practices.

2. Materials and Methods

The primary objective of this research was to investigate the effectiveness of integrating 10% pomegranate peel waste (PPW) into fired clay bricks for use in external walls within the Jazan region of Saudi Arabia. The research was conducted in four phases, with a graphical illustration of the research work flow presented in Figure 1.

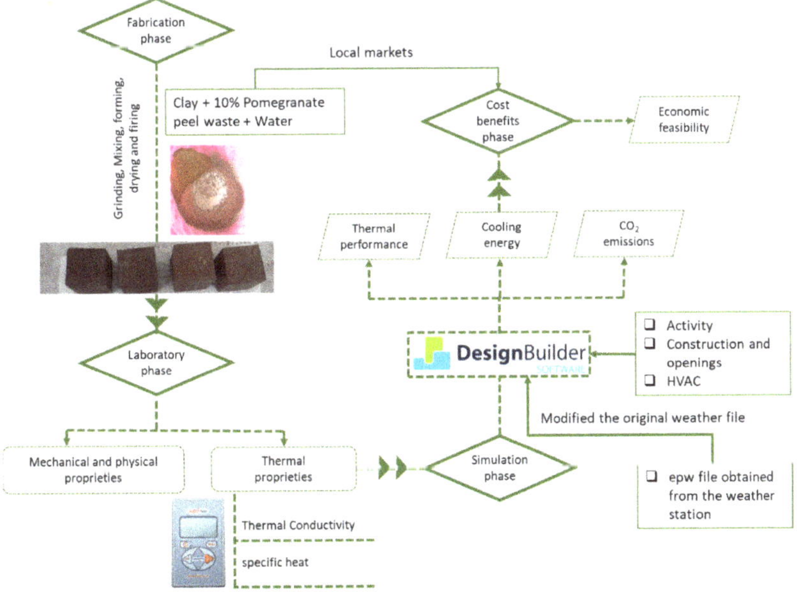

Figure 1. The methodological framework of the study.

The first phase involved the fabrication of brick samples at varying firing temperatures; the resulting samples are recorded in Table 1. In the second phase, the produced brick samples were subjected to rigorous mechanical, physical, and thermal testing. The third phase entailed the utilization of Design Builder simulation software to evaluate the potential energy savings that could be achieved using the developed brick samples. Finally, in the fourth phase, the cost-effectiveness of the proposed building materials was assessed.

Table 1. Brick samples and their firing temperatures.

Sample	Firing Temperature °C
S1	700
S2	800
S3	900

2.1. Study Area

Jazan City, the capital of the Jazan province in southwestern Saudi Arabia, is characterized by hot and humid climatic conditions, with average high temperatures reaching 33.5 °C in June. Conversely, January is the coldest month of the year, with average temperatures falling to as low as 25.7 °C. Precipitation levels in Jazan range from 1 mm to 19 mm, with the lowest precipitation level in June and the highest levels in the wettest month of the year.

Figure 2 depicts the location of Jazan City in KSA. The present study aims to investigate the potential benefits of incorporating pomegranate waste in clay bricks at a concentration of 10% to reduce energy consumption and enhance thermal performance in buildings located in Jazan. By utilizing clay bricks developed from pomegranate waste, this study seeks to address the issue of uninsulated walls in Saudi Arabian buildings and contribute to the promotion of sustainable building practices in the region.

Figure 2. Location of Jazan City.

2.2. Raw Materials and Fabrication Process

2.2.1. Materials

The main raw material used in this study for brick production was clay, obtained from the Aswan region of Upper Egypt. The studied clay exhibited similarities to the clay used in brick manufacturing in KSA. PPW waste was collected from local juice shops. After drying under normal sunlight, it was ground to achieve sufficient particle powder, then dried in an oven at 100 °C for two days to obtain fully dried powder before being mixed with clay. Table 2 explains the XRF of raw materials, which indicates high loss on ignition (LOI) of PPW, which is attributed to its high organic content. X-ray diffraction (XRD) was utilized to ascertain the mineral composition of the clay. Based on the XRD analysis, the predominant components of the clay material are silica and alumina. Figure 3 presents an analysis of the X-ray results of the utilized clay.

Table 2. XRF of raw materials.

Oxide Composition	Pomegranate Peel Waste wt.%	Clay wt.%
CaO	10.48	0.50
SiO_2	0.38	48.93
Al_2O_3	0.1	32.90
Fe_2O_3	0.91	1.19
SO_3	0.4	0.29
Na_2O	<0.01	0.09
P_2O_3	3.33	-
K_2O	11.68	0.014
MgO	4.01	0.09
MnO	0.01	-
TiO_2	0.1	5.92
CL	3.19	0.01
ZrO_2	-	0.46
Cr_2O_3	-	0.14
LOI	65.35	9.2

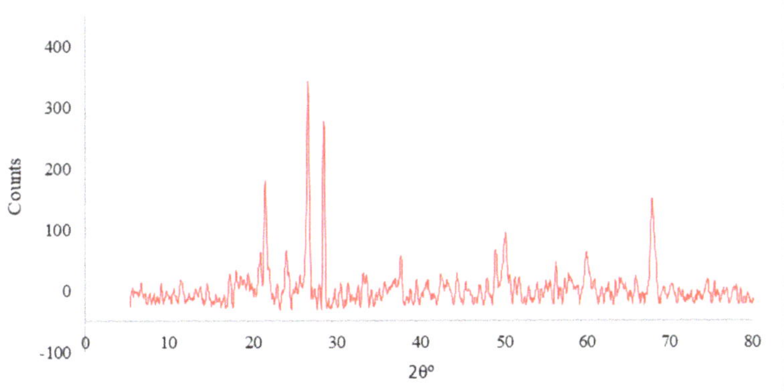

Figure 3. X-ray of the clay.

2.2.2. Brick Preparation

The brick samples were prepared according to our previous report [27]. Typically, brick samples were prepared using fine-dried PPW powder as a partial replacement for clay in the bricks and clay. The raw materials were combined to make a homogeneous dry mixture; then, three samples from each investigation were examined to find the average value of the investigated parameter. The clay was replaced by different percentages of

PPW (0–15 wt.% of PPW) [27]. The mixtures were mixed in dry conditions for 2 min to obtain a consistent dry mixture. The aforementioned mixtures were then blended again using an electric mixer with the addition of 20% water to improve the compression and consistency of the samples. The formed paste was molded in $50 \times 50 \times 50$ mm cubic steel molds. Then, a hydraulic pressure of 10 MPa with a 0.2 mm/min press rate was applied to press the samples. Next, to ensure total moisture content elimination, brick samples were oven-dried for 6 h at 120 °C, then cooled to room temperature. Finally, the produced brick samples were fired at different temperatures (700 °C, 800 °C, and 900 °C) for four hours at 10 °C/min.

2.2.3. Characterization

Characterization (compressive strength (CS), bulk density (BD), cold water absorption (WA), apparent porosity (AP), and thermal conductivity) of the fabricated brick membranes was conducted according to an identical procedure described in our recent publication [27].

2.3. Thermal Properties Tests

The thermal conductivity values of the brick specimens were determined using a KD2 Pro Thermal Properties Tester, which adheres to the guidelines outlined in ASTM D 5334 [28]. The transient heat conduction method was employed to obtain a digital recording of thermal conductivity. This method involves applying a heat pulse to the specimen and measuring the resulting temperature change over time using thermocouples. The temperature change over time was then analyzed to calculate the thermal conductivity of the specimens.

It is important to note that the ambient temperature at the time of testing is a crucial factor to consider in thermal conductivity measurements because temperature has a significant impact on thermal conductivity, and measurements taken at various temperatures can yield different results. To minimize the effect of temperature on the measurements, the specimens were conditioned to a standard temperature of 23 °C prior to testing. Additionally, each specimen was tested three times to ensure the accuracy and repeatability of the results.

2.4. Simulation Procedures

The daily schedule of the traditional Saudi Arabian lifestyle incorporates holidays, work hours, and other events. Several festivals are celebrated throughout the year in Saudi Arabia. As mandated by Saudi Arabian law, employers are required to abide by the country's standard work schedule, which runs from 8:00 a.m. to 3:00 p.m. from Sunday to Thursday. Recent demographic data [29] indicate that the average Saudi Arabian family comprises ten members, with building residents usually ranging from 15 to 65 years of age. The building schedule was designed with consideration of holidays and the typical weekday routine of building residents, who are expected to be at their workplaces during weekdays. Furthermore, during holidays, more people may be present in the building, which could lead to higher energy consumption for heating, cooling, and other activities. To simulate actual family routines, the simulation program was fed with relevant data; all the necessary input data for the simulation software are presented in Table 3.

Table 3. Study model input data.

Item	Specification
Building type	Residential building
Location	Jazan City—hot desert climate (Köppen: BSh)
Floor area (m^2)	240
No. of floors	Ground floor and one floor
Floor height (m)	3.6
Occupancy (persons per building)	10
Window glazing	6 mm single clear glazing
Window-to-wall ratio	10%
Lighting (Lux)	400
HVAC	4 split air conditioning units for each flat
Cooling setpoint (°C)	25
Heating setpoint (°C)	18

2.4.1. Model Description

Weather conditions such as temperature and humidity can impact the energy demands of buildings. In hot and arid climates, the energy requirements for cooling of buildings can be significant, which shows the importance of using materials with better insulation properties to reduce energy usage for both heating and cooling. As a result, this study aims to investigate the energy demand required to cool a family home located in the city of Jazan using a new insulated brick. The residential model under consideration features a two-story building with a total area of 240 m^2 designed to accommodate up to ten individuals comfortably. The architectural designs for this model are depicted in Figure 4. It is worth noting that this residential model can be found in various locations throughout the city of Jazan and shares similar attributes, including the use of concrete construction techniques that comply with the building thermal insulation guidelines outlined in the Saudi Building Code (SBC) [30].

(a)   (b)

Figure 4. The residential building model: (**a**) first floor; (**b**) second floor.

2.4.2. Weather Data Collection

The U.S. Department of Energy (DOE) website provides the 2002 EPW (energy plus weather) file for the Jazan climate zone in the modeling software Design Builder version (V.5.0.0.105). The EPW file is a textual CSV file containing hourly weather data for the research site throughout the year. As noted in prior research studies [31–34], we replaced the default weather data file in the software with data obtained from the weather station at Jazan University, which reflects the local climate conditions in the year 2022.

The objective was to create an environment that closely resembles the conditions observed in reality. To access the weather station's extracted data, the original EPW file had to be converted to a CSV file, which was a crucial step for obtaining accurate information. The dry-bulb temperature, relative humidity, global radiation, wind speed, and wind direction were used as inputs, with all measurements taken at the local weather station. The dew point and direct radiation were calculated using Element software Version (1.0.6). The newly generated EPW file was then utilized to import the updated CSV file into the Design Builder simulation software for the study.

2.4.3. Model Validation

To ensure the accuracy of the simulation results, a comparative analysis was carried out between the simulated data and the actual observations of air temperature. Specifically, field measurements were conducted in a selected bedroom located on the southwestern side of the investigated building on June 21; the first-floor bedroom air temperature was compared to verify the base case model.

The Hobo U12 data logger was utilized to record the temperatures and relative humidity of the selected bedroom throughout the day, which included work hours from 8:00 a.m. to 5:00 p.m., Sunday through Thursday, while the remaining hours were assumed to represent typical residential occupancy. The difference between the calculated and observed air temperatures in the studied bedroom varied from 0.08% to 4.61% throughout the day, which is within the acceptable margin of less than 5% [35]. The model validation is presented in Figure 5.

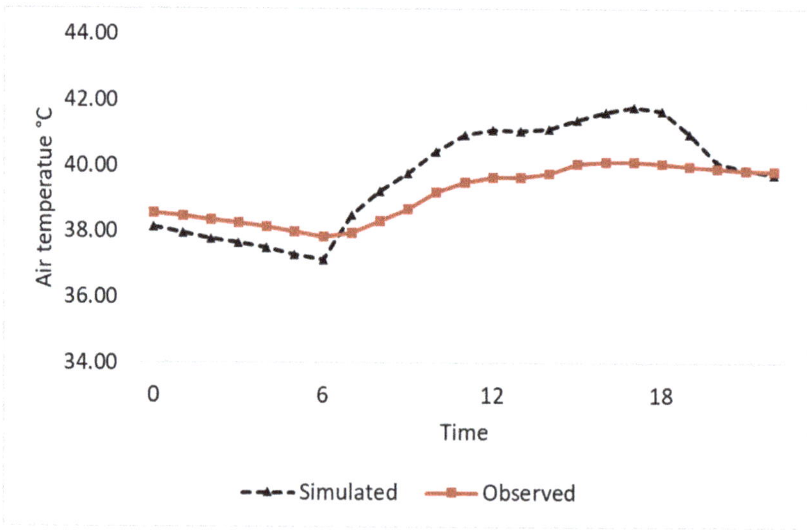

Figure 5. Validation of the investigated building model.

2.5. Electricity Prices

The monthly expenses associated with energy consumption are presented in Table 4, with the cost illustrated in Saudi Arabian riyal (SAR). The tariff utilized to calculate the energy consumption expenses is determined by the Saudi Electricity Company (SEC) for the residential sector. The SEC tariff system is designed to consider various factors, including the type of customer, the amount of energy consumed, and the time of day when the energy is utilized. The tariff system is periodically reviewed and updated to ensure its accuracy and alignment with the prevailing economic conditions and market dynamics. The use of the SEC tariff system to calculate the monthly energy consumption expenses in this study provides a reliable and consistent basis for evaluating the cost-effectiveness of the proposed building materials and energy efficiency measures.

Table 4. Residential electricity prices.

Bracket	Category (kWh)	Price (SAR)
First	0:6000	0.18
Second	more than 6000	0.30

3. Results and Discussions

The present study is designed to investigate the efficacy of utilizing bricks made from pomegranate peel in enhancing the thermal performance and cost-effectiveness of building structures. The study is divided into four distinct parts, each addressing a specific aspect of the building's thermal performance and energy efficiency.

The first part of the study focuses on evaluating the mechanical properties and physical characteristics of bricks. This involves assessing the compressive strength, water absorption, and density of the bricks, among other relevant parameters that can affect their performance in building applications. The second part of the study is concerned with investigating the thermal conductivity of bricks. This involves measuring the rate at which heat is transferred through the bricks and assessing their insulation properties.

The third part of the study presents simulations of the building's interior thermal performance, cooling energy consumption, and CO_2 emissions. This involves utilizing advanced modeling software to simulate the thermal behavior of the building under different scenarios and evaluate the potential benefits of incorporating pomegranate peel bricks in building construction. The fourth and final part of the study assesses the cost-effectiveness of utilizing bricks made from pomegranate peel to maintain appropriate indoor temperatures. This involves analyzing the cost implications of utilizing pomegranate peel bricks compared to conventional building materials and assessing the potential for long-term cost savings associated with improved thermal performance and reduced energy consumption.

3.1. Properties of Bricks: A Mechanical and Physical Assessment

This study focuses on the use of (PPW) as a replacement for clay in the fabrication of bricks. The study measured the effects of different percentages of PPW on the density, compressive strength, water absorption, and apparent porosity of the bricks.

Table 5 summarizes the density, compressive strength, water absorption, and apparent porosity of brick samples with different PPW percentages fired at 900 °C. As expected, the bulk density decreased when the PPW content increased in the brick samples. It is well demonstrated that the bricks fabricated with PPW have lower densities, varying between 1922 and 1348 kg/m^3 for 0% and 15% PPW contents, respectively, which corresponds to a decrease of about 29.9% in the density. A 17.2% decrease was obtained with the addition of 10% PPW. This may be due to the fact that during the firing of PPW bricks, more material was lost due to the decomposition of carbonaceous matter [36]. Replacing PPW with clay decreases the compressive strength of bricks in most low-rise buildings, and the acceptable compressive strength is about 8.6 MPa according to ASTM C 62 [37]. Thus, it can be easily seen that all the tested bricks overwhelmingly satisfy this limit, except those replaced with 15% PPW. Nevertheless, 10% PPW is the highest amount that can be added to achieve the specifications and may be the best choice among the single-component replacement groups.

Table 5. Average values for AP, CS, BD, and WA of bricks with different waste additives fired at 900 °C.

PPW (%)	AP (%)	CS (MPa)	BD (kg/m^3)	WA (%)
0	27.7	18.5	1922.02	13.9
5	29.9	13.9	1713.45	14.7
7.5	32.9	11.2	1607.1	16.2
10	33.03	10.3	1562.9	17.5
15	34.6	4.6	1348.3	20.1

Different amounts of PPW addition and firing at different temperatures impact the compressive strength of the produced bricks. This can be explained by the size of the capillary channels and voids that are created by porosity in the brick. Additionally, the amount of clay inside the brick sample decreases, which is a key reason why the compressive strength decreases as the PPW proportion rises and the firing temperature decreases. The samples with a compressive strength between 2.5 and 3.8 MPa, referred to as Earth Blocks Class 2 (EB2), can be used in low-height construction, where loads are lower. They can also be employed in secondary buildings in which they are adopted as insulating, exterior, or partition constructions. Test bricks known as Earth Blocks Class 3 (EB3), which have a mean compressive strength value between 3.8 and 5 MPa, can be deployed as non-load-bearing, self-supporting walls. Bricks with a compressive strength value above 5 MPa, referred to as Earth Blocks Class 4 (EB4), can be used as inner walls and load-bearing walls of low-rise and mid-rise buildings [38].

Higher percentages of PPW content resulted in an increase in water absorption: 20.1% for 15% PPW compared to 13.9% for 0% PPW, which confirms that PPW increased the pore volume. Furthermore, the apparent porosity increased from 27.7% to 34.6% with an increase in the PPW content from 0% PPW to 15% PPW. A similar trend was observed with the incorporation of biosolids in fired clay bricks. The increase in the percentage of biosolids resulted in an increase in the rate of absorption due to the creation of pores during the firing process [39]. Water absorption values signify the long-term durability performance of the bricks. Therefore, excessively high values can lead to cracking because of the increase in the volume.

Among the tested percentages, the results show that the fired brick samples made with 10% PPW have optimal characteristics. Thus, the sample composed of 10% PPW was fired at three different temperatures (700 °C, 800 °C, and 900 °C). Table 6 summarizes the results of density, compressive strength, water absorption, and apparent porosity of bricks samples with 10% PPW fired at different temperatures. In general, increasing the PPW content in the mixtures decreased the specimen weights. Replacing clay (dense materials) with PPW (light materials) resulted in a total volume increase, even after compaction at 10 MPa. Increases in the compacted mix volume resulted in decreases in specimen weights and densities. Increasing the firing temperature of 10% PPW causes a linear increase in bulk density; such an impact on raw samples may result from the low bulk density level of PPW relative to that of clay and from the change in particle packing of the clay mix originated by incorporating a lightening additive, as well as the formation of a glassy phase. Compared to the sample with 0% PPW content, the addition of 10% PPW reduced the bulk density by 19.04% at 700 °C, 19.6% at 800 °C, and 18.7% at 900 °C.

Table 6. Average values and standard deviations for AP, CS, BD, and CW of bricks with 10% PPW at different firing temperatures.

Firing Temperature (°C)	AP (%)	CS (MPa)	BD (kg/m^3)	WA (%)
700	35.80	6.7	1480.8	18.6
800	34.04	7.97	1507.7	17.5
900	33.03	10.3	1562.9	17.3

D. Eliche-Quesada and Yuecheng reported similar findings, revealing that the incorporation of coffee grounds and cigarette butts in the development of masonry components decreases their compressive strength and load-bearing ability [40]. Therefore, increased loss of organic material during firing leads to a higher level of total porosity in clay. Pursuant to current regulations, the bulk density in bricks may not be lower than 1050 kg/m^3. The bricks made with 10% PPW at 700 °C presented high porosity because the generation of gas from the decomposition of carbonaceous matter generates more porosity; in addition, the LOI of PPW bricks is higher than that of control bricks fired at 900 °C. This may be caused by the partial closure of open pores or by a decrease in the interconnectivity between pores due to the formation of a glassy phase at 900 °C.

Generally, the results show that as the percentage of PPW increased, the density of the bricks decreased, while the water absorption and apparent porosity increased. However, the compressive strength remained within acceptable limits, except for the bricks with 15% PPW. This study suggests that adding 10% PPW may be the best choice for achieving both the desired specifications and durability performance.

3.2. Thermal Conductivity Assessment of Bricks Incorporating Pomegranate Peel Waste

This study investigated the thermal conductivity values of bricks fabricated using pomegranate peel waste (PPW) as the primary ingredient and fired at different temperatures. The results show that the thermal conductivity values of the bricks decreased with increasing firing temperature. The decrease in thermal conductivity was attributed to the increased porosity of the bricks, which resulted from the development of larger pores in the brick structure when exposed to high temperatures. The highest thermal efficiency was observed in sample S3, which exhibited the lowest thermal conductivity value of 0.29 W/m.°C among all tested samples, as shown in Table 7. The rates of improvement in thermal conductivity values for S1, S2, and S3, when compared to conventional bricks, were 52.77%, 54.16%, and 59.72%, respectively, as shown in Figure 6.

Table 7. The characteristics of the investigated brick samples.

Sample	Firing Temperature (°C)	Thermal Conductivity W/m.°C	Specific Heat (J/kg^{-1}.K^{-1})	Density (kg/m^3)
Base case	900	0.72	800	1922
S1	700	0.34	2013	1480
S2	800	0.33	1800	1507
S3	900	0.29	1500	1562

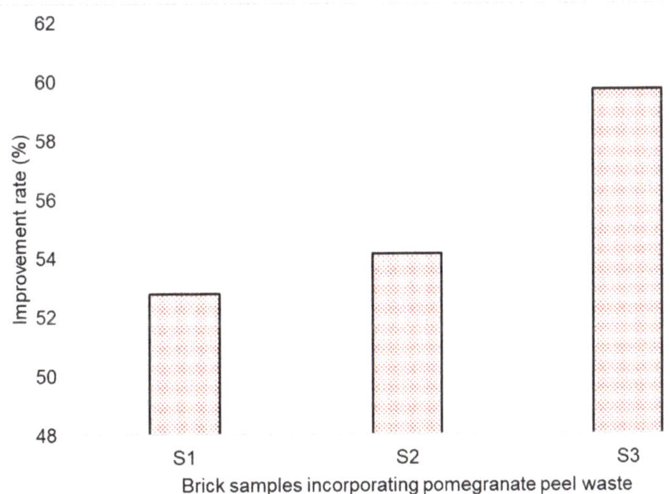

Figure 6. The rate of improvement in terms of thermal conductivity of the fabricated brick samples (S1–S3) compared to the conventional brick sample.

The findings suggest that utilizing PPW bricks in building construction can be an effective approach to reducing energy consumption and minimizing carbon emissions. Additionally, the use of sustainable materials in building construction can contribute to the achievement of environmental sustainability and a reduction in the carbon footprint of the construction industry.

Moreover, the results of this study emphasize the importance of firing temperature in determining the thermal conductivity and porosity of the fabricated bricks. Higher firing temperatures were found to increase the porosity and decrease the thermal conductivity of the bricks, leading to improved thermal efficiency. However, excessive porosity can compromise the structural integrity of the bricks, indicating the need for careful control of the firing process to obtain the desired properties.

The findings of this study have significant implications for the development of sustainable building materials and design. The use of PPW bricks in building construction has the potential to contribute to the reduction in energy consumption and carbon emissions while also promoting the use of sustainable materials. This study also highlights the need for further research to explore the full potential of PPW bricks and to develop new and innovative building materials for sustainable building design.

3.3. Thermal and Energy Performance of the Investigated Building

3.3.1. The Effect of the Investigated Bricks Made from Pomegranate Peel Waste on the Indoor Air Temperature

This study aimed to assess the insulation ability of bricks made from pomegranate peel waste against weather conditions. The tested brick wall samples were made from pomegranate peel waste and fired at different temperatures, then compared to a conventional brick case using Design Builder simulation software. The study was conducted on 21 June, the official longest day of the year in the northern hemisphere, as it is a day with high solar radiation and temperature fluctuations, as reported in previous studies [41]. The temperature of the air inside the building was monitored and recorded for conventional brick (BC) and the other brick wall samples under investigation. The results show that the installation circumstances of the pomegranate peel waste bricks significantly influenced the interior air temperature.

Figure 7a displays the thermal efficiency of the proposed brick samples made from pomegranate peel waste and shows the possible reductions, along with the conventional brick used as a comparison. It was found that incorporating pomegranate peel waste into exterior brick walls considerably reduces interior air temperature fluctuations throughout the day, which has significant implications for energy efficiency and indoor comfort.

The thermal performance of the proposed brick samples made from pomegranate peel waste was found to be highly comparable, despite their evident variances. Among all the tested samples, S3 showed the largest reduction in interior air temperature, with a potential decrease of 2.07 to 5.02 K, making it significantly different from BC. However, S1 exhibited the least desirable thermal performance, with an average reduction in air temperature of only 0.93 K to 3.34 K, indicating that it had fewer benefits than any other brick sample. The thermal performances of S1 and S2 tended to be similar, which could be attributed to their comparable thermal conductivity. The greatest rate of improvement regarding internal air temperature was observed during daylight hours, with an average improvement ranging between 13% and 15% for S3, 10% to 13% for S2, and 7% to 9% for S1. These outcomes are in line with previous studies that have reported a significant enhancement in internal thermal performance through the integration of agricultural waste in fired bricks [6,9,13,14,19].

To assess the building's thermal performance, Fanger's predicted mean vote (PMV) was employed [41], which is a popular statistic used in studies of thermal environmental modeling, design, assessment, and management. PMV takes into account various parameters, such as air temperature, humidity, clothing insulation, and metabolic rate to evaluate the thermal comfort of occupants. PMV results were extracted from Design Builder simulation software according to ASHRAE and depending on the following equation:

$$PMV = [(0.303 * e^{-0.036 * M} + 0.028) * (M - W) + (0.1 * W)] - 3.05 * 10^{-3} * M * (M - 58.15) - 0.42 \qquad (1)$$

where PMV is the predicted mean vote of thermal sensation, ranging from -3 (cold) to $+3$ (hot); M is the metabolic rate of the occupant, expressed in metabolic equivalents (met), with 1 met equaling the metabolic rate at rest (58.15 W/m^2); W is the external work rate of the occupant, expressed in W/m^2; and e is the vapor pressure of water vapor in the air, expressed in kPa.

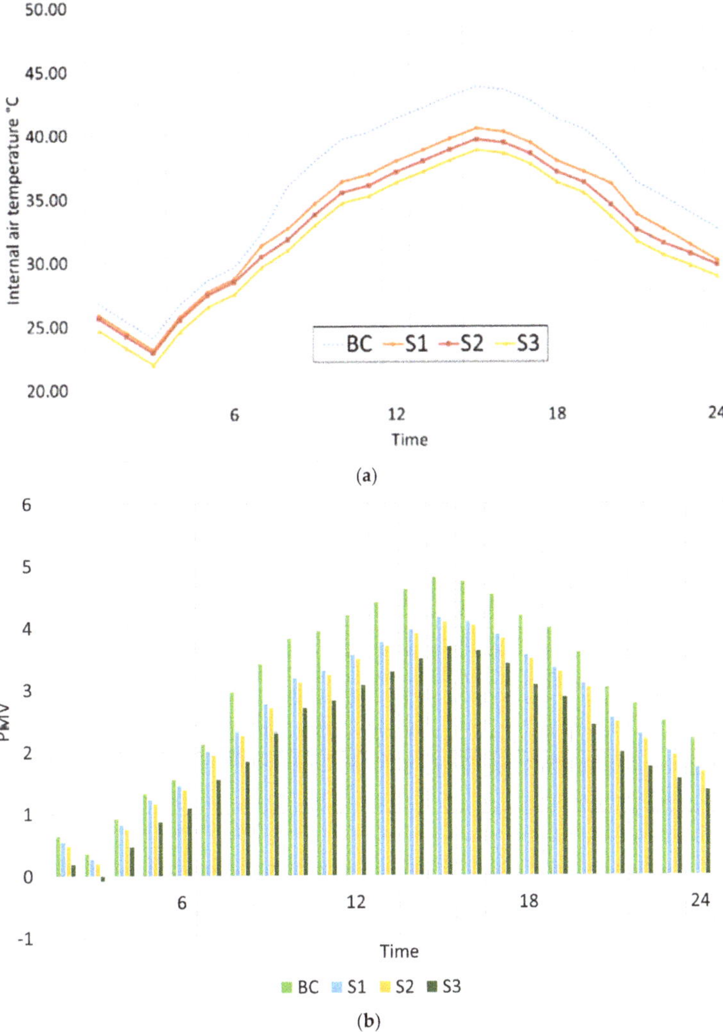

Figure 7. The simulated thermal conditions in the investigated building. (**a**) Indoor air temperature using varied brick samples; (**b**) thermal comfort using the PMV index.

The results show a wide variation in PMV prediction across all evaluated brick samples, with PMV following the same patterns as the indoor air temperature in the analyzed environments. Figure 7b displays a comparison of the tested brick samples for PMV. The PMV values of S3 were the lowest among all the tested brick samples, making it the most successful of the set. Specifically, at the morning and afternoon hours of 3:00, PMV levels in S3 were reported to be 0.1 and 3.72, respectively. S2 was the second-best choice after S3 for creating acceptable temperature conditions, with PMV levels between 0.2 and 4.09. According to these findings, the building's interior environment met the standards set out by the Saudi rating system (MOSTADAM).

It is important to note that this study is subject to some limitations. First, the study was conducted in a specific climatic region and may not be representative of other regions with different weather conditions. Additionally, the study focused only on the optimum percentage of pomegranate peel waste (PPW) according to the mechanical and physical properties of the bricks and did not consider other factors, such as the long-term durability of the bricks. Therefore, further research is needed to fully evaluate the potential of using pomegranate waste in brick production under different climatic conditions and to assess the long-term durability of the bricks. Nonetheless, the study's findings provide a valuable contribution to the field of sustainable building practices and offer a unique approach to reducing waste and promoting energy efficiency.

3.3.2. An Analysis of Cooling Energy Demands

This section presents an evaluation of the cooling energy requirements of buildings constructed with pomegranate peel waste (PPW) bricks using Design Builder modeling software. The simulation results illustrated in Figure 8 provide insights into how the firing temperature of PPW bricks affects the cooling energy requirements of the building. The results demonstrate that conventional bricks achieve poor cooling energy performance compared to PPW bricks, which can significantly reduce cooling energy usage.

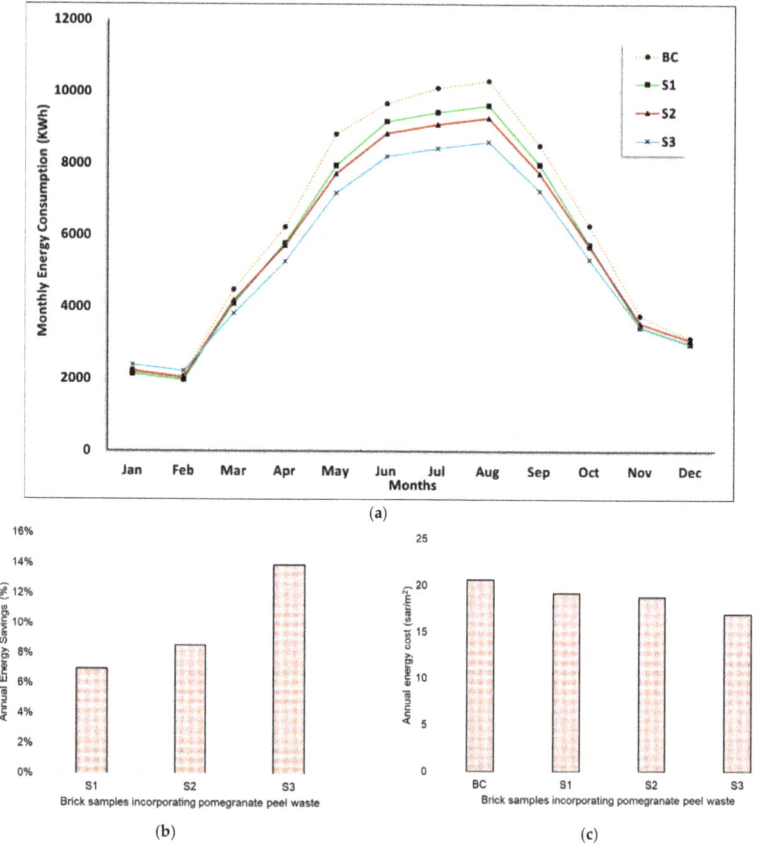

Figure 8. Simulation results for energy consumption attributed to the scenarios of polycarbonate windows: (**a**) monthly energy consumption; (**b**) annual savings; (**c**) annual costs.

It is noteworthy that the highest variation in energy use was observed during the summer months, between March and October, as shown in Figure 8a. During this period, cooling systems consume their annual maximum energy due to the high temperature in Jazan, which can reach up to 40 °C. Figure 8b shows that substituting conventional bricks with S3 can reduce the building's cooling energy consumption by 10,499.65 kWh, corresponding to an improvement rate of approximately 13.89% compared to using conventional bricks as a base case. Bricks made from PPW fired at 800 °C reduced cooling energy consumption by 6457.05 kWh, representing an improvement of 8.54%. Similarly, firing PPW bricks at 700 °C reduced cooling energy usage by 5296 kWh, with an improvement rate of 6.97%. According to Figure 8c, the monthly cost of energy in SAR for a building located in Jazan City is lowest for the S3 brick sample and highest for the BC brick sample. The cost of energy decreases as the firing temperature of brick samples increases. Abdel Hamid et al. reported that using varying percentages of PPW ranging between 2.5% and 12.5% in building bricks can result in energy savings of approximately 23.3% [42]. In comparison, this study found that incorporating 10% of PPW material in building construction can lead to energy savings of about 13.89%. The results from both studies suggest that PPW has the potential to significantly reduce energy consumption in buildings in various climatic regions.

The findings of this study are significant because space cooling accounts for a substantial portion of a building's overall energy demand, ranging from 37% to 42% [34,43]. Therefore, the use of PPW bricks in building construction could significantly reduce the energy usage of buildings, resulting in lower energy costs and reduced carbon emissions. The results of this study suggest that PPW bricks can be a viable alternative to conventional bricks in terms of reducing cooling energy consumption in buildings.

This study aimed to investigate the annual energy consumption and associated costs required for cooling in buildings constructed with different materials, including BC, S1, S2, and S3. The results demonstrate that the annual energy costs for cooling in the BC building were SAR $20.6/m^2$. In contrast, the buildings constructed with S1, S2, and S3 materials exhibited annual energy costs of SAR $19.2/m^2$, SAR $18.8/m^2$, and SAR $17/m^2$, respectively.

Comparison of the proposed building materials with the BC revealed that the use of S1, S2, and S3 materials resulted in significant reductions in energy consumption and associated costs for cooling. Specifically, the energy cost reductions were 7.02%, 8.97%, and 17.71% for S1, S2, and S3, respectively.

3.3.3. Analysis of Pomegranate Peel Waste Bricks' Potential Impact on Carbon Dioxide Emissions

The issue of increasing carbon dioxide (CO_2) emissions in buildings over their entire life cycle is a growing concern. Despite technological advancements that have helped reduce CO_2 emissions from building activities, embodied carbon dioxide in buildings has been on the rise in recent decades. CO_2 emissions from utilities in buildings, such as heating, cooling, ventilation, and lighting, are a major contributor to this trend [44]. Embodied carbon dioxide [45] is influenced by various factors, such as resource extraction, manufacturing, transportation, construction, maintenance, and demolition. The materials and components used in buildings play a significant role in increasing the amount of CO_2 in the atmosphere. Past research has linked CO_2 emissions to air conditioning use in hot desert buildings. Studies have investigated various building materials, including window layouts, roofing tiles, and wall bricks, to improve thermal performance, reduce cooling energy use, and mitigate operational CO_2 emissions. Such studies have highlighted the potential of using more sustainable building materials and design strategies to reduce the embodied carbon footprint of buildings. However, further research is needed to explore effective approaches for reducing embodied carbon dioxide in building construction and operation, especially in the context of changing climate conditions and evolving building codes and standards.

The results of this study show that CO_2 emissions increase significantly during the summer season due to rising energy consumption for cooling purposes. August, the hottest month in Jazan City, had the highest CO_2 emissions. This study evaluated the CO_2 emissions associated with S1, S2, and S3 brick samples made from pomegranate peel waste (PPW) at firing temperatures of 700 °C, 800 °C, and 900 °C, respectively, using Design Builder modeling software. Among all tested samples, S3 exhibited the most efficient performance, with an improvement rate of 12% in reducing CO_2 emissions. Samples S1 and S2 also showed a reduction in CO_2 emissions, with improvement rates of 4.85% and 6.07%, respectively, as shown in Figure 9.

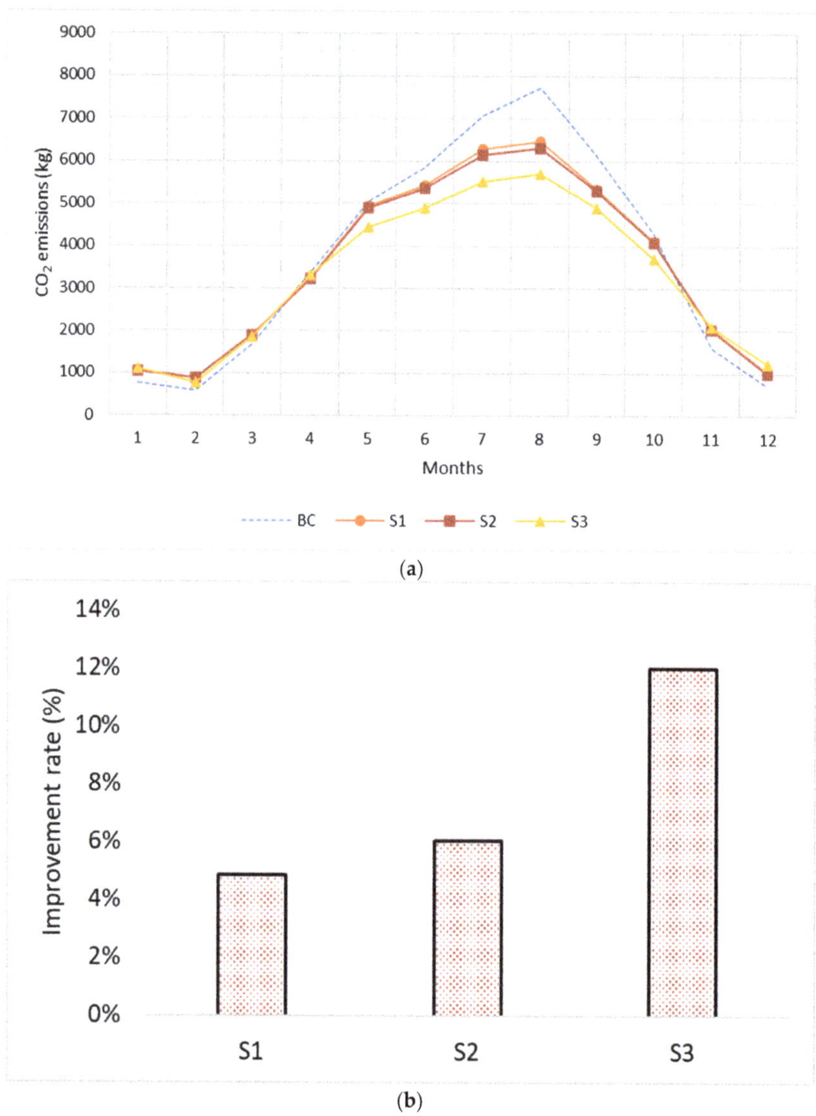

Figure 9. Simulation results: (**a**) monthly CO_2 emissions; (**b**) improvement rate.

These findings suggest that using more energy-efficient materials for cooling purposes can significantly reduce CO_2 emissions in buildings, particularly during the summer season. The use of PPW bricks fired at higher temperatures resulted in a lower embodied carbon footprint, which can be attributed to the reduction in cooling energy requirements. This study highlights the potential of using sustainable building materials, such as PPW bricks, to reduce the embodied carbon footprint of buildings.

However, it is important to note that this study only evaluated the CO_2 emissions associated with cooling energy consumption and did not consider other factors that contribute to embodied carbon dioxide, such as transportation and maintenance. Further research is needed to comprehensively evaluate the embodied carbon footprint of buildings and explore effective strategies for reducing it.

In conclusion, the use of sustainable building materials such as PPW bricks has the potential to significantly reduce the embodied carbon footprint of buildings. The study findings demonstrate that the firing temperature of PPW bricks plays a crucial role in reducing cooling energy requirements and associated CO_2 emissions during the summer season. This study highlights the importance of using more energy-efficient materials in building construction to mitigate the impact of buildings on the environment. The findings suggest that the use of PPW bricks in building construction can be an effective strategy for reducing the embodied carbon footprint of buildings, which is crucial in mitigating the impact of buildings on climate change. However, further research is required to fully evaluate the potential of using PPW bricks and other sustainable building materials in different situations and to explore effective approaches for reducing the embodied carbon footprint of buildings. This research is essential in promoting sustainable building practices and addressing the challenges presented by climate change and evolving building codes and standards.

3.4. The Advantages of Using Fabricated PPW Bricks: A Building Economic Study

The global trend towards energy conservation driven by high energy costs and environmental concerns has led to increased interest in sustainable building practices. While insulation of ceilings has been a common practice, exterior walls, despite their large surface area, have often been overlooked. This study aims to investigate the feasibility of using pomegranate waste as a substitute for clay in the production of conventional bricks, with the potential for significant cost savings. The objective is to explore the potential of this approach in contributing to energy conservation efforts and promoting sustainable building practices.

During the months of May to August, energy consumption exceeding 6000 kWh was observed in all studied samples, with the addition of two months (April and October) in BC. Consequently, all the aforementioned months were evaluated according to the second energy price category. Conversely, the remaining months were evaluated according to the first energy price category due to low energy consumption levels below 6000 kWh. The study involved the use of brick samples containing 10% pomegranate peel waste (PPW) with firing temperatures of 700 °C, 800 °C, and 900 °C for S1, S2, and S3, respectively. The impact of the firing temperature on production costs was deemed negligible, and therefore, the most efficient sample (S3) in terms of energy savings as compared to the conventional brick sample (BC) was investigated in terms of cost analysis.

The study findings indicate that S3, among all firing temperature options, has the potential to achieve energy savings of up to 13.89%. This promising result motivated us to explore the feasibility of producing bricks from pomegranate waste. While energy savings are an important factor, it is equally crucial to evaluate the financial benefits of such a production process. Therefore, a cost analysis was conducted, which involved determining the additional investment required for each suggested brick sample, as well as evaluation of the total annual energy cost savings in Saudi riyals (SAR).

This study assumed that the primary expenses for constructing both a conventional brick wall and the proposed PPW brick sample wall would be comparable, with a minor increase in the cost of the PPW brick sample due to the low cost of pomegranate peel waste. To determine the economic viability of the suggested PPW brick sample, we obtained production cost data (in SAR/m^2) for both the conventional brick wall and the manufactured S3 brick sample from the local market, as shown in Table 8. The initial cost analysis considered the costs of materials and workers. In addition, the cost of transporting pomegranate peel waste (PPW) from Medina, which is considered one of the major cities in pomegranate cultivation and pomegranate juice production, to Jazan, the area under investigation, was determined using local market data, with a cost of SAR 1000 per 1000 m^2 of PPW [46]. The labor costs per production rate (SAR/m^2) were determined by dividing the total investment in the production process (SAR) by the production amount (m^2). The data on labor costs were obtained from the statistics report provided by the Ministry of Human Resources and Social Development in KSA (HRSD) [47]. Based on the data presented in Table 9, we evaluated the annual energy costs, total construction costs, annual savings, and simple payback period (SPP) for each brick sample. The SPP was calculated as follows [48]:

$$\text{SPP} = \frac{\text{Additional Investment}}{\text{Annual Saving}}$$

Table 8. The material costs for both conventional and PPW brick samples.

Material	Material Unit Cost (SAR/m^2)
Clay (kaolin)	6
Pomegranate peel waste	1.2
Cement mortar	4

Table 9. The calculated simple payback period associated with the brick samples.

	Wall Cost (SAR)	Additional Investment (SAR)	Energy Cost (SAR/Year)	Annual Savings (SAR/Year)	SPP (Year)
BC	22,628.39	0	20,795.34	0	
S3	25,454.97	2826.58	16,476.54	4318.8	0.65

The study found that S3 demonstrated the potential to achieve substantial energy and financial savings, with a short SPP of 0.65 years. Notably, the additional investment required for the suggested brick sample was deemed minor, with a small increase in cost due to the low cost of pomegranate peel waste. The study's findings indicate that S3 is a promising option for building owners seeking to reduce energy costs and enhance sustainability aspects. Finally, this study provides valuable insights into the potential of using pomegranate waste in brick production to achieve energy and financial savings while promoting sustainable building practices.

4. Conclusions

The purpose of this study was to investigate the potential effects of incorporating pomegranate peel waste (PPW) into brick clay at a concentration of 10% and firing at varying temperatures on the thermal and energy performance of the produced bricks. The study focused on evaluating the thermal parameters of density, thermal conductivity, and specific heat. Additionally, the mechanical and physical characteristics of the bricks were extensively tested. Based on the study findings, the following conclusions were drawn:

First, the addition of PPW to the brick samples resulted in enhanced thermal insulation properties by promoting the formation of pores, which directly reduced thermal conductivity.

Secondly, among all the fabricated brick samples, the insulation brick made from S3 and fired at 900 °C exhibited the most efficient thermal performance. It displayed a density of 1562 kg/m^3, a specific heat of 1500 J/kg^{-1} K^{-1}, and a thermal conductivity of 0.29 Wm^{-1} K^{-1}.

Thirdly, S3 was found to be 13.89% more effective in preserving cooling energy, producing 12% less annual carbon dioxide emissions than conventional bricks. Additionally, it had the shortest simple payback period (SPP) of 0.65 years, indicating its cost-effectiveness. These findings are consistent with previous research [6,10,14] that has demonstrated a noteworthy enhancement in thermal properties, energy efficiency, and reduction in CO_2 emissions, resulting in the shortest simple payback period (SPP).

This study provides recommendations for reducing thermal, energy, and total energy expenses and proposes the use of an environmental brick sample composed of 10% PPW fired at 900 °C instead of conventional bricks. Additionally, the study recommends that national authorities utilize the findings to improve national energy codes and insulation criteria for building envelopes. Finally, the study highlights the potential use of pomegranate peel waste in fired bricks for construction in hot and arid regions and its generalizability to other regions with similar climates. However, several limitations should be considered, such as resource availability and cost, construction suitability, and policy and circular economic factors. Further research is necessary to fully evaluate the feasibility and potential impact of using recycled PPW bricks. This should include an assessment of their environmental and economic implications; durability; thermal and energy performance; and potential for reuse, recycling, or repurposing. Regulatory frameworks, incentives, and market demand must also be considered.

Author Contributions: Conceptualization, I.M.M.; Methodology, A.R., M.M.A. and A.Y.; Validation, N.Z.; Formal analysis, I.M.M.; Investigation, A.A.; Resources, N.Z.; Data curation, A.A.; Writing—original draft, A.R. and M.M.A.; Writing—review & editing, A.Y. All authors have read and agreed to the published version of the manuscript.

Funding: The authors extend their appreciation to the Deputyship for Research & Innovation, Ministry of Education in Saudi Arabia for funding this research work through project number ISP22-4.

Institutional Review Board Statement: Not applicable.

Informed Consent Statement: Not applicable.

Data Availability Statement: Not applicable.

Acknowledgments: The authors extend their appreciation to the Deputyship for Research & Innovation, Ministry of Education in Saudi Arabia, for funding this research work through project number ISP22-4.

Conflicts of Interest: The authors declare no conflict of interest.

References

1. Almasoud, A.H.; Gandayh, H.M. Future of solar energy in Saudi Arabia. *J. King Saud Univ. Eng. Sci.* **2015**, *27*, 153–157. [CrossRef]
2. Dincer, I. Renewable energy and sustainable development: A crucial review. *Renew. Sustain. Energy Rev.* **2000**, *4*, 157–175. [CrossRef]
3. Boretti, A.; Castelletto, S.; Al-Kouz, W.; Nayfeh, J. The energy future of Saudi Arabia. In Proceedings of the 2020 5th International Conference on Sustainable and Renewable Energy Engineering (ICSREE 2020), Paris, France, 6–9 May 2020; EDP Sciences: Les Ulis, France, 2020; Volume 181, p. 03005.
4. Junshu, F.; Xiaoqing, Y. Evolution and Inspiration of Saudi Arabian Energy Strategy. *IOP Conf. Ser. Earth Environ. Sci.* **2019**, *295*, 052012. [CrossRef]
5. Esmail, S.M.; Cheong, J.H. Studies on optimal strategy to adopt nuclear power plants into Saudi Arabian energy system using MESSAGE tool. *Sci. Technol. Nucl. Install.* **2021**, *2021*, 8818479. [CrossRef]
6. Ahmed, S.; El Attar, M.E.; Zouli, N.; Abutaleb, A.; Maafa, I.M.; Ahmed, M.; Yousef, A.; Ragab, A. Improving the Thermal Performance and Energy Efficiency of Buildings by Incorporating Biomass Waste into Clay Bricks. *Materials* **2023**, *16*, 2893. [CrossRef]
7. Tarek, D.; El-Naggar, M.; Sameh, H.; Yousef, A.; Ragab, A. Energy Efficiency Coupled with Lightweight Bricks: Towards Sustainable Building: A review. *SVU-Int. J. Eng. Sci. Appl.* **2023**, *4*, 1–28. [CrossRef]
8. Visco, A.; Scolaro, C.; Facchin, M.; Brahimi, S.; Belhamdi, H.; Gatto, V.; Beghetto, V. Agri-food wastes for bioplastics: European prospective on possible applications in their second life for a circular economy. *Polymers* **2022**, *14*, 2752. [CrossRef]
9. Anitha, K. Utilization of Agricultural Waste materials in Building Industry—An overview. *SPAST Abstracts* **2021**, *1*. Available online: https://spast.org/techrep/article/view/305 (accessed on 12 May 2023).

10. Ghorbani, M.; Dahrazma, B.; Saghravani, S.F.; Yousofizinsaz, G. A comparative study on physicochemical properties of environmentally-friendly lightweight bricks having potato peel powder and sour orange leaf. *Constr. Build. Mater.* **2021**, *276*, 121937. [CrossRef]
11. Rashad, A.M.; Gharieb, M.; Shoukry, H.; Mokhtar, M. Valorization of sugar beet waste as a foaming agent for metakaolin geopolymer activated with phosphoric acid. *Constr. Build. Mater.* **2022**, *344*, 128240. [CrossRef]
12. Ahmadi, R.; Souri, B.; Ebrahimi, M. Evaluation of wheat straw to insulate fired clay hollow bricks as a construction material. *J. Clean. Prod.* **2020**, *254*, 120043. [CrossRef]
13. Abd El-Hady, R.; Zayan, A.; Mohamed, A. Effect of Bio-Material in Thermal Insulation Case-Study: Energy Saving in Residential Building in Aswan City. *IOP Conf. Ser. Earth Environ. Sci.* **2022**, *1056*, 012014. [CrossRef]
14. Aravind, N.; Sathyan, D.; Mini, K. Rice husk incorporated foam concrete wall panels as a thermal insulating material in buildings. *Indoor Built Environ.* **2020**, *29*, 721–729. [CrossRef]
15. Mehrzad, S.; Taban, E.; Soltani, P.; Samaei, S.E.; Khavanin, A. Sugarcane bagasse waste fibers as novel thermal insulation and sound-absorbing materials for application in sustainable buildings. *Build. Environ.* **2022**, *211*, 108753. [CrossRef]
16. Javed, A.; Ahmad, A.; Tahir, A.; Shabbir, U.; Nouman, M.; Hameed, A. Potato peel waste-its nutraceutical, industrial and biotechnological applcations. *AIMS Agric. Food* **2019**, *4*, 807–823. [CrossRef]
17. *Statistical Yearbook*; Water Saudi Ministry of Environment, and Agriculture: Riyadh, Saudi Arabia, 2019.
18. Karim, S.; Alkreathy, H.M.; Ahmad, A.; Khan, M.I. Effects of methanolic extract based-gel from Saudi pomegranate peels with enhanced healing potential on excision wounds in diabetic rats. *Front. Pharmacol.* **2021**, *12*, 704503. [CrossRef]
19. Huy, N.S.; Tan, N.N.; Hang, M.T.N. Environmentally friendly unburnt bricks using raw rice husk and bottom ash as fine aggregates: Physical and mechanical properties. *J. Sci. Technol. Civ. Eng. HUCE* **2021**, *15*, 110–120. [CrossRef]
20. Hassan, A.M.S.; Abdeen, A.; Mohamed, A.S.; Elboshy, B. Thermal performance analysis of clay brick mixed with sludge and agriculture waste. *Constr. Build. Mater.* **2022**, *344*, 128267. [CrossRef]
21. Marques, B.; Tadeu, A.; Antonio, J.; Almeida, J.; de Brito, J. Mechanical, thermal and acoustic behaviour of polymer-based composite materials produced with rice husk and expanded cork by-products. *Constr. Build. Mater.* **2020**, *239*, 117851. [CrossRef]
22. Ramos, A.; Briga-Sá, A.; Pereira, S.; Correia, M.; Pinto, J.; Bentes, I.; Teixeira, C. Thermal performance and life cycle assessment of corn cob particleboards. *J. Build. Eng.* **2021**, *44*, 102998. [CrossRef]
23. Chee, M.C. Effects of natural fibers inclusion in clay bricks: Physic-mechanical properties. *J. Int. J. Civ. Environ. Eng.* **2011**, *1*, 51–57.
24. Uchechukwu Elinwa, A. Effect of addition of sawdust ash to clay bricks. *Civ. Eng. Environ. Syst.* **2006**, *23*, 263–270. [CrossRef]
25. Vėjelienė, J.; Gailius, A.; Vėjelis, S.; Vaitkus, S.; Balčiūnas, G. Development of thermal insulation from local agricultural waste. In Proceedings of the 8th International Conference Environmental Engineering, Vilnius, Lithuania, 19–20 May 2011.
26. Wu, H.; Hu, R.; Yang, D.; Ma, Z. Micro-macro characterizations of mortar containing construction waste fines as replacement of cement and sand: A comparative study. *Constr. Build. Mater.* **2023**, *383*, 131328. [CrossRef]
27. Maafa, I.M.; Abutaleb, A.; Zouli, N.; Zeyad, A.M.; Yousef, A.; Ahmed, M.M. Effect of agricultural biomass wastes on thermal insulation and self-cleaning of fired bricks. *J. Mater. Res. Technol.* **2023**, *24*, 4060–4073. [CrossRef]
28. *ASTM:D5334-08*; Standard Test Method for Determination of Thermal Conductivity of Soil Soft Rock by Thermal Needle Probe Procedure. ASTM International: West Conshohocken, PA, USA, 2008.
29. Aldubyan, M.; Krarti, M. Impact of stay home living on energy demand of residential buildings: Saudi Arabian case study. *Energy* **2022**, *238*, 121637. [CrossRef]
30. Aldersoni, A.; Albaker, A.; Alturki, M.; Said, M.A. The Impact of Passive Strategies on the Overall Energy Performance of Traditional Houses in the Kingdom of Saudi Arabia. *Buildings* **2022**, *12*, 1837. [CrossRef]
31. Abdelrady, A.; Abdelhafez, M.H.H.; Ragab, A. Use of insulation based on nanomaterials to improve energy efficiency of residential buildings in a hot desert climate. *Sustainability* **2021**, *13*, 5266. [CrossRef]
32. Tarek, D.; Ahmed, M.; Hussein, H.S.; Zeyad, A.M.; Al-Enizi, A.M.; Yousef, A.; Ragab, A. Building Envelope Optimization Using Geopolymer Bricks to Improve the Energy Efficiency of Residential Buildings in Hot Arid Regions. *Case Stud. Constr. Mater.* **2022**, *17*, e01657. [CrossRef]
33. Ragab, A.; Abdelrady, A. Impact of green roofs on energy demand for cooling in Egyptian buildings. *Sustainability* **2020**, *12*, 5729. [CrossRef]
34. Ragab Abdel Radi, A. The impact of phase change materials on the buildings enegy efficiency in the hot desert areas the annexed rooms of the traffic building in new aswan city as a case study. *JES J. Eng. Sci.* **2020**, *48*, 302–316. [CrossRef]
35. Rahman, M.; Rasul, M.; Khan, M. Energy conservation measures in an institutional building by dynamic simulation using DesignBuilder. In Proceedings of the 3rd IASME/WSEAS International Conference on Energy & Environment, Cambridge, UK, 23–25 February 2008.
36. Aouba, L.; Bories, C.; Coutand, M.; Perrin, B.; Lemercier, H. Properties of fired clay bricks with incorporated biomasses: Cases of olive stone flour and wheat straw residues. *Constr. Build. Mater.* **2016**, *102*, 7–13. [CrossRef]
37. *ASTM C62-04*; Standard Specification for Building Brick (Solid Masonry Units Made From Clay or Shale). ASTM: West Conshohocken, PA, USA, 2010.
38. Limami, H.; Manssouri, I.; Cherkaoui, K.; Khaldoun, A. Study of the suitability of unfired clay bricks with polymeric HDPE & PET wastes additives as a construction material. *J. Build. Eng.* **2020**, *27*, 100956.

39. Ukwatta, A.; Mohajerani, A.; Eshtiaghi, N.; Setunge, S. Variation in physical and mechanical properties of fired-clay bricks incorporating ETP biosolids. *J. Clean. Prod.* **2016**, *119*, 76–85. [CrossRef]
40. Eliche-Quesada, D.; Pérez-Villarejo, L.; Iglesias-Godino, F.J.; Martínez-García, C.; Corpas-Iglesias, F.A. Incorporation of coffee grounds into clay brick production. *Adv. Appl. Ceram.* **2011**, *110*, 225–232. [CrossRef]
41. Abdelhafez, M.H.H.; Aldersoni, A.A.; Gomaa, M.M.; Noaime, E.; Alnaim, M.M.; Alghaseb, M.; Ragab, A. Investigating the Thermal and Energy Performance of Advanced Glazing Systems in the Context of Hail City, KSA. *Buildings* **2023**, *13*, 752. [CrossRef]
42. Abdel Hamid, E.; Abadir, M.; Abd El-Razik, M.; El Naggar, K.; Shoukry, H. Performance assessment of fired bricks incorporating pomegranate peels waste. *Innov. Infrastruct. Solut.* **2023**, *8*, 18. [CrossRef]
43. Mahmoud, A.R. The Influence of Buildings Proportions and Orientations on Energy Demand for Cooling in Hot Arid Climate. *SVU-Int. J. Eng. Sci. Appl.* **2022**, *3*, 8–20. [CrossRef]
44. Abdelhafez, M.H.H.; Touahmia, M.; Noaime, E.; Albaqawy, G.A.; Elkhayat, K.; Achour, B.; Boukendakdji, M. Integrating Solar Photovoltaics in Residential Buildings: Towards zero energy buildings in hail city, ksa. *Sustainability* **2021**, *13*, 1845. [CrossRef]
45. De Wolf, C.; Yang, F.; Cox, D.; Charlson, A.; Hattan, A.S.; Ochsendorf, J. Material quantities and embodied carbon dioxide in structures. In *Institution of Civil Engineers-Engineering Sustainability*; Thomas Telford Ltd.: London, UK, 2015.
46. Ministry of Transport and Logistic Services. *Transportation Report 2022*; Ministry of Transport and Logistic Services: Riyadh, Saudi Arabia, 2022.
47. Ministry of Human Resources and Social Development (HRSD). *Labor Market Survey Report*; KSA: Ministry of Human Resources and Social Development (HRSD): Riyadh, Saudi Arabia, 2022.
48. Mujeebu, M.A.; Ashraf, N.; Alsuwayigh, A. Energy performance and economic viability of nano aerogel glazing and nano vacuum insulation panel in multi-story office building. *Energy* **2016**, *113*, 949–956. [CrossRef]

materials

Article

Development of Clay-Composite Plasters Integrating Industrial Waste

Andreea Hegyi [1,2], Cristian Petcu [3,*], Adrian Alexandru Ciobanu [4,*], Gabriela Calatan [5] and Aurelia Bradu [4]

[1] NIRD URBAN-INCERC Cluj-Napoca Branch, 117 Calea Florești, 400524 Cluj-Napoca, Romania; andreea.hegyi@incerc-cluj.ro
[2] Faculty of Materials and Environmental Engineering, Technical University of Cluj-Napoca, 103-105 Muncii Boulevard, 400641 Cluj-Napoca, Romania
[3] NIRD URBAN-INCERC Bucharest, 266 Șoseaua Pantelimon, 021652 Bucharest, Romania
[4] NIRD URBAN-INCERC Iași Branch, 6 Anton Șesan Street, 700048 Iași, Romania; bradu_aurelia@yahoo.com
[5] Office for Pedological and Agrochemical Studies Cluj-Napoca, 1 Fagului Str., 400483 Cluj-Napoca, Romania; gabi_kavida@yahoo.com
* Correspondence: cristian.petcu@yahoo.com (C.P.); adrian.ciobanu@incd.ro (A.A.C.)

Abstract: This research investigates the feasibility of developing clay composites using natural materials and incorporating waste by-products suitable for plastering diverse support structures. The study identified a versatile composition suitable for a wide range of support materials and explored the potential of revaluing industrial waste and by-products by reintegrating them into the Circular Economy. The experimental investigation outlines the process of evaluating the influence of different raw materials on the performance of the clay composite. The findings confirm that using limestone sludge and fly ash as additives to clay contributes to reducing axial shrinkage and increasing mechanical strengths, respectively. The optimal percentage of additives for the clay used are identified and provided. Using hydraulic lime as a partial substitute for clay reduces the apparent density of dried clay composites, axial shrinkage, and fissures formation while improving adhesion to the substrate. Introducing dextrin into this mix increases the apparent density of the hardened plaster while keeping axial shrinkage below the maximum threshold indicated by the literature. Mechanical strengths improved, and better compatibility in terms of adhesion to the support was achieved, with composition S3 presenting the best results and a smooth, fissure-free plastered surface after drying.

Keywords: clay plasters; drying shrinkage; plaster characterization; physical and mechanical properties; sustainable building materials

Citation: Hegyi, A.; Petcu, C.; Ciobanu, A.A.; Calatan, G.; Bradu, A. Development of Clay-Composite Plasters Integrating Industrial Waste. *Materials* **2023**, *16*, 4903. https://doi.org/10.3390/ma16144903

Academic Editor: Antonio Caggiano

Received: 22 May 2023
Revised: 3 July 2023
Accepted: 6 July 2023
Published: 9 July 2023

1. Introduction

Nowadays, there is a global interest in identifying sustainable solutions that are as environmentally friendly as possible for the construction industry. The highly polluting impact of the cement industry—currently the primary raw material used for developing and maintaining the built environment—is well known. Therefore, a series of research studies have turned to innovation based on traditional "heritage" using vernacular materials [1]. The results of this research, besides offering the possibility of preserving traditional identity, support the documented development of construction made from raw-clay-based materials in combination with various additives (mineral, plant, animal, derived from waste or industrial by-products). In addition to the great advantage of reduced environmental impact, buildings made from raw-clay-based materials have several benefits in terms of indoor air quality and, implicitly, the health of the population: water vapor permeability, the ability to regulate indoor air humidity, and high storage/heat release capacity, thus contributing to thermal comfort, increasing indoor air quality and energy efficiency. However, the difficulties these materials present are primarily in terms of mechanical strength and resistance to the action of climatic factors being lower compared to concrete buildings, as well as a reduced degree of compatibility with classic finishing materials available on the market [2–13]. Consequently, there is a need to develop plaster and finish materials that are compatible with the primary materials used for traditional construction (natural stone, burnt ceramic brick, and masonry elements based on raw clay, wood, and wood-based or other lignocellulosic materials).

Indoor finishing materials play a significant role in shaping the indoor climate due to their moisture-buffering capabilities. These capabilities arise from the sorption and diffusion properties of the materials, which help maintain a balanced indoor environment [2–5]. The inherent thermal properties of clay building elements and finishes contribute to their capacity to effectively regulate indoor temperature. This is particularly beneficial when faced with significant daily temperature fluctuations, as highlighted by [6]. Their substantial thermal mass facilitates the absorption and storage of heat throughout daylight hours, subsequently releasing it during cooler periods. This natural temperature modulation fosters a comfortable interior ambiance and impacts the building energy efficiency [7–12]. Additionally, clay's permeability to water vapors enables earthen walls to function as a moisture buffer, adeptly managing humidity levels. Minke shows that, when in equilibrium with saturated moisture air, clay blocks can absorb eight times more moisture than burnt bricks [13]. By absorbing excess water from the indoor air, then gradually releasing it when conditions are dry, clay walls maintain a balanced indoor environment, further enhancing their appeal in contemporary construction. In regions with naturally lower air humidity, such as northern climates, these aspects become particularly important. McGregor [14] shows that the nature of clay minerals has a significant impact on moisture capacity. Moisture storage capacity is affected by the clay particles' variable surface charge and size. Incorporating finer and more active clay minerals, like montmorillonite, can substantially boost moisture capacity. However, this may also introduce secondary effects, such as increased swelling and shrinkage when moisture levels change. Addressing the issue of increasing water permeability, researchers have discovered innovative methods to enhance the moisture buffering capabilities of clay plasters by incorporating cellulose from waste paper [15]. However, in a milder climate, clay alone may be sufficient to provide humidity-buffering properties. In a newly constructed German house, with both interior and exterior walls made of earth, measurements taken over an eight-year period revealed that the relative humidity inside the home remained consistently close to 50% all year round [13]. This natural regulation of temperature and humidity reduces the need for artificial heating and cooling systems and is beneficial in reducing the risk of respiratory illnesses, allergies, and many other issues.

The specialized literature showcases numerous studies that highlight the durability of buildings constructed using clay and indigenous techniques in diverse geographic and climatic conditions. These studies emphasize the adaptability and resilience of clay-based construction, which has been used for centuries across different cultures and environments [16–20]. The oldest example of using mudbricks in Europe is in Heuneburg, Germany, dating back to the 6th century BCE. Writings by Pliny the Elder indicate the construction of rammed earth forts in Spain at the end of the 100s BCE [20]. Bugini et al. [21] examines a clay plaster consisting of illite, chlorite, kaolinite, and fine quartz, which is believed to be the first-ever identified example in Roman Lombardy. This unique sample was discovered at the Santa Maria alla Porta site, located in the area of Milan's Imperial Palace, dating back to the first century CE.

In the past couple of decades, houses with earthen building elements have gained significant popularity and recognition. This surge in interest can be attributed to several key factors that make clay houses an appealing choice. However, an essential characteristic is that using natural, non-toxic materials in the building process significantly reduces the risk of indoor air pollution and other health hazards, making clay homes a safer and eco-friendlier option [22]. One important parameter that characterizes the indoor air quality is ozone (O_3). It readily reacts with various indoor materials and compounds in indoor air, leading to lower indoor ozone concentrations when outdoor air is the primary source.

However, the byproducts of indoor ozone reactions can be irritating or harmful to building occupants. Darling et al. [23] suggest that clay-based coatings could serve as efficient passive removal materials, exhibiting relatively low emission rates of by-products that diminish quickly within a two-month period. Another study [24], assessing clay plaster as a passive removal material (PRM) for improving air quality, found that clay plaster improved the perceived air quality and considerably decreased aldehyde concentrations. However, there are relatively few data in the literature on the characteristics of clay plasters [25].

Clay houses have become increasingly popular and well-regarded, with more and more specialists showing interest in this type of construction [26,27]. The results of studies conducted on the possibility of using natural materials in construction are presented in the specialized literature, but these results, in addition to their many benefits, also reveal some drawbacks of clay that are worth investigating for potential improvements [13,28 30]. These drawbacks include a high risk of fissuring during the drying process for clay plaster due to significant axial shrinkage and a high sensitivity to water. To be used in the manufacture of construction materials, clayey soil must contain at least 15–16% clay to achieve the necessary plasticity and workability for labor [29,30]. To avoid the appearance of fissures, a linear shrinkage between 3 and 12% for soft mixtures or between 0.4 and 2% for drier mixtures is acceptable [13,28–31].

Given that there is limited research on the addition of waste materials to clay composites, and considering that in Romania, 80% of such waste is stored in vast areas set up in nature [32,33], constituting a potential environmental hazard, this study aims to valorize these wastes as additives in clay mixtures. Composites based on clay with added ash and sludge have been studied for the production of masonry elements [34,35], and optimal recipes with good results have been developed, thus continuing the study of composites based on clay with added ash and sludge for plastering. The study continues by examining clay mixtures with added ash and sludge for plastering applications.

Clay plasters are known for their eco-friendly and sustainable characteristics [3,13,22–24,36] but also have certain drawbacks [13,37]. One of the main concerns is the shrinkage during the drying process. As the water content in the plaster decreases, the material contracts, which can lead to the formation of fissures or cracks. This affects the plaster's appearance and compromises its ability to adhere properly to the substrate, potentially reducing the structural integrity of the overall construction [13,36,37]. Furthermore, the compatibility of clay plasters with various substrates can be an issue. In construction projects, a wide range of materials may be used as substrates, including wood, brick, stone, and even modern materials like concrete or oriented strand boards (OSB). To simplify the application process and increase the adoption of clay plasters, it is crucial to develop composites that can bond effectively with a diverse array of substrates. This may involve the incorporation of additional binders, fibers, and other additives that can enhance adhesion without compromising the plaster's environmental profile.

The study focused on examining the potential for developing clay compositions suitable for plastering surfaces constructed from various support materials. The research highlights the impact of different raw materials on the overall performance and characteristics of the clay composite. As a result, the originality and value of the experimental research conducted lie in the following:

- Investigating the realm of eco-friendly materials with negligible environmental impact;
- Assessing the impact of additional or substitute raw materials on the performance of clay composites;
- The objective was to identify a composition that could be widely used across multiple support materials. This would eliminate the need for adapting the plaster composition every time the support layer changes. Ultimately, this would create a favorable framework for achieving a homogeneous and uniform appearance and reduce any delays in the plastering process.
- The research aimed to identify ways to reintroduce and revalue industrial waste and by-products. This would help create a favorable framework for implementing the

principles of the Circular Economy. By finding new uses for these materials, the research aimed to reduce waste and promote sustainability.

2. Materials and Methods

To achieve a low-embedded energy material, it is essential to prioritize the use of locally sourced raw materials. For preparing clay-based composite materials intended for plastering the surfaces of structures, the research focused on locally produced materials, such: clay extracted directly from the soil in Valea Draganului, Cluj Napoca, Romania (46°54′10″ N 22°49′53″ E); fly ash generated by the Mintia power-plant, Romania (a by-product of Romania's third largest power-plant); limestone sludge, a by-product of limestone processing industry in Hunedoara, Romania; hydraulic lime with the commercial name RÖFIX NHL5; dextrin purchased from a local producer in Cluj Napoca, Romania; and sodium chloride, NaCl, with 90% purity.

Clay is a fine-grained sedimentary rock composed of a blend of silica and fragments of materials such as quartz and mica [38] (X-ray diffraction (XRD) using a high-resolution Brucker D8 diffractometer). The particular clay used in this study, considered to be mostly of montmorillonite type, from the point of view of the particle size distribution, consists of approximately 38.15% clay minerals, 41.93% sand, and 19.92% silt, determined by the sieving and sedimentation method according to the Romanian standard STAS 1913-5 [39]. Based on its size distribution, it is classified as clay loam (Figure 1). It was characterized in terms of oxide composition by XRF analysis and the relevant data are provided in Table 1. From a mineralogical perspective (Figure 2), experimental tests indicated the presence of predominant minerals: quartz 66%, muscovite 32.4%, and vermiculite 1.6%. The tests were carried out in the research laboratories of the Technical University of Cluj-Napoca (XRD analysis) and the Babeş-Bolyai University of Cluj-Napoca (XRF analysis). According to the specialized literature [13,40,41], the mineralogical composition of this raw material indicates that it is appropriate for creating compositions that can be used in construction.

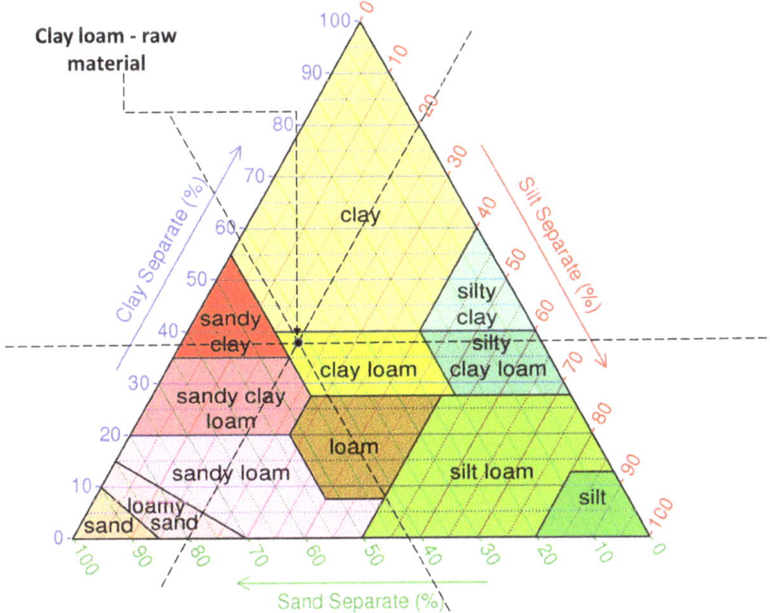

Figure 1. Characterization of the clay used as raw material—material placement in the ternary diagram.

Table 1. The oxide composition of the clay, determined by XRF analysis (X-ray fluorescence).

Oxides	SiO_2	Al_2O_3	Fe_2O_3	CaO	MgO	K_2O	Na_2O	TiO_2	PC
Content [%]	74.17	12.74	4.38	0.7	1.0	1.43	0.73	0.05	4.78

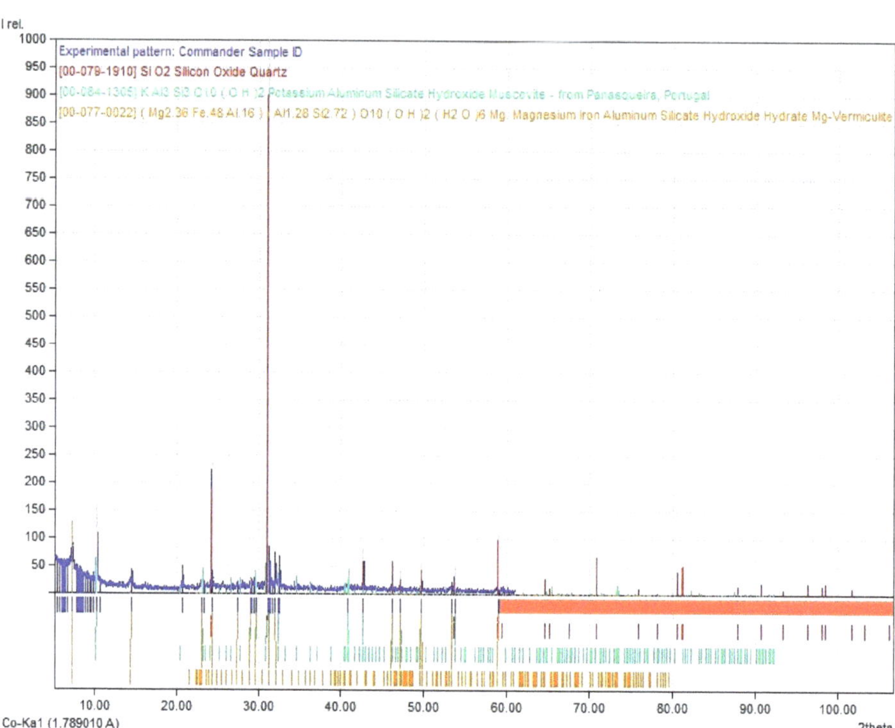

Figure 2. Characterization of the clay used as raw material—mineralogical analysis, determined by XRD analysis (X-ray diffraction).

The fly ash underwent a comprehensive analysis to determine its properties and potential applications. Its chemical characteristics were assessed by examining the oxide composition (as shown in Table 2), revealing the percentages of various oxides present in the ash. In addition, the fly ash's physical properties, such as fineness and bulk density, were evaluated and documented in Table 3. The fineness was determined by dry-sieving using a HELOS RODOS/L, R5 (dry dispersion in the free aerosol jet for laser diffraction), and the bulk density was determined as a ratio between mass and volume using the pycnometer method.

Table 2. The oxide composition of the fly ash, determined by XRF analysis (X-ray fluorescence).

Oxides	SiO_2	Al_2O_3	Fe_2O_3	CaO	MgO	SO_3	Na_2O	K_2O	P_2O_5	TiO_2	Cr_2O_3	Mn_2O_3	ZnO	SrO	PC
Content [%]	53.75	26.02	7.91	2.54	1.54	0.35	0.59	2.57	0.12	1.02	0.05	0.09	0.02	0.03	2.11

Table 3. Physical characteristics of ash samples.

Fly Ash from Mintia Power Plant	Fineness $R_{0.045}$	Apparent Density (Mg/m^3)
	39.20%	1.67

The limestone sludge underwent experimental analysis to determine its properties, focusing on particle size distribution and bulk density, which are presented in Table 4. This characterization provides insights into the sludge's potential applications and performance in various settings, particularly when used as an additive in construction materials.

Table 4. Physical characteristics of the limestone sludge.

Sieve mesh sizes (mm)	0.063	0.125	0.250	0.500	1
Passings through the sieve (%)	69	93	95	99	100
Apparent density (kg/m^3)			1780		

The hydraulic lime, introduced into the composite as a substitute for part of the clay, was characterized by experimentally determining the bulk density and compressive strength (Table 5).

Table 5. Physical characteristics of the hydraulic lime.

Characteristics	Apparent Density (kg/m^3)	Compression Strength (N/mm^2)
Hydraulic lime	520	5

Dextrin are constituted from water-soluble amorphous carbohydrates with a small molecule, obtained from the hydrolysis reaction of starch or glycogen, with the chemical formula $(C_6H_{10}O_5)_n$. Dextrin is available as a white, pure powder, soluble in water and poorly soluble in alcohol. When mixed with water, they form a very sticky syrup, which can be used as a substitute for gum arabic and gum tragacanth. Dextrin was characterized by experimentally determining its bulk density, 800 kg/m^3.

The water used for preparation was dosed so that the sum of the liquid raw material quantities was constant in relation to the sum of the dry material quantities, specifically, 33% by mass, ensuring a consistency of the fresh mixture of 95 ± 5 mm.

2.1. Preliminary Experimental Composites

In order to identify and analyze the performance of clay compositions intended for plastering work, a series of mixtures made from three or more of the previously presented raw materials were prepared under laboratory conditions. The compositional design was based on preliminary research regarding the influence of different raw materials on the performance of clay composites and considered a series of guidelines:

- Fly ash was considered a potential raw material due to its pozzolanic performance [32,42–54]. Preliminary experimental research analyzed the axial shrinkage of four clay composites prepared with fly ash used as an additive to a constant amount of clay, specifically 15 g, 20 g, 25 g, and 30 g of limestone sludge added to 50 g of clay, according to Table 6.
- Limestone sludge was selected based on preliminary experimental tests and literature reports that indicate beneficial effects [13,41,55–57] due to the additional calcium oxide contribution in reducing the volume variations of the composite. Thus, preliminary experimental research analyzed the axial shrinkage of four clay composites prepared with limestone sludge used as an additive to a constant amount of clay, precisely 15 g, 20 g, 25 g, and 30 g of limestone sludge added to 50 g of clay, according to Table 7.
- NaCl was identified in preliminary tests, and in accordance with the specifications of the specialized literature [13,41,58–61], as a beneficial additive in clay-based composites. It helps in reducing axial shrinkage and minimizes the risk of cracking by moderating and standardizing the drying process. In clay composites, NaCl was introduced as an 8% saline solution prepared by dissolving the salt in potable water.
- Hydraulic lime was chosen based on initial experimental tests and numerous literature findings that highlight its positive impact in reducing drying shrinkage and the likelihood of fissuring, due to the added calcium oxide content [62–71]. In this instance,

preliminary experimental research examined the effects of partially substituting clay with hydraulic lime in compositions consisting of clay, fly ash, and limestone sludge. The study focused on evaluating the risk of fissuring, adhesion to clay-based support, axial shrinkage, apparent density, and mechanical resistance in four clay composites, as outlined in Table 8. In all the mixtures, sodium chloride was incorporated as an 8% mass concentration aqueous solution. The amounts of fly ash and limestone sludge were kept constant. The ratios were determined based on prior experimental studies, keeping a constant mass proportion of additive materials (limestone sludge/fly ash) to clay. In instances where clay was partially replaced by hydraulic lime, a constant mass proportion of additive materials (limestone sludge/fly ash) to the combination of (clay + hydraulic lime) was maintained.

-   Dextrin was selected based on literature reports indicating beneficial effects of similar organic additives [3,37,72–74], which include improved paste workability, decreased drying rate, increased uniformity in the drying process, and, ultimately, a reduction in fissures within the dried plaster layer.

Table 6. Composition of the preliminary clay composites experimentally prepared to assess the influence on mechanical resistance of adding fly ash to clay.

Sample Code	PA1	PA2	PA3	PA4	PA5
Clay (g)	50	50	50	50	50
Fly ash (g)	-	15	20	25	30

Table 7. Composition of the preliminary clay composites experimentally prepared to assess the influence on axial shrinkage of adding limestone sludge to clay.

Sample Code	PS1	PS2	PS3	PS4	PS5
Clay (g)	50	50	50	50	50
Limestone sludge (g)	-	15	20	25	30

Table 8. Composition of the preliminary clay composites experimentally prepared to assess the influence of the partial substitution of clay with hydraulic lime.

Raw Materials / Clay Composition Code	Clay (g)	Hydraulic Lime (g)	Fly Ash (g)	Limestone Sludge (g)	Saline Solution NaCl, 8% (g)
PV1	50	0	25	25	17
PV2	30	20	25	25	17
PV3	25	25	25	25	17
PV4	20	30	25	25	17

2.2. Clay-Based Compositions Developed in Accordance with the Preliminary Research Findings

Following the preliminary experimental research results, the aim was to determine the impact of dextrin on the performance of clay composites for plastering surfaces in eco-traditional buildings. As a result, five different composites were developed, as shown in Table 9.

The chosen substrates for applying clay composite plasters were selected from commonly used construction materials in eco-traditional wall buildings. These include ceramic bricks, limestone, oriented strand boards (OSB), unfired clay masonry units, and masonry elements made from a composite material consisting of clay and cereal straw.

Table 9. Composition of the experimental clay composites (percentage composition reported to the total mass of the mixture).

Raw Materials Clay Composition Code	Clay (g)	Lime (g)	Fly Ash (g)	Limestone Sludge (g)	Saline Solution NaCl, 8% (g)	Dextrin (g)
S1	30	20	25	25	17	2
S2	30	20	25	25	17	4
S3	30	20	25	25	17	6
S4	30	20	25	25	17	8
S5	30	20	25	25	17	10

2.3. Experimental Testing Methods

For preliminary clay composites PA1–PA5, PS1–PS5, PV1–PV5, and dextrin-containing clay composites S1–S5, prismatic samples of 40 × 40 × 160 mm were produced by casting in metal molds. After maintaining for 24 h in constant laboratory conditions of 23 ± 2 °C and 65 ± 1% relative humidity, the samples were removed from the molds and stored for 40 days for curing/drying.

For the clay composites PV1–PV5 and S1–S5, which underwent experimental testing for adherence to the substrate, the process involved preparing flat samples using different substrates. These substrates included ceramic bricks, limestone, OSB boards, unfired clay masonry units, and masonry elements made of composite material based on clay and cereal straw. A 3–5 mm thick layer of the clay composite material was applied to the previously mentioned substrates. The surfaces of the substrate were treated with bone glue aqueous solution before applying the clay composites. For this purpose, an 8% concentration bone glue solution was prepared by dissolving bone glue in warm water. The application of the clay composites was carried out only after the drying of the support surfaces treated with the bone glue solution.

All samples were stored for 40 days until reaching constant mass. The constant mass of the sample was regarded as an indication of both material maturity and dryness. The samples were maintained under constant laboratory conditions of 23 ± 2 °C and 65 ± 1% relative humidity and were then experimentally tested under the same environmental conditions to analyze the following parameters:

- identification of fissures through visual analysis;
- axial shrinkage after 40 days of drying, following the testing method established by Romanian standard STAS 2634 [75], as a percentage reduction of the length of the specimen when it reaches the equilibrium humidity (40 days), in relation to the initial length recorded at de-molding;
- apparent density in hardened and dried state, 40 days after casting, following the testing method established by European standard EN 1015-10 [76], as a ratio between the mass and the volume of the specimen;
- mechanical strengths of the dried state, 40 days after casting, following the testing method established by European standard EN 1015-11 [77], using an automatic press. In order to determine the flexural strength, the concentrated load method was used halfway between the supports, positioned at a distance of 100 mm from each other, each at a distance of 20 mm from the ends of the prismatic specimen. Later, on the two halves of the prism resulting from the tensile testing by bending, investigation was carried out to determine the compressive strength as a ratio between the maximum load recorded at the time of breaking and the surface of the plates through which the compressive stress was applied (40 × 40 mm).
- adherence to substrate in the dried state, 40 days after casting, following the testing method established by European standard EN 1015-12 [78], by the pulling method, using an Elcometer pull-off device.

To ensure repeatability of the experimental tests, a minimum of 3 samples were prepared for each case. The reported values represent the arithmetic mean of individual values.

3. Results and Discussions

3.1. Preliminary Experimental Research Results

The preliminary experimental research revealed aspects that were later incorporated into the design of the tested clay compositions to determine the optimal variant for creating plaster material:

-   Using limestone slurry as an additive to clay, as in Figure 3, resulted in a 1.63–2.54% reduction in axial shrinkage compared to the control clay sample (without limestone slurry). The most suitable limestone slurry to clay ratio, which produced the desired effect of reducing axial shrinkage, was determined to be 1:2 parts by mass (sample code PS4).
-   Incorporating fly ash as an additive to clay increased mechanical strengths relative to the control clay sample (without limestone slurry). This aligns with the findings of other experimental studies in the field [79]. While compressive strength increases with higher levels of fly ash (as shown in Figure 4), flexural tensile strength was adversely affected when the ratio of fly ash to clay exceeded a certain percent (as demonstrated in Figure 5). Aggregating this information, the optimal ratio of fly ash to clay from the point of view of mechanical strengths was established as 1:2 parts by mass (sample code PA4).
-   Using hydraulic lime as a partial substitute for clay led to a significant reduction in the apparent density of dried clay composites, axial shrinkage, and mechanical strengths (Figures 6–9), especially when the amount of clay replaced by hydraulic lime was higher. In this context, it is important to weigh the advantages and disadvantages of using hydraulic lime as a substitute for clay in these composites, which are intended for plastering surfaces of eco-traditional construction. On one hand, the use of hydraulic lime leads to reduced axial shrinkage, which is a positive outcome. On the other hand, it results in decreased mechanical strengths, which is not advantageous. Therefore, the presence or absence of fissures and the adherence of the composites to clay support (clay masonry elements) are considered crucial factors in determining the appropriate quantity of hydraulic lime to be used. Visual evaluation of the samples indicated the absence of fissures for composites PV1–PV3. The PV4 composite detached from the support while drying, making visual evaluation inconclusive in this instance. However, the maximum adherence to support was recorded for composite PV2, in which clay was replaced by hydraulic lime at a ratio of 40% (mass percentages). Therefore, this composite is considered the most appropriate option to further experimental investigation.

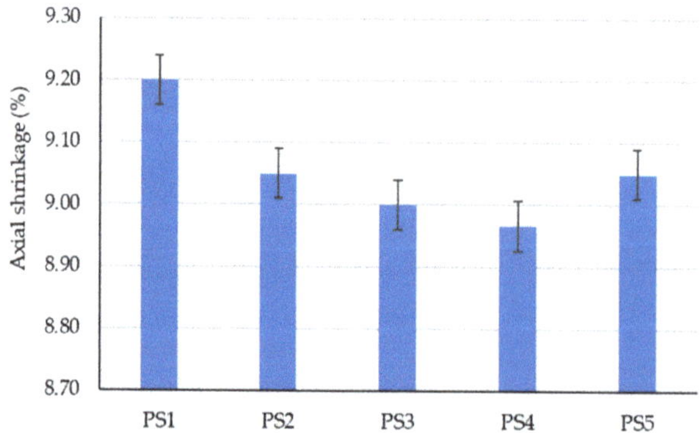

Figure 3. The influence of adding limestone slurry on the axial shrinkage of the clay material.

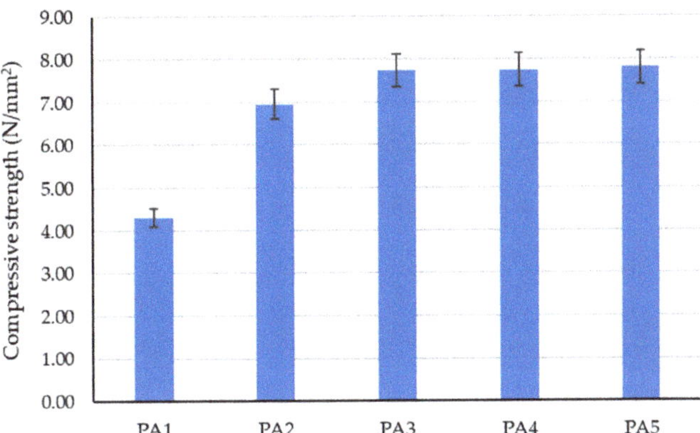

Figure 4. The influence of adding fly ash on the compressive strength of the clay material.

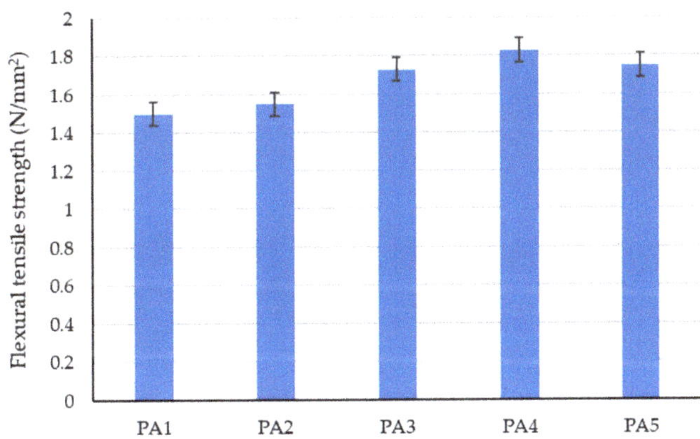

Figure 5. The influence of adding fly ash on the flexural tensile strength of the clay material.

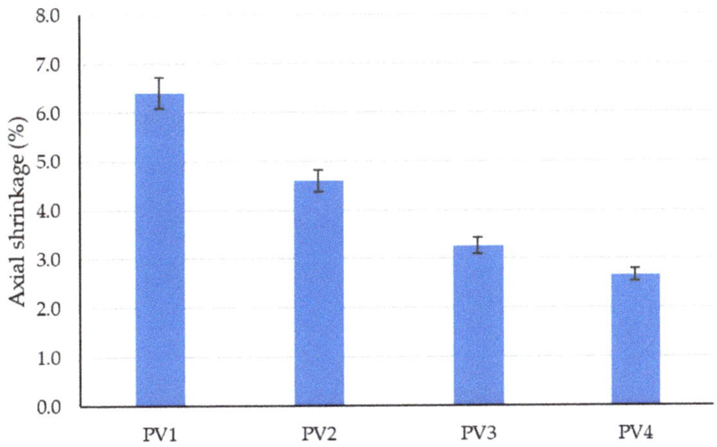

Figure 6. Influence of partial substitution of clay with hydraulic lime on the axial shrinkage of the clay material.

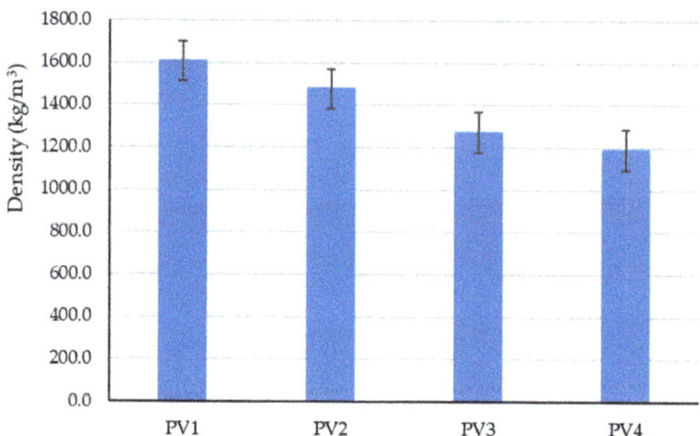

Figure 7. Influence of partial substitution of clay with hydraulic lime on the apparent density of the clay material.

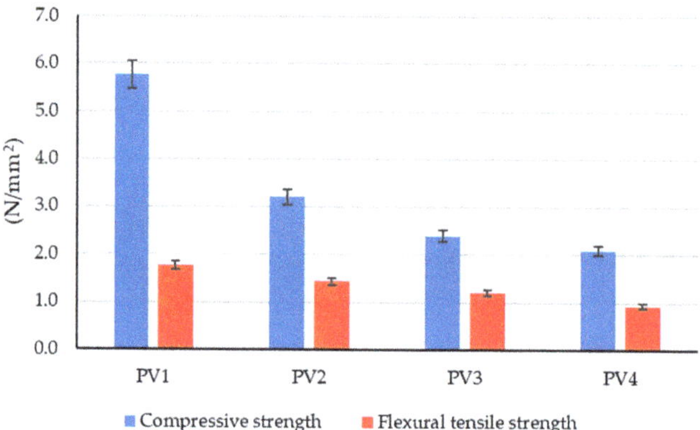

Figure 8. Influence of partial substitution of clay with hydraulic lime on the mechanical strengths of the clay material.

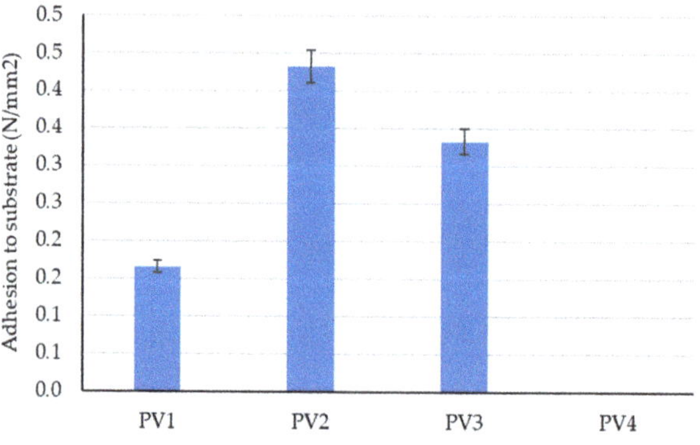

Figure 9. Influence of partial substitution of clay with hydraulic lime on the adhesion of the clay material to a clay-based masonry element support.

3.2. Results of Experimental Research for Clay-Based Compositions Developed Following Preliminary Research Results

The experimental results obtained are presented in Figures 10–13. Experimental outcomes for clay compositions PV1–PV4 and S1–S5 (Figure 10) showed axial shrinkage below the maximum allowable limit indicated by specialized literature (12%). Therefore, all the analyzed compositions satisfy this requirement from this viewpoint.

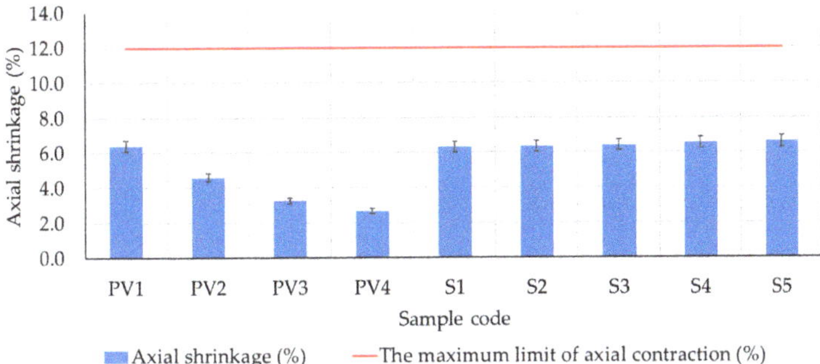

Figure 10. Axial shrinkage of clay mixtures after 40 days of maturation in constant laboratory conditions.

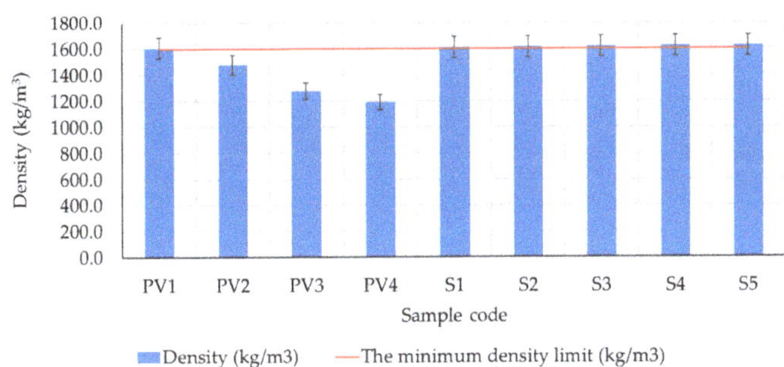

Figure 11. The apparent density of the clay composites after 40 days of maturation under constant laboratory conditions.

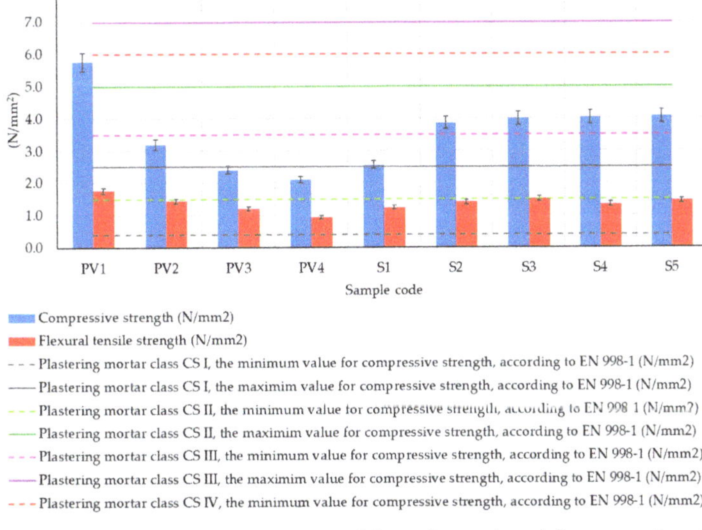

Figure 12. Resistance to compression and flexural extension of clay composites after 40 days of maturation under constant laboratory conditions.

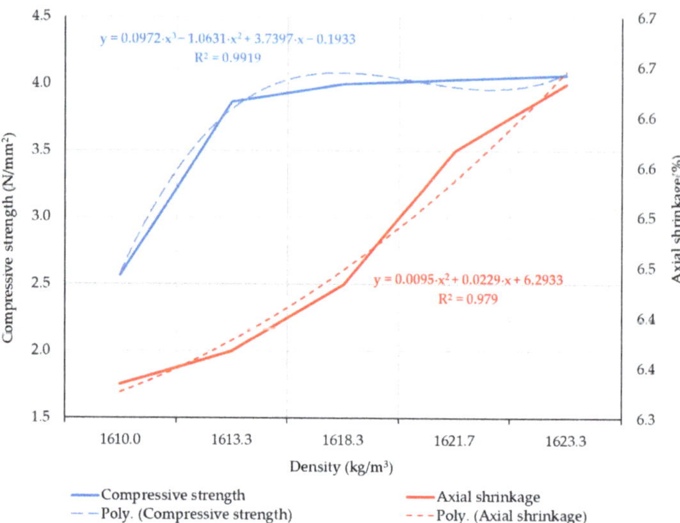

Figure 13. The evolution of compressive strength and axial shrinkage depending on the apparent density of the clay composite.

Considering that these mixtures are intended for plastering applications, it is important to attempt for reduced shrinkage levels. By doing so, the potential for cracking and fissuring can be minimized, ensuring a more durable and long-lasting plaster performance. Clay compositions PV3 and PV4, which contain a large amount of hydraulic lime, have the lowest axial shrinkage compared to the other samples. Moreover, analyzing the axial shrinkage recorded for clay compositions PV1–PV4 (compositions that do not contain dextrin), a decreasing trend of this indicator is observed as the amount of lime increases. Therefore, the beneficial effect of lime in reducing the axial shrinkage of the composite is confirmed. On the other hand, the introduction of dextrin into the composite matrix results in increased axial shrinkage. The values are within a range comparable to the axial shrinkage of sample PV1, which contains neither hydraulic lime nor dextrin. However, these compositions (S1–S5) possess a lower lime content compared to the compositions that demonstrated the best performance regarding axial shrinkage. When comparing the experimental results of samples S1–S5 to the performance of sample P2 (composition without dextrin, identified as the most suitable for designing composites S1–S5), there was a 37–44% increase in axial shrinkage. However, since these values fall within the limit specified by specialized literature (max. 12%), this increase was not considered a drawback for the development of clay composites with dextrin intended for plastering.

According to specialized literature, density values within the range of 1600 kg/m^3 and 2000 kg/m^3 provide the advantage of good thermal resistance for the material [2,8,13,14,19]. However, these indications from the literature mainly consider the behavior of masonry elements intended for constructing walls made of clay composites. In this case, the intended use of the analyzed clay composites is plastering. Therefore, although thermal resistance characteristics are not negligible, they are not as important a parameter in performance evaluation. Consequently, it was determined that several factors are more crucial for evaluating composite performance. These include a smooth surface free from fissuring risks, strong adherence to the supporting material, vapor permeability, and resistance to environmental agents.

The experimental results recorded for the density of the analyzed clay composites (Figure 11) indicated several interesting aspects. It can be observed that using hydraulic lime as a substitute for clay in clay composites PV2–PV4 leads to a reduction in apparent density by 8%, 20%, and 25%, respectively, compared to the apparent density of the clay composite prepared without hydraulic lime (PV1). However, this decreasing trend in apparent density is not maintained in the case of clay composites prepared with dextrin (compositions S1–S5). In this case, the parameter followed shows a slight increase, even above the 1600 kg/m^3 limit indicated by specialized literature as the minimum for good thermal resistance behavior. Thus, in the case of compositions with dextrin, apparent density values ranging from 1610 to 1623 kg/m^3 were recorded, increasing as the amount of dextrin used was higher. Therefore, it is considered that the use of dextrin in clay composites brings slight benefits, leading to an apparent density within a range of values favorable for good thermal resistance behavior.

Regarding the compressive strength of the analyzed clay composites (Figure 12), two clear trends were observed: on one hand, the partial substitution of clay with hydraulic lime leads to a decrease in compressive strength by approximately 45%, 58%, and 64% (samples PV2–PV4). On the other hand, the introduction of dextrin into the clay composite (samples S1–S5) slightly improves compressive strength, with the improvement increasing as the amount of dextrin used is higher. Comparing the experimental data to sample PV1, which lacks hydraulic lime, reveals a 30–33% reduction in compressive strength for composites made with partial substitution of clay with lime and added dextrin (samples S2–S5). This is seen as a minor improvement in performance when contrasted with the 45–64% decrease observed in samples without dextrin. Still, considering that samples with dextrin also have hydraulic lime, their compressive strength, although improved, does not reach the value recorded for sample PV1, the clay composite made without partially substituting clay with lime. In this case, when comparing the experimental results of samples S1–S5 with the performance of sample P2 (the composition without dextrin identified as the most suitable for designing compositions S1–S5), an increase in compressive strength for composites with a minimum content of 4 g dextrin per 50 g (clay + hydraulic lime) was observed. Examining the compressive strengths of compositions with dextrin (S1–S5) reveals that this additive may have a minor positive impact on the composite's mechanical performance. However, this beneficial effect is constrained, with a maximum benefit that cannot be surpassed.

On the other hand, analyzing the experimental results in the context of the classification limits indicated by EN 998-1 [80], a specific standard for industrially produced plaster and skimming mortars based on inorganic binders, it can be said that all tested clay composites fall within the minimum limits imposed by this standard (0.4 N/mm^2—minimum value for class CS I). Furthermore, all tested clay composites could be classified in the second class in terms of compressive strength, CS II, indicated by this standard, except for the clay composite PV1, which would fall into a higher class, namely CS III. This finding is particularly valuable in terms of the applicability of these composite materials. The ease of their in situ use increases when they can be compared to cement-based plaster or skimming mortar products, which are commonly employed in the construction industry.

In terms of flexural tensile strength, as shown in Figure 12, a pattern similar to compressive strength is observed. With the partial substitution of clay with hydraulic lime in samples PV2–PV4, the flexural tensile strength decreases by 18%, 32%, and 47%. This reduction is in comparison to the flexural tensile strength of the composite that does not contain lime (sample PV1). Upon introducing dextrin to the clay composition, there is a visible trend of reduced strength loss as the dextrin content increases. This trend continues up to a certain threshold, as seen in samples S1–S5. However, when the amount of dextrin is further increased, the mechanical performance and flexural tensile strength decline, as observed in samples S4–S5, falling below the maximum value recorded for sample S3. This behavior suggests that the flexural tensile strength can experience slight improvement with the addition of a small amount of dextrin in the clay composite, but the extent of this improvement is limited. This finding aligns with the observation made during the analysis of compressive strength performance. As a result, it can be inferred that while adding dextrin to the clay composition may provide some benefits to the mechanical performance, these positive effects are relatively reduced and limited.

Analyzing the compressive strength and axial shrinkage for the clay composites S1–S5, in relation to their density, it was observed that these parameters show a progression that can be mathematically expressed through functions of density (f(density)), as shown in Figure 13, with a sufficiently good accuracy indicated by the R^2 factor being greater than 0.9. Both compressive strength and axial shrinkage can be mathematically modeled using polynomial functions. In both cases, the existence of these functions allowing mathematical reproducibility with a high degree of accuracy is considered a useful tool for designing subsequent clay compositions.

Identifying mathematical functions for adhesion to the support in relation to density or compressive strength of clay composites is not as straightforward. In this case, mathematical functions satisfying the $R^2 > 0.9$ condition are polynomial functions of degree 3 or higher, indicating reduced accuracy and challenges in modeling the phenomenon. This implies that numerous factors influence the adhesion parameter to the support, and it does not solely depend on the clay composite's characteristics, as initially anticipated.

The key characteristics of plaster mortars involve their ability to adhere strongly to substrates and resist fissuring. A good plaster mortar should maintain a solid bond with the underlying surface while remaining fissure-free, ensuring durability and a visually appealing finish. These properties are crucial for the performance and longevity of plastered surfaces in construction applications. Regarding the adhesion to the clay composite masonry element substrate, Figure 14 shows an undeniable increase in this parameter with the introduction of dextrin in the composition (S1–S5 vs. PV1–PV4). It is important to note that the compositional design of samples S1–S5 was based on composition PV2. This further highlights the advantages of adding dextrin to the mixture, as it significantly improves the adhesion of the composites to their respective substrates.

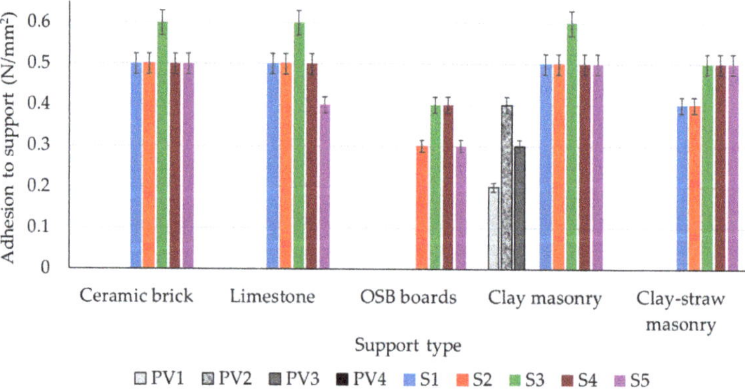

Figure 14. Adhesion to the support of clay mixtures after 40 days of maturation under constant laboratory conditions.

When examining the behavior of clay composites that incorporate dextrin (S1–S5), it becomes evident that composition S3 stands out, exhibiting the best performance irrespective of the underlying support layer. Concurrently, the type of support appears to play a role in influencing the adhesion of the clay composite. Different support materials might lead to varying levels of adhesion, which could affect the overall performance and stability of the clay composite when applied.

To enhance adhesion to the support, before the clay composites were applied, all support surfaces were pre-treated with an 8% aqueous solution of bone glue. Despite this treatment, the oriented strand board (OSB) surface seems to pose the greatest challenge when it comes to compatibility with clay-based plaster. This varying behavior of clay plasters in relation to different support types may be due to specific properties of the support surface, such as water absorption capacity and surface roughness. This hypothesis can also be supported by analyzing the presence or absence of fissures that appeared at the end of the drying period on the surface of clay plasters.

A visual analysis of the surface appearance of various clay composites, after drying, is provided, which can be found in Table 10. This analysis offers insights into the texture and appearance of these composites, which can clarify their suitability for different applications and their overall performance in terms of aesthetics and functionality. It can be observed that there are cases of micro-fissuring, fissuring, and cases where fissures or even detachment from the surface are identified (composition S1 on OSB support), depending on the type of composition and the nature of the support, but there are also cases where the dried plaster surface is smooth, without traces of fissuring or other defects. Composition S3 is identified as having the best compatibility with all analyzed supports. Figure 15 shows exemplary images of the samples made by applying clay composites S1–S5 on different support layers.

Table 10. Visual analysis of the clay composite layer applied to the support, after drying.

Clay Composite Code	The Appearance of Plaster Layer after Drying				
	Ceramic Brick	Limestone	OSB	Masonry Element Made of Unfired Clay	Masonry Element Made of Clay and Cereal Straw
S1	no fissures	no fissures	detachment	no fissures	fissures
S2	no fissures	no fissures	fissures	no fissures	fissures
S3	smooth appearance, without fissure	smooth appearance, without fissures	smooth appearance, without fissures	smooth appearance, without fissures	no fissures
S4	no fissures	micro-fissures	fissures	micro-fissures	micro-fissures
S5	fissures	fissures	Major fissures, cracks	fissures	deep fissures

Figure 15. Representative images presenting the application of clay composite S1–S5 on various support materials, illustrating their compatibility and appearance after drying. (**a**) Ceramic brick support; (**b**) limestone support; (**c**) clay support; (**d**) support made of composite material based on clay and cereal straw; (**e**) OSB support.

Examining the adherence performance of clay composites S1–S5 to various supports (as depicted in Figure 14), we find that the results are rather promising when compared with benchmarks found in the literature (Table 11). The minimum recorded value was 0.3 N/mm^2 and the maximum was 0.6 N/mm^2, suggesting the suitability of these composites for both finishing and decoration purposes in construction. When cross-reference the obtained data with other studies, as represented in Table 11, it is noticed that both limestone sludge and, more notably, hydraulic lime result in reduced axial contractions. These contractions are within the 12% limit suggested by the literature. The use of hydraulic lime, and particularly fly ash, in clay composites enhances mechanical strength, both in terms of compression and bending. Regarding the density of the clay composite, it is slightly lower than average but still aligns with references from the literature. Given these data, we can conclude that our results are reasonably in line with those from other studies. However, we must take into account two important considerations:

- the experimental findings cited in scholarly sources are influenced by various factors. These include the clay's compositional, oxide and mineralogical characteristics, along with the type and quantity of added materials [13,18,21,26,29,41,67,81]. These characteristics can vary based on the region where the clay is sourced. Hence, any comparison of these results should be general, focusing on the overall trends in physical and mechanical properties, rather than exact numerical similarities;
- From the point of view of adherence to the support layer of the composite material based on clay, performance improvements can be achieved. This improvement was notably demonstrated in the later developed composites, namely S1–S5.

Table 11. Benchmarking against previous research findings.

Bibliographic Reference/Sample Code		Axial Shrinkage [%]	Compression Strength [N/mm^2]	Flexural Strength [N/mm^2]	Density [kg/m^3]	Adherence to Support [N/mm^2]
Clay from Dallgow-Döberitz area, Germany [13]		15.0	3.30	0.63	1748	-
Clay from Cluj-Napoca area, Romania [19]		-	2.40	-	1960	-
Four clay types from Toulouse area, France [25]		1.5–2.1	1.30–2.10	0.49–0.64	-	0.06–0.8
Clay from Albi, France [25]		2.5	1.70	0.57	-	0.06
Clay from Poland [37]		2.2	1.34–1.50	0.49–0.58	1802–1853	
Clay from Oluvil, Sri Lanka [41]		-	2.8-8.9		-	0.15-0.44
Clay from Chom Thong, Thailand [82]		17.1	0.32	-	1050	-
Clay from Middle Belt, Nigeria [83]		25.0	10.2	-	1700	-
Clay from Guimaraes, Portugal [84]		-	1.50	-	1748	0.10
Clay from Bath area, UK [85]		-	2.50		1960	0.11–0.28
Sandy-loam soils from Kôdéni, Burkina Faso [86]		3.13	1.80	0.57	1720	-
Clay used for the developed composites		9.20	4.30	1.55	1600	-
Composites PS2–PS5		8.97–9.05	-	-	-	-
Composites PA2–PA5		-	6.95–7.80	1.55–1.83	-	-
Composites PV1–PV4		2.70–6.40	2.1–5.8	0.9–1.8	1190–1606.7	0.0–0.4
Composites S1–S5	Masonry element made of unfired clay	6.3–6.6	2.6–4.1	1.2–1.5	1610–1623	0.5–0.6
	Ceramic brick					0.5–0.6
	Masonry element made of clay and cereal straw					0.4–0.5
	Limestone					0.4–0.6
	OSB					0.0–0.4

4. Conclusions

The experimental findings are remarkable and practical because the research focused on identifying and highlighting the impact of alternative or additive raw materials, such as clay, and by-products from other industries on the performance of clay composites.

The experimental research investigated the potential of using clay composites as plasters for various surfaces commonly found in eco-traditional construction. These surfaces included ceramic brick, limestone, OSB boards, masonry elements made of clay, and masonry elements made of composite material based on clay and cereal straw. Based on this research, the following observations can be made:

- The research found that adding limestone sludge to clay helped reduce axial shrinkage. For the best results, the recommended ratio of limestone sludge to clay was 1:2 by mass;
- It is demonstrated that adding fly ash to a clay composite resulted in increased mechanical strengths, particularly compressive strength. For the best results, the recommended ratio of fly ash to clay was 1:2 by mass;
- Utilizing hydraulic lime as a partial substitute for clay leads to a decrease in the apparent density of dried clay composites, axial shrinkage, and mechanical strengths. This reduction is more significant when a larger amount of clay is replaced by hydraulic lime. Additionally, this partial substitution has a positive impact on minimizing fissures formation and enhancing adhesion to the substrate. As a result of the experimental findings, the preliminary composition PV2 was determined to be the most appropriate for designing and developing dextrin-containing clay composites;
- The addition of dextrin to clay composites with fly ash, limestone sludge, and hydraulic lime resulted in a higher apparent density of the hardened mortar. This increase surpassed the minimum threshold of 1600 kg/m^3, which is considered to provide good thermal resistance according to the specialized literature. Furthermore, clay composites with dextrin exhibited axial shrinkage below the maximum threshold of 12% mentioned in the literature. These composites also showed improved mechanical strengths and better adhesion compatibility to the support. Composition S3 demonstrated the most favorable outcomes, displaying a smooth and fissure-free surface after drying.

The experimental research, on the one hand, demonstrates the possibility of making clay-based plastering mortars even at an industrial level. On the other hand, it underlines the importance of a detailed analysis of the clayey raw material whose characteristics differ depending on the place of extraction. Therefore, each case must adapt and customize the recipe and the production process. These observations and the need for customized design need to be extended to fly ash, whose characteristics differ from one power plant to another.

Author Contributions: Conceptualization, A.H., C.P., A.A.C., G.C. and A.B.; methodology, A.H. and G.C.; formal analysis, C.P., A.A.C. and A.B.; investigation, A.H., C.P., A.A.C., G.C. and A.B.; data curation, A.H. and C.P.; writing—original draft preparation, A.H. and C.P.; writing—review and editing, C.P. and A.A.C.; project administration, A.A.C. All authors have read and agreed to the published version of the manuscript.

Funding: This research was funded by the Romanian Government Ministry of Research Innovation and Digitization, project No. PN 23 35 03 01 "Integrated system of development and scientific research of constructions and vital infrastructures to extreme environmental, seismic and climatic actions and the exploitation of sustainable resources of materials and energy".

Institutional Review Board Statement: Not applicable.

Informed Consent Statement: Not applicable.

Data Availability Statement: Not applicable.

Conflicts of Interest: The authors declare no conflict of interest.

References

1. Sandu, I.; Deak, G.; Ding, Y.; Ivashko, Y.; Sandu, A.V.; Moncea, M.-A.; Sandu, I.G. New Materials for Finishing of Ancient Monuments and Process of Obtaining and Applying. *Int. J. Conserv. Sci.* **2021**, *12*, 1249–1258.
2. Altmäe, E.; Ruus, A.; Raamets, J.; Tungel, E. Determination of Clay-Sand Plaster Hygrothermal Performance: Influence of Different Types of Clays on Sorption and Water Vapour Permeability. In *Cold Climate HVAC 2018: Sustainable Buildings in Cold Climates*; Springer International Publishing: Cham, Switzerland, 2019; pp. 945–955. [CrossRef]
3. Lagouin, M.; Laborel-Préneron, A.; Magniont, C.; Aubert, J.-E. Development of a high clay content earth plaster. *IOP Conf. Ser. Mater. Sci. Eng.* **2019**, *660*, 012068. [CrossRef]
4. Cascione, V.; Lim, D.; Maskell, D.; Shea, A.; Walker, P. The response of clay plaster to temperature and RH sinusoidal variations. *MATEC Web Conf.* **2019**, *282*, 02005. [CrossRef]
5. Vares, O.; Ruus, A.; Raamets, J.; Tungel, E. Determination of hygrothermal performance of clay-sand plaster: Influence of covering on sorption and water vapour permeability. *Energy Procedia* **2017**, *132*, 267–272. [CrossRef]
6. Rajapaksha, M.; Halwatura, P.D.R. Dynamic, Clay, Secondary Walling for Heat Reduction, in Tropical Indoors. *Engineer* **2022**, *55*, 89–98. [CrossRef]
7. El Fgaier, F.; Lafhaj, Z.; Chapiseau, C.; Antczak, E. Effect of sorption capacity on thermo-mechanical properties of unfired clay bricks. *J. Build. Eng.* **2016**, *6*, 86–92. [CrossRef]
8. Kumar, D.; Alam, M.; Doshi, A.J. Investigating the Influence of Thermal Conductivity and Thermal Storage of Lightweight Concrete Panels on the Energy and Thermal Comfort in Residential Buildings. *Buildings* **2023**, *13*, 720. [CrossRef]
9. Jiang, Y.; Phelipot-Mardele, A.; Collet, F.; Lanos, C.; Lemke, M.; Ansell, M.; Lawrence, M. Moisture buffer, fire resistance and insulation potential of novel bio-clay plaster. *Constr. Build. Mater.* **2020**, *244*, 118353. [CrossRef]
10. Su, B.; Jadresin Milic, R.; McPherson, P.; Wu, L. Thermal performance of school buildings: Impacts beyond thermal comfort. *Int. J. Environ. Res. Public Health* **2022**, *19*, 5811. [CrossRef]
11. Sun, H.; Calautit, J.K.; Jimenez-Bescos, C. Examining the regulating impact of thermal mass on overheating, and the role of night ventilation, within different climates and future scenarios across China. *Clean. Eng. Technol.* **2022**, *9*, 100534. [CrossRef]
12. Yu, S.; Hao, S.; Mu, J.; Tian, D. Optimization of Wall Thickness Based on a Comprehensive Evaluation Index of Thermal Mass and Insulation. *Sustainability* **2022**, *14*, 1143. [CrossRef]
13. Minke, G. *Building with Earth: Design and Technology of a Sustainable Architecture*, 4th ed.; Birkhäuser Verlag GmbH: Basel, Switzerland, 2021.
14. McGregor, F.; Heath, A.; Shea, A.; Lawrence, M. The moisture buffering capacity of unfired clay masonry. *Build. Environ.* **2014**, *82*, 599–607. [CrossRef]
15. Nutt, N.; Kubjas, A.; Nei, L. Adding waste paper to clay plaster to raise its ability to buffer moisture. *Proc. Estonian Acad. Sci.* **2020**, *69*, 179–185. [CrossRef]
16. Silveira, D.; Varum, H.; Costa, A.; Martins, T.; Pereira, H.; Almeida, J. Mechanical properties of adobe bricks in ancient constructions. *Constr. Build. Mater.* **2012**, *28*, 36–44. [CrossRef]
17. Bui, Q.-B.; Morel, J.-C.; Reddy, B.V.V.; Ghayad, W. Durability of rammed earth walls exposed for 20 years to natural weathering. *Build. Environ.* **2009**, *44*, 912–919. [CrossRef]
18. Kumar, A. Vernacular practices: As a basis for formulating building regulations for hilly areas. *Int. J. Sustain. Built Environ.* **2013**, *2*, 183–192. [CrossRef]
19. Cioca, G.T.C.; Horvath, I.B. The new vernacular based architecture. *J. Appl. Eng. Sci.* **2011**, *1*, 27–34.
20. Cioca, G.T.C.; Horvath, I.B. Modular Building Using Rammed Earth. *Acta Tech. Napoc. Civ. Eng. Archit.* **2012**, *55*, 173–181.
21. Bugini, R.; Corti, C.; Folli, L.; Rampazzi, L. Roman Wall Paintings: Characterisation of Plaster Coats Made of Clay Mud. *Heritage* **2021**, *4*, 889–905. [CrossRef]
22. Diviš, J.; Růžička, J. The Influence of Clay Structures to the Hygrothermal Component of the Indoor Environment. *Materials* **2022**, *15*, 1744. [CrossRef]
23. Darling, E.; Corsi, R.L. Field-to-laboratory analysis of clay wall coatings as passive removal materials for ozone in buildings. *Indoor Air* **2017**, *27*, 658–669. [CrossRef]
24. Darling, E.K.; Cros, C.J.; Wargocki, P.; Kolarik, J.; Morrison, G.C.; Corsi, R.L. Impacts of a clay plaster on indoor air quality assessed using chemical and sensory measurements. *Build. Environ.* **2012**, *57*, 370–376. [CrossRef]
25. Delinière, R.; Aubert, J.E.; Rojat, F.; Gasc-Barbier, M. Physical, mineralogical and mechanical characterization of ready-mixed clay plaster. *Build. Environ.* **2014**, *80*, 11–17. [CrossRef]
26. Suciu, M.C.; Suciu, N. Dezvoltarea sustenabila—Problema cheie a secolului XXI. *AGIR Bullet.* **2007**, *1*, 124–125.
27. Moquin, M. Ancient Solutions for Future Sustainability: Building with Adobe, Rammed Earth, and Mud. In Proceedings of the First International Conference of CIB TG 16, Tampa, FL, USA, 6–9 November 1994.
28. Bui, Q.B.; Morel, J.C.; Hans, S.; Meunier, S. Compression behaviour of non-industrial materials in civil engineering by three scale experiments: The case of rammed earth. *Mater. Struct.* **2009**, *42*, 1101–1116. [CrossRef]
29. Kiroff, L.; Roedel, H. Sustainable Construction Technologies: Earth Buildings in New Zealand. In Proceedings of the Second International Conference of Sustainable Construction Materials and Technologies, Ancona, Italy, 29–30 June 2010; Volume 1.
30. Vural, N.; Vural, S.; Engin, N.; Sumerkan, M.R. Eastern Black Sea Region. A sample of modular design in the vernacular architecture. *Build. Environ.* **2007**, *42*, 2746–2761. [CrossRef]

31. Jayasinghe, C.; Kamaladasa, N. Compressive strength characteristics of cement stabilized rammed earth walls. *Constr. Build. Mater.* **2007**, *21*, 1971–1976. [CrossRef]

32. Lazarescu, A.; Szilagyi, H.; Baera, C.; Hegyi, A. *Betoane Alternative-Betonul Geopolimer. Cercetari si Oportunitati Emergente*; Editura Napoca Star: Cluj-Napoca, Romania, 2020; pp. 144–152.

33. Feuerborn, H.-J.; Harris, D.; Heidrich, C. Global Aspects on Coal Combustion Products. In Proceedings of the EUROCOALASH 2019 Conference, Dundee, UK, 10–12 June 2019; University of Dundee: Dundee, UK, 2019; pp. 1–17.

34. Calatan, G.; Hegyi, A.; Dico, C.; Szilagyi, H. Opportunities regarding the use of adobe-bricks within contemporary architecture. *Procedia Manuf.* **2020**, *46*, 150–157. [CrossRef]

35. Călătan, G.; Hegyi, A.; Grebenisan, E.; Mircea, A.C. Possibilities of Recovery of Industrial Waste and by-Products in Adobe-Brick-Type Masonry Elements. *Proceedings* **2020**, *63*, 1. [CrossRef]

36. La Noce, M.; Lo Faro, A.; Sciuto, G. Clay-Based Products Sustainable Development: Some Applications. *Sustainability* **2021**, *13*, 1364. [CrossRef]

37. Brzyski, P.; Grudzińska, M. Influence of Linseed Oil Varnish Admixture on Glauconite Clay Mortar Properties. *Materials* **2020**, *13*, 5487. [CrossRef] [PubMed]

38. Andres, D.M.; Manea, D.L.; Fechete, R.; Jumate, E. Green plastering mortars based on clay and wheat straw. *Procedia Technol.* **2016**, *22*, 327–334. [CrossRef]

39. *STAS 1913-5*; Foundation Soil. Determination of Particle Size Distribution. (In Romanian: Teren de Fundare. Determinarea Granulozității). The National Organization for Standardization from Romania ASRO: Bucharest, Romania, 1985.

40. Bleam, W.F. Clay Mineralogy and Clay Chemistry. In *Soil and Environmental Chemistry*; Elsevier: Amsterdam, The Netherlands, 2012; pp. 85–116.

41. Thamboo, J.A. Material characterisation of thin layer mortared clay masonry. *Constr. Build. Mater.* **2020**, *230*, 116932. [CrossRef]

42. Chen, R.; Hao, D.X.; Li, D.Z. Mechanical property of coal ash mixed with clay. In *Applied Mechanics and Materials*; Trans Tech Publications Ltd.: Zurich, Switzerland, 2013; Volume 357, pp. 854–857. [CrossRef]

43. Da Silva, S.R.; Andrade, J.J.D.O. A review on the effect of mechanical properties and durability of concrete with construction and demolition waste (CDW) and fly ash in the production of new cement concrete. *Sustainability* **2022**, *14*, 6740. [CrossRef]

44. Jaskulski, R.; Jóźwiak-Niedźwiedzka, D.; Yakymechko, Y. Calcined clay as supplementary cementitious material. *Materials* **2020**, *13*, 4734. [CrossRef] [PubMed]

45. Kwon, Y.H.; Kang, S.H.; Hong, S.G.; Moon, J. Acceleration of intended pozzolanic reaction under initial thermal treatment for developing cementless fly ash based mortar. *Materials* **2017**, *10*, 225. [CrossRef]

46. Sun, X.J.; Zhao, Y.S.; Sang, Z.W. Influence of Fly Ash on Clay Liner. In *Advanced Materials Research*; Trans Tech Publications Ltd.: Zurich, Switzerland, 2012; Volume 518, pp. 2543–2546. [CrossRef]

47. Turan, C.; Javadi, A.A.; Vinai, R.; Russo, G. Effects of Fly Ash Inclusion and Alkali Activation on Physical, Mechanical, and Chemical Properties of Clay. *Materials* **2022**, *15*, 4628. [CrossRef]

48. Turan, C.; Javadi, A.A.; Vinai, R. Effects of Class C and Class F Fly Ash on Mechanical and Microstructural Behavior of Clay Soil—A Comparative Study. *Materials* **2022**, *15*, 1845. [CrossRef]

49. Wiśniewski, K.; Rutkowska, G.; Jeleniewicz, K.; Dąbkowski, N.; Wójt, J.; Chalecki, M.; Siwiński, J. The Impact of Fly Ashes from Thermal Conversion of Sewage Sludge on Properties of Natural Building Materials on the Example of Clay. *Sustainability* **2022**, *14*, 6213. [CrossRef]

50. Sandu, A.V. Obtaining and Characterization of New Materials. *Materials* **2021**, *14*, 6606. [CrossRef]

51. Jamaludin, L.; Razak, R.A.; Abdullah, M.M.A.B.; Vizureanu, P.; Bras, A.; Imjai, T.; Sandu, A.V.; Abd Rahim, S.Z.; Yong, H.C. The Suitability of Photocatalyst Precursor Materials in Geopolymer Coating Applications: A Review. *Coatings* **2022**, *12*, 1348. [CrossRef]

52. Zailan, S.N.; Mahmed, N.; Abdullah, M.M.A.B.; Rahim, S.Z.A.; Halin, D.S.C.; Sandu, A.V.; Vizureanu, P.; Yahya, Z. Potential Applications of Geopolymer Cement-Based Composite as Self-Cleaning Coating: A Review. *Coatings* **2022**, *12*, 133. [CrossRef]

53. Azimi, E.A.; Abdullah, M.M.A.B.; Vizureanu, P.; Salleh, M.A.A.M.; Sandu, A.V.; Chaiprapa, J.; Yoriya, S.; Hussin, K.; Aziz, I.H. Strength Development and Elemental Distribution of Dolomite/Fly Ash Geopolymer Composite under Elevated Temperature. *Materials* **2020**, *13*, 1015. [CrossRef] [PubMed]

54. Luhar, I.; Luhar, S.; Abdullah, M.M.A.B.; Razak, R.A.; Vizureanu, P.; Sandu, A.V.; Matasaru, P.-D. A State-of-the-Art Review on Innovative Geopolymer Composites Designed for Water and Wastewater Treatment. *Materials* **2021**, *14*, 7456. [CrossRef] [PubMed]

55. Alnuaim, A.; Dafalla, M.; Al-Mahbashi, A. Enhancement of Clay–Sand Liners Using Crushed Limestone Powder for Better Fluid Control. *Arabian J. Sci. Eng.* **2020**, *45*, 367–380. [CrossRef]

56. Courard, L.; Michel, F.; Pierard, J. Influence of clay in limestone fillers for self-compacting cement based composites. *Constr. Build. Mater.* **2011**, *25*, 1356–1361. [CrossRef]

57. Ibrahim, H.H.; Alshkane, Y.M.; Mawlood, Y.I.; Noori, K.M.G.; Hasan, A.M. Improving the geotechnical properties of high expansive clay using limestone powder. *Innov. Infrastruct. Solut.* **2020**, *5*, 112. [CrossRef]

58. Abood, T.T.; Kasa, A.B.; Chik, Z.B. Stabilization of silty clay soil using chloride compounds. *J. Eng. Sci. Technol.* **2007**, *2*, 102–110.

59. Barman, D.; Dash, S.K. Stabilization of expansive soils using chemical additives: A review. *J. Rock Mech. Geotech. Eng.* **2022**, *14*, 1319–1342. [CrossRef]

60. Musso, G.; Scelsi, G.; Della Vecchia, G. Chemo-Mechanical Behaviour of Non-Expansive Clays Accounting for Salinity Effects. *Géotechnique*, 2022; *ahead of print*. [CrossRef]
61. Ying, Z.; Cui, Y.J.; Benahmed, N.; Duc, M. Salinity effect on the compaction behaviour, matric suction, stiffness and microstructure of a silty soil. *J. Rock Mech. Geotech. Eng.* **2021**, *13*, 855–863. [CrossRef]
62. Alhokabi, A.; Hasan, M.; Amran, M.; Fediuk, R.; Vatin, N.I.; Alshaeer, H. The Effect of POFA-Gypsum Binary Mixture Replacement on the Performance of Mechanical and Microstructural Properties Enhancements of Clays. *Materials* **2022**, *15*, 1532. [CrossRef]
63. Jiang, P.; Zhou, L.; Zhang, W.; Wang, W.; Li, N. Unconfined compressive strength and splitting tensile strength of lime soil modified by nano clay and polypropylene fiber. *Crystals* **2022**, *12*, 285. [CrossRef]
64. Kamaruddin, F.A.; Anggraini, V.; Kim Huat, B.; Nahazanan, H. Wetting/drying behavior of lime and alkaline activation stabilized marine clay reinforced with modified coir fiber. *Materials* **2020**, *13*, 2753. [CrossRef] [PubMed]
65. Maheri, M.R.; Maheri, A.; Pourfallah, S.; Azarm, R.; Hadjipour, A. Improving the durability of straw-reinforced clay plaster cladding for earthen buildings. *Int. J. Archit. Herit.* **2011**, *5*, 349–366. [CrossRef]
66. Meddah, A.; Goufi, A.E.; Pantelidis, L. Improving Very High Plastic Clays with the Combined Effect of Sand, Lime, and Polypropylene Fibers. *Appl. Sci.* **2022**, *12*, 9924. [CrossRef]
67. Onyelowe, K.; Van, D.B.; Igboayaka, C.; Orji, F.; Ugwuanyi, H. Rheology of mechanical properties of soft soil and stabilization protocols in the developing countries-Nigeria. *Mater. Sci. Energy Technol.* **2019**, *2*, 8–14. [CrossRef]
68. Schwantes, G.; Dai, S.B. Preliminary results for using micro-lime-clay soil grouts for plaster reattachment on earthen support. In *Structural Analysis of Historical Constructions: Anamnesis, Diagnosis, Therapy, Controls*; CRC Press: Boca Raton, FL, USA, 2016; pp. 1392–1398.
69. Shaker, A.A.; Al-Shamrani, M.A.; Moghal, A.A.B.; Vydehi, K.V. Effect of confining conditions on the hydraulic conductivity behavior of fiber-reinforced lime blended semiarid soil. *Materials* **2021**, *14*, 3120. [CrossRef] [PubMed]
70. Shirmohammadi, S.; Ghaffarpour Jahromi, S.; Payan, M.; Senetakis, K. Effect of lime stabilization and partial clinoptilolite zeolite replacement on the behavior of a silt-sized low-plasticity soil subjected to freezing–thawing cycles. *Coatings* **2021**, *11*, 994. [CrossRef]
71. Messis, M.; Benaissa, A.; Bouhamou, N. Hydromechanical Characterization of Raw Earth Mortar–Stabilizing Cement and Lime. *J. Appl. Eng. Sci.* **2022**, *12*, 203–208. [CrossRef]
72. Hamard, E.; Morel, J.-C.; Salgado, F.; Marcom, A.; Meunier, N. A procedure to assess the suitability of plaster to protect vernacular earthen architecture. *J. Cult. Herit.* **2013**, *14*, 109–115. [CrossRef]
73. Santos, T.; Nunes, L.; Faria, P. Production of eco-efficient earth-based plasters: Influence of composition on physical performance and bio-susceptibility. *J. Clean. Prod.* **2017**, *167*, 55–67. [CrossRef]
74. Stazi, F.; Nacci, A.; Tittarelli, F.; Pasqualini, E.; Munafò, P. An experimental study on earth plasters for earthen building protection: The effects of different admixtures and surface treatments. *J. Cult. Herit.* **2016**, *17*, 27–41. [CrossRef]
75. *STAS 2634*; Common Mortars for Masonry and Plastering. Test Methods (In Romanian: Mortare Obişnuite Pentru Zidărie şi Tencuieli. Metode de Încercare). The National Organization for Standardization from Romania ASRO: Bucharest, Romania, 1980.
76. *EN 1015-10*; Methods of Test for Mortar for Masonry—Part 10: Determination of Dry Bulk Density of Hardened Mortar. The National Organization for Standardization from Romania ASRO: Bucharest, Romania, 1999.
77. *EN 1015-11*; Methods of Test for Mortar for Masonry—Part 11: Determination of Flexural and Compressive Strength of Hardened Mortar. The National Organization for Standardization from Romania ASRO: Bucharest, Romania, 2019.
78. *EN 1015-12*; Methods of Test for Mortar for Masonry—Part 12: Determination of Adhesive Strength of Hardened Rendering and Plastering Mortars on Substrates. The National Organization for Standardization from Romania ASRO: Bucharest, Romania, 2016.
79. Koňáková, D.; Čáchová, M.; Vejmelkova, E.; Keppert, M.; Jerman, M.; Bayer, P.; Rovnaníková, P.; Černý, R. Lime-based plasters with combined expanded clay-silica aggregate: Microstructure, texture and engineering properties. *Cem. Concr. Compos.* **2017**, *83*, 374–383. [CrossRef]
80. *EN 998-1*; Specification for Mortar for Masonry—Part 1: Rendering and Plastering Mortar. The National Organization for Standardization from Romania ASRO: Bucharest, Romania, 2016.
81. Lima, J.; Faria, P.; Santos Silva, A. Earth plasters: The influence of clay mineralogy in the plasters' properties. *Int. J. Archit. Herit.* **2020**, *14*, 948–963. [CrossRef]
82. Lertwattanaruk, P.; Choksiriwanna, J. The physical and thermal properties of adobe brick containing bagasse for earth construction. *Int. J. Build. Urban Inter. Landsc. Technol. (BUILT)* **2011**, *1*, 57–66. [CrossRef]
83. Okunade, E.A. The effect of wood ash and sawdust admixtures on the engineering properties of a burnt laterite-clay brick. *J. Appl. Sci.* **2008**, *8*, 1042–1048. [CrossRef]
84. Faria, P.; Santos, T.; Silva, V. Earth-based mortars for masonry plastering. In Proceedings of the 9th International Masonry Conference, Guimarães, Portugal, 7–9 July 2014.
85. Lawrence, M.; Heath, A.; Walker, P. Mortars for thin unfired clay masonry walls. In Proceedings of the LEHM 5th International Conference on Building with Earth, Koblenz, Germany, 9–12 October 2008; pp. 66–73.
86. Bamogo, H.; Ouedraogo, M.; Sanou, I.; Aubert, J.E.; Millogo, Y. Physical, Hydric, Thermal and Mechanical Properties of Earth Renders Amended with Dolomitic Lime. *Materials* **2022**, *15*, 4014. [CrossRef]

Article

Fabrication of Thermal Insulation Bricks Using *Pleurotus florida* Spent Mushroom

Sally A. Ali [1], Marwa Kamal Fahmy [2], Nasser Zouli [3,*], Ahmed Abutaleb [3], Ibrahim M. Maafa [3], Ayman Yousef [3,4,*] and M. M. Ahmed [4]

[1] Department of Botany and Microbiology, Faculty of Science, Helwan University, Cairo 11795, Egypt; sally_ali@science.helwan.edu.eg

[2] Department of Architecture, Faculty of Engineering at Mataria, Helwan University, Cairo 11718, Egypt; marwa_fakhry@m-eng.helwan.edu.eg

[3] Department of Chemical Engineering, Faculty of Engineering, Jazan University, Jazan 45142, Saudi Arabia; azabutaleb@jazanu.edu.sa (A.A.); imoaafa@jazanu.edu.sa (I.M.M.)

[4] Department of Mathematics and Physics Engineering, Faculty of Engineering at Mataria, Helwan University, Cairo 11718, Egypt; marwa_elnagar77@yahoo.com

* Correspondence: nizouli@jazanu.edu.sa (N.Z.); aymanyousef84@gmail.com (A.Y.)

Abstract: This study explores the potential for making lightweight bricks via the use of dry, pulverized spent mushroom materials (SMM) as a thermal insulator. There are five distinct replacement proportions of SMM that are used, and they range from 0% to 15% of the weight of the clay. The firing of the fabricated bricks at temperatures of 700, 800, and 900 °C led to the development of pores on the interior surface of the bricks as a consequence of the decomposition of SMM. The impact of SMM on the physicomechanical characteristics of fabricated bricks is assessed based on standard codes. Compressive strength, bulk density, and thermal conductivity decreased as the SMM content increased, reaching up to 8.7 MPa, 1420 kg/m^3, and 0.29 W/mK at 900 °C and 15% substitution percentage. However, cold water absorption, boiling water absorption, linear drying shrinkage, linear firing shrinkage, and apparent porosity increased with the increase in SMM, reaching 23.6%, 25.3%, and 36.6% at 900 °C and 15% substitution percentage. In the study simulation model, there was a significant improvement in energy consumption, which reached an overall reduction of 29.23% and 21.49% in Cario and Jazan cities, respectively.

Keywords: lightweight bricks; spent mushroom materials; bio-based materials; thermal insulation; energy efficiency buildings

Citation: Ali, S.A.; Fahmy, M.K.; Zouli, N.; Abutaleb, A.; Maafa, I.M.; Yousef, A.; Ahmed, M.M. Fabrication of Thermal Insulation Bricks Using *Pleurotus florida* Spent Mushroom. *Materials* **2023**, *16*, 4905. https://doi.org/10.3390/ma16144905

Academic Editor: Carlos Leiva

Received: 30 May 2023
Revised: 2 July 2023
Accepted: 3 July 2023
Published: 9 July 2023

1. Introduction

Due to the overexploitation of numerous natural resources, there is a lack of natural resources in the world for manufacturing conventional bricks [1]. Huge amounts of raw materials are consumed in the brick industry [2,3]. To overcome this issue, several attempts have been made to incorporate different waste materials into the brick-making process, including natural fibers, textile laundry wastewater sludge, foundry sand, granite sawing waste, perlite, processed waste tea, sewage sludge, structural glass waste, fly ash, sugar cane bagasse ash, organic residue, steel dust, bottom ash, rice husk ash, silica fume, and municipal solid waste incineration fly ash [4–7]. An innovative biotechnological method for recycling lignocellulosic waste involves growing oyster mushrooms. Mushroom substrates have a limited ability to condition and fertilize soil due to their high salt, nutrient, and alkaline contents [8]. Cellulosic substrates, including cotton wastes, maize cob wastes, bean straws, crushed bagasse, molasses wastes, coffee husks, paper wastes, industrial cardboard wastes, tree sawdust, and rice straws can be used to grow mushrooms. The majority of commercially grown Agaricus mushrooms are grown on straw or hay substrates. Rice straw mushrooms (RSM) are regarded as a nutritious food (USITC 2010) [9,10]. Growing edible mushrooms meets the dual goals of treating rice straw, which acts as a substrate for the production of food (mushrooms), and as a source of food for the mushroom-spent bedding. One of the biggest environmental problems in the mushroom production process that utilizes rice straw as soil is the vast amounts of spent mushroom material (SMM) generated as solid waste by-products—1 kg of mushrooms produces around 5 kg of solid waste [11]. SMM waste is frequently disposed in landfills [12]. About 27 billion kg of cultivated mushrooms were produced worldwide in 2012 [13]. Governments and researchers are examining the potential uses of these leftover materials as a result of careful environmental management of this by-product of mushroom production. Today, around 50% of the population lives in cities, using a lot of energy resources and emitting more than 70% of the world's carbon emissions. This is mainly because of urban development, which has been identified as the cause of issues with sustainability [14]. One approach for addressing this is using sustainable materials such as SMM in the production of construction materials. This has some beneficial effects on the environment and the economy and, in addition to optimizing the thermal performance of the urban block form, is necessary for the management of daily wastes that cause significant environmental problems [15]. Beyond the context of the surrounding, the building's shape and the windows' parameters have an impact on energy consumption [16]. Using sustainable materials with thermally isolated behavior is considered the most effective parameter for achieving sustainability. Inefficient thermal materials, which cause high energy consumption, have environmental and economic consequences during the building's life [17,18]. Hence, avoiding or even reducing as much thermal gain as possible in buildings decreases the internal temperature and improves the indoor thermal quality, thus directly decreasing the need for air conditioning, resulting in a decrease in the energy consumed in the building and the city. The main distinguishing objectives and novelty of this study are managing waste materials by producing lightweight bricks made by replacing clay with SMM materials in various ratios that meet the obligatory values of the physicomechanical characteristics assigned by standards. It was also critical for manufacturing lightweight bricks with effective thermal insulation for controlling energy usage. These bio-based, lightweight bricks can both help with trash disposal and resource conservation because they are made from rice straw waste that has undergone the mushroom biodegradation process. Thus, it makes economic and environmental sense to utilize SMM as a clay body addition. Table S1 contains the physicomechanical and thermal characteristics of bricks substituted with different waste materials.

2. Experimental

2.1. Materials and Methodology

Clay was obtained from Aswan, Egypt, and spent mushroom materials (SMM) was grown in the laboratory. *Pleurotus florida* spawn (culture) was obtained from the Agricultural Research Centre, Ministry of Agriculture, Giza, Egypt.

2.2. Preparation of Spent Mushroom Materials from Cultivated Pleurotus florida

Spent mushroom materials (SMM) of *Pleurotus florida*—also known as the oyster mushroom or the white oyster mushroom—were prepared after cultivation (Figure 1). The SMM was prepared in the laboratory to show the process of mushroom growth. The preparation was carried out following the protocol by Stoknes et al. [19], with slight modifications as follows: To ensure that water completely soaked through the rice straws, the straws were cut into 20–30 cm lengths, soaked in water that was three times as heavy as the straw, left to soak for 24 h, and then air dried. After soaking, the rice straws were placed and sealed in sterile plastic bags. The bags were subjected to a 15-min sterilization process at 121 °C and then cooled at room temperature for 24 h. Each bag was filled with 50 g of *Pleurotus florida* spawn (culture) that had already been prepared and cooled; the bag was then sealed. The inoculated rice straw was incubated at 25 °C with light requirements of 1000–1500 (2000 lux) for not less than 30 days, until fruit body development. The first harvest was carried out after 30–40 days, and the yield during the first harvest was 250 g. The raw materials of this edible fungus, as well as the cultivation and harvesting conditions for spent mushroom material preparation, are shown in Figure 1a–k. After autoclaving for 20 min and drying at 60 °C for 48 h, each piece of residual *Pleurotus florida* spent mushroom material was crushed using a pulverizer (Elaraby, Cairo, Egypt, MX 900/2). Each substrate's powder was then separated into two parts using a sieve (through a mesh with a 2-mm pore size sieve). The sieving process is performed to separate the impurities and ungrounded particles, and only the finer part was used for the experiments.

Figure 1. Preparation of SMM from cultivated *Pleurotus florida* (**a**) chopped rice straws, (**b**) soaked rice straw, (**c**) inoculation of rice straws with 50 g *Pleurotus florida* spawn, (**d**) sealed, inoculated, and sterilized bags, (**e**) fruitful body development, (**f**) first harvest, (**g**) formation of SMM, (**h**) SMM after autoclaving, (**i**) SMM after heat drying, (**j**) A pulverizer, and (**k**) powdered spent mushroom materials in a petri dish.

2.3. Characterization of Clay and Spent Mushroom Materials

In this investigation, X-ray fluorescence (XRF) was used to assess the chemical composition of Aswan clay and SMM after they had been ground into powder using an AXIOS, Panalytical 2005 (Malvern Panalytical, Malvern, UK), and a wavelength dispersive (WD) XRF sequential spectrometer (Malvern Panalytical, Malvern, UK). By utilizing the Brukur (Billerica, MA, USA) D8 advanced computerized X-ray diffractometer apparatus with monochromatized CuK radiation operated at 40 kV and 40 mA, it was possible to establish the mineralogical composition of the raw material. The acquired SMM powder was used directly without further processing. The ASTM C136 standard sieve analytical approach was used to calculate the mean particle size [20].

2.4. Preparation and Testing of Samples

The clay was substituted with six different percentages of SMM ranging from 0% to 15% by weight. In order to obtain a homogenous mixture, the raw components were stirred at a speed of 50 rpm for 10 min. The pastes were produced by adding 18 wt.% of water to each of the aforementioned mixtures to get the desired consistency. They were molded into steel cube molds ($50 \times 50 \times 50$ mm^3) and compressed under 10 MPa pressure. The formed samples were left out in the open for a period of 12 h where they were exposed to direct sunlight. The samples were then de-molded and dried overnight in an oven at a temperature of 120 °C. In the final stage, the samples were sintered at three different temperatures: 700, 800, and 900 °C. The samples were kept in the furnace for a total of 4 h at each temperature. Figure 2 shows the preparation procedure for the bricks. The percentages of linear drying shrinkage for the dried samples were measured. Additionally, linear firing shrinkage, cold and boiling water absorption, apparent porosity, bulk density, compressive strength, and thermal conductivity tests were used to analyze the characteristics of fired samples based on ASTM standard codes and similar to our previous work [3]. The samples were tested in triplicate, and the average value of the parameter under investigation was computed. Figure 3 shows the photo-image of the fabricated bricks.

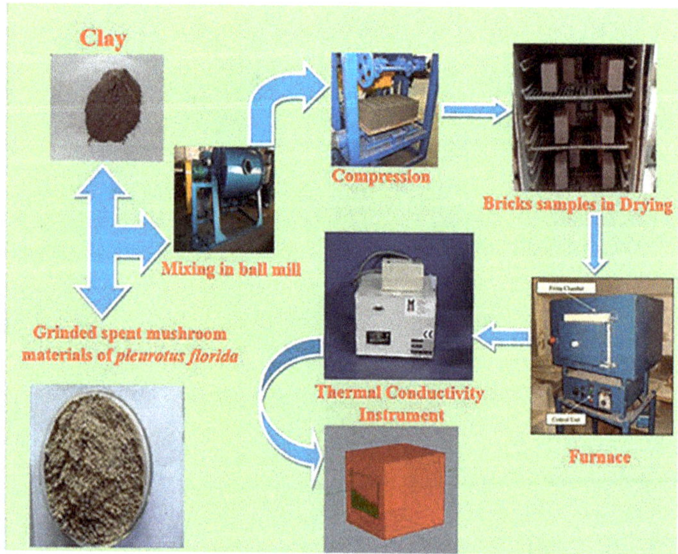

Figure 2. Schematic of the brick production process.

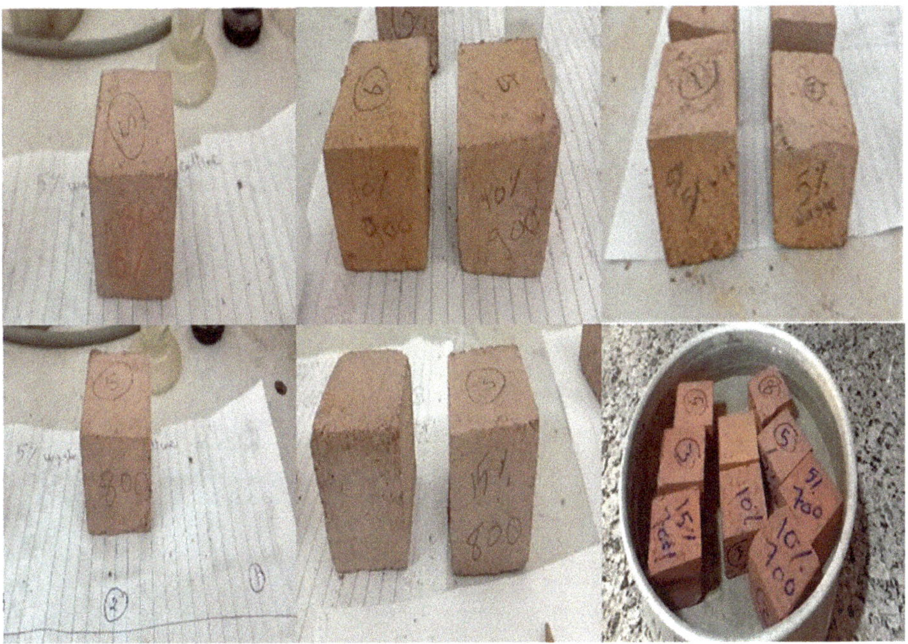

Figure 3. Photo-image of the fabricated bricks.

2.5. Raw Material Characterization

Figure 4 shows the XRD of the raw materials. As seen in the figure, silica and alumina were the two primary compounds present in the clay (Figure 4a), while SMM is composed of sodium aluminum silicate, davidsmithite, trinepheline, sodium aluminum silicate hydroxide, and wollastonite (Figure 4b). Table 1 displays the XRF analysis of the raw materials. The results indicate that SiO_2 and Al_2O_3 are the major compounds in clay, with compositions of between 20% and 50% of SiO_2 and between 10% and 20% of Al_2O_3, which were within the recommended ranges [68, 69].

Table 1. Chemical composition of raw materials.

Composition	Clay wt. (%)	SMM wt. (%)
Al_2O_3	32.906	0.07
SiO_2	48.931	30.2
Na_2O	0.094	0.38
K_2O	0.014	0.66
CaO	0.505	1.58
MgO	0.09	--
TiO_2	5.918	--
Fe_2O_3	1.193	0.1
SO_3	0.291	--
F	--	--
Cl	0.011	--
Cr_2O_3	0.138	--
ZrO_2	0.465	--
LOI	9.2	32
TOTAL	99.756	64.99

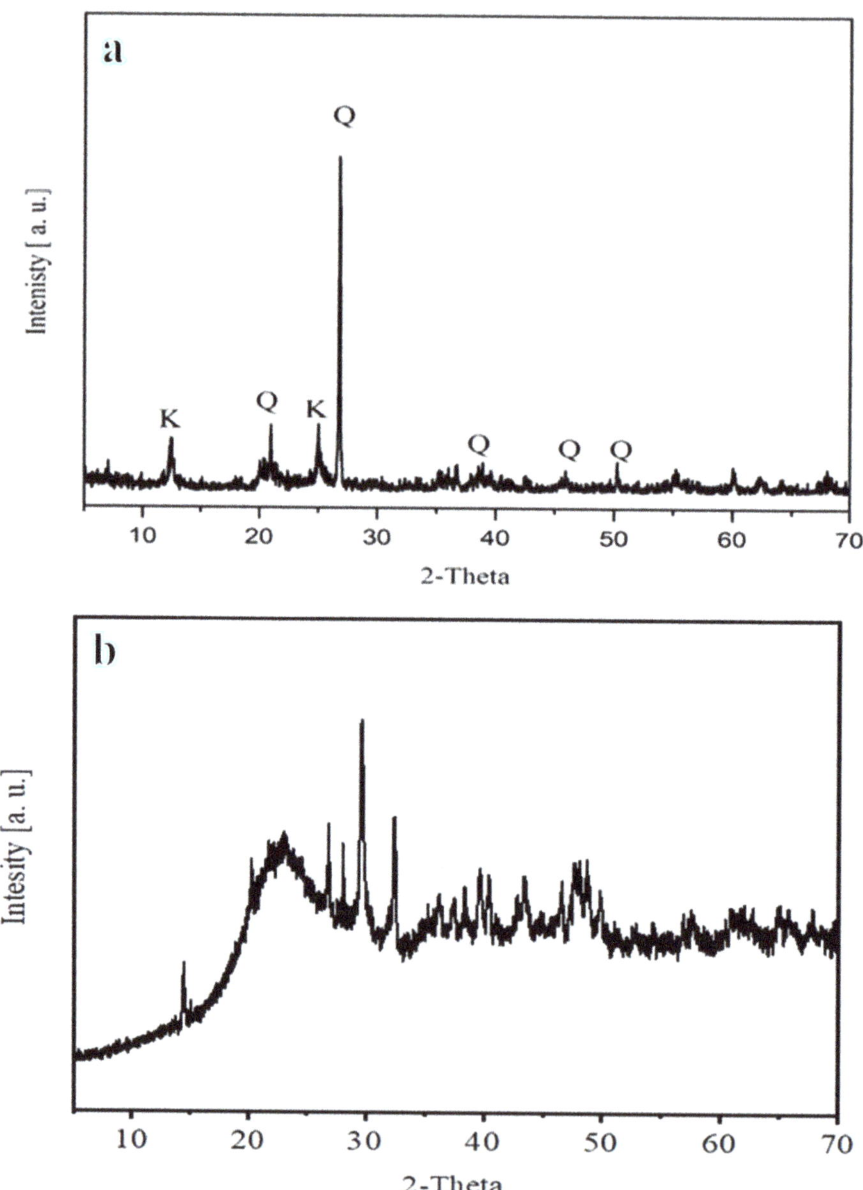

Figure 4. XRD analysis of clay (**a**) and SMM (**b**).

SiO_2 and Al_2O_3 were present in lower quantities in SMM. However, it contained more calcium oxide (CaO) and potassium oxide (K_2O) than clay. Furthermore, its high organic content is thought to be the cause of its significant loss on ignition (LOI) [21]. Kaolin dehydroxylation is considered the primary cause of the LOI and raising thermal conductivity [22]. Figure 5 displays the particle size distribution of the raw materials. This graph demonstrates the differences in their median sizes (D50), which equal 0.26 mm for clay and 0.6 mm for SMM, demonstrating that clay is finer than SMM.

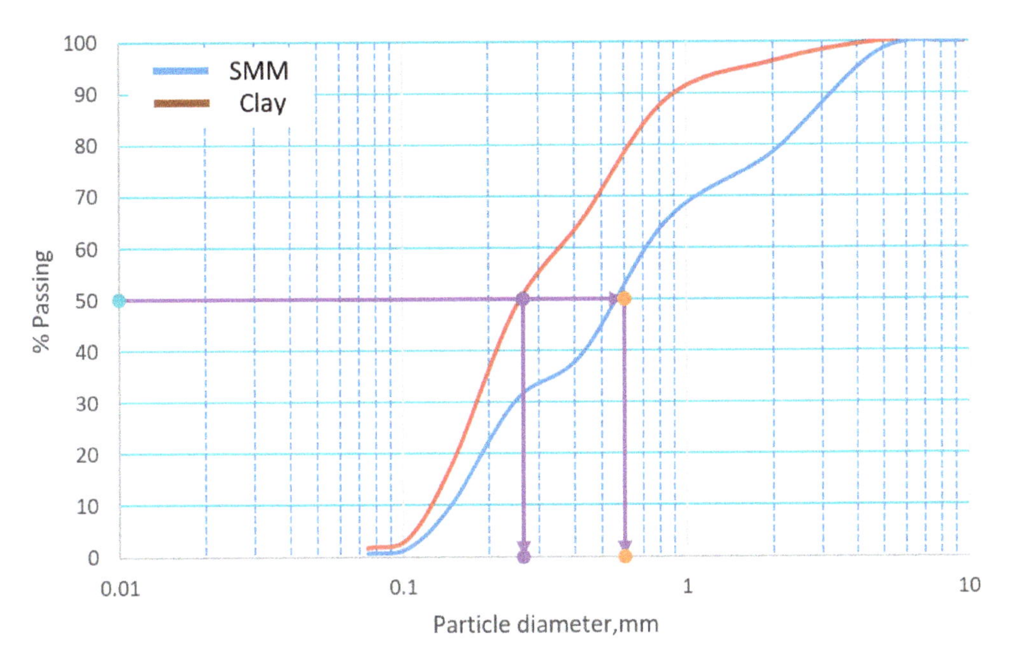

Figure 5. Distribution of the raw materials' particle sizes.

2.6. Simulation Procedure

The study optimized the building design, aiming to achieve optimum thermal performance. First, the model was built in Grasshopper (Figure 6). Then, the thermal model was written in Grasshopper using Honeybee, ladybug, and climate studio (Grasshopper plugins). The plugins work as engines for EnergyPlus, Radiance, Daysim, and OpenStudio [11]. Different materials were identified and inserted into the model as exterior wall materials. The materials were made from Energy Plus 23.1.0 materials. Optimization began with simulation of different types of bricks in a room as a case study. The simulation included studying cooling loads and CO_2 emissions.

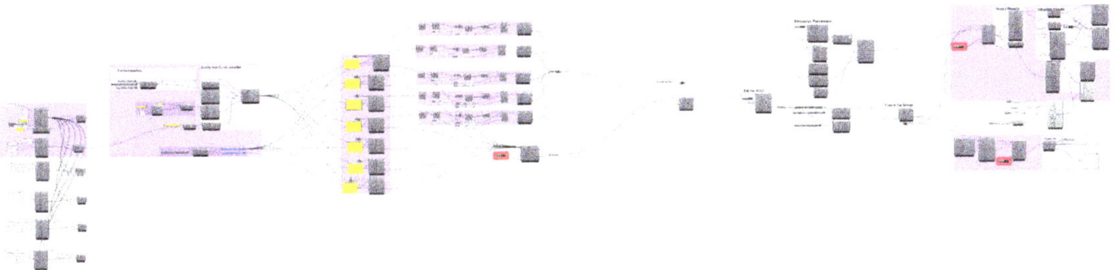

Figure 6. Definition made in Grasshopper.

2.7. Description of the Study Model

The case study consisted of one unit with dimensions of 4.0 m width, 6.0 m length, and 3.0 m height, located in Cairo, Egypt (Figure 7a). Figure 7b shows the average temperature in Cairo, Egypt. The impact of the new wall section made of different fabricated bricks was compared with the base case. The energy saved using different fabricated bricks was calculated. The exterior walls were 0.25 m thick and contained three layers (plaster–brick–plaster). The material used was EnergyPlus Material. For the HVAC system, the Honeybee plugin acted as an engine for EnergyPlus, and the air conditioning system was set to 24 °C. The building operated 24 h/day as it is a residential unit. Energy loads were simulated using Honeybee as an engine for Radiance, Daysim, OpenStudio, and EnergyPlus, and energy performance was evaluated through cooling loads. The building's annual energy consumption (kWh/m^2) was computed for easier comparison between various materials.

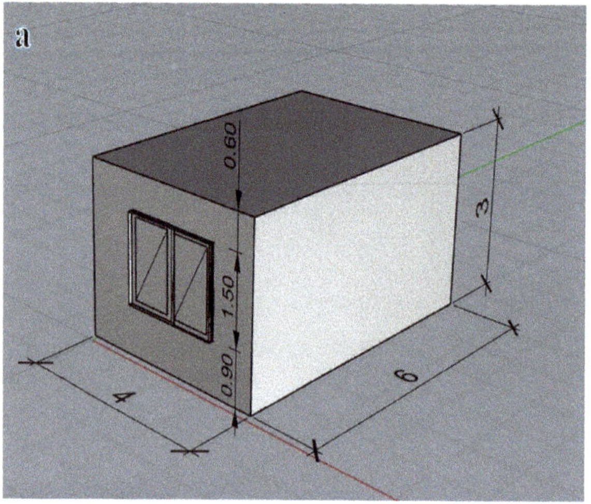
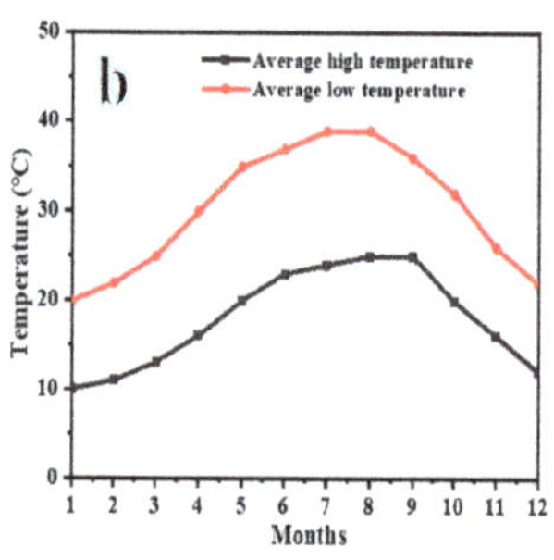

Figure 7. Study model (**a**) and average temperature (**b**).

3. Results and Discussion

3.1. Linear Shrinkage

Due to water loss, the clay particles became compact as a result of shrinkage. The increase in the SMM content led to an increase in linear drying shrinkage (Figure 8a). Compared with the control sample, which displayed a linear drying shrinkage of 3.17%, the sample composed of 15% SMM had a maximum linear drying shrinkage of 4.9%. This effect can be attributed to the reduction in quartz content, which helped to increase the volumetric stability of the mixture [23]. The degree of densification during a fire is often determined in part by shrinkage. SMM particles fuse during sintering at high temperatures, resulting in more closely packed particles that improve linear firing shrinkage. Firing temperature is a crucial factor influencing the degree of shrinkage, and large shrinkage could be problematic since it might result in fissures and dimensional flaws. To reduce shrinkage, the firing temperature must be managed during the sintering process. It is worth mentioning that the mechanical performance of the bricks can be maintained after sintering if the linear firing shrinkage is less than 8%. As shown in Figure 8b, firing shrinkage increases with the increase in the amount of SMM and the firing temperature. The findings revealed that shrinkage in the fired clay brick samples ranged from 6.6% to 7.6% with the addition of 15% SMM, while firing shrinkage in the control fired clay bricks ranged from 0.29% to 1.7% without the addition of any SM; thus, shrinkage is influenced by the concentration of the added SMM. All brick samples' shrinkage after firing fell within the permitted range for commercial manufacturing, as stated in ASTM C326 [24]. However, the linear shrinkage of all the evaluated additions was under 8%. The sample composed of 15 wt.% SMM showed 7.6% linear firing shrinkage at 900 °C.

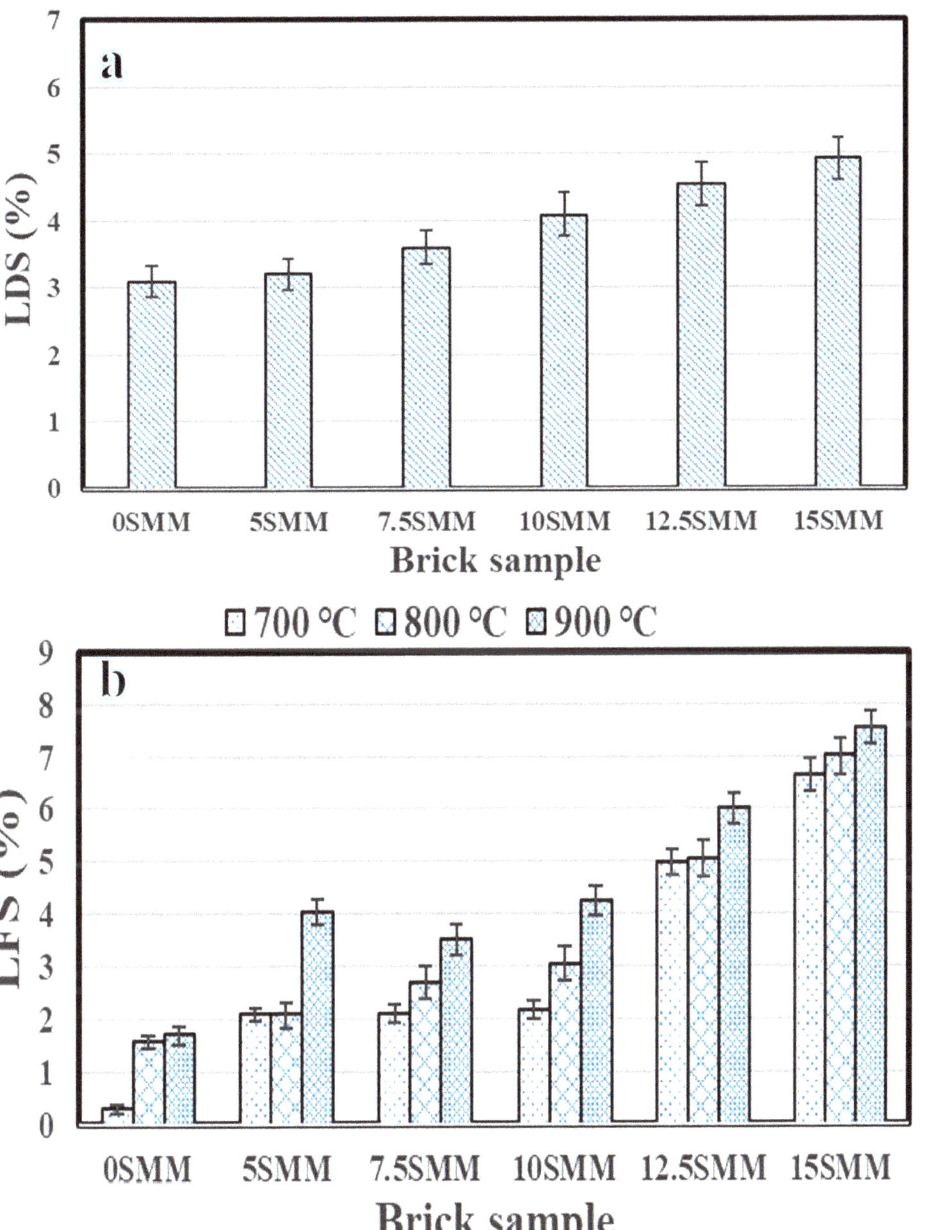

Figure 8. Linear drying shrinkage (a) and firing shrinkage (b).

3.2. Bulk Density

The specific gravity of the raw material source, the manufacturing process, and the firing temperature can all influence the bulk density of fired clay bricks. Figure 9 shows the influence of SMM and firing temperature on the bulk density of the fired clay bricks. The findings indicate that the maximum replacement percentage (15% SMM) produced the lowest bulk density of 1385.2, 1420.8, and 1419.99 kg/m^3 when the samples were subjected to firing temperatures of 700, 800, and 900 °C, respectively. Under the same firing temperatures of 700, 800, and 900 °C, the bulk density was 1828.7, 1874.2, and 1922 kg/m^3, respectively, for the brick composed of 0% SMM. Accordingly, the bulk density reduced as SMM content increased, possibly due to the increased formation of pores that occur as a result of the degradation of carbonaceous matter in SMM during the sintering process [25]. On the other hand, bulk density increased slightly as the firing temperature increased at each percentage of SMM, possibly due to the verification or higher consolidation between body particles during the sintering process. Based on the results of the bulk density measurement, the brick can be categorized as lightweight if it has a density lower than 1680 kg/m^3 in accordance with the requirements of the standard ASTM C90 [26,27]. The bricks met the lightweight standard at all SMM percentages analyzed.

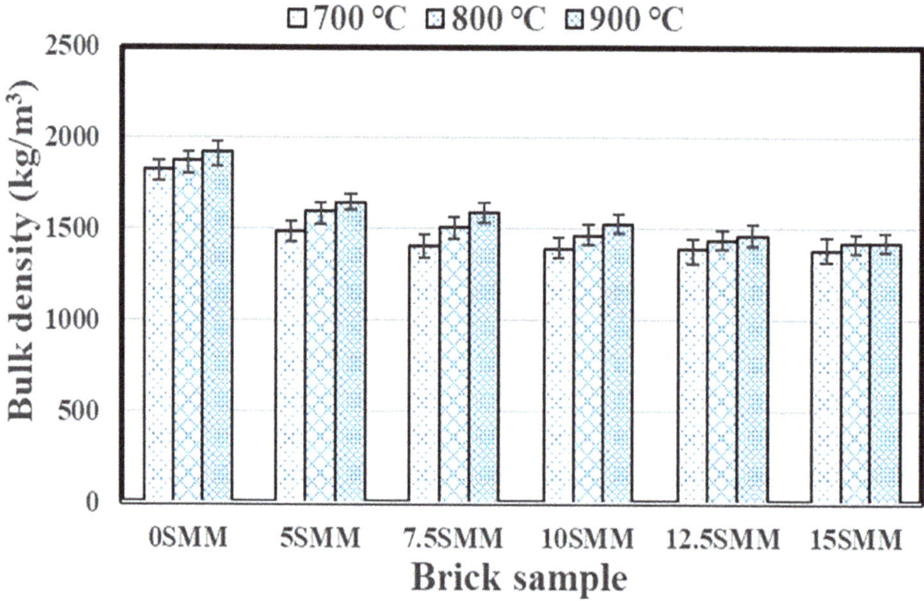

Figure 9. Average values of the bulk density of the samples.

3.3. Water Absorption

The findings of water absorption by fabricated bricks at 700, 800, and 900 °C, and the clay replacement percentages varying from 0% to 15% SMM are shown in Figure 10. Brick durability can be assessed by measuring its ability to absorb water or by its physical characteristics. As shown in the figure, water absorption by bricks is linearly correlated with the percentage of SMM present, indicating that SMM serves as a pore-generating agent. When clay was substituted with 15 wt.% of SMM, the rate of absorption of cold water rose to 34.7%, 29.6%, and 23.6% at 700, 800, and 900 °C, respectively, compared with 16.4%, 14.5%, and 13.9% at 0% SMM substitution (Figure 10a). On the other hand, boiling water absorption was 37.6%, 34.5%, and 25.3% when 15% SMM was used as a substitute for clay, up from 19.3%, 18.2%, and 17.2% at 700, 800, and 900 °C, respectively, when 0% SMM substitution was used (Figure 10b). Furthermore, the bricks demonstrated the lowest water absorption percentage and pore volume at 900 °C in accordance with the requirements of the standard ASTM C62 [6,28], likely due to the increase in the formation of a glassy phase at high firing temperatures. In general, water absorption increased with the increase in the

SMM in the mixtures and decreased with increasing firing temperature. Brick samples with high water absorbency had greater total porosity, improving the insulating characteristics of the clay bricks.

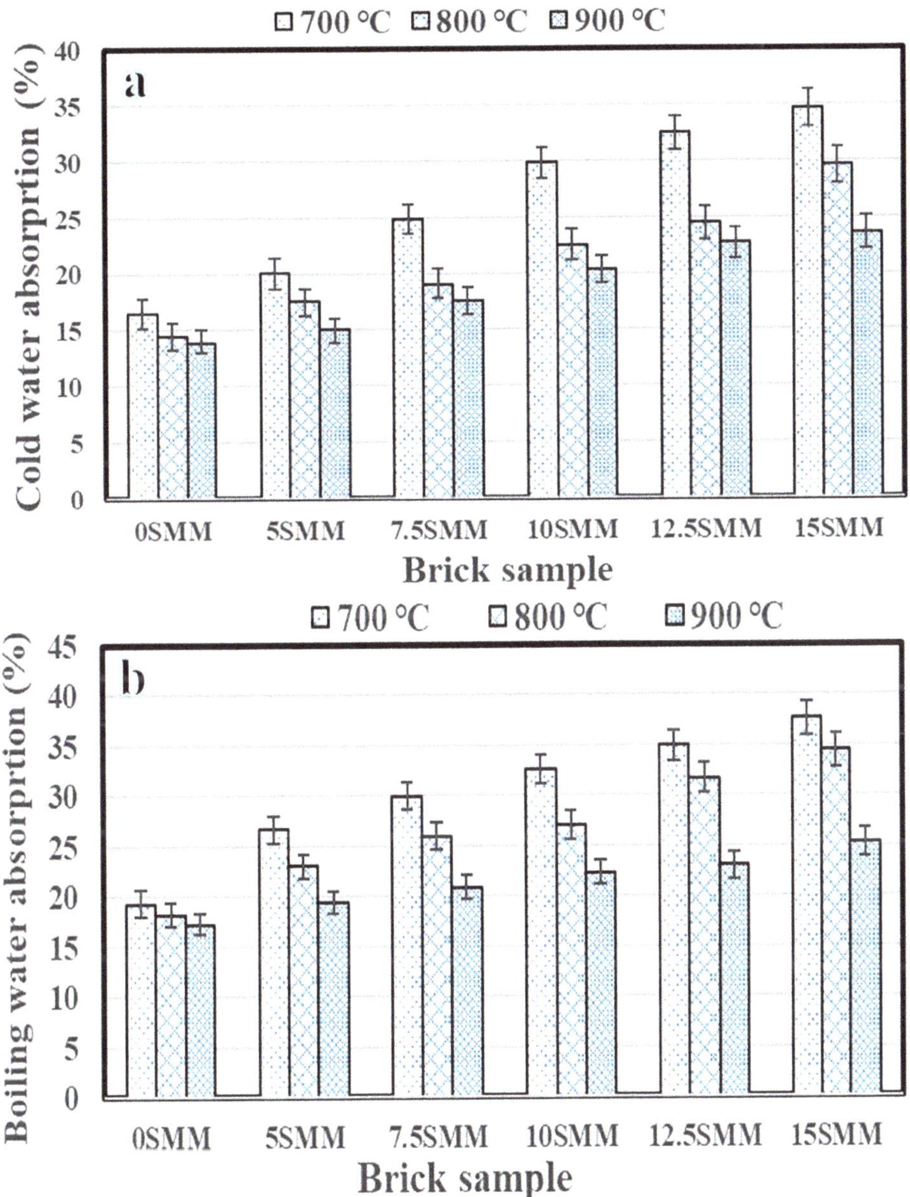

Figure 10. Cold water absorption (**a**) and boiling water absorption (**b**).

3.4. Apparent Porosity

Figure 11 illustrates the findings of apparent porosity tests conducted on brick samples burnt at temperatures of 700, 800, and 900 °C using SMM clay-replacement percentages varying from 0 to 15%. It was noted that the brick samples containing SMM had more porosity than the control bricks at different temperatures. This could be due to the gases resulting from the decomposition of carbonaceous materials and the fact that SMM has a higher LOI than clay [23]. According to Sutcu and Akkurt [29], an increase in overall porosity of over 50% is effective at improving the insulating qualities of fired clay bricks. Overall, total porosity increased almost linearly as the percentage of SMM rise, demonstrating the effectiveness of SMM as a pore-forming agent. The results demonstrated an increase in porosity of 29.6, 27.9, and 24.2% when clay was substituted with 15 wt.% SMM at firing temperatures of 700, 800, and 900 °C, respectively. Brick samples fired at 900 °C exhibited the lowest porosity, likely due to the formation of the glassy phase at high temperatures. Although the formation of pores in the brick structure is the cause of decreased compressive strength, higher water absorption, and total porosity, the increased porosity has a positive impact on the performance of bricks because these pores make it easier for water vapor to move inside the brick skeleton, thereby preventing cracks from spreading, especially in humid climate zones [30].

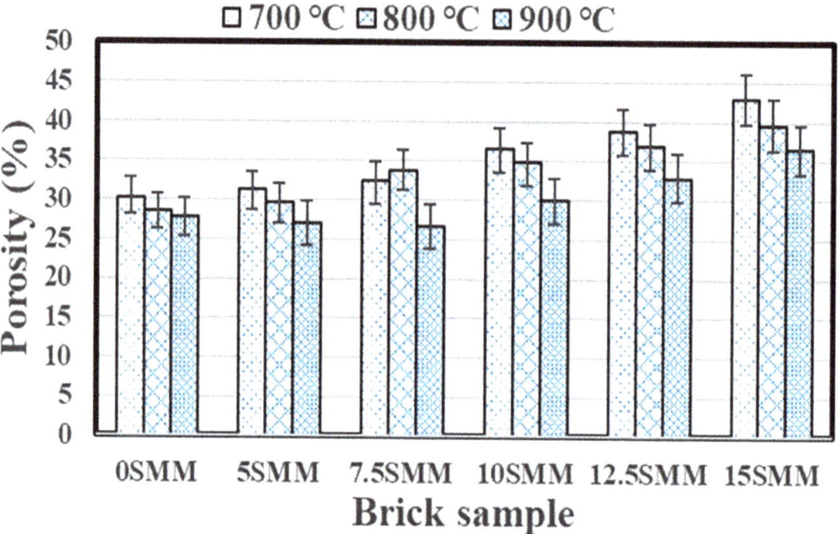

Figure 11. Apparent porosity results.

3.5. Compressive Strength

Figure 12 indicates the findings of compression tests conducted on fired brick specimens. The compressive strength of the samples was determined using the equation below; the burnt samples were tested by applying a perpendicular force that is consistent with ISO 9652 [31].

$$C = \frac{P}{A}, \left(\frac{N}{mm^2} \right)$$

(1)

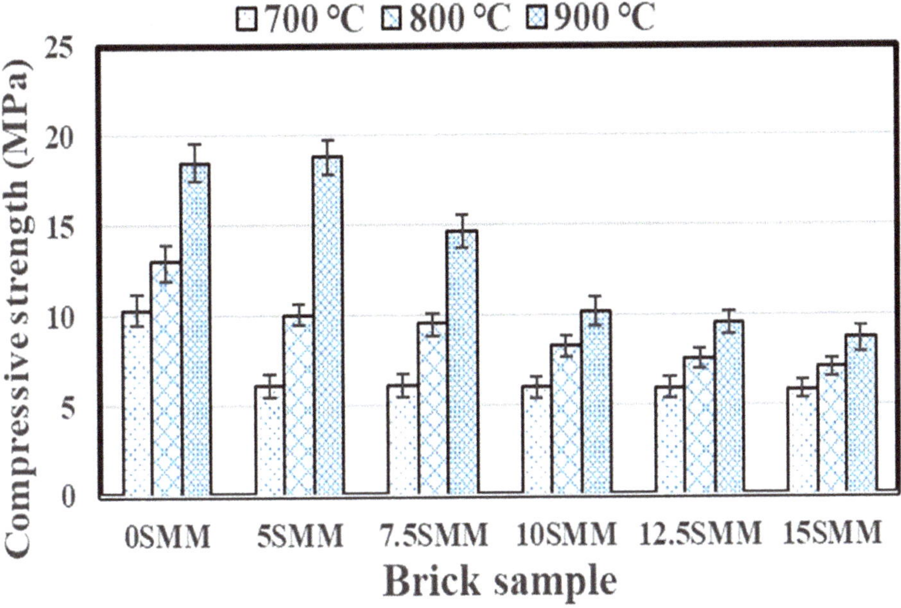

Figure 12. Compressive strength results.

The variable C denotes the compressive strength of fired samples, measured in units of N/mm^2 (MPa). The variable "P" denotes the maximum load, while the variable "A" represents the average cross-sectional area of the fired samples. The mean compressive strength of the specimen was determined by computing the average of the compressive strengths of three distinct brick specimens. Based on the findings in the figure, the addition of SMM and firing temperature had a significant impact on the compressive strength of fired clay bricks. While the addition of SMM did enhance some physical features, it had a detrimental effect on some mechanical characteristics. After sintering the fabricated bricks at 700 °C, the compressive strength of the brick specimens decreased from 10.23 MPa to 6.06 MPa for 5 SMM, 6.04 MPa for 7.5 SMM, 6 MPa for 10 SMM, 5.89 MPa for 12.5 SMM, and 5.8 MPa for 15 SMM. Due to the significant number of pores and cavities produced by the decomposition and combustion of organic matter, the integration of SMM, as anticipated, decreased the compressive strength [32]. The greater the pore size due to the decomposition of SMM, the smaller the exposed area of the compressive-resistant brick section, suggesting that this may be the cause of the drop in compressive strength. However, increasing the brick firing temperature to 800 °C increased the compressive strength to 12.9 MPa for the 0 SMM sample, 10 MPa for the 5 SMM sample, 9.5 MPa for the 7.5 SMM sample, 8.2 MPa for the 10 SMM sample, 7.6 MPa for the 12.5 SMM sample, and 7 MPa for the 15 SMM sample. In addition, increasing the brick firing temperature to 900 °C substantially improved the compressive strengths of the 0 SMM, 5 SMM, 7.5 SMM, 10 SMM, 12.5 SMM, and 15 SMM samples to 18.4 MPa, 18.8 MPa, 14.6 MPa, 10.13 MPa, 9.5 MPa, and 8.6 MPa, respectively. By sealing off internal pores, raising the firing temperature significantly accelerated verification and improved the development of strength. This is consistent with other studies that found that bricks produced at higher firing temperatures have higher densities and mechanical strengths [33,34]. Many studies [19] have shown that when firing temperatures rise, the bulk density of the brick framework increases, leading to greater compressive strength in the brick. The compressive strength requirements for non-load-bearing clay bricks are set at 3.5 N/mm^2 by the IS 1077-1992 standards [35]. Therefore, bricks having 15% SMM sintered at 800 or 900 °C still meet the required standards. The ASTM C62 [28] minimum suggested a compressive strength value of 8.6 MPa, which is consistent with brick samples fired at 800 and 900 °C. Furthermore, bricks containing 5% SMM and fired at 900 °C produced the highest compressive strength of 18.9 MPa.

3.6. Thermal Conductivity

Figure 13 shows the variance in thermal conductivity of bricks fired at 900 °C as a function of the SMM content. As shown in the figure, thermal conductivity reduced as the SMM content increased. The brick-free SMM had the greatest thermal conductivity (0.77 W/mK), with the brick being denser than other formulations; on the other hand, bricks composed of 15% SMM had the lowest thermal conductivity (0.293 W/mK). As is known, heat is transferred through solid materials by free electrons and/or lattice vibration waves (phonons). In ceramic materials, thermal conduction results from phonons. The presence of pores in the structure of ceramic materials facilitates phonon scattering, which may reduce the thermal conductivity of ceramic materials [36]. The bricks fired at 900 °C, including 5 SMM, 10 SMM, 12.5 SMM, and 15 SMM, showed significant reductions of 0.481, 0.43, 0.37, 0.32, and 0.29 (W/mK), respectively, in thermal conductivity compared with 0.77 (W/mK) of 0 SMM. The development of pores during sintering as a result of the production of gases that arise from the decomposition of carbonate materials and the high LOI in the SMM may be a possible explanation for the decrease in heat conductivity that occurs with the increase in the proportion of SMM in the material.

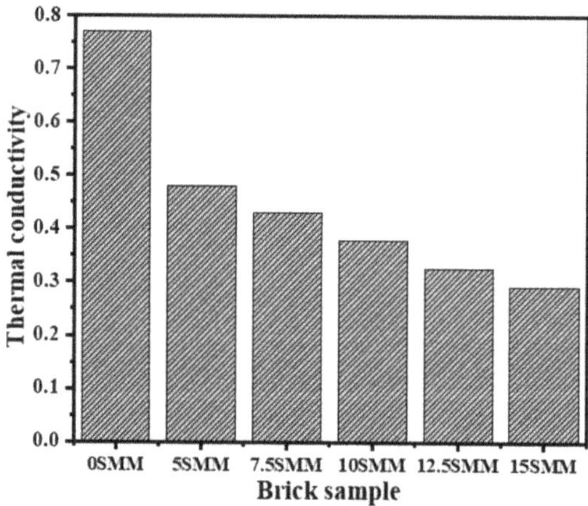

Figure 13. Thermal conductivity results.

3.7. SEM Microstructure Analysis

Figure 14 shows the microstructural characteristics of bricks that were burned at 900 °C and contained either 5% or 15% SMM. It is plain to see that the bricks that were infused with SMM produced a sizeable number of micropores, all of which were dispersed throughout the matrix of the hardened bricks. The pores that were visible took on a number of different shapes, including spheres and ovals. When the SMM percentage was increased, both the number of micropores and their diameters also increased. The average pore size increased from 2.92 to 5.71 μm when the percentage of SMM increased from 5% to 15%. This helps to explain the decrease in thermal conductivity, compressive strength, and bulk density, as well as the rise in water absorption and apparent porosity.

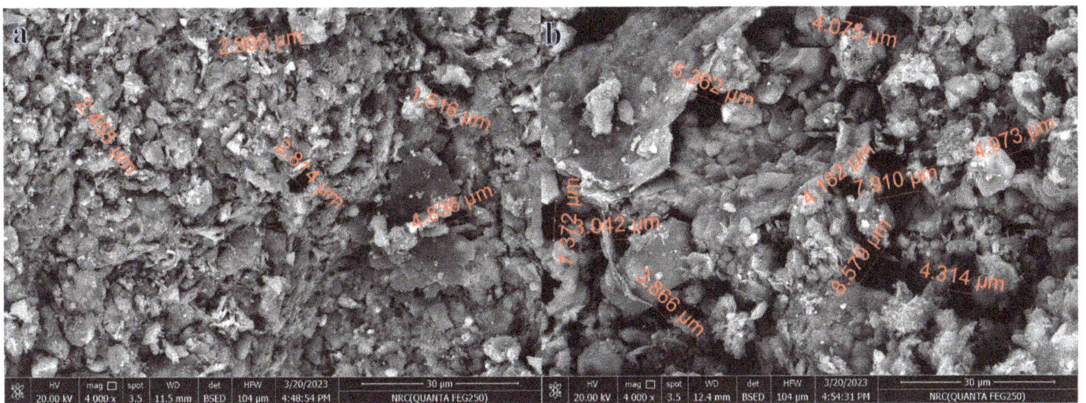

Figure 14. SEM images of 5% (**a**) and 15% (**b**) SMM bricks fired at 900 °C.

3.8. Estimation of Annual Energy Consumption for Cooling

The effect of bricks with different SMM proportions on a case study's cooling demands was evaluated using Grasshopper plugins Rhin 7 thermal simulation software. The application starts by simulating a room as a basic case to identify the optimal SMM ratio in terms of energy efficiency. As can be seen in Figure 15, the monthly variation in energy use was largest during the summer months of April through October. In addition, traditional bricks consumed the highest energy for cooling (Figure 15a,b) in both Cairo and Jazan cities. Samples with 15 SMM bricks used substantially less energy for cooling than their standard counterparts (Figure 15c,d) in Cairo and Jazan cities. The highest monthly average for cooling energy used occurred in July, the warmest month of the year; this was particularly true in the mornings. Figure 16 shows the simulation results, which indicate that different SMM brick compositions affect cooling energy consumption. The replacement of standard bricks with a 15 SMM brick sample reduced the building's cooling needs by 129.514 kWh, which is approximately 29.23%, compared with the 0 SMM (183 kWh) bricks (Figure 16a). The total amount of energy that was utilized for cooling reduced by 134.606 kWh, which is an improvement of 26.44%, when 12.5 SMM was used. Furthermore, cooling energy consumption can be reduced by 141.898 kWh with 10 SMM, 148.611 kWh with 7.5 SMM, and 154.63 kWh with 5 SMM, representing improvement rates of 22.46%, 18.79%, and 15.50%, respectively, (Figure 16b). This significantly reduces the building's energy consumption, as space cooling accounts for 37–42% of its overall energy requirement [37,38]. The suggested manufactured brick samples were examined for thermal and energy efficiency in the city of Jazan in the Kingdom of Saudi Arabia (Figure 16a,b), which has a climate characterized by hot, arid desert conditions. The sample bricks were compared with a standard brick for the purpose of establishing a reference point. The hypothesis posits that a model identical to that implemented in Cairo should be utilized in Jazan. Using 15 SMM resulted in a reduction in energy consumption of 21.49%. Utilization of clay bricks derived from SMM exhibited promising prospects of enhancing thermal efficiency and energy conservation and reducing the carbon footprint in arid and high-temperature settings.

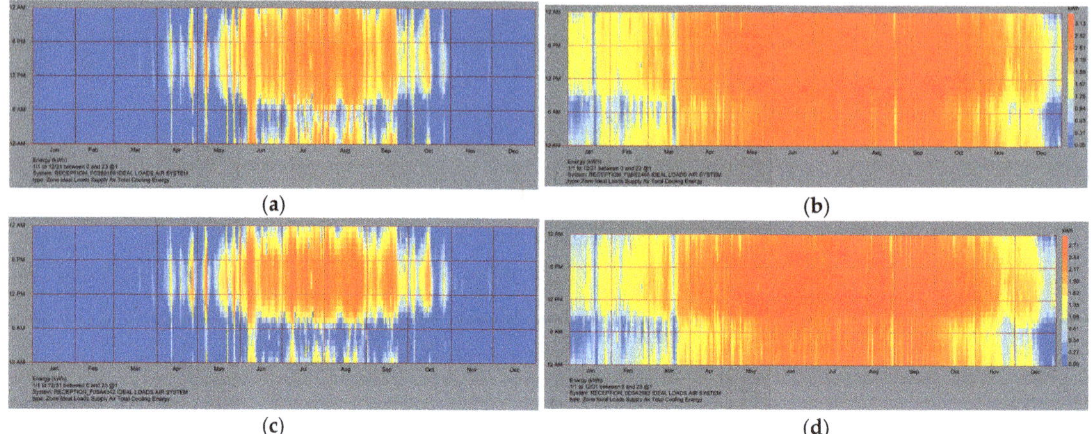

Figure 15. Monthly required cooling load for 0 SMM in Cairo (**a**) and Jazan (**b**) and for 15 SMM in Cairo (**c**) and Jazan (**d**).

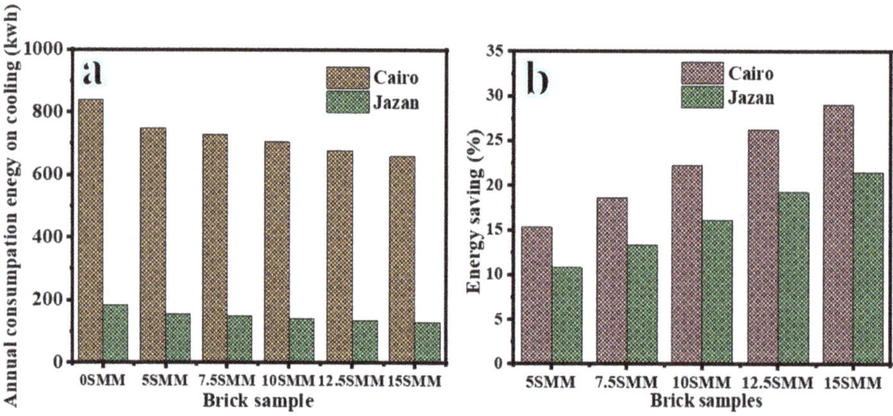

Figure 16. Annual energy consumed for cooling (**a**) and percentage of energy saved (**b**) using different percentages of SMM.

3.9. Influence of the Proposed Fabricated Bricks on CO_2 Emissions

Embodied CO_2 has become an increasingly significant contributor to the entire life-cycle of construction in the last few decades. This is the case despite the fact that developments in technology have led to a decrease in operational CO_2 emissions. The main sources of operational CO_2 emissions are HVAC and lighting; the major sources of embodied CO_2 emissions are construction materials [39]. The results of this study suggest a link between CO_2 emissions from building operations and the increasing prevalence of air conditioning in hot climates. The construction industry likely contributes to the rising atmospheric concentrations of embodied and operational CO_2. Construction materials have been the subject of extensive studies and developments. As a result, buildings' thermal performance has improved, the amount of energy required to keep them cool has decreased, and CO_2 emissions have decreased. The results of this research suggest that SMM bricks may benefit building occupants' health by decreasing their dependence on air conditioning and slowing the rate at which heat travels through walls. As a result, indoor temperatures will be more pleasant. In addition, it is possible that cutting down CO_2 emissions would be possible thanks to the utilization of SMM in bricks. Figure 17 shows the massive amounts of CO_2 emissions produced in order to provide cooling. It was demonstrated that using 5 SMM might result in an annual decrease of 2.2% in CO_2 emissions.

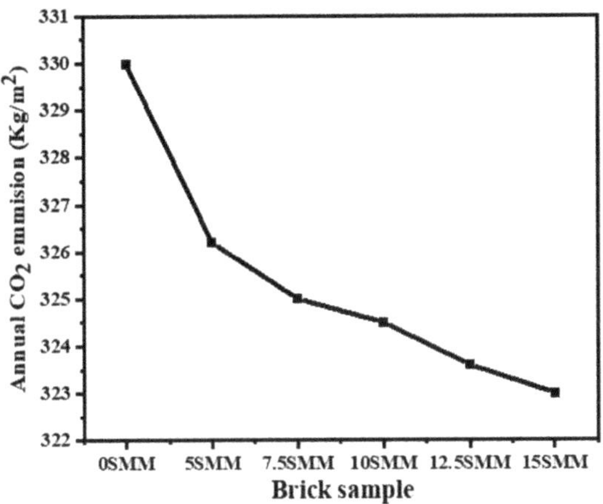

Figure 17. Annual CO_2 emission of the fabricated bricks.

4. Conclusions

The following are inferences drawn from the study's practical findings:

- The use of SMM as a pore-forming material allows for the production of fired bricks with low heat transmission and sufficient compressive strength. Linear firing shrinkage is adversely affected by the replacement of SMM waste. When the operating temperature is raised, there is a corresponding rise in firing shrinkage; the bricks produced at 900 °C with 15% SMM showed a major increase (7.6%) in firing shrinkage.
- The results demonstrate that increasing the quantity of SMM as a substitute for clay leads to a loss in compressive strength. The most modest compressive strength, 8.7 MPa at 900 °C, was achieved at the highest SMM % and may be adequate to fulfill the standards for load-bearing blocks in ordinary constructions in accordance with ASTM C62 [28].
- The bulk densities of all samples reduced as the percentage of SMM added as a replacement for clay increased, and samples containing SMM were less dense compared with the control sample.
- Because of the correlation between SMM and water absorption and porosity, decreased compressive strength, bulk density, and thermal conductivity are observed when SMM is increased. Regardless of firing temperature, higher SMM concentrations result in lower thermal conductivity by the brick specimens. When burnt at 900 °C, brick samples containing 15 wt.% SMM exhibited the smallest thermal conductivity (0.29 W/mK).
- SEM analysis of the SMM bricks indicated that the number and diameters of micropores increased as the SMM ratio increased.
- Thermal efficiency is maximized in bricks containing 15% SMM. In the case study provided, a 2.2% decrease in CO_2 emissions and a 29.23% decrease in energy consumption were achieved.

Supplementary Materials: The following supporting information can be downloaded at: https://www.mdpi.com/article/10.3390/ma16144905/s1, Table S1: Demonstrates physco-mechanical and thermal characteristics of bricks substituted with different waste material. References [40–52] are cited in the supplementary materials.

Author Contributions: Methodology, S.A.A., M.K.F. and M.M.A.; Validation, N.Z. and I.M.M.; Formal analysis, A.A.; Investigation, N.Z. and A.A.; Data curation, M.K.F. and I.M.M.; Writing—original draft, S.A.A., M.K.F. and M.M.A.; Writing—review & editing, A.Y. All authors have read and agreed to the published version of the manuscript.

Funding: The authors extend their appreciation to the Deputyship for Research & Innovation, Ministry of Education in Saudi Arabia for funding this research work through project number ISP22-4.

Institutional Review Board Statement: Not Applicable.

Informed Consent Statement: Not Applicable.

Data Availability Statement: The data presented in this study are available from the corresponding authors upon reasonable request.

Acknowledgments: The authors extend their appreciation to the Deputyship for Research & Innovation, Ministry of Education in Saudi Arabia for funding this research work through project number ISP22-4.

Conflicts of Interest: The authors declare no conflict of interest.

References

1. Tarek, D.; El-Naggar, M.; Sameh, H.; Yousef, A.; Ragab, A. Energy Efficiency Coupled with Lightweight Bricks: Towards Sustainable Building: A review. *SVU-Int. J. Eng. Sci. Appl.* **2023**, *4*, 1–28. [CrossRef]
2. Ahmed, M.M.; El-Naggar, K.; Tarek, D.; Ragab, A.; Sameh, H.; Zeyad, A.M.; Tayeh, B.A.; Maafa, I.M.; Yousef, A. Fabrication of thermal insulation geopolymer bricks using ferrosilicon slag and alumina waste. *Case Stud. Constr. Mater.* **2021**, *15*, e00737. [CrossRef]
3. Maafa, I.M.; Abutaleb, A.; Zouli, N.; Zeyad, A.M.; Yousef, A.; Ahmed, M.M. Effect of agricultural biomass wastes on thermal insulation and self-cleaning of fired bricks. *J. Mater. Res. Technol.* **2023**, *24*, 4060–4073. [CrossRef]
4. Chee, M.C. Effects of natural fibers inclusion in clay bricks: Physic-mechanical properties. *J. Int. J. Civ. Environ. Eng.* **2011**, *1*, 51–57.
5. Raut, S.P.; Ralegaonkar, R.V.; Mandavgane, S.A. Development of sustainable construction material using industrial and agricultural solid waste: A review of waste-create bricks. *Constr. Build. Mater.* **2011**, *25*, 4037–4042. [CrossRef]
6. Al-Fakih, A.; Mohammed, B.S.; Liew, M.S.; Nikbakht, E. Incorporation of waste materials in the manufacture of masonry bricks: An update review. *J. Build. Eng.* **2019**, *21*, 37–54. [CrossRef]
7. Ahmed, M.M.; Abadir, M.F.; Yousef, A.; El-Naggar, K.A.M. The use of aluminum slag waste in the preparation of roof tiles. *Mater. Res. Express* **2021**, *8*, 125501. [CrossRef]
8. Belewu, M.A.; Belewu, K.Y. Cultivation of mushroom (*Volvariella volvacea*) on banana leaves. *Afr. J. Biotechnol.* **2005**, *4*, 1401–1403.
9. Feeney, M.J.; Miller, A.M.; Roupas, P. Mushrooms—Biologically distinct and nutritionally unique: Exploring a "third food kingdom". *Nutr. Today* **2014**, *49*, 301. [CrossRef]
10. Ahlawat, O.P.; Tewari, R.P. *Cultivation Technology of Paddy Straw Mushroom (Volvariella volvacea)*; National Research Centre for Mushroom: Solan, India, 2007.
11. Finney, K.N.; Ryu, C.; Sharifi, V.N.; Swithenbank, J. The reuse of spent mushroom compost and coal tailings for energy recovery: Comparison of thermal treatment technologies. *Bioresour. Technol.* **2009**, *100*, 310–315. [CrossRef]
12. Sánchez, C. Cultivation of *Pleurotus ostreatus* and other edible mushrooms. *Appl. Microbiol. Biotechnol.* **2010**, *85*, 1321–1337. [CrossRef]
13. Royse, D.J. A global perspective on the high five: Agaricus, Pleurotus, Lentinula, Auricularia & Flammulina. In Proceedings of the 8th International Conference on Mushroom Biology and Mushroom Products (ICMBMP8), New Delhi, India, 19–22 November 2014; pp. 1–6.
14. Bibri, S.E.; Krogstie, J. ICT of the new wave of computing for sustainable urban forms: Their big data and context-aware augmented typologies and design concepts. *Sustain. Cities Soc.* **2017**, *32*, 449–474. [CrossRef]
15. Sanaieian, H.; Tenpierik, M.; Van Den Linden, K.; Seraj, F.M.; Shemrani, S.M.M. Review of the impact of urban block form on thermal performance, solar access and ventilation. *Renew. Sustain. Energy Rev.* **2014**, *38*, 551–560. [CrossRef]
16. Khani, A.; Khakzand, M.; Faizi, M. Multi-objective optimization for energy consumption, visual and thermal comfort performance of educational building (case study: Qeshm Island, Iran). *Sustain. Energy Technol. Assess.* **2022**, *54*, 102872. [CrossRef]
17. Bakmohammadi, P.; Noorzai, E. Optimization of the design of the primary school classrooms in terms of energy and daylight performance considering occupants' thermal and visual comfort. *Energy Rep.* **2020**, *6*, 1590–1607. [CrossRef]
18. Tarek, D.; Ahmed, M.M.; Hussein, H.S.; Zeyad, A.M.; Al-Enizi, A.M.; Yousef, A.; Ragab, A. Building envelope optimization using geopolymer bricks to improve the energy efficiency of residential buildings in hot arid regions. *Case Stud. Constr. Mater.* **2022**, *17*, e01657. [CrossRef]

19. Stoknes, K.; Scholwin, F.; Krzesiński, W.; Wojciechowska, E.; Jasińska, A. Efficiency of a novel "Food to waste to food" system including anaerobic digestion of food waste and cultivation of vegetables on digestate in a bubble-insulated greenhouse. *Waste Manag.* **2016**, *56*, 466–476. [CrossRef]

20. *ASTM C136/C136M-14*; Standard Test Method for Sieve Analysis of Fine and Coarse Aggregates. American Society for Testing and Materials: West Conshohocken, PA, USA, 2014.

21. Patil, R.C.; Patil, U.P.; Jagdale, A.A.; Shinde, S.K.; Patil, S.S. Ash of pomegranate peels (APP): A bio-waste heterogeneous catalyst for sustainable synthesis of α, α'-bis (substituted benzylidine) cycloalkanones and 2-arylidene-1-tetralones. *Res. Chem. Intermed.* **2020**, *46*, 3527–3543. [CrossRef]

22. Mphahlele, R.R.; Pathare, P.B.; Opara, U.L. Drying kinetics of pomegranate fruit peel (cv. Wonderful). *Sci. Afr.* **2019**, *5*, e00145. [CrossRef]

23. Maza-Ignacio, O.T.; Jiménez-Quero, V.G.; Guerrero-Paz, J.; Montes-García, P. Recycling untreated sugarcane bagasse ash and industrial wastes for the preparation of resistant, lightweight and ecological fired bricks. *Constr. Build. Mater.* **2020**, *234*, 117314. [CrossRef]

24. *ASTM C326*; Standard Test Method for Drying and Firing Shrinkages of Ceramic Whiteware Clays. ASTM International: West Conshohocken, PA, USA, 2018.

25. Aouba, L.; Bories, C.; Coutand, M.; Perrin, B.; Lemercier, H. Properties of fired clay bricks with incorporated biomasses: Cases of olive stone flour and wheat straw residues. *Constr. Build. Mater.* **2016**, *102*, 7–13. [CrossRef]

26. *ASTM:C90*; Standard Specification for Loadbearing Concrete Masonry Units. ASTM International: West Conshohocken, PA, USA, 2014.

27. Assaffii, I.A.; Muslat, M.M.; Al-Issawy, K.J. Improve the local components of the substrate to increase the production oyster mushroom *Pleurotus ostreatus* and post-spent quality. *Anbar J. Agric. Sci.* **2011**, *9*, 113–129.

28. *ASTM C62*; Standard Specification for Building Brick (Solid Masonry Units Made From Clay or Shale). ASTM International: West Conshohocken, PA, USA, 2017.

29. Ölmez, H.; Erdem, E. The effects of phosphogypsum on the setting and mechanical properties of Portland cement and trass cement. *Cem. Concr. Res.* **1989**, *19*, 377–384. [CrossRef]

30. Rashad, A.M.; Gharieb, M.; Shoukry, H.; Mokhtar, M.M. Valorization of sugar beet waste as a foaming agent for metakaolin geopolymer activated with phosphoric acid. *Constr. Build. Mater.* **2022**, *344*, 128240. [CrossRef]

31. *ISO:9652-4*; Test Methods for Masonry Units. International Organization for Standardization (ISO): Geneva, Switzerland, 2000.

32. Singh, L.P. Investigation of Physical Properties of Bricks Utilizing Fly Ash, Lime and Gypsum. Ph.D. Dissertation, Kanpur University, Kanpur, India, 1994.

33. Ambalavanan, R.; Roja, A. Feasibility studies on utilization of waste lime and gypsum with flyash. *Indian Concr. J.* **1996**, *70*, 611–615.

34. Viana, C.E.; Dias, D.P.; Holanda, J.N.F.D.; Paranhos, R.P.D.R. The use of submerged-arc welding flux slag as raw material for the fabrication of multiple-use mortars and bricks. *Soldag. Inspeção* **2009**, *14*, 257–262. [CrossRef]

35. *NF ISO 5017*; Dense Shaped Refractory Products-Determination of Bulk Density, Apparent Porosity and True Porosity-Produits Réfractaires Façonnés Denses. AFNOR: Saint-Denis, France, 2013.

36. Klemens, P.G. Phonon scattering by oxygen vacancies in ceramics. *Phys. B Condens. Matter* **1999**, *263*, 102–104. [CrossRef]

37. Mahmoud, A.R. The Influence of Buildings Proportions and Orientations on Energy Demand for Cooling in Hot Arid Climate. *SVU-Int. J. Eng. Sci. Appl.* **2022**, *3*, 8–20. [CrossRef]

38. Ragab Abdel Radi, A. The impact of phase change materials on the buildings enegy efficiency in the hot desert areas the annexed rooms of the traffic building in new aswan city as a case study. *JES J. Eng. Sci.* **2020**, *48*, 302–316. [CrossRef]

39. De Wolf, C.; Yang, F.; Cox, D.; Charlson, A.; Hattan, A.S.; Ochsendorf, J. Material quantities and embodied carbon dioxide in structures. In *Proceedings of the Institution of Civil Engineers-Engineering Sustainability*; Thomas Telford Ltd.: London, UK, 2015.

40. Galán-Arboledas, R.J.; Cotes-Palomino, M.T.; Bueno, S.; Martínez-García, C. Evaluation of spent diatomite incorporation in clay based materials for lightweight bricks processing. *Constr. Build. Mater.* **2017**, *144*, 327–337. [CrossRef]

41. Eliche-Quesada, D.; Felipe-Sesé, M.A.; López-Pérez, J.A.; Infantes-Molina, A. Characterization and evaluation of rice husk ash and wood ash in sustainable clay matrix bricks. *Ceram. Int.* **2017**, *43*, 463–475. [CrossRef]

42. Abdrakhimov, V.Z.; Abdrakhimova, E.S. Promising use of waste coal in the production of insulating material without the use of traditional natural materials. *Inorg. Mater. Appl. Res.* **2017**, *8*, 788–794. [CrossRef]

43. Sutcu, M.; Ozturk, S.; Yalamac, E.; Gencel, O. Effect of olive mill waste addition on the properties of porous fired clay bricks using Taguchi method. *J. Environ. Manag.* **2016**, *181*, 185–192. [CrossRef] [PubMed]

44. Eliche-Quesada, D.; Leite-Costa, J. Use of bottom ash from olive pomace combustion in the production of eco-friendly fired clay bricks. *Waste Manag.* **2016**, *48*, 323–333. [CrossRef] [PubMed]

45. Velasco, P.M.; Ortiz, M.P.M.; Giró, M.A.M.; Melia, D.M.; Rehbein, J.H. Development of sustainable fired clay bricks by adding kindling from vine shoot: Study of thermal and mechanical properties. *Appl. Clay Sci.* **2015**, *107*, 156–164. [CrossRef]

46. Bories, C.; Aouba, L.; Vedrenne, E.; Vilarem, G. Fired clay bricks using agricultural biomass wastes: Study and characterization. *Constr. Build. Mater.* **2015**, *91*, 158–163. [CrossRef]

47. Sutcu, M.; Alptekin, H.; Erdogmus, E.; Er, Y.; Gencel, O. Characteristics of fired clay bricks with waste marble powder addition as building materials. *Constr. Build. Mater.* **2015**, *82*, 1–8. [CrossRef]

48. Sánchez-Martínez, J.; Sesé, F.; Ángel, M.; Infantes-Molina, A. Silica–Calcareous Non Fired Bricks Made of Biomass Ash and Dust Filter from Gases Purification. 2017. Available online: https://reunir.unir.net (accessed on 20 May 2022).
49. Raut, A.N.; Gomez, C.P. Development of thermally efficient fibre-based eco-friendly brick reusing locally available waste materials. *Constr. Build. Mater.* **2017**, *133*, 275–284. [CrossRef]
50. Holmes, N.; O'Malley, H.; Cribbin, P.; Mullen, H.; Keane, G. Performance of masonry blocks containing different proportions of incinator bottom ash. *Sustain. Mater. Technol.* **2016**, *8*, 14–19. [CrossRef]
51. Sakhare, V.V.; Ralegaonkar, R.V. Use of bio-briquette ash for the development of bricks. *J. Clean. Prod.* **2016**, *112*, 684–689. [CrossRef]
52. Ismail, S.; Anas, Z.A.H.; Yaacob, Z. Mechanical and thermal properties of brick produced using recycled fine aggregate. *Key Eng. Mater.* **2016**, *706*, 112–116. [CrossRef]

materials

MDPI

Review

A Study on Sustainable Concrete with Partial Substitution of Cement with Red Mud: A Review

Hisham Jahangir Qureshi [1,*], Jawad Ahmad [2], Ali Majdi [3], Muhammad Umair Saleem [4], Abdulrahman Fahad Al Fuhaid [1] and Md Arifuzzaman [1]

1 Department of Civil and Environmental Engineering, College of Engineering, King Faisal University, Al-Ahsa 31982, Saudi Arabia
2 Department of Civil Engineering, Swedish College of Engineering, Wah Cantt 47070, Pakistan
3 Department of Building and Construction Techniques Engineering, Al-Mustaqbal University College, Hillah 51001, Iraq
4 Service Stream Limited Co., Chatswood, NSW 2067, Australia
* Correspondence: hqureshi@kfu.edu.sa

Abstract: Every year, millions of tons of red mud (RDM) are created across the globe. Its storage is a major environmental issue due to its high basicity and tendency for leaching. This material is often kept in dams, necessitating previous attention to the disposal location, as well as monitoring and maintenance during its useful life. As a result, it is critical to develop an industrial solution capable of consuming large quantities of this substance. Many academics have worked for decades to create different cost-effective methods for using RMD. One of the most cost-effective methods is to use RMD in cement manufacture, which is also an effective approach for large-scale RMD recycling. This article gives an overview of the use of RMD in concrete manufacturing. Other researchers' backgrounds were considered and examined based on fresh characteristics, mechanical properties, durability, microstructure analysis, and environmental impact analysis. The results show that RMD enhanced the mechanical properties and durability of concrete while reducing its fluidity. Furthermore, by integrating 25% of RDM, the environmental consequences of cumulative energy demand (CED), global warming potential (GWP), and major criteria air pollutants (CO, NO_X, Pb, and SO_2) were minimized. In addition, the review assesses future researcher guidelines for concrete with RDM to improve performance.

Keywords: red mud; eco-friendly concrete; durability; slump; mechanical strength

Citation: Qureshi, H.J.; Ahmad, J.; Majdi, A.; Saleem, M.U.; Al Fuhaid, A.F.; Arifuzzaman, M. A Study on Sustainable Concrete with Partial Substitution of Cement with Red Mud: A Review. *Materials* **2022**, *15*, 7761. https://doi.org/10.3390/ma15217761

Academic Editors: Stefano Guarino and José Barroso de Aguiar

Received: 9 September 2022
Accepted: 17 October 2022
Published: 3 November 2022

Publisher's Note: MDPI stays neutral with regard to jurisdictional claims in published maps and institutional affiliations.

1. Introduction

To make constant strides toward sustainable growth, a massive revolution in the cement and concrete sector is required to decrease environmental pollution, particularly carbon dioxide [1]. Concrete manufacturing has increased in recent decades, and it is currently one of the top concerns of scholars who are interested in sustainable development [2–5]. The utilization of waste materials in concrete manufacturing has been shown to minimize natural resource use [6–9]. Construction uses more raw materials and energy than any other economic activity on the planet today. Concrete is a common construction material used all over the globe. Cement, being a fundamental component of concrete, is produced via an energy-intensive process. Cement manufacturing produces a lot of greenhouse gas emissions, which contribute to global warming [5,10]. The cement factories are the second-largest industrial carbon dioxide emitter, accounting for 5 to 7% of total CO_2 emissions [11,12]. The use of various waste materials as cement substitutes has been researched for CO_2 reduction [13–15]. In contemporary civilizations, using garbage as a secondary material in the building sector is a cost-effective and environmentally friendly means of disposing of waste [16,17]. Sustainability in the construction industry has become crucial, and numerous solutions have emerged to lessen the environmental effect of present building operations [18,19]. Rising energy supply costs, decreasing CO_2 ejections, and the delivery of unrefined, low-quality ingredients all pose risks to the cement business [20]. As a result, alternative suppliers must be sought instead of cement. According to research, concrete created from waste materials such as plastic waste is one of the waste disposal alternatives [21]. Furthermore, in this century, most researchers are concentrating on developing sustainable concrete by incorporating various industrial wastes such as waste glass [22], waste marble [23], silica fume [24], copper slag [25], waste foundry sand [26], cellulosic materials [27], recycled aggregate concrete [28], as well as fly ash [29,30]. In addition to utilizing different waste materials cementitious materials in concrete, RMD is also a good option.

1.1. RMD

Bauxite residue is an insoluble byproduct of the Bayer alumina manufacturing process. The high alkalinity of the liquid phase isolated from the RMD slurry is a major environmental concern. The challenges related to bauxite waste formation were first overlooked and ignored throughout the discovery and usage of bauxite for profit. However, when the world's population exploded in the mid-twentieth century, resources began to deplete and waste products began to accumulate on the planet's surface [31]. As a consequence, by the end of the twentieth century, the use of bauxite waste had become a worldwide issue. Every year, roughly 150 million tons of bauxite residue are generated worldwide [32]. In general, 1–2.5 tons of bauxite residue are created for every ton of alumina produced, with the amount varying depending on the kind of bauxite ore utilized [33]. The RMD collected from an alumina refinery is shown in Figure 1.

Figure 1. RMD [34].

Bauxite is a mineral composed of hydrated aluminium oxides and mixtures of other elements, such as iron. The aluminium mixtures in the bauxite may be found in many kinds of aluminium parts, and contaminations may influence extraction conditions. Aluminium oxides and hydroxides are amphoteric, which means they are acidic and fundamental at the same time. Al_3 solubility in water is minimal, although it increases dramatically with high or low PH. Bauxite metal is warmed in a weight vessel beside a sodium hydroxide setup at a temperature of 150 to 200 °C in the Bayer technique. In an extraction technique, the aluminium is broken down as sodium aluminate (basically $[Al(OH)_4]$) at these temperatures. Gibbsite is accelerated when the fluid is cooled and seeded with fine-grained aluminium hydroxide precious stones from previous extractions after partitioning the accumulation by sifting. The RMD process is seen in Figure 2 [35].

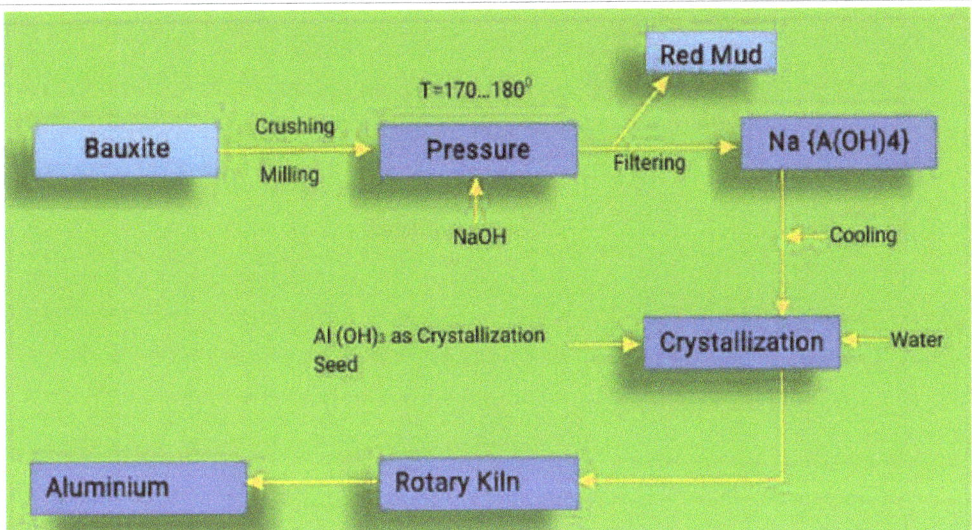

Figure 2. Manufacturing Process of RMD [35].

The chemical difference in every circumstance, applying RMD is a challenge. However, detecting the RMD features for each location where it is created and establishing a template as a foundation is achievable. Two variables influence the quantity of RMD produced. The first pertains to the ore quality, while the second refers to the processing conditions. The amount of RMD produced has been approximated by many writers. The reported values vary from 0.3 to 2.5 tons per ton of poorer grade bauxite treated [36]. Patents for RMD applications were created between 1964 and 2008, according to research [37]. It is expected that 33 percent of them are used in civil building projects, as shown in Figure 3. Although RMD has a variety of uses, as seen in Figure 3, its primary use is in the building industry.

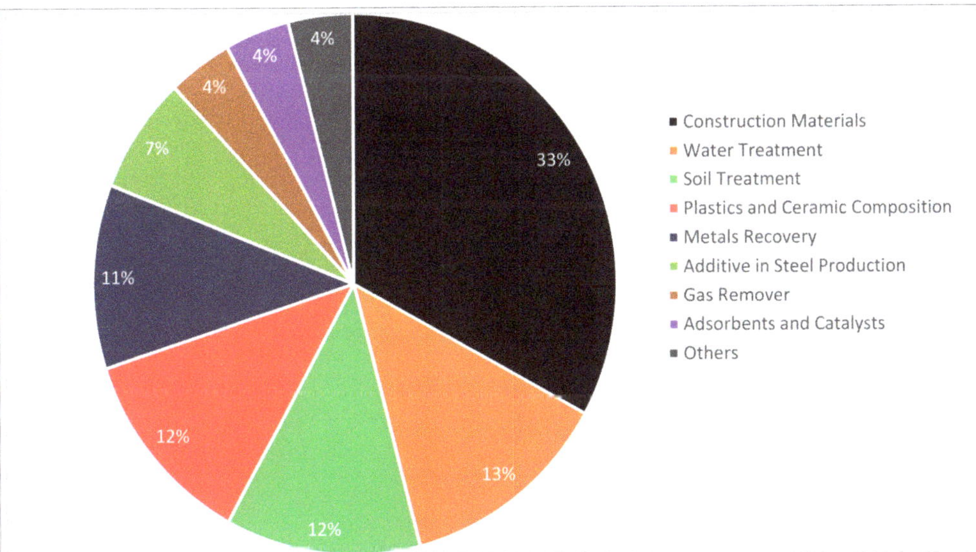

Figure 3. RMD Application (1964 to 2008): Data source [38].

Figure 4 depicts data on bauxite residue applications research publications by companies, academic institutions, and research organizations during a 20-year period. The biggest number of research articles have been published in the field of building and construction. The largest volume of RMD consultation materials was released from 1971 to 1990, and subsequently, quickly declined, as seen in Figure 4. As a result, a concise review is needed to highlight current developments and prospective applications of RDM in the construction industry.

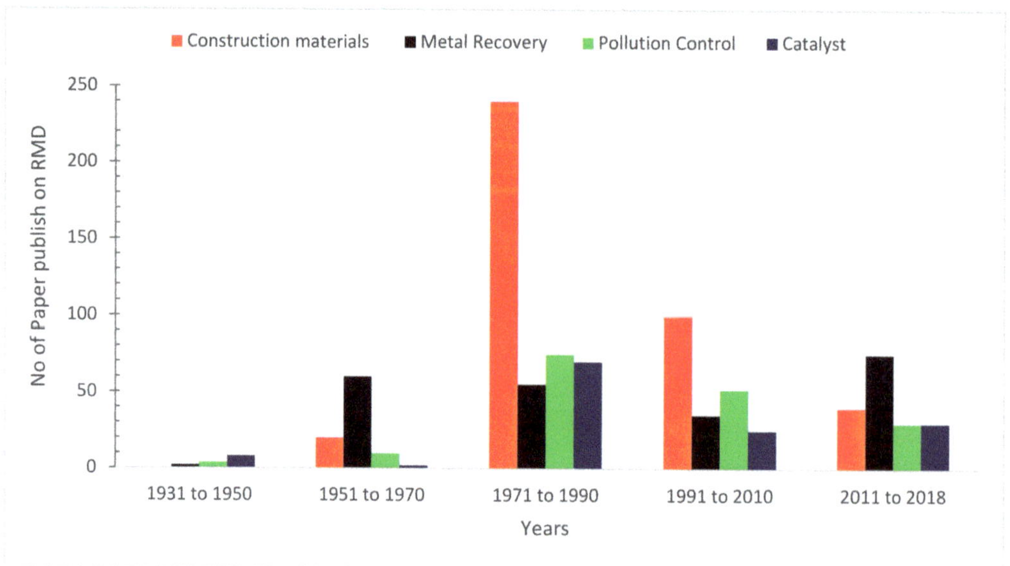

Figure 4. No. of Publication on RMD (1931 to 2018): Data Source [36].

1.2. Weaknesses of RMD

RMD affects agricultural landscapes due to its strong alkalinity, hence, it should be cleaned numerous times before usage. There is a chance that the RMD lake may leak. If it is organized underground, it pollutes the ground water table. Based on the writing overview, RMD fills in as an excellent folio material and has proven to be a decent cementation material. The tiny pores of cement are diminished by RMD, and as the amount of RMD in solid increments increases, the amount of water swallowed decreases. A basic combination of 70% RMD and 30% CaO yields a product with a compressive quality of 7 MPa. $Ca(OH)_2$, C_4AH_{13} and C_4AH_{11} are the hydrates framed after 4 days. Later investigations have confirmed these findings. When compared to fly powder, RMD does not impart much compressive quality, although flexural quality and resistance to porousness have been seen in RMD bond concrete. Despite the fact that RMD is less accessible than flies' fiery remnants, it is critical to use or reuse RMD since it has several harmful ecological effects [39].

1.3. Chemical Composition

Table 1 shows the chemical components of RDM while Figure 5 shows the XRD of RDM. The primary components are alumina and iron oxide, as assumed, but the relative proportions of SiO_2 and Na_2O are also important. In addition to aluminium hydroxide and a complicated $Na_5Al_3CSi_3O_{15}$ phase, XRD detects several of these oxides. Different researchers have reported different chemical compositions, as can be shown in Figure 5. It is likely that the chemical composition of RDM varies as the source changes. However, the accumulation of different ingredients such as silica, alumina, and iron reported by each author is higher than 50%, indicating that RDM has the potential to be employed as binding material. According to ASTM [29], the sum of the three principal oxide constituents, namely, silica, alumina, and iron, must be at least 50% for a material to be classed as pozzolanic. According to Table 1, all the RDM samples used in the various research projects may be classified as pozzolanic as per ASTM [29].

Table 1. Chemical Composition of RMD.

Authors	[40]	[41]	[42]	[43]	[44]
SiO_2	14.7	9.0	14.88	45.76	17.60
Al_2O_3	17.7	12.0	23.53	40.69	43.43
Fe_2O_3	27.6	37	36.48	2.85	0.65
MgO	1.7	-	1.61	0	-
CaO	14.7	6.0	1.83	4.98	2.87
Na_2O	5.4	5.0	9.41	0	10.55
K_2O	0.1	-	-	0.45	2.0

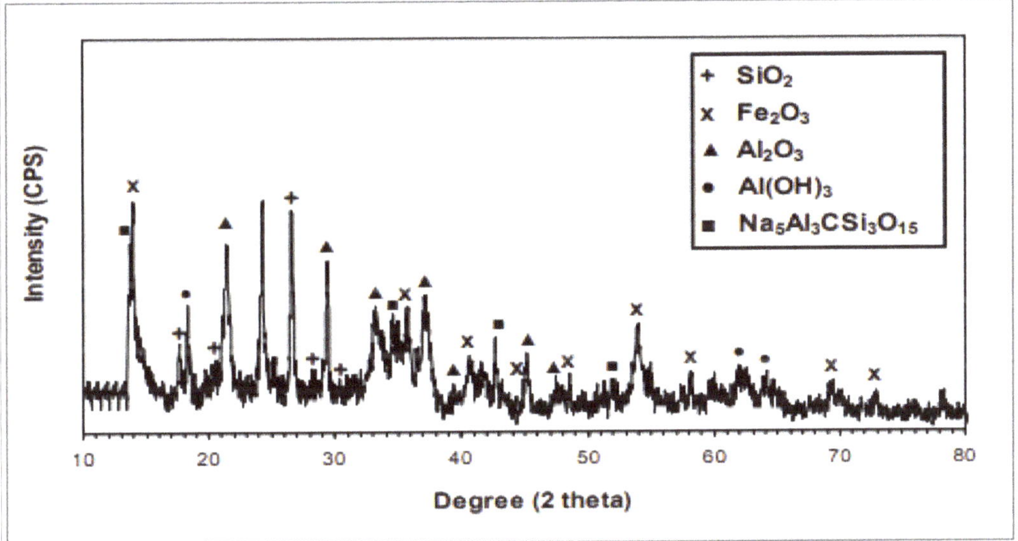

Figure 5. XRD of RDM [45].

1.4. Significances

The use of RMD in the manufacturing of concrete as a cement substitute has a number of environmental and economic advantages, including minimizing soil and groundwater contamination, reducing dust pollution, conserving natural resources for cement clinker, and lowering the cost of concrete production. The purpose of this paper is to present a detailed review of current progress on the use of RMD in concrete production, as well as to clearly point out four directions for using RMD in concrete production, namely, fresh properties, mechanical properties, durability aspects, and environmental aspects. Furthermore, the review also evaluates the future researcher guideline for better performance.

2. Fresh Properties

2.1. Workability

As indicated in Figure 6, the slump flow diminishes dramatically as the quantity of RMD supplied boosts. The rise in adsorbed water generated by the porous nature and the greater specific surface area of RMD causes the reduction in workability. Although the addition of RMD may cause water between particles to relax, it also raises the packing density, which squeezes the free water between the particles and improves workability [46]. This might be due to the porous RMD which enhanced water absorption ability [47]. RDM, which has smaller particles than cement, reduces the fluidity of concrete by absorbing more moisture in the fresh concrete condition. However, since water made up around 48 percent of the total mass of the RDM employed in this investigation, adequate moisture could be provided in the fresh concrete phase [48]. However, the larger surface area of the seawater-neutralized bauxite refinery residence is most likely to blame for the decrease in a slump. Furthermore, unlike water added to natural sand during concrete production, water in seawater-neutralized bauxite refinery reside containing mix may not be as readily available to lubricate the mix, as it may be held within the fine-particle aggregates or chemically bound with the hygroscopic seawater-neutralized bauxite refinery reside [49]. A study [42] explored the slump flow of self-compacting concrete as per standard [50] and claimed that RMD decreased the slump flow of concrete.

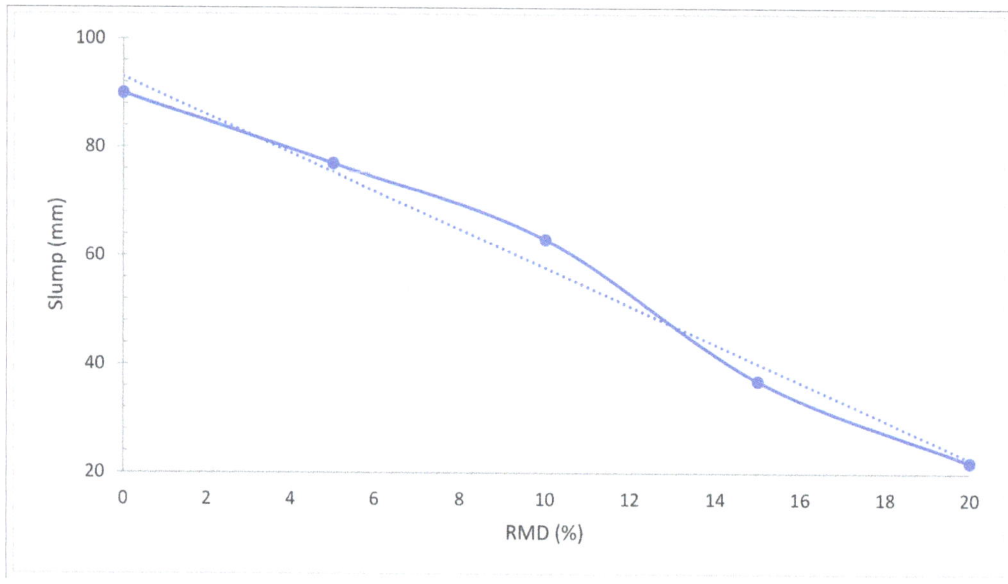

Figure 6. Slump Flow: Data Source [51].

2.2. Fresh Density

Figure 7 depicts the changes in the fresh density of concrete after 28 days in relation to RDM cement replacement. The dry density ranges from 1685 to 1789 kg/m^3, whereas the fresh density ranges from 1752 to 1844 kg/m^3. When RDM concentration in concrete is raised, the specific gravity of the concrete drops. The specific gravity of the concrete specimen having 25% RDM is around 5% lower than the reference specimen with 0% RDM, according to the findings [40]. In contrast, a study claimed that the fresh density of concrete was enhanced due to the micro filling effect of pozzolanic material which gives more dense concrete. Furthermore, due to the pozzolanic reaction, the binding properties of concrete paste also contribute to the improved density of concrete. The combined micro filling voids and pozzolanic reaction have a positive influence on the density of concrete [52]. According to research, due to improved particle packing, there was an initial rise in density and a reduction in porosity due to micro filling voids in concrete aggregates. However, at 20% RMD addition, the behaviour becomes restrained because of additional challenges in moulding and shaping samples due to lack of flowability [45].

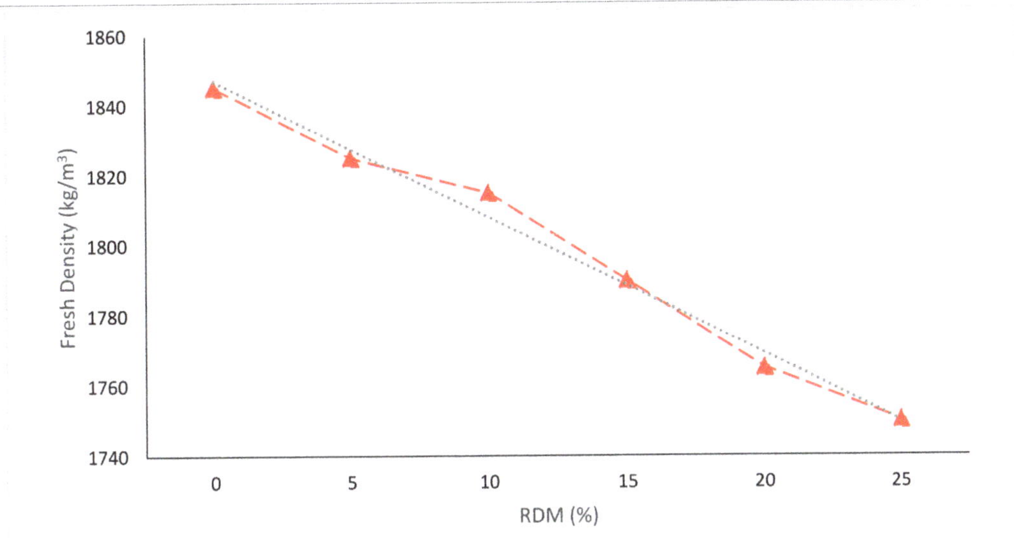

Figure 7. Fresh Density: Data Source [40].

2.3. Setting Time

The addition of RMD has the effect of speeding up the setting process, as shown in Figure 8. For mortars without RMD and those containing 20% waste, the end of the setting time changes from 345 to 300 min.

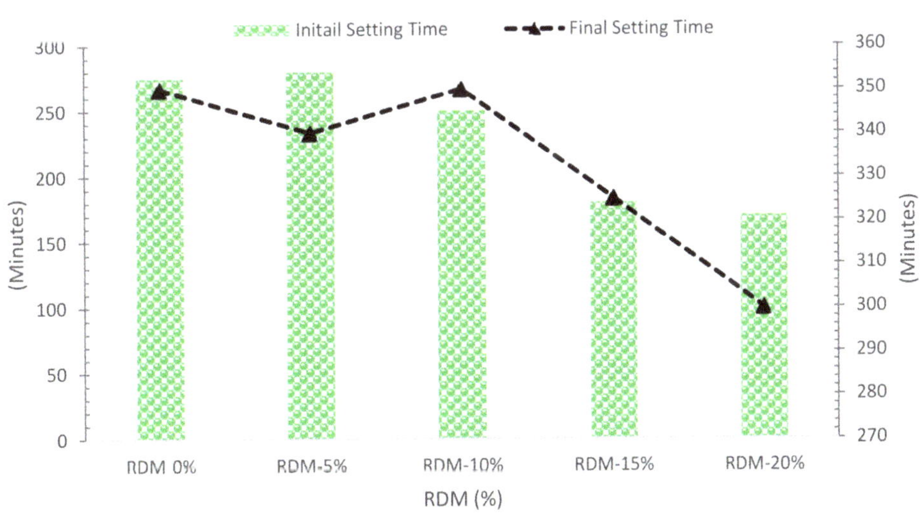

Figure 8. Setting Time [45].

The presence of aluminium and sodium hydroxides, which are known as accelerators [53], might explain this impact, in the mud, and also because of its high alkaline content. The fineness of waste particles might potentially help to retain water by competing with cement. As all formulas have the same amount of water, the leftover free fraction, which may be coupled with cement particles, will be consumed quickly [54]. In contrast, another study suggests that the pozzolanic materials decrease the setting time of cement paste as the reaction proceeds slowly [55]. The change in behavior might be possible due to a change in chemical composition.

3. Mechanical Properties

3.1. Compressive Strength (CMS)

Figure 9 depicts the CMS of concrete with various RDM replacement ratios. It can be shown that with 10% RDM replacement, maximum CMS is obtained. According to research, samples with a low level of RDM had good CMS [56]. The characteristics of RMD cementitious content change when 20% of the RMD is replaced with cement by weight. As a result, more than 20% RMD substitution reduces the concrete solid mass's compressive quality and stiffness [57]. Increases the load-bearing capability of RMD-based concrete samples by up to 10% by increasing the RMD component [58]. The self-compacting concrete (SCC) mixes created with RMD inclusion showed equivalent CMS to the control at 28 days, according to the findings. However, the impacts of red dirt content on CMS were more significant at 56 and 90 days. At 56 days, the strength of the SCC mixture samples containing 30% and 40% RMD was 89.4 MPa and 90.1 MPa, respectively, which were 8% and 9% higher than the control sample [43]. The addition of RMD had no effect on the hydration process, but when the RMD content was more than 20% (by weight of cement), the hydration of cement paste was reduced [54]. The result shows that the CMS of 20 percent partly substituted RMD concrete is higher than that of conventional concrete. This is due to an increase in the cement's pressure quality and bond with the aggregate. As the contribution of bonding between cement and aggregate is reduced as compared to the former, increasing the amount of RMD in concrete does not result in an increase in CMS values [59]. The biggest changes were noted in the 7-day strength improvements, with 13% disparities between the 12.5 percent and 50 percent replacement levels. The major explanation is most likely owing to a high concentration of hatrurite in the cement matrix, which will be explained later. Enhanced RMD content increased CMS somewhat at 56 days, with a maximum difference of less than 4%. As a result, it is possible that RMD concentration has no effect on compressive strength, however, RMD SCC mixes in general had greater CMS than control concrete [48]. The impact of RMD on the properties of cement mortars in terms of setting time, pozzolanic activity, and mechanical strength was explored in the research. They discovered that adding RMD sped up the setting process and lowered the pozzolan reaction and that the CMS fell as the quantity of RMD increased. It was discovered that RMD may be used to partly substitute cement in non-structural mortars and concrete [60]. The compressive strength and TS of a cement mortar, in which RDM substituted up to 50% of the cement, were observed to decline when the RDM concentration was increased [48]. With an increasing amount of replacements, RMD accelerates the curing time of concrete and lowers CMS [61]. According to research, replacing 6% of RMD in concrete boosted CMS by 6%. This might be attributed to microstructure densification, as seen by the decrease in calcium hydroxide concentration after RMD replacement [62]. The strength of the concrete was reduced after 10% RMD substitution, however, it was not lower than regular concrete mixtures. CMS was lowered at 15% and 20% RMD replacement in concrete owing to inadequate cement hydration due to the presence of increased RMD content at higher replacement levels [63]. A similar justification was given by Cheng et al. [64], the large specific surface area of RMD absorbs more water in the concrete mix, resulting in a lack of water for adequate cement hydration. Table 2 shows a summary of CMS with partial substitution of RDM.

Figure 9. Compressive Strength: Data Source [44].

Table 2. Summary of Compressive strength.

Reference	RDM Replacement with Cement	Compression Strength (MPa)
[6]	0%, 2.5%, 5%, 7.5%, and 10%	28 Days: 42, 45, 43, 35, and 30. 56 Days: 45, 47, 45, 40, and 33.
[40]	0%, 5%, 10%, 15%, 20%, and 25%	7 Days: 21, 21, 20, 19, 18, and 17. 28 Days: 28, 27, 26, 25, 22, and 17.
[65]	0%, 10%, 20%, 30%, 40%, 50%, and 60%	28 Days: 62, 40, 110, 92, 78, 50, and 60.
[43]	0%, 10%, 20%, 30%, and 40%	28 Days: 80, 81, 81, 82, and 83. 56 Days: 82, 84, 85, 90, and 90. 90 Days: 90, 91, 92, 100, and 97.
[66]	0%, 1.0 F% + 10 B%, 1.0 F% + 20 B%, and 1.0 F% + 30 B%	7 Days: 26.95, 26, 26.35, and 25.87. 28 Days: 38.87, 35.72, 36.87, and 36. 90 Days: 47.51, 46.21, 46.58, and 44.48.
[44]	0%, 2.5%, 5%, 7.5%, 10%, 12.5%, and 15%	7 Days: 30, 31, 33, 35, 40, 36, and 32. 28 Days: 42, 45, 49, 50, 53, 49, and 45.
[34]	0%, 20%, and 40%	28 Days: 35, 29, and 18. 56 Days: 38, 32, and 18. 90 Days: 41, 35, and 19.
[51]	0%, 5%, 10%, 15%, and 20%	56 Days: 48, 48, 55, 46, and 45. 180 Days: 37, 33, 41, 41, 26, and 28.
[58]	0%, 2.5%, 5%, 7.5%, 10%, 12.5%, and 15%	UTRM: 40, 40, 43, 40, 35, 34, and 32. TRM: 40, 42, 43, 41, 45, 43, and 40.
[59]	0%, 10%, 20%, and 30%	7 Days: 20, 21, 22, and 23. 14 Days: 27, 28, 27, and 27. 28 Days: 32, 31, 35, and 33.

Table 2. *Cont.*

Reference	RDM Replacement with Cement	Compression Strength (MPa)
[46]	0%, 20%, 40%, and 60%	28 Days: 159.7, 139.8, 129.8, and 107.3
[63]	0%, 5%, 10%, 15%, and 20%	7 Days: 30.16, 33.65, 35.67, 34.54, and 31.27. 28 Days: 43.55, 45.34, 48.1, 46.05, and 44.09. 90 Days: 48.2, 55.61, 59.75, 53.79, and 52.18. 150 Days: 49.5, 61.1, 67.5, 60.2, and 57.5.
[67]	0%, 5%, 10%, 15%, and 20%	28 Days: 42, 43, 47, 44, and 41. 56 Days: 45, 46, 49, 46, and 43.
[68]	0, 96, 115, 144, and 192 kg	28 Days: 58, 77, 91, 71, and 70.
[48]	0%, 12.5%, 25%, and 50%	7 Days: 30, 35, 32, and 35. 28 Days: 47, 53, 50, and 50. 56 Days: 57, 58, 60, and 61.
[39]	0%, 10%, 20%, and 30%	7 Days: 21.09, 25.36, 24.64, and 21.53. 28 Days: 33.50, 36.47, 34.26, and 31.78.
[69]	0%, 2%, 4%, 6%, 8%, 10%, and 12%	28 Days: 40, 45, 40, 36, 34, 33, and 31.

Fiber = F; Bauxite = B; RMD = RDM; Treated RMD = UTRM; Untreated RMD = TRM.

Figure 10 shows a relative examination of concrete CMS with various amounts of RDM. For testing, the optimal dosage of RDM (10%) is used. The reference concrete is the control concrete's CMS after 7 days. The CMS of RDM concrete with a 10% replacement is 33 percent higher than reference concrete after 7 days of curing. The CMS of 10% RDM replacement is 76 percent higher than reference concrete CMS after 28 days of curing. Furthermore, after 7 and 28 days of curing, all the RDM replaced mix had higher CMS than reference concrete (control concrete CMS at 7 and 28 days).

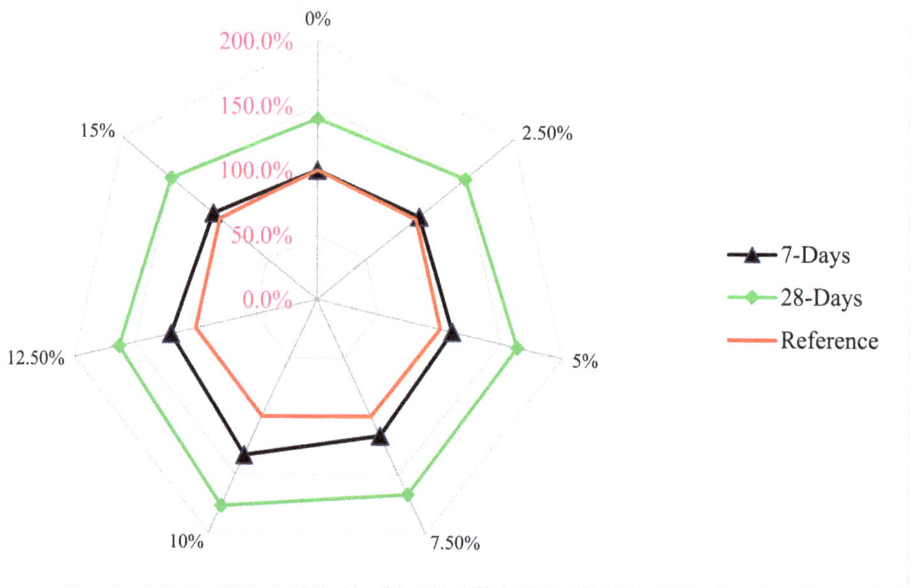

Figure 10. Relative Strength: Data Source [44].

3.2. Flexural Strength (FLS)

Figure 11 depicts the FLS of concrete with various RDM replacement ratios. In a similar approach to compressive strength, maximum FLS is reached at 10% RDM replacement. However, according to one study, FLS diminishes as the amount of RDM increases [40]. The FLS of concrete dropped by 1.6, 10.7, 17, 17.6 and 28.4% for RDM levels of 5, 10, 15, 20, and 25% replacement, respectively, compared to the reference sample with no RDM [40]. Tang et al. [70] found that increasing the RDM concentration as a cement substitute reduced flexural strength. According to research [71], increasing the RDM content of a hybrid composite produced using RDM as a filler and sisal fiber as the reinforcement in a polyester matrix boosts the tensile and impact strengths by up to 20%. Under tensile and impact loadings, the composite showed good resistance to fracture formation and propagation. Another study found that adding RDM to a polyester composite reinforced with sisal and banana fibers improved impact and flexural strength, making it ideal for applications requiring high load-bearing capability [72]. Ganeshan et al. [73] discovered that adding RDM to natural fiber–polyester composites significantly boosted the flexural strengths of the polyester composites while lowering the TSs. The findings of research on self-compacted concrete using a mixture of 10% iron ore and 2% RDM by cement weight indicated that the flexural, tensile, and compressive strengths were raised [74]. According to research, when RDM concentration grows, compressive and FLS falls, yet, 5% of RDM addition produces superior results [75]. A study explored using RDM as a substitute for fly ash in self-compacting mortar and concrete. RDM shows lowered the flowability of mortar and concrete but increased their compressive and flexural strengths [76]. The 28-day FLS of ultra-high-performance concrete (UHPC) with 20%, 40%, 60%, and 70% RDM (by volume and mass) is 41.6 MPa, 37.6 MPa, 35.3 MPa, 19.6 MPa, 40.7 MPa, and 32.4 MPa, respectively. Due to the toughening effect of steel fiber, the drop in FLS is not as noticeable as the decrease in compressive strength. Steel fiber toughening is the most important factor in flexural strength. As a consequence, the FLS loss caused by RMD is less noticeable than the CMS deterioration [46]. The FLS of concrete with 30% RDM substitution is greater than the other test samples [59]. Table 3 shows a summary of FLS with partial substitution of RDM. It can also be noticed that fewer studies consider FLS of concrete with RDM.

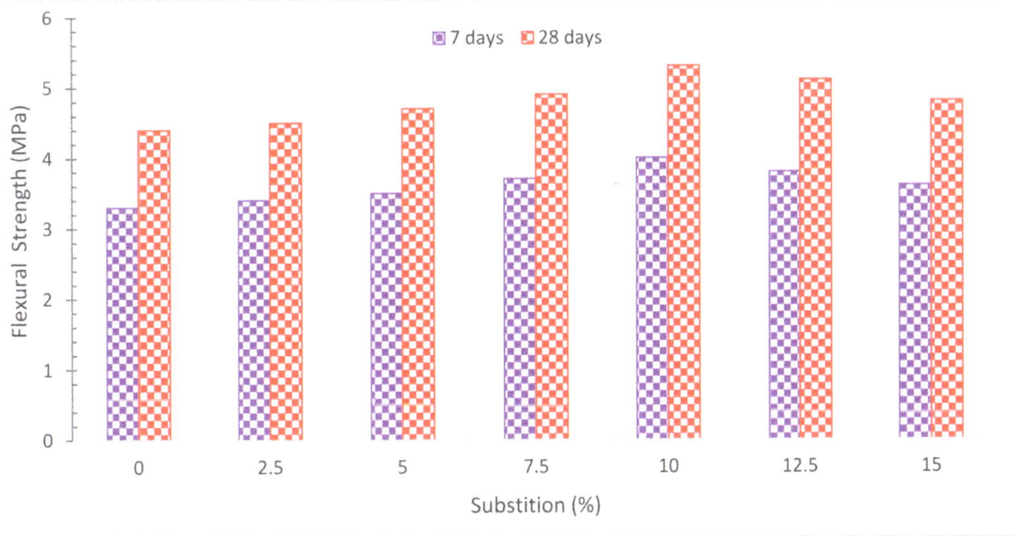

Figure 11. Flexural Strength: Data Source [44].

Table 3. Summary of Flexural strength.

Reference	RDM Replacement with Cement	Flexure Strength (MPa)
[40]	0%, 5%, 10%, 15%, 20%, and 25%	28 Days: 8.0, 8.0, 7.1, 6.8, 6.7, and 6.0.
[65]	0%, 10%, 20%, 30%, 40%, 50%, and 60%	28 Days: 52, 30, 22, 26, 15, 13, and 24.
[44]	0%, 2.5%, 5%, 7.5%, 10%, 12.5%, and 15%	7 Days: 3.3, 3.4, 3.5, 3.7, 4.0, 3.8, and 3.6. 28 Days: 4.4, 4.5, 4.7, 4.9, 5.3, 5.1, and 4.8.
[51]	0%, 5%, 10%, 15%, and 20%	28 Days: 5.0, 5.5, 6.2, 5.8, and 6.1.
[58]	0%, 2.5%, 5%, 7.5%, 10%, 12.5%, and 15%	UTRM: 4.4, 4.3, 4.5, 4.4, 4.1, 4.0, and 3.9. TRM: 4.4, 4.4, 4.4.4.5, 4.5, 4.4, and 4.4.
[59]	0%, 10%, 20%, and 30%	28 Days: 6.2, 4.3, 6.3, and 7.0.
[46]	0%, 20%, 40%, and 60%	42.43, 41.6, 37.6, and 35.5.

Figure 12 depicts the relationship between compressive and FLS with various RDM replacement ratios at 7 and 28 days of curing. With an R square value larger than 0.90, a substantial connection between CMS and FLS may be noticed. As a result, the linear regression equation may be used to forecast the FLS of concrete.

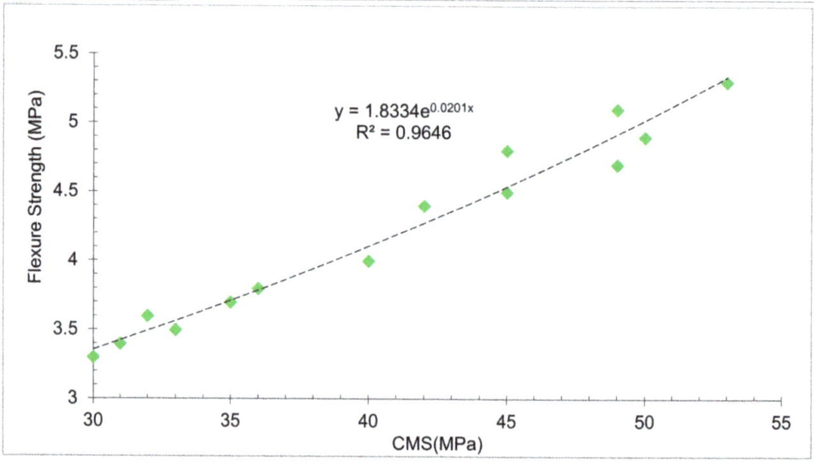

Figure 12. Correlation between Flexural and Compressive Strength: Data Source [44]. Green diamond is the data point.

3.3. Split Tensile Strength (TS)

The TS of concrete with different substitution ratios of RDM is shown in Figure 13. It can be observed that maximum TS is achieved at 10% substitution of RDM in a similar way to compressive strength. According to research [63], the strength of concrete improved up to 10% substitution of RDM, and more substations of RDM after 10% caused the strength to decrease. Another study found that when up to 2.5 percent of cement was replaced with RDM, the splitting TS rose before declining as the RDM content was increased [6]. When compared to normal concrete, the strength parameters of concrete, such as compressive strength and TS, improve with a 20% substitution of RDM [59]. Increases in RDM content seem to result in minor increases in CMS and elastic modulus, as well as a little loss in TS [48]. According to research, the SCC with 2% RMD and 10% iron ore tailings had the greatest compressive, tensile and flexural strengths [77]. Furthermore, the internal curing of the RDM might be the cause of TS growth at higher curing ages [78]. There was no significant variation in TS across any of the samples, particularly after 56 days. The inclusion of RMD had no significant effect on the TS of SCC, according to the findings [48].

For 28 and 56 days, the RMD concrete had lower TS than the control, but for 90 days, the RMD concrete had higher TS than the control. The porosity of RMD was credited with increasing its tensile and compressive strength. It was said that RMD absorbed water and then released it later to help with hydration [43]. The optimal value of split TS was reached by replacing 10% of the cement with neutralized RMD and adding 5% of hydrated lime [39]. Furthermore, Table 4 shows summary of TS with partial substitution of RDM.

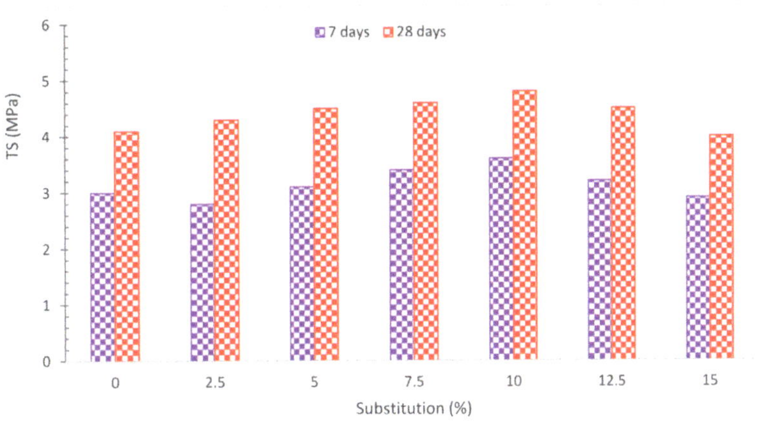

Figure 13. Tensile Strength: Data Source [44].

Table 4. Summary of TS.

Reference	RDM Replacement with Cement	Split TS (MPa)
[6]	0%, 2.5%, 5%, 7.5%, and 10%	28 Days: 4.8, 5.2, 4.8, 3.9, and 3.5. 56 Days: 5.1, 5.3, 5.1, 4.3, and 3.8.
[40]	0%, 5%, 10%, 15%, 20%, and 25%	7 Days:1.8, 1.7, 1.6, 1.6, 1.6, and 1.5. 28 Days: 2.8, 2.7, 2.5, 2.3, 2.2, and 1.8.
[65]	0%, 10%, 20%, 30%, 40%, 50%, and 60%	28 Days: 30, 20, 13, 08, 10, 07, and 04.
[43]	0%, 10%, 20%, 30%, and 40%	28 Days: 5.2, 5.2, 5.0, 5.5, and 5.8. 56 Days: 5.7, 5.4, 5.2, 5.8, and 5.9. 90 Days: 6.0, 6.0, 6.0, 7.2, and 7.3.
[66]	0%, 1.0 F% + 10 B%, 1.0 F% + 20 B%, and 1.0 F% + 30 B%	28 Days: 1.73, 2.16, 2.34, and 1.99. 56 Days: 2.42, 2.76, 3.0, and 2.37. 90 Days: 2.97, 3.30, 3.59, and 2.75.
[44]	0%, 2.5%, 5%, 7.5%, 10%, 12.5%, and 15%	7 Days: 3.0, 2.8, 3.1, 3.4, 3.6, 3.2, and 2.9. 28 Days: 4.1, 4.3, 4.5, 4.6, 4.8, 4.5, and 4.0.
[58]	0%, 2.5%, 5%, 7.5%, 10%, 12.5%, and 15%	UTRM: 3.8, 4.1, 4.3, 4.1, 3.8, 3.7, and 3.7. TRM: 3.8, 4.2, 4.2, 4.2, 4.5, 4.2, and 4.1.
[59]	0%, 10%, 20%, and 30%	28 Days: 2.9, 2.5, 3.3, and 3.0.
[67]	0%, 5%, 10%, 15%, and 20%	28 Days: 4.2, 4.8, 5.0, 4.6, and 4.4. 56 Days. 4.6, 5.0, 5.2, 4.8, and 4.6.
[68]	0 kg, 96 kg, 115 kg, 144 kg, and 192 kg	28 Days:7.5, 9.5, 10.3, 9.7, and 8.7.
[48]	0%, 12.5%, 25%, and 50%	28 Days: 4.6, 4.7, 4.4, and 4.6. 56 Days: 4.8, 4.8, 5.0, and 4.9.
[39]	0%, 10%, 20%, and 30%	7 Days: 2.1, 2.6, 2.2, and 2.1. 28 Days: 2.1, 2.6, 2.2, and 2.1.

Figure 14 shows the correlation between the CMS and TS with different substitution ratios of RDM at 7 and 28 days' curing. It can be seen that a strong correlation between CMS and TS exists with an R square value greater than 0.90. Therefore, the equation developed based on linear regression can be used to predict the TS of concrete.

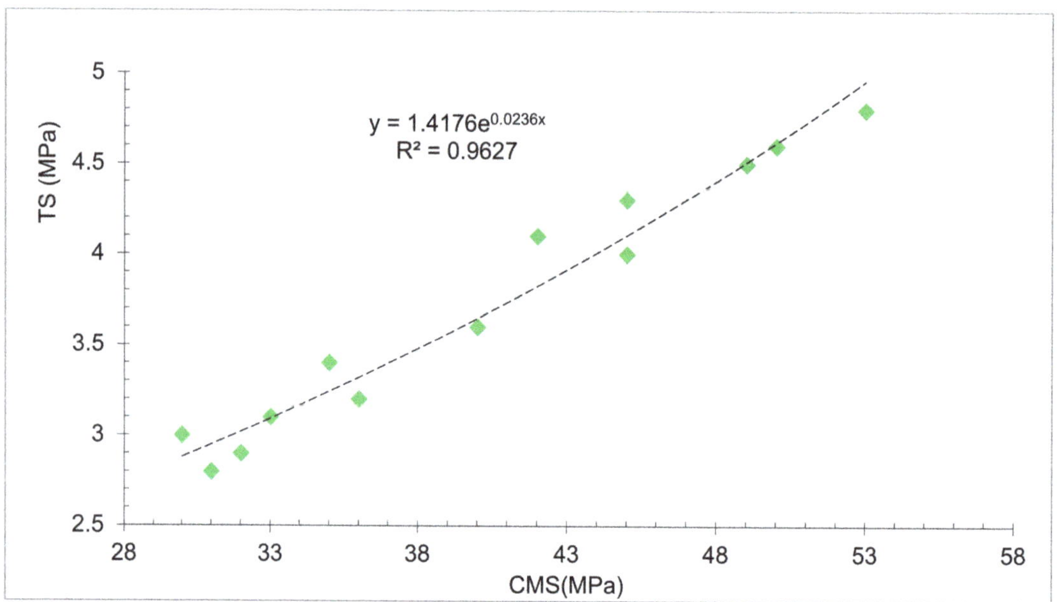

Figure 14. Correlation between Tensile and Compressive Strength: Data Source [44]. Green diamond is data point.

4. Durability

4.1. Water Absorption

As indicated in Figure 15, increasing the amount of RDM replacement reduces the percentage of water absorption. The water absorption test was performed after a certain curing age, such as 7, 28, 90, and 150 days. The goal is to determine the water absorption resistance of RMD concrete as hydration progresses. The findings show that when the curing age increases and the replacement amount of RDM increases, the water absorption values decrease. The micro filling effect of pozzolanic material which gives more dense concrete by filling voids results in decreases in water absorption. Furthermore, due to the pozzolanic reaction, the binding properties of concrete paste also contribute to the improved density of concrete leading to lower water absorption. The combined micro filling voids and pozzolanic reaction have a positive influence on the water absorption of concrete [52]. RMD promotes pozzolanic activity at a later age, resulting in fewer connections between pores. The fineness of RMD particles (average particle size 14 microns) is another cause of less water absorption; all micro-cracks and holes in the concrete are filled. As a result of its enormous specific surface area, RMD may help concrete absorb less water [58]. The greater $Ca(OH)_2$ crystals were fractured into multiple tiny crystals and less orientated in the RMD-based cement hydration process, which leads to minimizing pore connections and water absorption, according to Manfroi et al. 2014 [79]. Due to the existence of multiple "pits" and "folds" on the surface of the RM particles, increasing the RDM concentration had a negative influence on the water absorption of SCC, and therefore the capacity of absorbing water increased [43]. Overall, it can be claimed that as the RDM ratio grew, the ability of concrete to absorb water improved. A concrete containing 2.5 percent RDM, on the other hand, behaved similarly to the control mix [6]. The results revealed that using

15, 20, and 25% RDM as a cement substitution increased water absorption by 23 and 30%, respectively, as compared to a control specimen with no RDM. According to other research, the key explanation for the enhanced water absorption is the increased porosity caused by the usage of RDM [60,80] in terms of quality, it divides concrete into three categories: bad, average, and excellent, with water absorption of 0–3 percent, 3–5 percent, and 5% or more, respectively. Water absorption levels of over 3% and below 5% were found in all concrete specimens evaluated in this study, putting the concrete in the average absorption category. According to another study [81], high-quality concrete has a water absorption rate of less than 5%.

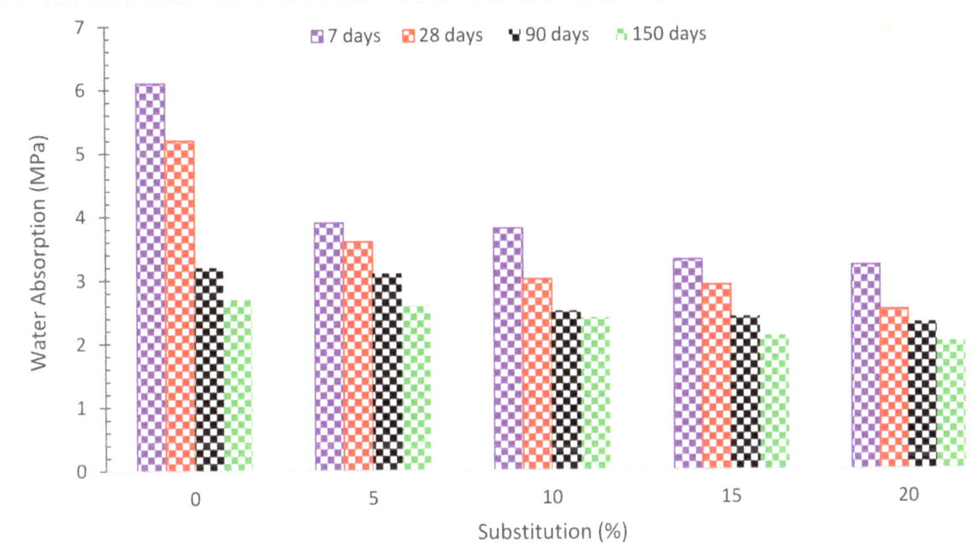

Figure 15. Water Absorption [63].

4.2. Chloride Permeability

The resistance of the samples to chloride ion penetration was determined by testing the chloride permeability of the RDM-based concrete. The RCPT was performed in accordance with ASTM C 1202 [82] and the charge that travelled through the samples was measured in coulombs. Figure 16 shows the total charges that travelled through concrete samples made RDM and cured for 28 and 56 days, respectively. With increasing curing age, the resistance to chloride-ion penetration definitely increased. The reason for this is that as curing age increases, hydration products develop. The RDM-based concrete had superior chloride-ion penetration resistance than the control samples. As RDM is alkaline, it increased the resistance of the concrete to chloride ion penetration and hence reduced the total charge transferred. Thus, substituting RDM for cement in the concrete reduced the charge transmitted, indicating increased resistance to chloride ion penetration. Research indicated that tiny RM particles are responsible for the decrease of chloride ions penetration and carbonation depth [83]. The combined pozzolanic and filling voids, RDM improved the resistance of concrete against chloride-ion penetration.

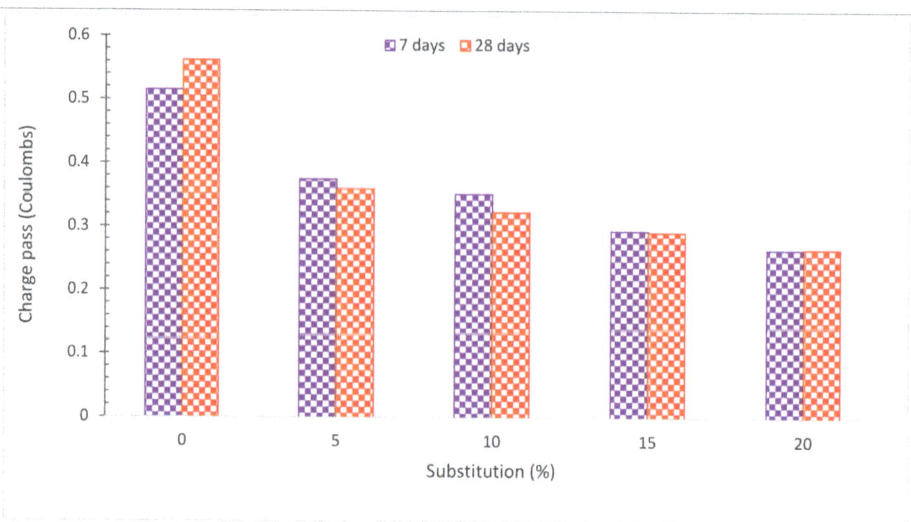

Figure 16. Chloride Permeability [67].

4.3. Sorptivity Test

With an increase in the degree of RMD replacements in concrete, the sorptivity values displayed in Figure 17 decreased. The sorptivity values are lowered after 28 days of curing from 0.562 mm/min$^{0.5}$ (control) to 0.266 mm/min$^{0.5}$ (20% RDM) which is ascribed to the fineness of RDM, which makes the concrete surface extremely thick by filling all the spaces. This filling property would aid in the development of a continuous pozzolanic reaction between RMD and cement, improving the strength and durability of concrete over time. As C–S–H filled all of the capillary holes during hydration, microstructure analysis indicated that replacing RMD in concrete lowered sorptivity values. This occurs because RMD fills holes and cracks in concrete, which results in lower sorptivity [83].

Figure 17. Sorptivity of Concrete [63].

4.4. Surface Resistivity Test

The surface resistivity test may be performed to determine the electrical resistivity of water-saturated concrete and provide a quick indication of its resistance to chloride ion penetration. The resistance to current leakage along the surface of insulating material is characterized by the substance's surface resistivity. A voltage is transmitted between the two electrodes and a test specimen is placed between them. A resistivity meter is used to determine the value [84]. The surface resistivity of RMD-replaced concrete specimens is similar to that of normal concrete, according to test findings. The test specimens' resistivity values are shown in Figure 18. Less information is available on the surface resistivity of concrete with RDM, and detailed investigation is required.

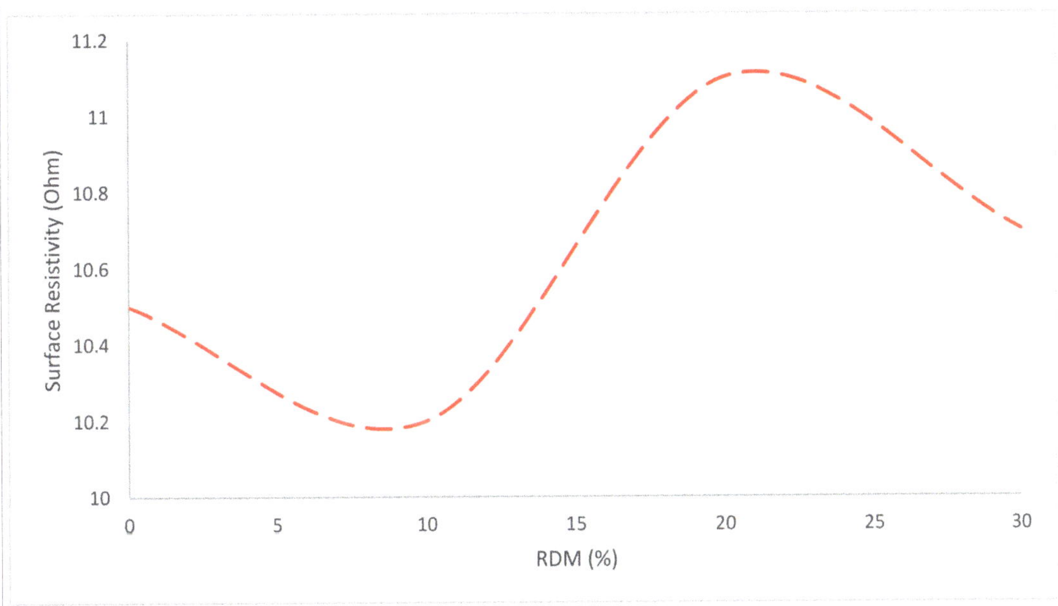

Figure 18. Surface Resistivity [59].

5. Microstructure Analysis

5.1. Interfacial Transition Zone

The area between aggregate and paste is known as the interfacial transition zone (ITZ). When a porous fracture separates the two locations, the ITZ is at its weakest. The ITZ is highly established and must correspond to strong strength if there is just a little fracture evident or it virtually seems like one uniform surface. Figure 19 shows SEM images of ITZ of reference and RMD concrete. The dark fissures in (0% RDM) ITZ were used to identify the porosity. As demonstrated in Figure 19a, the fly ash did not bind effectively with the cement paste or aggregate, resulting in a loss in compressive and TS. RDM 12.5%, RDM 25%, and RDM 50% seem to have the same porosity ITZ as the control. The cracks of RDM 12.5%, RDM 25%, and RDM 50% were comparable in breadth to the controlled crack. In terms of porosity and microfractures, RMD concrete samples bonded similarly to control concrete samples. It can be seen from a comparison of the SEM pictures of the different mixes that there was no significant variation in ITZ for all mixes in terms of penetrability and fracture size. All blends may have identical compressive and TS [48].

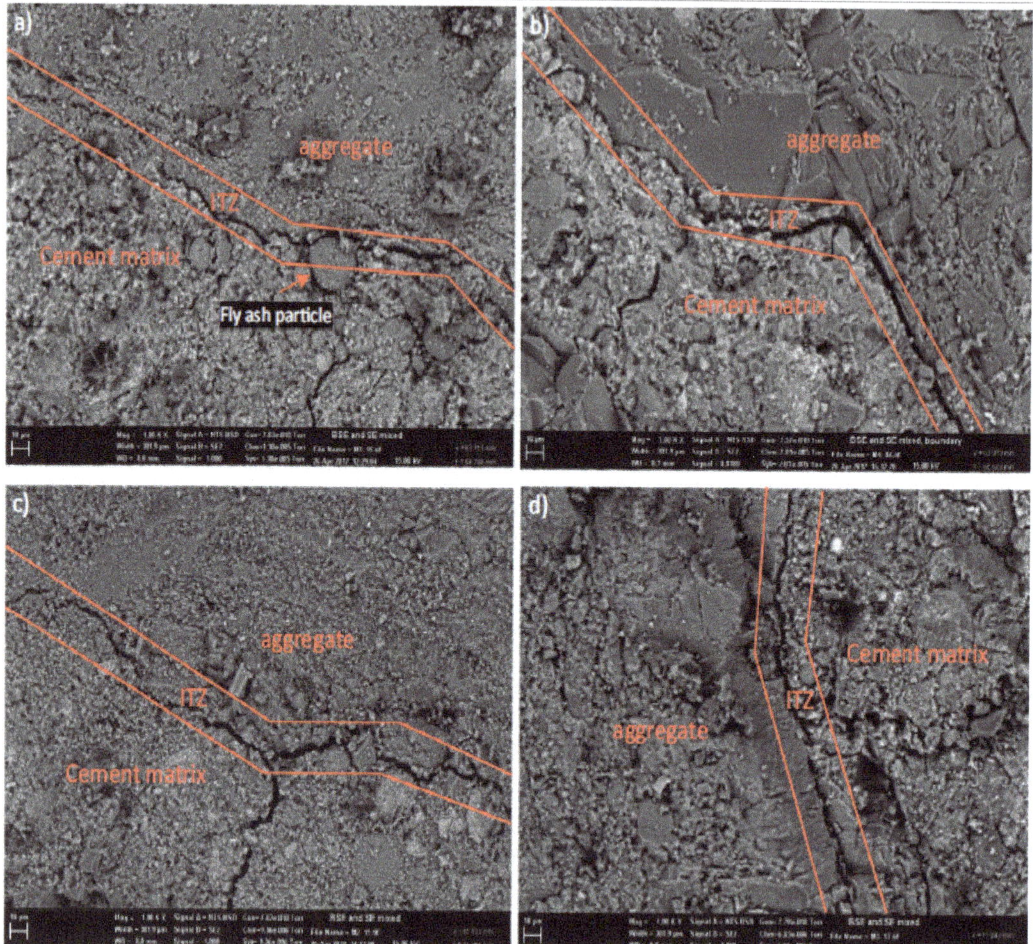

Figure 19. Interfacial Transition Zone (ITZ): As per Elsevier Permission [48]. (**a**) 0% (**b**) 12.5% (**c**) 25% and (**d**) 50% RMD.

When cement is hydrated, quartz and calcium oxide react, assisting in the development of CSH gel. The presence of larnite and hatrurite in the red mud concrete provides conclusive proof that CSH gel was formed, giving the concrete strength qualities. According to research [63], the Ca/Si ratios are 1.13, 0.99, 0.95, 1.01, and 1.08 for red mud replacement levels of 0%, 5%, 10%, 15%, and 20%, respectively. Based on the findings, concrete with 10% red mud replacement has a lower Ca/Si ratio than other mixtures. The creation of CSH gels increases with decreasing Ca/Si ratios and decreases with increasing Ca/Si ratios [67], which accounts for the increased strength shown in mixes with 10% red mud replacement. The findings demonstrate that RMD has a good pozzolanic activity that is comparable to that of fly ash [43].

5.2. Energy Dispersive Spectroscopy (EDS)

EDS was employed on the hardened cement paste samples during the SEM investigation. This would allow researchers to establish what compounds appear and whether the structure is modified significantly when RMD concentration increased compared to the control. Since calcium hydroxide intensity impacts the Ca/Si ratio of the CSH and the creation of the CSH network, the calcium/silicon (Ca/Si) ratio was tested to see how effectively the cement had formed [85]. CSH is critical for the development of strength as it works as a binder [86]. According to research, there was no significant variance in ITZ for all mix in terms of penetrability and fracture size when comparing SEM images of the various mixtures. As a result, the compressive and TSs of all mixtures may be similar [48]. The sites of the EDS analyses on the RMC0 simple are displayed in Figure 20. Table 5 reveals the quantitative findings of the elemental. The EDS data show that there was a high concentration of calcium and silicate, with a Ca/Si ratio of 1.44. C-S-H values for concrete was typically about 1.7, with CH being somewhat higher [87].

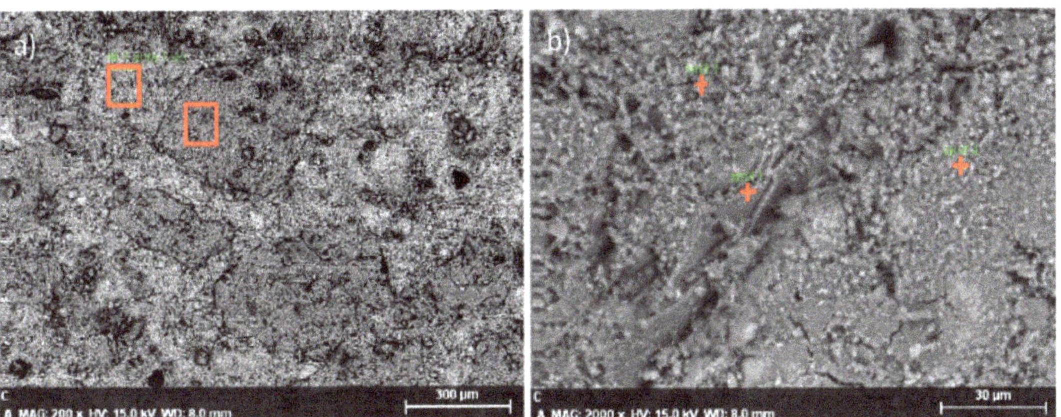

Figure 20. EDS Results: As per Elsevier Permission [48]. (**a**) areas (**b**) spots.

Table 5. Energy dispersive spectroscopy (EDS) Results [48].

Chemical Name	Control	RDM-12.5%	RDM-25%	RDM-50%
Na	0.74	1.70	1.43	5.53
Mg	0.69	0.57	0.87	1232.47
Al	4.09	12.82	4.99	13.03
Si	37.38	33.98	50.05	28.03
P	-	-	-	0.08
S	0.67	0.08	0.91	2.39
CI	-	-	-	0.28
K	0.86	3.40	1.30	1.60
Ca	53.96	42.87	38.28	37.76
Ti	-	1.51	-	0.92
Fe	1.61	3.07	2.17	8.20

6. Environmental Impact Analysis

Figure 21 depicts the effects of CO_2 vs. RDM replacement in concrete. As can be shown in Figure 21, increasing the RDM concentration from 0% to 25% reduces CO_2 from 556.8 to 409.9 kg·m^3. As a result, CO_2 rises in proportion to the amount of cement. In general, it can be said that cement is the source of the most CO_2 releases. As a result, even when delivery distances are much larger than the cement delivery distance, the usage of RDM is viable.

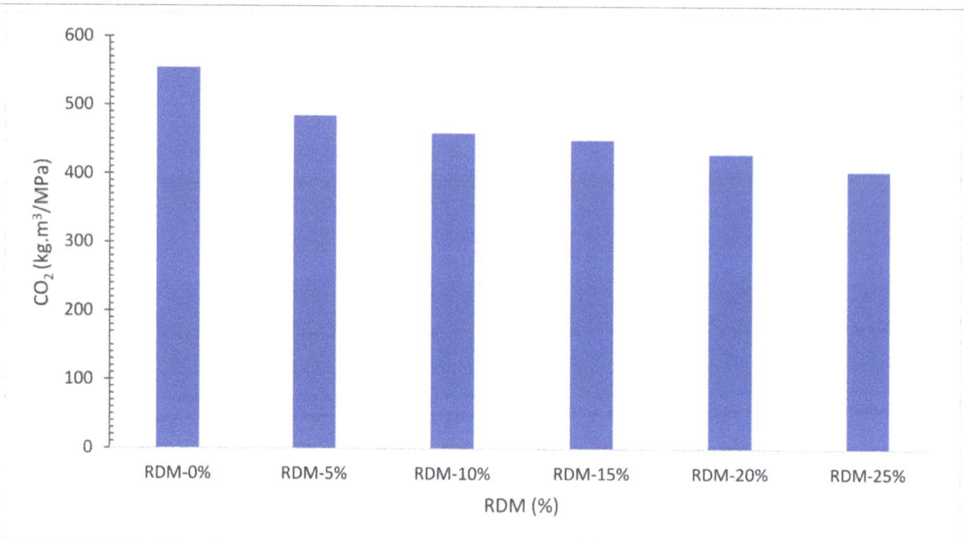

Figure 21. GWP during Concrete Production: Data Source [40].

The concentration of CO_2 discharges per unit volume of concrete per 28-day strength characteristics (compressive and tensile) for the concrete employed in the research are displayed in Figure 22. As this index allows for the consideration of both performance (strength) and contribution of concrete to GWP per unit volume and strength, it is characterized as a good alternative for assessing the various effects of concrete usage [88]. As can be observed, the concentration of CO_2 discharges normalized by various strength qualities including RDM up to 20% substitute is less than the intensity of CO_2 emissions for a reference specimen with no RDM.

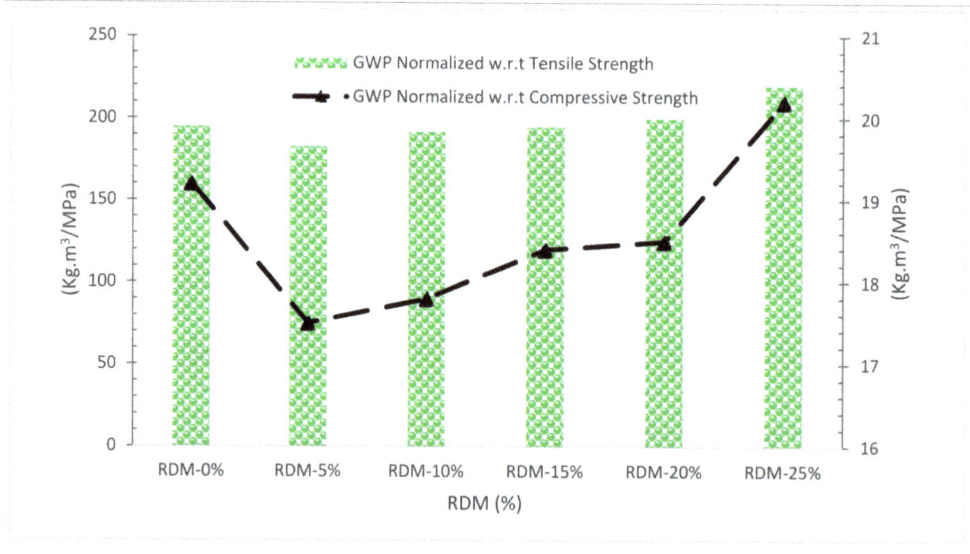

Figure 22. Normalized GWP W.r.t Compressive strength and TS: Data Source [40].

The energy needed throughout the life cycle of lightweight concrete is shown by CED in this research. Figure 23 shows a graphical evaluation of normalized CED in concrete with various RDM contents. As can be observed, improving the RDM from 0 to 1% reduces the value of CED by around 31%.

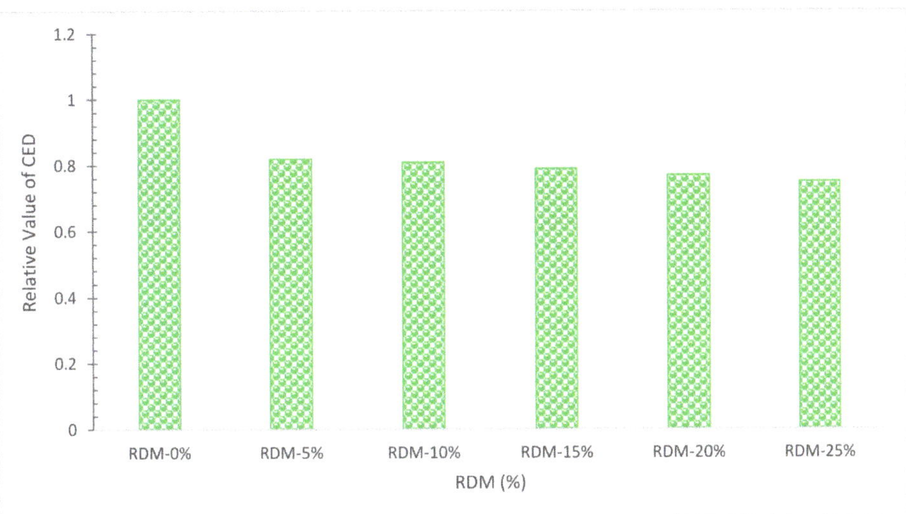

Figure 23. CED for Concrete with various percentages of RDM: Data Source [40].

The comparison of normalized main criterion air pollutants in concrete with varying RDM concentrations is displayed in Figure 24. It can be examined, that as the RDM content boosts, the amounts of CO, NO_X, Pb, and SO_2 decreased by roughly 32.5 percent, 17.1 percent, 31.8 percent, and 22.4 percent, respectively, as compared to 0 percent RDM. In general, the primary criterion air pollutants of CO, NO_X, Pb, and SO_2 rise with increased cement concentration owing to fuel-burning during the pyro processing step.

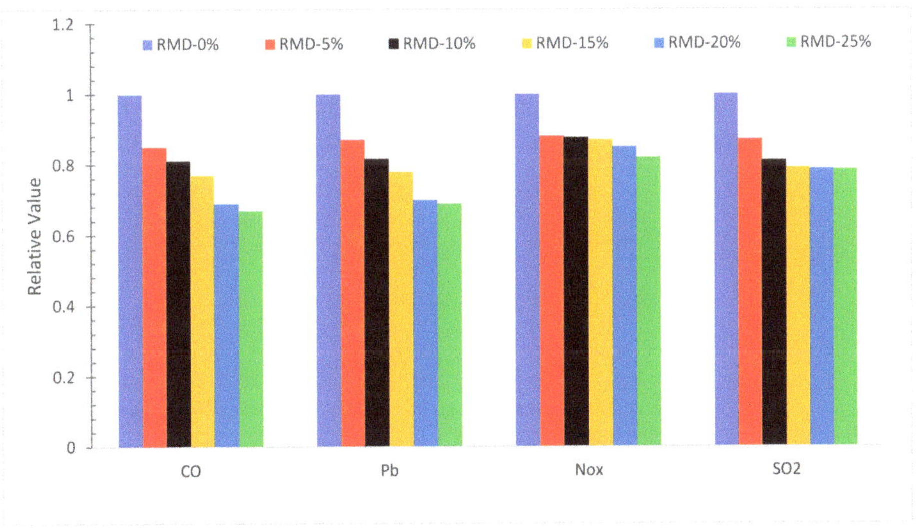

Figure 24. Air Pollution for Concrete with various percentages of RDM: Data Source [40].

7. Conclusions

The use of RMD as a cement substitute provides a variety of environmental and economic benefits, including reduced soil and groundwater contamination, reduced dust pollution, conservation of natural resources for cement clinker, and lower concrete construction costs. The goal of this analysis is to provide a comprehensive overview of current progress in the use of RMD in concrete production, as well as to clearly identify four directions for using RMD in concrete production: fresh properties, mechanical properties, durability aspects, and environmental aspects. The detailed conclusion is given below.

- The chemical composition of RDM shows that it can be used as pozzolanic material.
- The flowability of concrete decreased as the substitution ratio of RDM increased due to its porous nature.
- Mechanical performance such as compressive strength, flexural strength and TS improved with the substation of RDM up to a certain level. Maximum CMS was achieved at 10% substitution of RDM which is 43% higher than reference concrete compressive strength. Further, the substitution of RDM results in the decreased mechanical performance of concrete. Therefore, finding an optimum is important for maximum performance. Different researchers recommend a different optimum dose of RDM. This might be possible due to different sources of RDM. However, the majority of researchers recommend 10 to 15% substitution of RDM as an optimum dose.
- Water absorption and chloride permeability decreased concrete considerably with substitution RDM. However, less information is available in this regard.
- The velocity of ultrasonic waves is reduced when the RDM concentration is increased.
- SEM results show that the substation of RDM improved the interfacial transition zone (ITZ). It is a result of the micro filling effect of RDM which fills the crack (ITZ), leading to more dense concrete.
- EDX results ensure the pozzolanic activity, creating additional compounds (CSH), which enhanced the cementitious properties of the paste.
- CO_2, a rate of worldwide warming, reduces from 556.8 to 409.9 $Kg \cdot m^3$ as RDM content rises from 0% to 25%, showing that concrete sustainability improves as RDM content increases.
- When the RMD content is increased from 0% to 25%, the amounts of CED, CO, NO_X, Pb, and SO_2 are reduced by around 31%, 32.5 percent, 31.8 percent, 17.1 percent, and 22.4 percent, respectively.

Overall, the analysis reveals that RDM has the potential to be used as a cement substitute. Decreased soil and groundwater contamination, reduced dust pollution, conservation of ecological assets, and cheaper concrete production costs are some of the environmental and economic advantages of using RMD as a cement alternative in concrete manufacturing.

8. Recommendation

- Fewer data are accessible on durability aspects particularly dry shrinkage and creeps. Therefore, this review recommends a detailed investigation of dry shrinkage and creep properties of concrete with RDM.
- Different methods, such as thermal activation or alkali activation, should be applied to improve the pozzolanic activity of RDM.
- Thermal conductivity and heat insulation characteristics with RMD should be investigated in detail.
- Although RDM improved the strength of concrete, but concrete is still weak in tension. Therefore, this review also recommends fibers in RDM-based concrete to enhance the tensile capacity.

Author Contributions: Writing—original draft preparation, H.J.Q. and J.A.; Conceptualization, J.A., A.M. and H.J.Q.; methodology, J.A.; software, H.J.Q. and A.M.; resources, A.M.; writing—review and editing, M.U.S., A.F.A.F. and M.A.; project administration, M.U.S. and M.A.; funding acquisition, M.A., A.M. and M.U.S. All authors have read and agreed to the published version of the manuscript.

Funding: Deanship of Scientific Research, Vice Presidency for Graduate Studies and Scientific Research, King Faisal University, Saudi Arabia (Project No: GRANT 1020).

Institutional Review Board Statement: Not applicable.

Informed Consent Statement: Not applicable.

Data Availability Statement: All the data available in main text.

Acknowledgments: This work was supported by the Deanship of Scientific Research, Vice Presidency for Graduate Studies and Scientific Research, King Faisal University, Saudi Arabia (Project No: GRANT 1020). The authors would like to acknowledge the technical and instrumental support they received from King Faisal University, Saudi Arabia.

Conflicts of Interest: The authors declare no conflict of interest.

References

1. Junaid, M.F.; Rehman, Z.U.; Kuruc, M.; Medved', I.; Bačinskas, D.; Čurpek, J.; Čekon, M.; Ijaz, N.; Ansari, W.S. Lightweight concrete from a perspective of sustainable reuse of waste byproducts. *Constr. Build. Mater.* **2021**, *319*, 126061. [CrossRef]
2. Kim, S.K.; Kang, S.T.; Kim, J.K.; Jang, I.Y. Effects of particle size and cement replacement of LCD glass powder in concrete. *Adv. Mater. Sci. Eng.* **2017**, *2017*, 3928047. [CrossRef]
3. Kurad, R.; Silvestre, J.D.; De Brito, J.; Ahmed, H. Effect of incorporation of high volume of recycled concrete aggregates and fly ash on the strength and global warming potential of concrete. *J. Clean. Prod.* **2017**, *166*, 485–502. [CrossRef]
4. Pacheco-Torgal, F.; Cabeza, L.F.; Labrincha, J.; De Magalhaes, A. Assessing the Environmental Impact of Conventional and "green" Cement Production. In *Eco-Efficient Construction and Building Materials: Life Cycle Assessment (LCA), Eco-Labeling and Case Studies*; Woodhead Publishing: Sawston, UK, 2013; ISBN 9780857097675.
5. Shekhawat, B.S.; Aggarwal, V. Utilisation of waste glass powder in concrete-a literature review. *Int. J. Innov. Res. Sci. Eng. Technol.* **2007**, *3297*, 2319–8753.
6. Ghalehnovi, M.; Roshan, N.; Hakak, E.; Shamsabadi, E.A.; De Brito, J. Effect of red mud (bauxite residue) as cement replacement on the properties of self-compacting concrete incorporating various fillers. *J. Clean. Prod.* **2019**, *240*, 118213. [CrossRef]
7. Smirnova, O.M.; De Navascués, I.M.P.; Mikhailevskii, V.R.; Kolosov, O.I.; Skolota, N.S. Sound-absorbing composites with rubber crumb from used tires. *Appl. Sci.* **2021**, *11*, 7347. [CrossRef]
8. Smirnova, O. Compatibility of shungisite microfillers with polycarboxylate admixtures in cement compositions. *ARPN J. Eng. Appl. Sci.* **2019**, *14*, 600–610.
9. Alvee, A.R.; Malinda, R.; Akbar, A.M.; Ashar, R.D.; Rahmawati, C.; Alomayri, T.; Raza, A.; Shaikh, F.U.A. Experimental study of the mechanical properties and microstructure of geopolymer paste containing nano-silica from agricultural waste and crystalline admixtures. *Case Stud. Constr. Mater.* **2022**, *16*, e00792. [CrossRef]
10. Jani, Y.; Hogland, W. Waste glass in the production of cement and concrete—A review. *J. Environ. Chem. Eng.* **2014**, *2*, 1767–1775. [CrossRef]
11. He, Z.; Zhu, X.; Wang, J.; Mu, M.; Wang, Y. Comparison of CO_2 emissions from OPC and recycled cement production. *Constr. Build. Mater.* **2019**, *211*, 965–973. [CrossRef]
12. Nelson, J.; Grayson, D. World business council for sustainable development (WBCSD). In *Corporate Responsibility Coalitions*; Routledge: London, UK, 2017; pp. 300–317.
13. Meyer, C.; Xi, Y. Use of recycled glass and fly ash for precast concrete. *J. Mater. Civ. Eng.* **1999**, *11*, 89–90. [CrossRef]
14. Siddique, R. Utilization of industrial by-products in concrete. *Procedia Eng.* **2014**, *95*, 335–347. [CrossRef]
15. Althoey, F.; Farnam, Y. The effect of using supplementary cementitious materials on damage development due to the formation of a chemical phase change in cementitious materials exposed to sodium chloride. *Constr. Build. Mater.* **2019**, *210*, 685–695. [CrossRef]
16. Brito, J.; Silva, R. Use of waste materials in the production of concrete. In proceedings of the key engineering materials. *Trans. Tech. Publ.* **2015**, *634*, 85–96.
17. Althoey, F. Compressive strength reduction of cement pastes exposed to sodium chloride solutions: Secondary ettringite formation. *Constr. Build. Mater.* **2021**, *299*, 123965. [CrossRef]
18. Isler, J.W. *Assessment of Concrete Masonry Units Containing Aggregate Replacements of Waste Glass and Rubber Tire Particles*; University of Colorado at Denver: Denver, CO, USA, 2012; ISBN 1267297336.
19. Agency, I.E.; Agency, I.E. *Tracking Industrial Energy Efficiency and CO_2 Emissions*; OECD: Paris, France, 2007; ISBN 9789264030169.
20. Benhelal, E.; Zahedi, G.; Shamsaei, E.; Bahadori, A. Global strategies and potentials to curb CO_2 emissions in cement industry. *J. Clean. Prod.* **2013**, *51*, 142–161. [CrossRef]

21. Faraj, R.H.; Ali, H.F.H.; Sherwani, A.F.H.; Hassan, B.R.; Karim, H. Use of recycled plastic in self-compacting concrete: A comprehensive review on fresh and mechanical properties. *J. Build. Eng.* **2020**, *30*, 101283. [CrossRef]
22. Ahmad, J.; Martínez-García, R.; De-Prado-Gil, J.; Irshad, K.; El-Shorbagy, M.A.; Fediuk, R.; Vatin, N.I. Concrete with partial substitution of waste glass and recycled concrete aggregate. *Materials* **2022**, *15*, 430. [CrossRef]
23. Amin, S.K.; Allam, M.E.; Garas, G.L.; Ezz, H. A study of the chemical effect of marble and granite slurry on green mortar compressive strength. *Bull. Natl. Res. Cent.* **2020**, *44*, 19. [CrossRef]
24. Shannag, M. High strength concrete containing natural pozzolan and silica fume. *Cem. Concr. Compos.* **2000**, *22*, 399–406. [CrossRef]
25. Naganur, C.J. Effect of copper slag as a partial replacement of fine aggregate on the properties of cement concrete. *Int. J. Res.* **2014**, *1*, 8.
26. Singh, G.; Siddique, R. Effect of waste foundry sand (WFS) as partial replacement of sand on the strength, ultrasonic pulse velocity and permeability of concrete. *Constr. Build. Mater.* **2012**, *26*, 416–422. [CrossRef]
27. Salman, N.M.; Ma, G.; Ijaz, N.; Wang, L. Importance and potential of cellulosic materials and derivatives in extrusion-based 3D concrete printing (3DCP): Prospects and challenges. *Constr. Build. Mater.* **2021**, *291*, 123281. [CrossRef]
28. Behera, M.; Bhattacharyya, S.; Minocha, A.; Deoliya, R.; Maiti, S. Recycled aggregate from C&D waste & its use in concrete—A breakthrough towards sustainability in construction sector: A review. *Constr. Build. Mater.* **2014**, *68*, 501–516. [CrossRef]
29. Faraj, R.H.; Sherwani, A.F.H.; Jafer, L.H.; Ibrahim, D.F. Rheological behavior and fresh properties of self-compacting high strength concrete containing recycled PP particles with fly ash and silica fume blended. *J. Build. Eng.* **2021**, *34*, 101667. [CrossRef]
30. Alobaidi, Y.M.; Hilal, N.N.; Faraj, R.H. An experimental investigation on the nano-fly ash preparation and its effects on the performance of self-compacting concrete at normal and elevated temperatures. *Nanotechnol. Environ. Eng.* **2020**, *6*, 2. [CrossRef]
31. Thakur, R.S.; Das, S.N. *International Series on Environment-Red Mud Analysis and Utilization*; PID Wiley East. Ltd.: New Delhi, India, 1994.
32. Evans, K. The history, challenges, and new developments in the management and use of bauxite residue. *J. Sustain. Met.* **2016**, *2*, 316–331. [CrossRef]
33. Paramguru, R.K.; Rath, P.C.; Misra, V.N. Trends in red mud utilization–A review. *Miner. Process. Extr. Met. Rev.* **2004**, *26*, 1–29. [CrossRef]
34. Choe, G.; Kang, S.; Kang, H. Mechanical properties of concrete containing liquefied red mud subjected to uni-axial compression loads. *Materials* **2020**, *13*, 854. [CrossRef]
35. Saravanan, B.; Vijayan, D.S. Status review on experimental investigation on replacement of red-mud in cementitious concrete. *Mater. Today Proc.* **2020**, *33*, 593–598. [CrossRef]
36. Rai, S.; Bahadure, S.; Chaddha, M.J.; Agnihotri, A. Disposal practices and utilization of red mud (bauxite residue): A review in indian context and abroad. *J. Sustain. Metall.* **2020**, *6*, 1–8. [CrossRef]
37. Shinomiya, L.D.; Gomes, J.O.; Alves, J.O. Planejamento de cenários para uso de resíduos industriais: Aplicação para lama vermelha. *Rev. Gestão Em Eng.* **2015**, *2*, 43–66.
38. Lima, M.S.S.; Thives, L.P.; Haritonovs, V.; Bajars, K. Red mud application in construction industry: Review of benefits and possibilities. In Proceedings of the IOP Conference Series: Materials Science and Engineering, Birmingham, UK, 13–15 October 2017; IOP Publishing: Bristol, UK, 2017; Volume 251, p. 12033.
39. Gaur, M.; Pandey, A.; Ashish, A. Performance of concrete utilizing red mud as a partial replacement of cement with hydrated lime. *Int. J. Res. Eng. Sci. Manag.* **2018**, *1*, 12.
40. Nikbin, I.M.; Aliaghazadeh, M.H.; Charkhtab, S.; Fathollahpour, A. Environmental impacts and mechanical properties of lightweight concrete containing bauxite residue (red mud). *J. Clean. Prod.* **2018**, *172*, 2683–2694. [CrossRef]
41. Ortega, J.M.; Cabeza, M.; Tenza-Abril, A.J.; Real-Herraiz, T.; Climent, M.; Sánchez, I. Effects of red mud addition in the microstructure, durability and mechanical performance of cement mortars. *Appl. Sci.* **2019**, *9*, 984. [CrossRef]
42. Tang, W.C.; Wang, Z.; Donne, S.W.; Forghani, M.; Liu, Y. Influence of red mud on mechanical and durability performance of self-compacting concrete. *J. Hazard. Mater.* **2019**, *379*, 120802. [CrossRef]
43. Liu, R.-X.; Poon, C.-S. Utilization of red mud derived from bauxite in self-compacting concrete. *J. Clean. Prod.* **2016**, *112*, 384–391. [CrossRef]
44. Shaik, A.B.; Kommineni, H.R. Experimental investigation on strength and durability properties of concrete using bauxite residue and metakaolin. *Mater. Today Proc.* **2020**, *33*, 583–586. [CrossRef]
45. Ribeiro, D.V.; Labrincha, J.A.; Morelli, M.R. Use of red mud as addition for Portland cement mortars. *J. Mater. Sci. Eng.* **2010**, *4*, 1–8.
46. Hou, D.; Wu, D.; Wang, X.; Gao, S.; Yu, R.; Li, M.; Wang, P.; Wang, Y. Sustainable use of red mud in ultra-high performance concrete (UHPC): Design and performance evaluation. *Cem. Concr. Compos.* **2020**, *115*, 103862. [CrossRef]
47. Ahmadi, B.; Shekarchi, M. Use of natural zeolite as a supplementary cementitious material. *Cem. Concr. Compos.* **2010**, *32*, 134–141. [CrossRef]
48. Tang, W.C.; Wang, Z.; Liu, Y.; Cui, H.Z. Influence of red mud on fresh and hardened properties of self-compacting concrete. *Constr. Build. Mater.* **2018**, *178*, 288–300. [CrossRef]
49. Clark, M.W.; McConchie, D.; Ryffel, T. Trace Metals in Brisbane River Estuary Sediments and Port of Brisbane Corporation Reclamation Paddocks. Ph.D. Thesis, Southern Cross University, Lismore, Australia, 1997.

50. EFNARC Specification. Guidelines for self-compacting concrete. *Rep. EFNARC* **2002**, *44*, 32.
51. Barbhuiya, S.A.; Basheer, P.A.M.; Clark, M.W.; Rankin, G.I.B. Effects of seawater-neutralised bauxite refinery residue on properties of concrete. *Cem. Concr. Compos.* **2011**, *33*, 668–679. [CrossRef]
52. Zaid, O.; Ahmad, J.; Siddique, M.S.; Aslam, F.; Alabduljabbar, H.; Khedher, K.M. A step towards sustainable glass fiber reinforced concrete utilizing silica fume and waste coconut shell aggregate. *Sci. Rep.* **2021**, *11*, 12822. [CrossRef]
53. Pruckner, F.; Gjørv, O.E. Effect of $CaCl_2$ and NaCl additions on concrete corrosivity. *Cem. Concr. Res.* **2004**, *34*, 1209–1217. [CrossRef]
54. Ribeiro, D.V.; Labrincha, J.A.; Morelli, M.R. Potential use of natural red mud as pozzolan for Portland cement. *Mater. Res.* **2011**, *14*, 60–66. [CrossRef]
55. Ahmad, J.; Aslam, F.; Martinez-Garcia, R.; El Ouni, M.H.; Khedher, K.M. Performance of sustainable self-compacting fiber reinforced concrete with substitution of marble waste (MW) and coconut fibers (CFs). *Sci. Rep.* **2021**, *11*, 23184. [CrossRef]
56. Hajjaji, W.; Andrejkovičová, S.; Zanelli, C.; Alshaaer, M.; Dondi, M.; Labrincha, J.A.; Rocha, F. Composition and technological properties of geopolymers based on metakaolin and red mud. *Mater. Des.* **2013**, *52*, 648–654. [CrossRef]
57. Rathod, R.R.; Suryawanshi, N.T.; Memade, P.D. Evaluation of the properties of red mud concrete. *IOSR J. Mech. Civ. Eng.* **2013**, *1*, 31–34.
58. Venkatesh, C.; Nerella, R.; Chand, M.S.R. Comparison of mechanical and durability properties of treated and untreated red mud concrete. *Mater. Today Proc.* **2019**, *27*, 284–287. [CrossRef]
59. Viyasun, K.; Anuradha, R.; Thangapandi, K.; Kumar, D.S.; Sivakrishna, A.; Gobinath, R. Investigation on performance of red mud based concrete. *Mater. Today Proc.* **2021**, *39*, 796–799. [CrossRef]
60. Senff, L.; Hotza, D.; Labrincha, J. Effect of red mud addition on the rheological behaviour and on hardened state characteristics of cement mortars. *Constr. Build. Mater.* **2011**, *25*, 163–170. [CrossRef]
61. Ribeiro, D.V.; Silva, A.S.; Labrincha, J.A.; Morelli, M.R. Rheological properties and hydration behavior of Portland cement mortars containing calcined red mud. *Can. J. Civ. Eng.* **2013**, *40*, 557–566. [CrossRef]
62. Yang, X.; Zhao, J.; Li, H.; Zhao, P.; Chen, Q. Recycling red mud from the production of aluminium as a red cement-based mortar. *Waste Manag. Res.* **2017**, *35*, 500–507. [CrossRef]
63. Venkatesh, C.; Nerella, R.; Chand, M.S.R. Role of red mud as a cementing material in concrete: A comprehensive study on durability behavior. *Innov. Infrastruct. Solut.* **2021**, *6*, 13. [CrossRef]
64. Cheng, X.; Yang, X.; Zhang, C.; Gao, X.; Yu, Y.; Mei, K.; Guo, X.; Zhang, C. Effect of red mud addition on oil well cement at high temperatures. *Adv. Cem. Res.* **2021**, *33*, 28–38. [CrossRef]
65. Adi, M.; Abu-Jdayil, B.; Ghaferi, F.A.; Yahyaee, S.A.; Jabri, M.A. Seawater-neutralized bauxite residue—Polyester composites as insulating construction materials. *Buildings* **2021**, *11*, 20. [CrossRef]
66. Habeeb, K.; Rawi, A. Effect of Adding Sisal Fiber and Iraqi Bauxite on Some Properties of Concrete. *Tech. Inst. Babylon* **2009**. Available online: https://www.iasj.net/iasj/download/c67e401b64b06dc0 (accessed on 16 October 2022).
67. Venkatesh, C.; Ruben, N.; Chand, M.S.R. Red mud as an additive in concrete: Comprehensive characterization. *J. Korean Ceram. Soc.* **2020**, *57*, 281–289. [CrossRef]
68. Haque, M.A.; Chen, B.; Liu, Y. The role of bauxite and fly-ash on the water stability and microstructural densification of magnesium phosphate cement composites. *Constr. Build. Mater.* **2020**, *260*, 119953. [CrossRef]
69. Venkatesh, C.; Chand, M.S.R.; Nerella, R. A state of the art on red mud as a substitutional cementitious material. In *Proceedings of the Annales de Chimie: Science des Materiaux*; Dunod: Paris, France, 2019; Volume 43, pp. 99–106.
70. Tang, L. *Study of the Possibilities of Using Red Mud as an Additive in Concrete and Grout Mortar*; Svensk Kärnbränslehantering AB Swedish Nuclear Fuel and Waste Management Co.: Stockholm, Sweden, 2014.
71. Vigneshwaran, S.; Uthayakumar, M.; Arumugaprabu, V. Development and sustainability of industrial waste-based red mud hybrid composites. *J. Clean. Prod.* **2019**, *230*, 862–868. [CrossRef]
72. Fedroff, D.; Ahmad, S.; Savas, B.Z. Mechanical properties of concrete with ground waste tire rubber. *Transp. Res. Rec.* **1996**, *1532*, 66–72. [CrossRef]
73. Ganeshan, P.; Raja, K. Improvement on the mechanical properties of madar fiber reinforced polyester composites. *Int. J. Adv. Engg. Tech.* **2016**, *7*, 261–264.
74. Hu, W.; Nie, Q.; Huang, B.; Shu, X.; He, Q. Mechanical and microstructural characterization of geopolymers derived from red mud and fly ashes. *J. Clean. Prod.* **2018**, *186*, 799–806. [CrossRef]
75. Vangelatos, I.; Angelopoulos, G.; Boufounos, D. Utilization of ferroalumina as raw material in the production of ordinary Portland cement. *J. Hazard. Mater.* **2009**, *168*, 473–478. [CrossRef]
76. Liu, R.; Poon, C. Effects of red mud on properties of self-compacting mortar. *J. Clean. Prod.* **2016**, *135*, 1170–1178. [CrossRef]
77. Shetty, K.K.; Nayak, G.; Vijayan, V. Use of red mud and iron tailings in self compacting concrete. *Int. J. Res. Eng. Technol.* **2014**, *3*, 111–114.
78. Rana, A.; Kalla, P.; Csetenyi, L. Sustainable use of marble slurry in concrete. *J. Clean. Prod.* **2015**, *94*, 304–311. [CrossRef]
79. Manfroi, E.P.; Cheriaf, M.; Rocha, J.C. Microstructure, mineralogy and environmental evaluation of cementitious composites produced with red mud waste. *Constr. Build. Mater.* **2014**, *67*, 29–36. [CrossRef]
80. Kang, S.-P.; Kwon, S.-J. Effects of red mud and alkali-activated slag cement on efflorescence in cement mortar. *Constr. Build. Mater.* **2017**, *133*, 459–467. [CrossRef]

81. Kosmatka, S.H.; Panarese, W.C.; Kerkhoff, B. *Design and Control of Concrete Mixtures*; Portland Cement Association: Skokie, IL, USA, 2002; Volume 5420.
82. Shi, C. Effect of mixing proportions of concrete on its electrical conductivity and the rapid chloride permeability test (ASTM C1202 or ASSHTO T277) results. *Cem. Concr. Res.* **2004**, *34*, 537–545. [CrossRef]
83. Raja, R.R.; Pillaib, E.P.; Santhakumarc, A.R. Effective utilization of redmud bauxite waste as a re-placement of cement in concrete for environmental conservation. *Ecol. Env. Conserv.* **2013**, *19*, 247–255.
84. Ashok, P.; Sureshkumar, M.P. Experimental studies on concrete utilising red mud as a partial replacement of cement with hydrated lime. *J. Mech. Civ. Eng.* **2014**, *4*, 1–10.
85. Ioannidou, A. Precipitation, Gelation and Mechanical Properties of Calcium-Silicate-Hydrate Gels. Ph.D. Thesis, ETH Zurich, Zürich, Switzerland, 2014.
86. Thomas, J.; Jennings, H. Calcium-Silicate-Hydrate (CSH) gel. *OSR J. Mech. Civ. Eng. (IOSR-JMCE)* **2015**, 1–10.
87. Richardson, I.G. The nature of C-S-H in hardened cements. *Cem. Concr. Res.* **1999**, *29*, 1131–1147. [CrossRef]
88. Yang, K.-H.; Song, J.-K.; Song, K.-I. Assessment of CO_2 reduction of alkali-activated concrete. *J. Clean. Prod.* **2013**, *39*, 265–272. [CrossRef]

Article

Effect of Biochar from Oat Hulls on the Physical Properties of Asphalt Binder

Camila Martínez-Toledo [1], Gonzalo Valdés-Vidal [1,*], Alejandra Calabi-Floody [1], María Eugenia González [2] and Oscar Reyes-Ortiz [3]

[1] Department of Civil Engineering, University of La Frontera, Temuco 4811230, Chile
[2] Department of Chemical Engineering, University of La Frontera, Temuco 4811230, Chile
[3] Department of Civil Engineering, Military University of Nueva Granada, Bogotá 111711, Colombia
* Correspondence: gonzalo.valdes@ufrontera.cl

Abstract: The purpose of this study was to verify the feasibility of using biochar from oat hulls (BO) as a potential bio-modifier to improve the physical properties of conventional asphalt binder. The BO and asphalt binder were characterized by confocal (fluorescence) laser microscopy, scanning electron microscopy and Fourier transform infrared spectroscopy. Then, an asphalt binder modification procedure was established and modifications with 2.5, 5.0 and 7.5% of BO on the weight of the asphalt binder were evaluated, using a particle size < 75 μm. The physical properties of the evaluated modified asphalt binder with BO were: rotational viscosity in original and aged state, aging index, Fraass breaking point, softening point, penetration, penetration rate and storage stability. The results indicated that the BO has a porous structure, able to interact with the asphalt binder by C=O and C=C bonds. In addition, modification of the asphalt binder with BO increases the rotational viscosity related to high-temperature rutting resistance. The results obtained from the Fraass breaking point and softening point indicated that the use of BO extends the viscoelastic range of the asphalt binder. In addition, the evaluated modifications present low susceptibility to aging and good storage stability.

Keywords: biochar; asphalt binder; physical properties

Citation: Martínez-Toledo, C.; Valdés-Vidal, G.; Calabi-Floody, A.; González, M.E.; Reyes-Ortiz, O. Effect of Biochar from Oat Hulls on the Physical Properties of Asphalt Binder. *Materials* **2022**, *15*, 7000. https://doi.org/10.3390/ma15197000

Academic Editors: Stefano Guarino and Flaviana Tagliaferri

Received: 15 July 2022
Accepted: 18 August 2022
Published: 9 October 2022

Publisher's Note: MDPI stays neutral with regard to jurisdictional claims in published maps and institutional affiliations.

1. Introduction

Nowadays, material engineers and scientists are becoming increasingly interested in modifying the properties of conventional asphalt binders using waste from an industry that causes tremendous pollution. This interest is based mainly on increasing the service life of asphalt pavements, generating a lighter environmental impact [1].

The performance of asphalt binder under different climatic and traffic conditions causes the appearance of certain failure modes that affect the structure of the pavement. One of the important characteristics in the performance of asphalt binder is its work range, which is the ability to change from a solid-elastic to a viscous state (viscoelastic range) according to the temperature changes it undergoes [2]. This, added to its response to the application of loads at different speeds, conditions the performance or rheological behavior of the asphalt binder, which affects the performance in a pavement structure [3].

One of the alternatives used to achieve more resistant and durable pavements is the use of different types of modified asphalt binder, which can improve such properties as rutting resistance, fatigue cracking, thermal cracking and other types of deterioration. In this context, the use of commercial polymer has contributed to the performance of asphalt binder, with SBS (styrene-butadiene-styrene) being one of the most used polymers in industry for the modification of asphalt binder. However, its use also involves a considerable increase in the cost of the end product compared to a conventional asphalt binder [4]. Therefore, various studies have been conducted of late that seek to assess other material additives that can improve the rheological properties of the asphalt binder. One of the lines of study is the addition of nano-modifiers, such as carbon nanotubes, graphene oxide, nanosilica and others [5–8]. Although these nano-modifiers have yielded good results, their performance and production costs mean that their industrial use is not yet economically viable as an asphalt binder-modifying additive [9,10]. On the other hand, micro-sized modifying materials (<150 μm) or micro-materials have been used with good results to improve the performance-related properties of the asphalt binder. Such is the case of asphaltite, graphite and activated carbon [11–13].

In this same line of enquiry, an alternative potential to modify asphalt binder may be biochar. Biochar is a solid, carbon-rich by-product obtained from pyrolysis [14]. Pyrolysis is a thermochemical process carried out between 300 and 1000 °C in the absence of oxygen to produce biofuels and transform waste biomass into biochar [15,16]. This technique reduces emissions caused by biomass incineration and reduces the use of landfills for waste disposal [17]. In addition, the biochar produced can be used as a tool for reducing greenhouse effect gases (GEG) and sequestering CO_2 in the air [18–20].

Research into the use of biochar from different types of biomass of plant origin, with particle sizes < 75 μm to modify asphalt binder, have shown that it can increase the anti-aging properties, maintain resistance to thermal cracking and improve rutting resistance [1,14,21,22]. In addition, it could be a cost-competitive modifier in relation to other modifiers with similar characteristics, such as graphite and activated carbon [23–26]. However, there is a very limited number of studies on the effects of biochar in asphalt binder [27], and studies of biochar from oat hulls in their application as an asphalt binder modifier do not exist. Biochar from oat hulls has a high content of carboxylic groups, low heavy metal content and high carbon content [28,29]. Additionally, some countries have states or regions that have oats as a large surface of their agro-industrial crops. For example, the region of La Araucanía in Chile has the widest distribution of its crops with oats, of which 62.1% of the surface is dedicated to agriculture [30]. These crops generate a large amount of residual biomass which mainly comes from oat hulls. Oats are estimated to have a hull:grain ratio being 30:70% by weight [31].

Accordingly, the aim of the present study is to verify the feasibility of the use of biochar from oat hulls (BO) obtained by slow pyrolysis at 300 °C as a potential bio-modifier to improve the physical properties of asphalt binder.

2. Materials and Methods

2.1. Materials

The asphalt binder used in this study is a conventional CA-24 asphalt binder (according to Chilean specifications), and its characterization is in Table 1.

Table 1. Physical properties of reference asphalt binder (CA-24).

Tests	Specs. [32]	CA-24
Original viscosity at 60 °C (P)	Min. 2400	2940
Penetration at 25 °C, 100 g, 5 s (dmm)	Min. 40	63
Ductility at 25 °C, 5 cm/min (cm)	Min. 100	100
Trichloroethylene solubility (%)	Min. 99	99.8
Flash point (°C)	Min. 232	310
Softening point (°C)	-	52.2
Penetration index	−1.0 to + 1.0	−0.1
RTFOT		
Mass loss (%)	Max. 0.8	0.08
Viscosity at 60 °C (P)	-	7860
Ductility at 25 °C, 5 cm/min (cm)	Min. 100	100
Durability index	Max. 3.5	2.7

The biochar selected is a by-product of the slow pyrolysis of oat hulls for 2 h of residence time, with a heating rate of 3.6 °C/min and a pyrolysis treatment temperature

(PTT) of 300 °C. Figure 1 shows the schematic diagram of the pyrolysis used, where the reactor is stainless steel and has a capacity of 0.117 m^3, approximately. This reactor is equipped with an inert gas connection (N$_2$) with a flow of 0.001 m^3/min to purge the oxygen generated during the reaction. The furnace parameters were programmed using a programmable logic controller (PLC). The oven is heated by the resistances installed inside, the temperature of which is controlled by a K-type thermocouple sensor (Ni-Cr-Ni). According to the procedure, the volatile by-products present in the synthesis gas were condensed by the circulation of water at room temperature; the reception of these condensable gases (bio-oil) took place in a TAR container, while the gaseous fraction was released into the atmosphere.

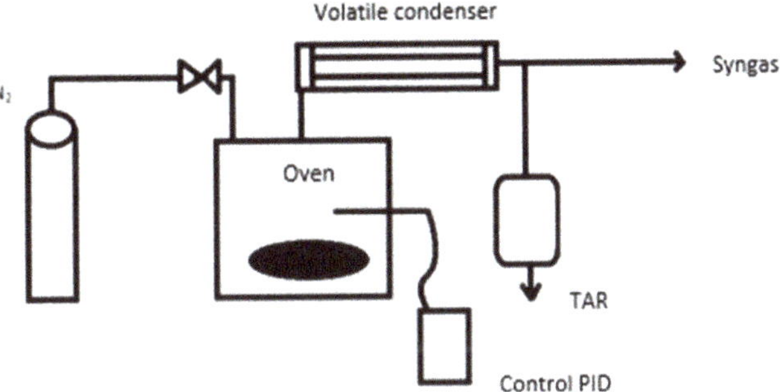

Figure 1. Schematic diagram of pyrolysis.

When the reaction was complete, a 44% BO yield was obtained. The BO obtained was subjected to a mid-sized reduction process using a 180 W grinder with a processing time of 30 s. Then, to obtain a particle size <75 µm, a sieve with a 0.075 mm mesh (N° 200) was used. Figure 2 shows the BO production sequence.

Figure 2. BO production sequence.

2.2. Asphalt Binder Modification Procedure

To establish the procedure of asphalt binder modification with BO, a matrix was created, composed of: three mixing times (30, 60, 120 min), three modification temperatures (160, 170, 180 °C) and a single BO content, equivalent to 5.0% in weight of the asphalt binder. Modification temperatures higher than the mixing temperature of the asphalt binder were considered because the addition of BO can increase the viscosity of asphalt. For each configuration, the modification of 1000 g of asphalt binder was carried out. Initially, the asphalt binder was heated in an oven to 130 °C for 30 min and then BO was added and the mixing process began, which was carried out using an electronic stirrer at 350 rpm. The BO was added gradually. The modification temperature was verified every 5 min using a digital thermometer with an accuracy to 0.1 °C until the mixing time was complete. Then, the samples were analyzed by confocal laser microscopy to evaluate the distribution and integration of the BO with the asphalt binder. The results determined that, with a constant temperature of 160 °C and 30 min of mixing, a homogenous sample between the BO and the asphalt binder is obtained without clumping.

2.3. Experimental Plan

For the temperature and selected mixing time, the additions of 2.5%, 5.0% and 7.5% of BO (<75 μm) on the weight of the asphalt binder were evaluated, along with a reference sample of asphalt binder (CA-24). Figure 3 provides the experimental plan to assess the effect of modification with BO on the physical properties of the asphalt binder at different stages. This experimental plan used physicochemical characterization procedures from scanning electron microscopy (SEM), energy-dispersive x-ray spectroscopy (EDS), confocal (fluorescence) laser microscopy, infrared spectroscopy (FTIR) and assay methods to evaluate physical properties of the asphalt binder related to its mechanical behavior and industrial use.

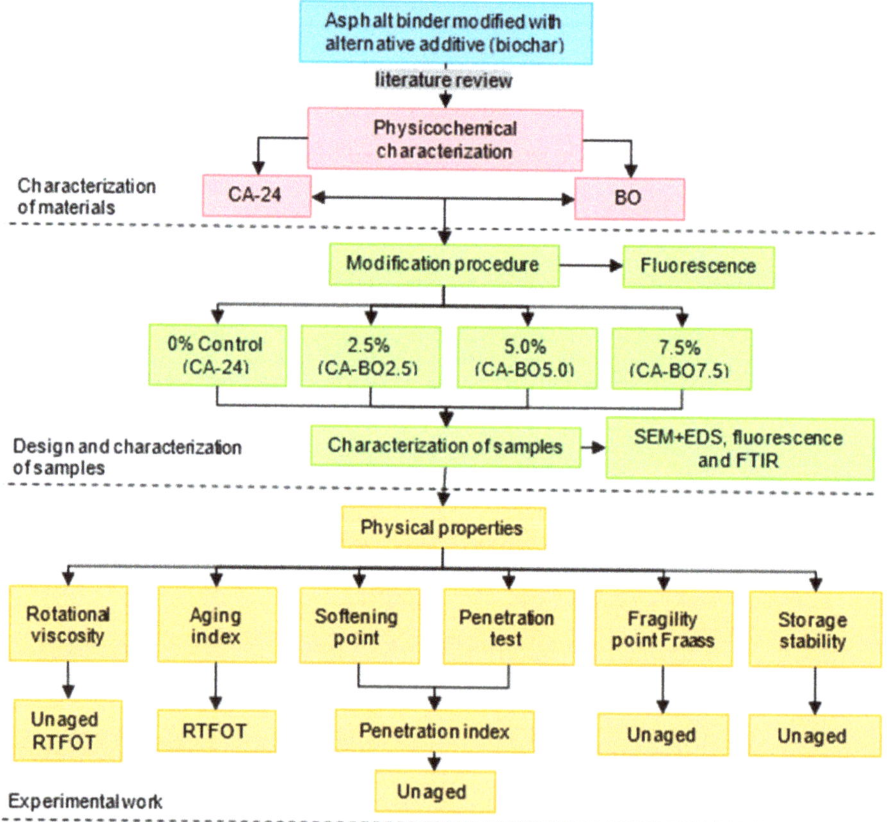

Figure 3. Experimental plan.

2.4. Test Methods

2.4.1. Physicochemical Characterization of Raw Materials and Samples

The SEM and EDS assays made possible an advanced analysis of the surface of the materials included in this study. In this way, the microscopic morphology, particle size and elemental chemical composition of the surface of the materials were obtained using a SU3500 scanning electron microscope from Hitachi High-Technologies Corporation, Tokyo, Japan, equipped with an energy-dispersive X-ray detector.

To verify the distribution and integration of the BO in the asphalt binder after the modification process, a FV1000 Olympus confocal laser (fluorescence) microscope was used in the excitation/emission spectra in 2 wavelengths: 488/590 nm and 530/590 nm, with a 20× magnification and a 30 μm depth in steps of 2 μm/slide.

To detect the functional groups present on the surface of the carbon materials, FTIR was carried out using Perkin Elmer Spectrum Two infrared spectrometer that includes an ATR (attenuated total reflectance) system. The FTIR assay was executed between 3400 cm^{-1} and 400 cm^{-1}, with a resolution of 4 cm^{-1} and a laser repetition rate of 20 at 20 °C.

2.4.2. Evaluation of the Physical Properties of Modified Asphalt

The rotational viscosity test (RV) was applied to measure the flow resistance and the workability of the samples at high operating temperatures of 52, 58, 60, 64, 70 and 76 °C, and at working temperatures of 135 and 165 °C, in both original and short-term aged samples using a rolling thin-film oven. The testing procedure was done according to AASHTO T316-19 using a Brookfield RVDV-III ULTRA rotational viscometer at a shear speed of 20 rpm.

The rolling thin-film oven test (RTFOT) was performed to simulate the short-term aging that occurs in asphalt during the manufacturing process of the mixture and its spreading in the pavement, subjecting the samples to the effect of heat and air at 163 °C for 85 min according to the procedure described in AASHTO T240-13 using a Controls Group 81-PV1612 thin-film rolling oven.

The aging index ($I_{ag}{}^{r}$) is a parameter used to determine the susceptibility to short-term aging of the asphalts evaluated, using Equation (1):

$$I_{ag}{}^{r} = V_{RTFOT} / V_{original} \tag{1}$$

where,

$I_{ag}{}^{r}$ = aging rate of the asphalt binder,
RV_{RTFOT} = rotational viscosity of the asphalt binder aged by RTFOT (poises),
$V_{original}$ = rotational viscosity of the asphalt binder in original state (poises).

The Fraass breaking point test was performed to evaluate the behavior of the asphalt at low operating temperatures since it determines the transition temperature at which asphalt goes from a viscoelastic state to an elastic state [33]. The testing procedure was done according to EN 12593:2007 using the Control Breaking Point apparatus, subjecting a thin film of asphalt to successive bending cycles at decreasing temperatures at a cooling rate of 1 °C/min.

The softening point test (SP) was conducted to determine the softening point temperature of asphalts evaluated according to the procedure described in ASTM D36-76 using the ring and ball apparatus. The softening point temperature obtained defines the transition of the asphalt from a viscoelastic state to a purely viscous state [34]. The lowest temperature at which the asphalt sample, suspended in a horizontal ring, was forced to fall due to the weight of a steel ball while it was heated at 5 °C/min in a water bath was recorded.

The penetration test (Pen) described in ASTM D5-13 was used to determine the hardness of the asphalts according to the penetration depth of a stainless-steel needle on their surface, the greatest values of which indicate softer consistencies of the material and vice versa. The procedure was done using a B057 automatic penetrometer, applying 100 g of weight for 5 s at 25 °C.

The penetration index (PI) or Pfeiffer index is a parameter to describe the thermal susceptibility of asphalt to temperature changes. In addition, it offers an indication of its colloidal structure and rheological behavior [32,33]. Table 2 provides a description of the characteristics of the asphalt binders based on the PI values.

Table 2. PI classification of asphalt binders [32].

IP Value	Description
PI > +1	Asphalts that are not very susceptible to temperature and show non-Newtonian flow behavior, with certain elasticity and thixotropy.
PI < −1	Asphalts that are highly susceptible to temperature and exhibit Newtonian flow behavior.
−1 < PI < +1	Asphalts that have rheological and flow characteristics intermediate between the two previous cases. Most of the asphalt binder used in paving has these characteristics.

This parameter can be calculated using Equation (2), with the results of the penetration tests at 25 °C and softening point.

$$PI = \frac{1952 \ - \ 500 \cdot \log(\text{Pen}) - 20SP}{50 \cdot \log(\text{Pen}) - SP - 120} \tag{2}$$

where,

PI = penetration index of the asphalt binder,
Pen = result of the penetration test at 25 °C,
SP = result of the softening point test using the ring and ball apparatus.

The storage stability test was conducted according to the procedure described in ASTM D58-92. Although this assay evaluates the storage stability of asphalts modified with polymers, like SBS, it was also used to evaluate the asphalt modification with BO because the test results offer the data required for the analysis of this study. The modified asphalt binder samples were poured into 1″ diameter tubes and conditioned in an oven at 163 ± 5 °C for 48 h. The samples were then conditioned in a freezer at -6.7 ± 5 °C for 4 h. Next, the tubes were cut into three sections of equal length, and the central part of each tube was discarded. Finally, these samples were poured into appropriately marked rings to perform the softening point test by means of the ring and ball apparatus according to ASTM D36-76. To consider storage stability good, the difference in the results of the softening point between the upper and lower sections of the sample cannot exceed 5 °C [33].

3. Results

3.1. Physicochemical Characterization of Raw Materials and Samples

3.1.1. Scanning Electron Microscopy (Sem) and Energy Dispersive X-ray Spectroscopy (EDS)

The image of oat hulls in Figure 4a obtained by SEM shows a great predominance of large particles ($\leq$1 mm) with heterogenous geometry, whereas the micrograph of the BO in Figure 4b shows particles of different sizes with an irregular surface and a certain pore development, with few cases of vitreous surfaces.

(a)   (b)

Figure 4. SEM micrograph: (**a**) Oat hulls and (**b**) BO.

Table 3 shows the results of SEM+EDS micro elemental analysis of the evaluated samples. It is observed that the predominant chemical element is carbon. Sulfur as well as oxygen are elements in common among all BO-modified asphalt binder samples. Thus, Fisher's least significant difference (LSD) method with a 95% confidence level was used for multiple comparisons of sample means. It was determined that there is a significant difference between the amount of carbon (C) in the reference asphalt binder (CA-24) and the BO-modified asphalt binders. However, among the modified asphalt binders, there is no significant difference in the amount of this chemical element (C), despite increasing the percentage of BO in the sample. With regard to the amount of sulfur (S) and oxygen (O), there are no significant differences between the samples of asphalt binder modified with BO, but the latter element (O) decreases significantly compared to the modifier (BO). It is also possible to observe other chemical elements in the samples modified with BO, such as silicon, potassium and phosphorus. These are inherent to the modifier used (BO). The higher the percentage of BO, the higher the presence of these elements.

Table 3. Micro elemental analysis of the samples.

Chemical Element	BO		CA-24		CA-BO2.5		CA-BO5.0		CA-BO7.5	
	Content (%)	σ	Content (%)	σ	Content (%)	σ	Content (%)	σ	Content (%)	σ
Carbon (C)	70.51	3.09	92.60	3.93	91.39	3.13	86.54	8.39	87.03	5.10
Sulfur (S)	-	-	3.80	0.15	3.36	0.26	3.08	1.60	3.58	0.64
Oxygen (O)	22.72	0.48	2.92	1.69	3.82	2.95	8.24	6.73	5.16	4.01
Calcium (Ca)	1.38	0.29	-	-	0.67	0.39	-	-	-	-
Nitrogen (N)	-	-	7.87	4.54	-	-	4.31	2.49	7.35	4.24
Silicon (Si)	3.29	3.25	-	-	-	-	1.62	0.96	-	-
Potassium (K)	1.89	0.53	-	-	-	-	-	-	3.51	2.03
Phosphorus (P)	0.61	0.35	-	-	-	-	-	-	1.22	0.70

Note: σ: standard deviation.

3.1.2. Confocal (Fluorescence) Laser Microscopy

The images in Figure 5 show that for all the percentages of modification with BO a good distribution in the asphalt binder is achieved, without the presence of clusters. This indicates that there is a homogenous integration between the two materials, which is why the modification procedure used is the one suitable for the asphalt binder modification with up to 7.5% BO. On the other hand, it is observed that the maximum sizes of BO particles are around 30 µm despite having separated the BO fraction using a 75 µm sieve. This result is attributed to the downsizing process used, because it was highly efficient in the production of small particles due to the characteristics of the grinder and the BO processing time.

3.1.3. Fourier Transform Infrared Spectroscopy (FTIR)

Figure 6 shows the spectra of the BO, CA-24 and the asphalt binder modified with BO. In the BO, the main infrared signals are in the band 3005.0 cm^{-1} corresponding to C-H, 1540.5 cm^{-1} corresponding to C=C type rings and 1077.61 cm^{-1} corresponding to C=C or C-O-C bonds. These peaks can be attributed to the lignin, cellulose and hemicellulose of the biochar used (BO), which have a content of 7.5%, 34.3% and 26.0% respectively [28,29]. The spectrum of CA-24 shows more noise than that of other studies but has characteristic signals similar to them [35]. The stretches at 2919.38 cm^{-1} and 2850.99 cm^{-1} correspond to methylene CH$_2$, the peak at 1591.9 cm^{-1} corresponds to the C=C rings of benzene, the peak at 1454.51 cm^{-1} corresponds to CH$_2$, the peak at 1375.33 cm^{-1} corresponds to methyl CH$_3$ and the peak at 1023.8 cm^{-1} corresponds to C-O-C or C-O. The peaks at 810.18 cm^{-1} and 723.56 cm^{-1} correspond to C-H. Several peaks are noted in the spectra of the modifications that corroborate the interaction between the functional groups present on the surface of the BO and the asphalt binder. The main interactions observed are C−O and C=C bonds in the band of 1021 cm^{-1} (represented by a vertical dotted line). When the percentage of BO in the asphalt binder is increased, a trend reduction in the transmittance is observed, confirming the interaction between the two matrices, however more trials are needed to quantify this reduction. The CA-BO2.5 and CA-BO5.0 samples present equal intensity of interaction. By contrast, the CA-BO7.5 sample is the one with the greatest interaction between the polymeric mixture, observing a greater reduction in the transmittance. In addition, the modifications of the asphalt binder bands suggest an interaction with the BO through functional groups present on the surface of the BO, forming C=O and C=C bonds without there being large differences in the fixations between the mixtures of 2.5%, 5.0% and 7.5% of BO.

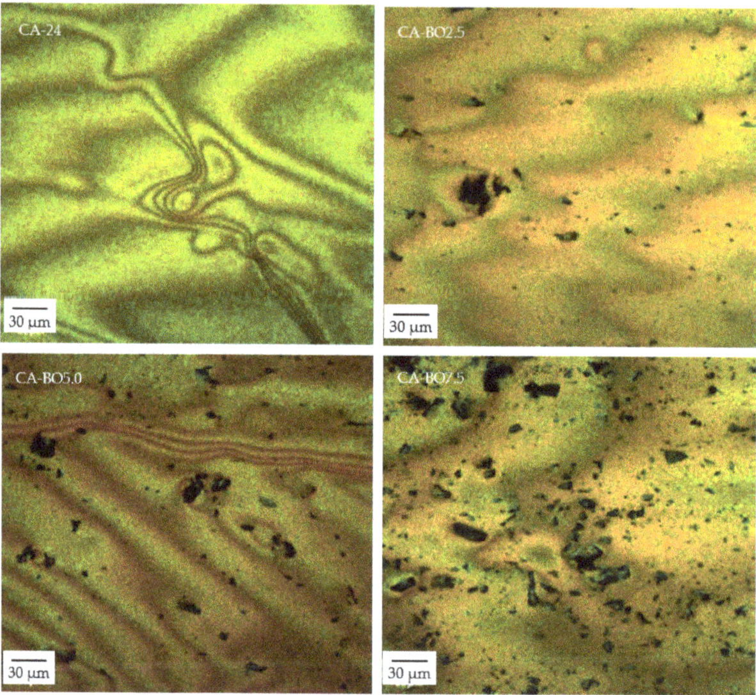

Figure 5. Fluorescence images of samples.

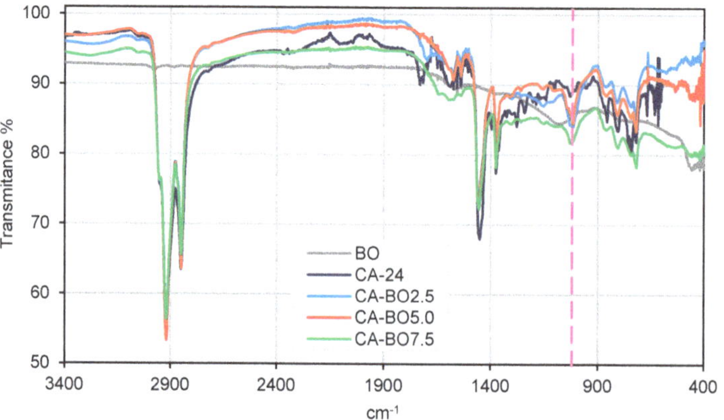

Figure 6. FTIR spectra of the samples analyzed.

3.2. Analysis of Physical Properties of Modified Asphalt

3.2.1. Rotational Viscosity (RV)

The results shown in Figure 7a indicate that the RV of the asphalt binder modified with BO increased for all the addition percentages considered and temperatures evaluated in relation to the CA-24. For example, at 60 °C the viscosities increased by 27.5%, 32.1% and 47.1% for the CA-BO2.5, CA-BO5.0 and CA-BO7.5 samples, respectively. These results show that as the amount of BO in the asphalt binder increases, a growing trend is registered in the increase in RV. This behavior indicates that the modification with BO can increase the viscosity of the asphalt binder in a high range of operating temperatures, which could aid in improving rutting resistance of the wearing course of a pavement [36]. These results are consistent with other studies, such as the study by Muhammed et al. [37], who modified asphalt binder with biochar from the pyrolysis of ground walnut shells and apricot seed shell granules, determining that the increase in the concentration of these bio-modifiers also increased the rigidity of the asphalt binder. This effect also was identified by Walters et al. [38], who used a thermo-chemical process to convert pig dung into bio-oil using biochar obtained to modify asphalt binder. The results indicated that the viscosities tended to increase as the addition percentages of biochar in the asphalt binder increased. In relation to the particle size of BO used in the modification of the asphalt binder (<75 µm), the effect on the RV is considered high, agreeing with the report by Zhang, et al. in 2018 [21], who described the particles of BO < 75 µm as registering higher RV than those modifications with larger particles (between 75 and 150 µm). This effect is because a greater number of small particles can be located in the same surface area. In addition, the porous structure and surface area that the BO possesses can produce a better adhesion and interaction with the asphalt binder, reducing its fluidity and increasing its viscosity. In this same context, the structure of the BO would allow a possible absorption of the lightest components of the asphalt binder, thereby increasing its RV [35].

Figure 7b shows that the RV of the reference asphalt binder and the RV of the asphalt binder modified with BO increased compared to the results obtained from these same samples in their original state due to the oxidation process that the asphalt binders underwent after aging by RTFOT. It is also worth noting that the RV of the asphalt binder modified with BO is greater than the RV of the reference asphalt binder (CA-24). This increase in the RV is proportional to the amount of BO added to the asphalt binder. In this sense, at 60 °C the CA-BO2.5, CA-BO5.0 and CA-BO7.5 samples increased their RV by 28.4%, 42.1% and 57.6% compared to the RV of the CA-24, respectively. These increases are greater than those determined in the original state of the asphalt binder as shown in Figure 7a.

In Figure 8 it is noted that both the optimal temperature to obtain the recommended mixture viscosity (~2 poises) and the optimum temperature to obtain the compaction viscosity (~3 poises) of the asphalt binder modified with BO increased compared to the temperatures of the reference asphalt binder, at 5 °C and 7 °C respectively. The highest temperatures corresponded to CA-BO7.5 (164 and 158 °C) and the lowest temperatures to the reference asphalt binder (159 and 151 °C). Within this range are the optimum mixture and compaction temperatures of CA-BO2.5 and CA-BO5.0, respectively. According to the data shown in Figure 8, CA-BO5.0 increases 38% and 44% more with the RV at 135 °C than the biochar from ground apricot seed shell and ground walnut shell, respectively. By contrast, at 165 °C, the RV increases by 23% and 26% more than those same modifiers, respectively [37].

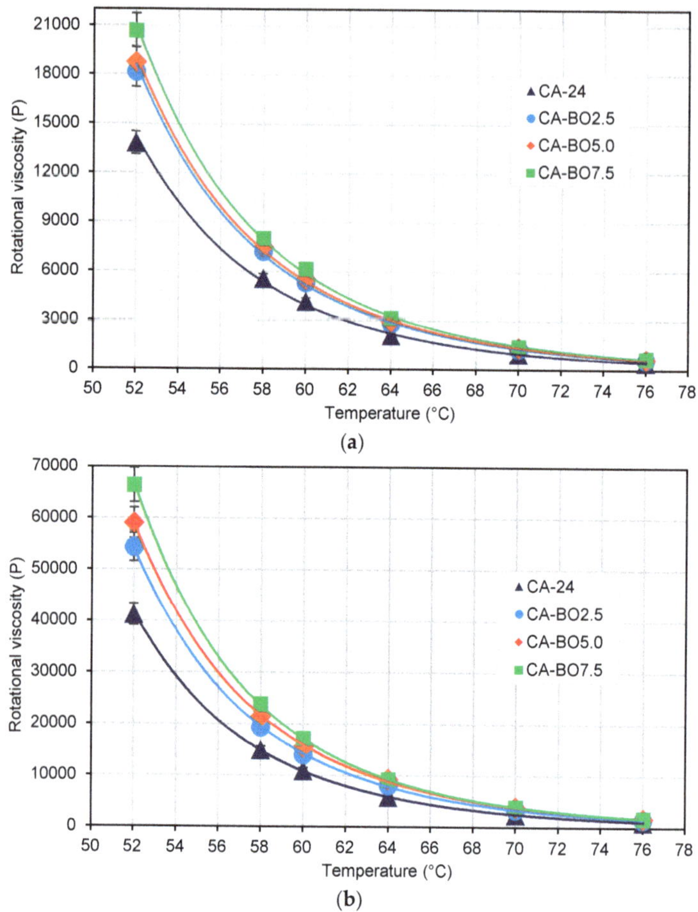

(a)

(b)

Figure 7. Rotational viscosity between 52 and 76 °C of asphalt binder samples analyzed at: (**a**) original state and (**b**) aged by RTFOT.

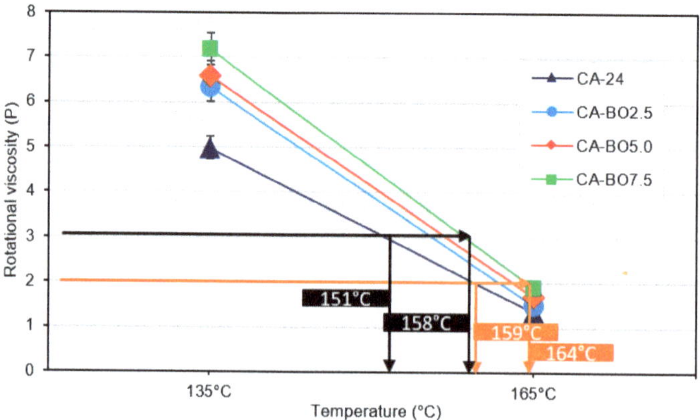

Figure 8. In black, mixing temperature (2 P) and in orange, compaction temperature (3 P) of the asphalt binder samples analyzed.

3.2.2. Aging Index ($I_{ag}{}^r$)

Figure 9 shows the evolution of $I_{ag}{}^r$ for different evaluated temperatures. A trend is observed when the $I_{ag}{}^r$ decreases as the temperature increases. This indicates that the aging effect in the reference asphalt binder and the asphalt binder modified with BO decreases as the evaluated temperature increases. On the other hand, for the highest contents of BO in the modification (CA-BO5.0 and CA-BO7.5), greater values are obtained in the parameter $I_{ag}{}^r$, with increases that vary in the range of 5% and 8% compared to the reference asphalt binder. However, these fluctuations do not generate significant differences between the samples with 5.0% and 7.5% of BO and $I_{ag}{}^r$ of the reference asphalt binder (CA-24), consistent with the results from the Kruskal–Wallis test, where a significant value of 0.323 was obtained. With respect to the CA-BO2.5 sample, this presents a behavior similar to the reference asphalt binder, showing equal variances according to Levene's test with a significant value >0.05, whereas the Student *t*-test results indicated that there is no significant difference between the means of CA-BO2.5 and the reference asphalt binder with a significant value equal to 0.680. These results could be due to the morphology of the BO particles, which are characterized by their heterogenous and porous shapes that could adsorb the asphalt binder, causing a reduction in the exposure of the asphalt binder to heat during the oxidation process, which would reduce the effects of aging [14,21]. On the other hand, the fluctuations obtained in the $I_{ag}{}^r$ parameter may be due to the loss of light compounds that the BO absorbed during the modification of the asphalt binder [35]. Nevertheless, the values obtained for the parameter $I_{ag}{}^r$ of the modifications evaluated in this study are lower than those specified in different standards, as in the case of the standard for Chile [32], which specifies for 60 °C a maximum $I_{ag}{}^r$ value of 4.43% higher than the maximum $I_{ag}{}^r$ value obtained at that temperature in this study.

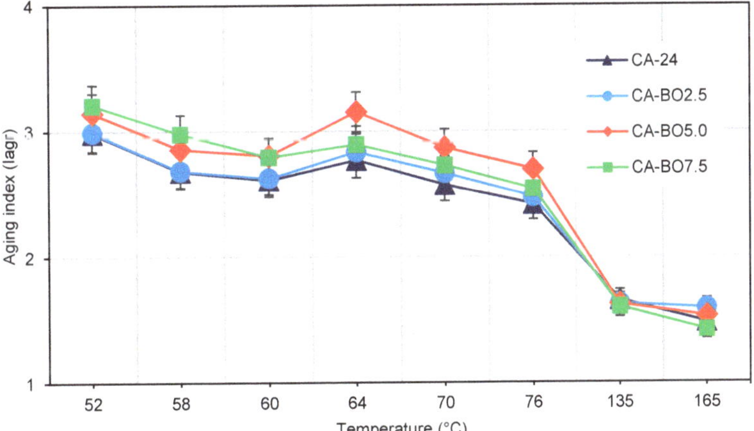

Figure 9. Aging index ($I_{ag}{}^r$) of the samples.

3.2.3. The Fraass Breaking Point vs. the Softening Point (SP)

The results show that the use of BO as an asphalt binder modifier can extend the viscoelastic range of the asphalt binder once the Fraass breaking point is reduced and the softening point is increased (Figure 10). In this respect, it is observed that the Fraass breaking point decreases as the BO content in the asphalt binder increases. For example, the CA-BO2.5, CA-BO5.0 and CA-BO7.5 samples achieved a breaking temperature of −8.5 °C, −9.5 °C and −10.0 °C respectively, being 2.0 °C, 3.0 °C and 3.5 °C lower than the breaking temperature of the CA-24. These results point to the asphalt binder modified with BO reducing the temperature at which the asphalt binder reaches the critical rigidity value after its fracture, and therefore its cracking [33]. On the other hand, it is observed that the softening point increases according to the increase in the amount of BO in the asphalt binder. For example, the CA-BO2.5, CA-BO5.0 and CA-BO7.5 samples reached a softening temperature of 51.0 °C, 53.8 °C and 55.3 °C respectively, increasing the softening temperature by 1.0 °C, 3.8 °C and 5.3 °C compared to the CA-24. The results show that the addition of BO increases the temperature range at which the asphalt binder can vary its behavior from a purely fragile state to a purely viscous state. Thus, the addition of this bio-modifier enables the asphalt binder to present a less fragile behavior at the same temperature as the reference asphalt binder. These results could justify an improvement in its response to cracking at low operating temperatures. By contrast, at the other extreme, they could show a greater capacity to resist higher temperatures, since its transition from a viscoelastic material to a viscous material occurs at a higher temperature, where the most common failures are rutting resistance. In that sense, an increase in the evaluated temperature ranges could mean an improvement in the resistance of the asphalt binder to the failures produced at low and high operating temperatures. According to the data in Figure 10, 5% of BO increases the SP of the asphalt binder by 8%, 12% and 16% more than the biochar DS-510F [39] and biochars from ground apricot seed shell and ground walnut shell, respectively [37]. Meanwhile, 7.5% of BO increases the SP of the asphalt binder by approximately 7% more than the biochar DS-510F [35,39].

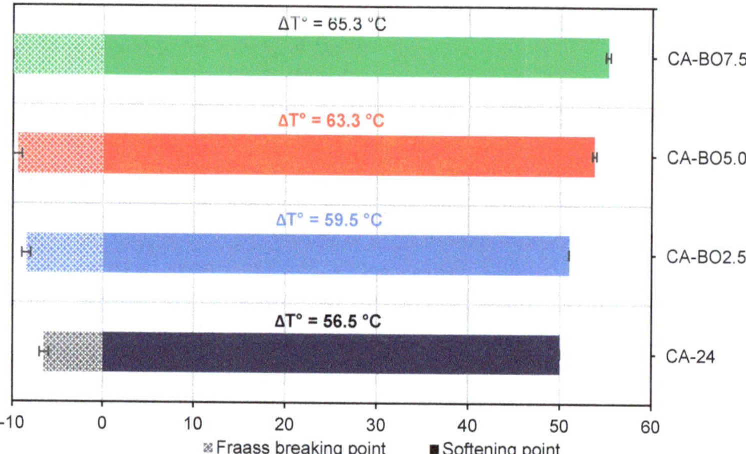

Figure 10. Viscoelastic range of the samples.

3.2.4. The Penetration (Pen)

The results provided in Figure 11 indicate that the addition of BO increases the hardness of the asphalt binder at 25 °C. The reduction in the penetration depth as the BO concentrations in the asphalt binder is increased is more obvious in the CA-BO5.0 and CA-BO7.5 samples, the results of which show similar values, with a decrease of 29.4% and 29.9%, respectively compared to the CA-24. However, the CA-BO2.5 presented a 16.1% reduction compared to the penetration of the CA-24. These results indicate that the asphalt binder modified with the different amounts of BO make it possible to increase the rigidity of the asphalt binder at the intermediate temperature, fulfilling the minimum penetration of 40 dmm by some standards that classify their asphalt binder by degree of viscosity, like the Chilean standard for example [32]. When using 5% or 7.5% of BO, the penetration of the asphalt binder decreases about 14% more than when using the same percentages of other modifiers, such as biochar DS-510F [35,39].

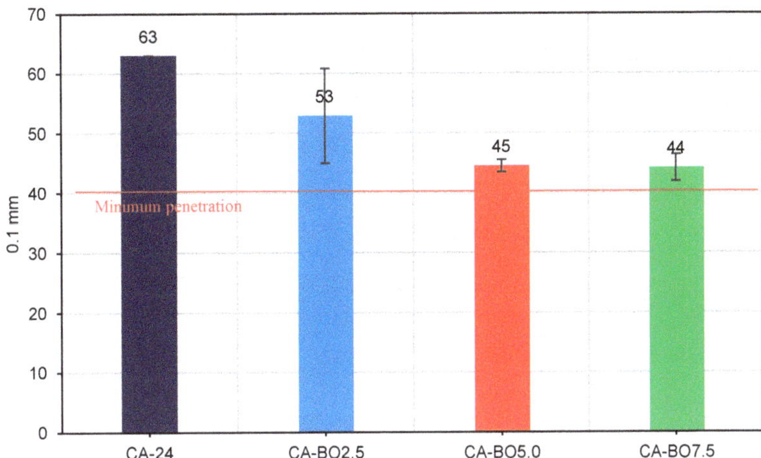

Figure 11. Penetration of the samples together with the required minimum penetration at 25 °C.

3.2.5. Penetration Index (PI)

From the data provided in Table 4, the PI values of the asphalt binder modified with BO are similar to the CA-24 up to the content of 5.0%. For a BO addition of 7.5%, an increase in the PI is recorded, indicating a reduction in thermal susceptibility compared to the CA-24. For this BO content, the effect on the reduction of the Pen results was observed, and at the same time on the increase in the SP values, which made it possible to reduce thermal susceptibility. This effect on IP reduction was also observed in other studies where biochar from ground apricot seed shell and ground walnut shell was used as an asphalt binder modifier [37]. Additionally, the classification of all the asphalt binder modified with BO corresponds to the category of intermediate thermal susceptibility, understood as a penetration index of between −1 and +1 [32].

Table 4. Penetration index of the samples.

Sample	CA-24	CA-BO2.5	CA-BO5.0	CA-BO7.5
Penetration index	−0.7	−0.8	−0.6	−0.3

3.2.6. Storage Stability

The results in Figure 12 indicate that the material corresponding to the upper superior of the sample records a lower SP than the lower section, which is observed for all the analyzed samples. With respect to the results obtained for both CA-BO2.5 and the CA-BO5.0, a difference is noted between the SP of the upper and lower sections of 1.5 °C and 2.5 °C, respectively. By contrast, the CA-BO7.5 sample shows a difference close to 4 °C. Differences in the SP in the range from 2 to 5 °C between the upper and lower sections of the sample are considered good storage stability [25]. In this sense, the results indicate that all the samples of asphalt binder modified with BO present good storage stability, demonstrating that the BO contents evaluated in the study would have good compatibility and interaction with asphalt binder, and that the modification process used enabled good bonding between phases. However, studies on asphalt binder modification report that to achieve a high homogeneity of the modified asphalt binder, the differences in the SP between the upper and lower sections of the sample should be ≤2.5 °C [40,41]. In that case, only the additions of 2.5% and 5.0% of BO fulfill this condition of high storage stability.

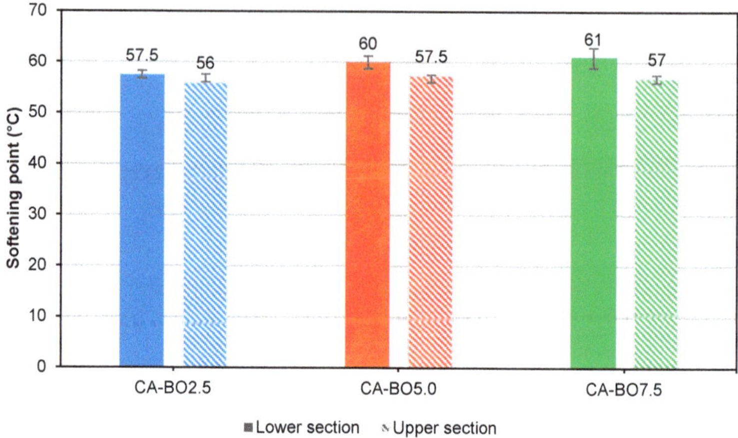

Figure 12. Storage stability of samples.

Figure 13 is a summary of the effects of the BO on the physical properties of the asphalt binder.

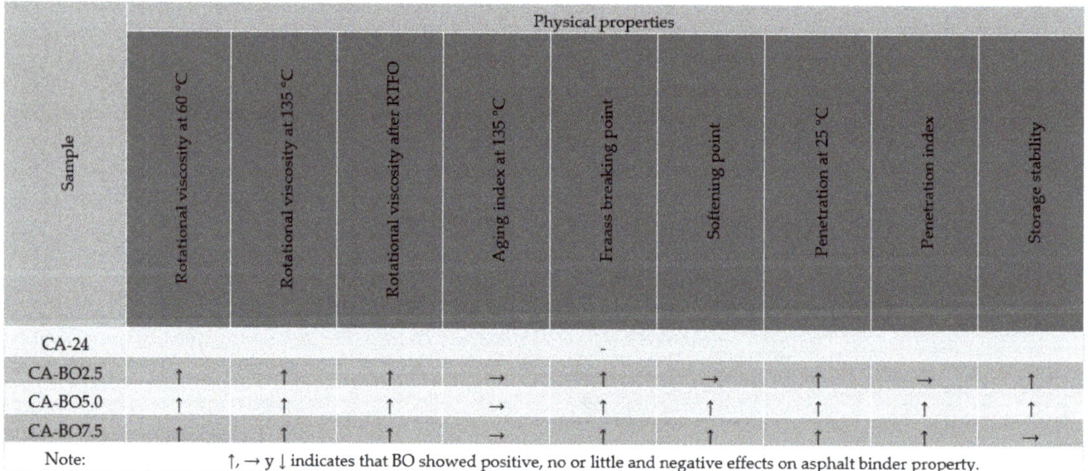

Sample	Physical properties								
	Rotational viscosity at 60 °C	Rotational viscosity at 135 °C	Rotational viscosity after RTFO	Aging index at 135 °C	Fraass breaking point	Softening point	Penetration at 25 °C	Penetration index	Storage stability
CA-24				–					
CA-BO2.5	↑	↑	↑	→	↑	→	↑	→	↑
CA-BO5.0	↑	↑	↑	→	↑	↑	↑	↑	↑
CA-BO7.5	↑	↑	↑	→	↑	↑	↑	↑	→
Note:	↑, → y ↓ indicates that BO showed positive, no or little and negative effects on asphalt binder property.								

Figure 13. Summary of the effects of BO on the physical properties of asphalt binder.

4. Discussion

The present study was conducted to verify the feasibility of the use of biochar from oat hulls (BO) as a potential bio-modifier of the physical properties of conventional CA-24 asphalt binder.

- It is determined that the asphalt binder and BO interact positively due to C=O and C=C bonds of the functional groups present on the surface of both materials.
- It is shown that the BO can be distributed homogenously in the asphalt matrix, in all the addition percentages considered, without causing clusters.
- The rotational viscosity of the asphalt binder in the original and short-term aged states increased with the addition of BO. This increase was directly proportional to the amount of BO added to the asphalt binder.

- The resistance to aging of the asphalt binder was maintained with the addition of BO. The values obtained for the parameter of aging of the modifications evaluated were lower than the regulatory requirements.
- The viscoelastic range of the asphalt binder can be extended with the addition of BO, being proportional to the increase in the modifying content, and being able to reduce the thermal susceptibility of the asphalt binder.
- The use of BO increases the consistency at an intermediate temperature, reducing the penetration of the asphalt binder.
- The reduction of the penetration, the increase in the softening point and the increase in viscosity demonstrate that BO improves the performance-related properties of the asphalt binder at high temperatures.
- With up to 7.5% modification with BO, good storage stability of the asphalt binder is obtained.
- Future studies are suggested to assess the effect of different PTT and residence times on the properties of the BO as a modifying additive. In addition, the effect of a smaller particle size and the effect of an additional digestion time after the asphalt binder modification stage should also be evaluated.

5. Conclusions

BO can be considered a potential bio-modifier of the physical properties of asphalt binder because it shows positive effects on asphalt binder properties at high temperatures, such as: rotational viscosity, softening point and penetration. These properties can contribute to improve the rutting resistance of the asphalt pavement. In relation to low temperatures, the benefits of BO as a modifier are discrete, but contribute to increasing the viscoelastic range of asphalt binder.

Author Contributions: Conceptualization, G.V.-V., C.M.-T. and A.C.-F.; methodology, G.V.-V., C.M.-T., A.C.-F. and M.E.G.; validation, G.V.-V., A.C.-F., M.E.G. and O.R.-O.; formal analysis, C.M.-T., G.V.-V., A.C.-F., M.E.G. and O.R.-O.; investigation, C.M.-T.; resources, C.M.-T., G.V.-V. and A.C.-F.; writing—original draft preparation, C.M.-T. and G.V.-V.; writing—review and editing, C.M.-T. and G.V.-V.; supervision, G.V.-V.; project administration, G.V.-V.; funding acquisition, G.V.-V. All authors have read and agreed to the published version of the manuscript.

Funding: This research was funded by Chilean Economic Development Agency (CORFO) conducted within the framework of the MACROFACULTAD Project, grant N° 14ENI2-26866, as well as the APC.

Institutional Review Board Statement: Not applicable.

Informed Consent Statement: Not applicable.

Data Availability Statement: Not applicable.

Conflicts of Interest: The authors declare no conflict of interest.

References

1. Gan, X.; Zhang, W. Application of biochar from crop straw in asphalt modification. *PLoS ONE* **2021**, *16*, e0247390. [CrossRef] [PubMed]
2. Holý, M.; Remišová, E. Characterization of Bitumen Binders on the Basis of Their Thermo-Viscous Properties. *Slovak J. Civ. Eng.* **2019**, *27*, 25–31. [CrossRef]
3. Ahed, Y.H.; Al-Haddad, A.H.A. Temperature Susceptibility of Modified Asphalt Binders. *IOP Conf. Series Mater. Sci. Eng.* **2020**, *671*, 012121. [CrossRef]
4. Behnood, A.; Gharehveran, M.M. Morphology, rheology, and physical properties of polymer-modified asphalt binders. *Eur. Polym. J.* **2019**, *112*, 766–791. [CrossRef]
5. Mamun, A. Arifuzzaman Nano-scale moisture damage evaluation of carbon nanotube-modified asphalt. *Constr. Build. Mater.* **2018**, *193*, 268–275. [CrossRef]
6. Yang, Q.; Li, X.; Zhang, L.; Qian, Y.; Qi, Y.; Kouhestani, H.S.; Shi, X.; Gui, X.; Wang, D.; Zhong, J. Performance evaluation of bitumen with a homogeneous dispersion of carbon nanotubes. *Carbon* **2020**, *158*, 465–471. [CrossRef]
7. Moreno-Navarro, F.; Sol-Sánchez, M.; Gámiz, F.; Rubio-Gámez, M. Mechanical and thermal properties of graphene modified asphalt binders. *Constr. Build. Mater.* **2018**, *180*, 265–274. [CrossRef]

8.   Leiva-Villacorta, F.; Aguiar-Moya, J.; Villegas, R.; Salazar, J.; Loría-Salazar, L. *Nano-Materiales En El Desempeño Del Asfalto*; Laboratorio Nacional de Materiales y Modelos Estructurales Lanamme UCR: San José, Costa Rica, 2014; p. 10.

9.   Calderón, C. Vías Más Duraderas Gracias a Los Nanomateriales. 2018. Available online: https://www.javeriana.edu.co/pesquisa/vias-mas-duraderas-gracias-a-los-nanomateriales/ (accessed on 12 May 2021).

10. Vicente Pérez, J.; Pérez Infante, J.I.; García, A.; Baselga Llidó, J.; Agzenai Ben Salem, Y.; Pozuelo de Diego, J.; Sanz Feito, J. Mezclas asfálticas con betunes modificados con nanotubos de carbono. Diseño y propiedades. *Carreteras* **2016**, *4*, 20–30.

11. Themeli, A.; Chailleux, E.; Farcas, F.; Chazallon, C.; Migault, B.; Buisson, N. Molecular structure evolution of asphaltite-modified bitumens during ageing; Comparisons with equivalent petroleum bitumens. *Int. J. Pavement Res. Technol.* **2017**, *10*, 75–83. [CrossRef]

12. Pan, P.; Wu, S.; Xiao, Y.; Wang, P.; Liu, X. Influence of graphite on the thermal characteristics and anti-ageing properties of asphalt binder. *Constr. Build. Mater.* **2014**, *68*, 220–226. [CrossRef]

13. Seyrek, E.; Yalçin, E.; Yilmaz, M.; Kök, B.V.; Arslanoğlu, H. Effect of activated carbon obtained from vinasse and marc on the rheological and mechanical characteristics of the bitumen binders and hot mix asphalts. *Constr. Build. Mater.* **2020**, *240*, 117921. [CrossRef]

14. Zhao, S.; Huang, B.; Ye, X.P.; Shu, X.; Jia, X. Utilizing bio-char as a bio-modifier for asphalt cement: A sustainable application of bio-fuel by-product. *Fuel* **2014**, *133*, 52–62. [CrossRef]

15. Bekchanova, M.; Campion, L.; Bruns, S.; Kuppens, T.; Jozefczak, M.; Cuypers, A.; Malina, R. Biochar's effect on the ecosystem services provided by sandy-textured and contaminated sandy soils: A systematic review protocol. *Environ. Évid.* **2021**, *10*, 7. [CrossRef]

16. Głąb, T.; Gondek, K.; Mierzwa–Hersztek, M. Biological effects of biochar and zeolite used for remediation of soil contaminated with toxic heavy metals. *Sci. Rep.* **2021**, *11*, 6998. [CrossRef]

17. Gupta, S.; Kua, H.W.; Low, C.Y. Use of biochar as carbon sequestering additive in cement mortar. *Cem. Concr. Compos.* **2018**, *87*, 110–129. [CrossRef]

18. Lee, J.; Hawkins, B.; Day, D.; Reicosky, D.C. Sustainability: The capacity of smokeless biomass pyrolysis for energy production, global carbon capture and sequestration. *Energy Environ. Sci.* **2010**, *3*, 1695–1705. [CrossRef]

19. Woolf, D.; Amonette, J.E.; Street-Perrott, F.A.; Lehmann, J.; Joseph, S.G. Sustainable biochar to mitigate global climate change. *Nat. Commun.* **2010**, *1*, 56. [CrossRef]

20. Roberts, K.G.; Gloy, B.A.; Joseph, S.; Scott, N.R.; Lehmann, J. Life Cycle Assessment of Biochar Systems: Estimating the Energetic, Economic, and Climate Change Potential. *Environ. Sci. Technol.* **2010**, *44*, 827–833. [CrossRef]

21. Zhang, R.; Dai, Q.; You, Z.; Wang, H.; Peng, C. Rheological Performance of Bio-Char Modified Asphalt with Different Particle Sizes. *Appl. Sci.* **2018**, *8*, 1665. [CrossRef]

22. Zhang, R.; Wang, H.; Ji, J.; Wang, H. Viscoelastic Properties, Rutting Resistance, and Fatigue Resistance of Waste Wood-Based Biochar-Modified Asphalt. *Coatings* **2022**, *12*, 89. [CrossRef]

23. Gonzalez, M.E.; Gonzalez, A.; Toro, C.; Cea, M.; Sepulveda, N.; Díez, M.C.; Navia, R. Biochar as a Renewable Matrix for the Development of Encapsulated and Immobilized Novel Added-Value Bioproducts. *J. Biobased Mater. Bioenergy* **2012**, *6*, 237–248. [CrossRef]

24. Brown, T.R.; Wright, M.M.; Brown, R.C. Estimating profitability of two biochar production scenarios: Slow pyrolysis vs fast pyrolysis. *Biofuels Bioprod. Biorefining* **2011**, *5*, 54–68. [CrossRef]

25. Babel, S.; Kurniawan, T.A. Low-cost adsorbents for heavy metals uptake from contaminated water: A review. *J. Hazard. Mater.* **2003**, *97*, 219–243. [CrossRef]

26. US Geological Survey. Mineral Commodity, Summaries 2005. 2005. Available online: http://minerals.usgs.gov/minerals/pubs/mcs/2005/mcs2005.pdf (accessed on 12 August 2022).

27. Alexander, H.; Reyes-lizcano, F.A.; Baudilio, S.; Gabriel, J.; Alfonso, C. Use of Biochar in Asphalts: Review. *Sustainability* **2022**, *14*, 4745. [CrossRef]

28. González, M.E.; Cea, M.; Sangaletti, N.; González, A.; Toro, C.; Diez, M.C.; Moreno, N.; Querol, X.; Navia, R. Biochar Derived from Agricultural and Forestry Residual Biomass: Characterization and Potential Application for Enzymes Immobilization. *J. Biobased Mater. Bioenergy* **2013**, *7*, 724–732. [CrossRef]

29. González, M.E.; Romero-Hermoso, L.; Hidalgo, P.; Meier, S.; Navia, R.; Cea, M. Effects of Pyrolysis Conditions on Physicochemical Properties of Oat Hull Derived Biochar. *BioResources* **2017**, *12*, 2040–2057. [CrossRef]

30. ODEPA. El Mercado De La Avena Blanca En Chile. 2017. Available online: https://www.odepa.gob.cl/wp-content/uploads/2017/12/Avena.pdf (accessed on 13 August 2022).

31. ICIgroup. Ficha Técnica De Producto Avena Entera Con Cáscara. 2013, p. 9. Available online: http://icigroup.gt/wp-content/fichas/Avena_Entera_con_cascara.pdf (accessed on 8 February 2021).

32. Dirección de Vialidad de Chile. Volumen 8. Especificaciones y métodos de muestreo, ensaye y control. In *Manual de Carreteras*; Ministerio de Obras Publicas de Chile, Ed.; Ministerio de Obras Públicas: Santiago, Chile, 2021.

33. Hunter, R.N.; Self, A.; Read, J. *The Shell Bitumen Handbook*, 6th ed.; Shell Bitumen: London, UK, 2015.

34. Lesueur, D. The colloidal structure of bitumen: Consequences on the rheology and on the mechanisms of bitumen modification. *Adv. Colloid Interface Sci.* **2009**, *145*, 42–82. [CrossRef]

35. Ma, F.; Dai, J.; Fu, Z.; Li, C.; Wen, Y.; Jia, M.; Wang, Y.; Shi, K. Biochar for asphalt modification: A case of high-temperature properties improvement. *Sci. Total Environ.* **2022**, *804*, 150194. [CrossRef]
36. Khiavi, A.K.; Ghanbari, A.; Ahmadi, E. Physical and High-Temperature Rheological Properties of PHEMA-Modified Bitumen. *J. Mater. Civ. Eng.* **2020**, *32*, 4020010. [CrossRef]
37. Ertugrul, M.; Mehmet, Y.; Kök, B.V.; Yalçin, E. Effects of various biochars on the high temperature performance of bituminous binder. 2017. Available online: https://www.researchgate.net/publication/312326052_Effects_of_various_biochars_on_the_high_temperature_performance_of_bituminous_binder (accessed on 13 August 2022).
38. Walters, R.C.; Fini, E.H.; Abu-Lebdeh, T. Enhancing asphalt rheological behavior and aging susceptibility using bio-char and nano-clay. *Am. J. Eng. Appl. Sci.* **2014**, *7*, 66–76. [CrossRef]
39. Dong, W.; Ma, F.; Li, C.; Fu, Z.; Huang, Y.; Liu, J. Evaluation of Anti-Aging Performance of Biochar Modified Asphalt Binder. *Coatings* **2020**, *10*, 1037. [CrossRef]
40. Wen, G.; Zhang, Y.; Zhang, Y.; Sun, K.; Fan, Y. Improved properties of SBS-modified asphalt with dynamic vulcanization. *Polym. Eng. Sci.* **2002**, *42*, 1070–1081. [CrossRef]
41. Al-Layla, M.M.; Hussien, A.K.; Mjthab, E.I. Evaluation of the properties and storage stability of EVA polymer modified asphalt. *J. Educ. Sci.* **1999**, *24*, 14–20. [CrossRef]

Review

Concrete Made with Dune Sand: Overview of Fresh, Mechanical and Durability Properties

Jawad Ahmad [1,*], Ali Majdi [2], Ahmed Farouk Deifalla [3,*], Hisham Jahangir Qureshi [4], Muhammad Umair Saleem [5], Shaker M. A. Qaidi [6] and Mohammed A. El-Shorbagy [7]

[1] Department of Civil Engineering, Military College of Engineering, Risalpur 4707, Pakistan
[2] Department of Building and Construction Technologies Engineering, Al-Mustaqbal University College, Hillah 51001, Iraq
[3] Structural Engineering Department, Faculty of Engineering and Technology, Future University in Egypt, New Cairo 11845, Egypt
[4] Department of Civil and Environmental Engineering, College of Engineering, King Faisal University, Al-Ahsa 31982, Saudi Arabia
[5] Service Stream Limited Co., Chatswood, NSW 2067, Australia
[6] Department of Civil Engineering, University of Duhok, Duhok 42001, Iraq
[7] Department of Mathematics, College of Science and Humanities in Al-Kharj, Prince Sattam bin Abdulaziz University, Al-Kharj 11942, Saudi Arabia
* Correspondence: jawadcivil13@scetwah.edu.pk (J.A.); ahmed.deifalla@fue.edu.eg (A.F.D.)

Citation: Ahmad, J.; Majdi, A.; Deifalla, A.F.; Qureshi, H.J.; Saleem, M.U.; Qaidi, S.M.A.; El-Shorbagy, M.A. Concrete Made with Dune Sand: Overview of Fresh, Mechanical and Durability Properties. *Materials* **2022**, *15*, 6152. https://doi.org/10.3390/ma15176152

Academic Editors: Stefano Guarino and Flaviana Tagliaferri

Received: 26 June 2022
Accepted: 22 July 2022
Published: 5 September 2022

Abstract: According to the authors' best information, the majority of research focuses on other waste materials, such as recycling industrial waste (glass, silica fume, marble and waste foundry sand), etc. However, some researchers suggest dune sand as an alternative material for concrete production, but knowledge is still scarce. Therefore, a comprehensive review is required on dune sand to evaluate its current progress as well as its effects on the strength and durability properties of concrete. The review presents detailed literature on dune sand in concrete. The important characteristics of concrete such as slump, compressive, flexural, cracking behaviors, density, water absorption and sulfate resistance were considered for analysis. Results indicate that dune sand can be used in concrete up to 40% without any negative effect on strength and durability. The negative impact of dune sand on strength and durability was due to poor grading and fineness, which restricts the complete (100%) substation of dune sand. Furthermore, a decrease in flowability was observed. Finally, the review highlights the research gap for future studies.

Keywords: dune sand; compressive strength; tensile strength; fine aggregate; durability

1. Introduction

All construction and development initiatives require concrete as their base since it is a widely used building material worldwide [1–4]. Finding lower priced cement made from local natural resources has emerged as a primary objective in order to make up for the scarcity in cement manufacturing [3,5–7]. Since 1970, there has been a great deal of research on the application of cementing additives as a partial substitute for Portland cement. The byproducts of other industry or natural sources are these additions [8]. The practical binding activity that the supply of the additive is influenced is by the volume, fineness, mineralogical composition and kind of cement, which in turn enhances the strength [9]. The density of the mortar might be raised by the creation of the secondary cementitious material calcium silicate hydrates (CSH). The latter is produced by including minute siliceous particles that serve a particular pozzolanic purpose and increase the strength properties of concretes [10–12].

Each of the most important ingredients of concrete has an impact on the environment to varied degrees. Numerous sustainability issues are brought up by the massive amounts of concrete utilized globally [13–15]. There has been considerable concern due to a rise in the amount of riverbed sand and gravel used in concrete [16]. Due to the extensive usage of concrete brought on by the surge in urbanization and industrialization, more natural sand has been removed from riverbeds. A few of the negative effects include increased riverbed distance, a drop in the water table, the discovery of bridge substructures, a significant impact on rivers, deltas, coastal ecosystems and marine ecosystems, land loss due to river or coastal erosion and a reduction in the quantity of deposit sources [17]. Furthermore, limits on sand removal from a river, which have led to an increase in sand charges, have seriously impacted the ability of the building sector to survive [18].

One of the essential ingredients for producing mortar and concrete is a fine aggregate, which also plays a crucial part in the design mix [19]. Fine aggregate is a key component of concrete, and the amount and kind of sand used to create a particular concrete mix will define its qualities. It significantly affects the flowability, resiliency to the effects of the environment, strength and dry shrinkage of concrete. In comparison to cement, sand makes up a bigger portion of the mixture. Another element that adds to concrete strength is that sand may fill up any pores or spaces in the material. Sand offers a mass of particles that can resist the action of applied stresses and endure longer than mortar alone. Fine aggregate also lowers volume changes brought on by the setting and hardening processes. Sand is thus essential for concrete's capacity to consolidate and give the necessary strength. In place of natural sand, there are other alternate sand options, such as foundry sand [20], marble waste [21], waste glass [22], recycled concrete aggregate [23], recycled ceramics aggregate [24], copper slag [25], iron ore tailing [26], plastic waste [27] and crush stone dust [28] that may be utilized as sand in concrete.

It is generally known that many construction projects, including building construction in recent years, water delivery channels, oil exploitation platforms, and building construction, are to be constructed in the desert. River sand is a crucial primary component for the increasing infrastructure projects. However, owing to the great distance and overuse of the river channel in several desert areas, the decline in river sand and the rise in need have come to be an insurmountable challenge. Despite being widely distributed across the desert, dune sand is seldom used in engineering construction, owing to its high prevalence of very small particle sizes and inability to conform to particle size grading standards.

Researchers utilize clay to make cement clay mortar as a masonry binder to construct buildings, reducing the need for cement, saving money and protecting the environment [29]. If dune sand and clay, which are effortlessly obtained nearby, can be technically utilized in place of natural fine aggregate and a binder as a building material, that will not only satisfy the demand for infrastructure creation, but also significantly reduce the charges of construction projects, making waste profitable. Many findings have been reported on the use of dune sand as a fine aggregate to create concrete, particularly in desert regions, in order to satisfy the criteria of engineering construction utilizing local supplies and decreasing the shipping expenses [30,31].

In light of these conditions, there is growing interest worldwide in the idea of substituting dune sand and clay for natural fine aggregate and cement in the creation of concrete and mortar. Some scholars have recently looked at the characteristics of mortar and concrete. The rheological characteristics of mortar including crushed sand were investigated by Mikael et al. [32]. The effect of natural river sand on the properties of mortar containing suspensions of coarse particles was assessed. According to the experiments, fine aggregate has a significant impact on both the water need and the mortar usability. The features of concrete made from dune sand from Maowusu sandy terrain have been studied by Jin et al. [30].

The issue of cementitious materials created using dune sand has sparked growing attention because of the environmental advantages. Numerous studies on the features of dune sand and the qualities of cementitious materials using it have been performed over the last ten years. Dune sand differs from regular engineering sand in that it has a poor gradation and high salinity. The varied mechanical characteristics of cementitious composites are considerably impacted by a high replacement rate of dune sand for construction sand [33]. The concrete made with a dune sand to fine aggregate ratio of 10% showed the highest compressive strength [34]. By adding ultrafine fibers [35] and mineral fillers [36] to cementitious composites or by regulating fractures and permeability utilizing ultrahigh performance materials at the macro level, the strength of dune sand-integrated cementitious composites may be increased.

The impact of very fine particles on the flowability and strength of concrete created with dune sand from the Australian desert was studied by Fu Jia Luo et al. [37]. The findings revealed that the metal fibers alter concrete properties by various mechanisms depending on the concentration of the sand–cement ratio and have no adverse effects on flowability. Additionally, dune sand concrete strength is on par with or even exceeds that of natural fine aggregate concrete. By valuing dune sand and pneumatic waste metal fibers, Allaoua Belferrag et al. [38] researched how to improve the compressive capacity of cement in dry conditions. The impact of adding a novel class of metal fibers made from recycled tires on the compressive strength of dune sand concrete has been researched. In comparison to concrete without fibers, the findings obtained reveal an increase in the compressive capacity for metal fiber-reinforced sand dune concrete. According to Krobba et al. [39], the characteristics of mortar, including density, strength, shrinkage, elasticity and bonding strength, are affected by the addition of natural microfibers. In comparison to mortar made from beach sand (without Alfa natural microfibers) and tested under identical circumstances, the findings obtained demonstrated that mortar containing natural microfibers had improved mechanical and physical qualities [39].

Brief literature shows that dune sand can be used as an alternative material for the production of concrete. However, knowledge is still scarce, which restricts the use of dune sand in concrete, and a comprehensive review is required to identify the positive and negative effects of dune sand on concrete performance. Furthermore, according to the authors' best information, no detailed review on dune sand concrete has been considered up to now. Therefore, this review presents the detailed literature on the characteristics of concrete with dune sand. For analysis, the important properties of concrete such as slump, compressive strength, flexural strength, cracking behaviors, density, water absorption and sulfate resistance were taken into account. The successful review provides multiple benefits, including the current progress of dune sand in concrete, the effect of dune sand flowability, strength and durability performance of concrete. The review also suggests future guidelines for the upcoming generation. Figure 1 represents a different section of the review.

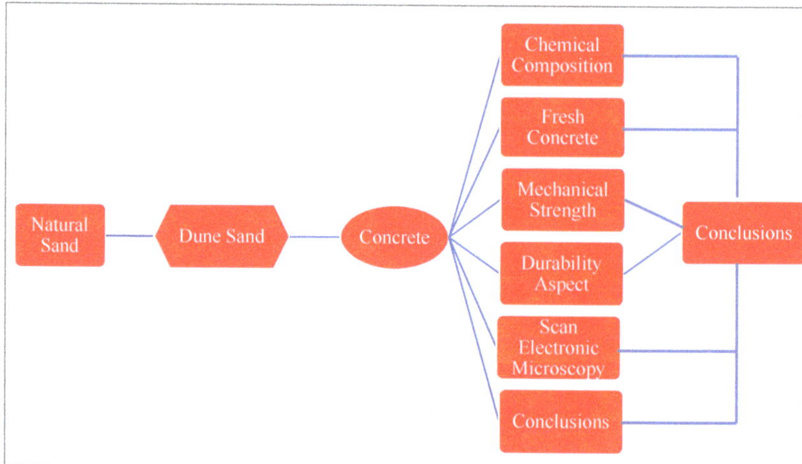

Figure 1. Different sections of the review.

2. Physical and Chemical Properties

Dune sand is very fine sand with a maximum particle size of 1.18 mm [40]. The diameters of more than 90% of its components are between 0 and 0.4 mm [41]. Dune sand typically has a fineness modulus of less than 1.5 or average particle size of less than 0.25 mm, making it superfine sand [42]. Dune sand is classified as a D1 soil type by the GTR 2000 Soil Classification [43]. It is described as porous, irregular and poorly graded dirt. This implies that dune sand by itself will not be adequately compact, and as a result, its immediate bearing index is insufficient [44]. Therefore, it will be crucial to treat this sand using hydraulic binders as correctors.

The largest size of the coarse grains is 0.5 mm, while the diameter of the tiniest particles is in the range of 0.04 mm. It is fine golden sand. The curvature coefficient (Cc) is around 0.96, while the uniformity coefficient (Cu) is in the range of 2.0. Therefore, it is extremely fine, improperly graded sand. The most often calculated component for fine aggregates is the fineness modulus, which is used to find the aggregate gradation degree of consistency. The examined dune sand samples had fineness modulus values ranging from 1.07 to 1.30. These findings show that the investigated dune sands fall short of the fine aggregate gradation criteria. Therefore, to achieve an appropriate degree of gradation, it is important to increase the gradation of these dune sands by combining them with well-graded crushed fine aggregates of ceramic waste [45]. Furthermore, physical properties are presented in Table 1.

Table 1. Physical properties of dune sand.

Authors	[45]	[46]	[47]	[48]	[49]
Dry-rodded density kg/m^3	-	1663	-	-	-
Water Absorption (%)	-	-	3.84	-	-
Fineness Modulus	1.07	1.45	1.44	-	0.65
Sand equivalency (%)	87%	-	-	-	69%
Bulk density (kg/m^3)	2525	-	1560	-	-
Specific Gravity	-	2.77	-	-	2.64
Surface area cm^2/g	-	116.8	-	3000	-
Apparent specific gravity (kg/m^3)	1434	-	-	-	-

Dune sand is siliceous, made up mostly of SiO_2 (silica), with traces of calcium and magnesium species, according to the chemical analyses reported in Table 2. X-ray diffraction of the dune is shown in Figure 2. High percentages of quartz and traces of illite and calcite are found, according to X-ray examination [50].

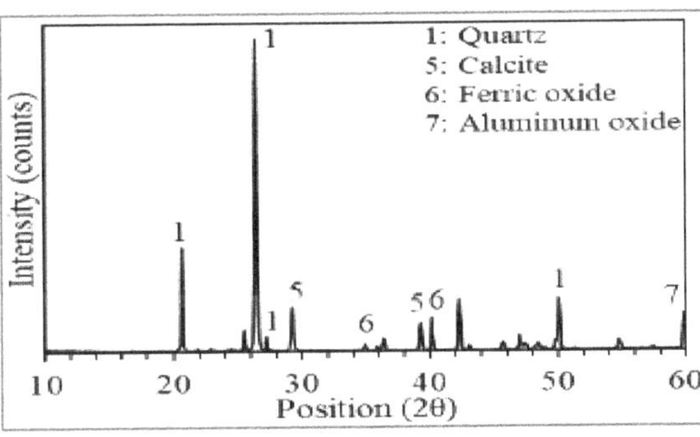

Figure 2. XRD of dune sand [51].

Table 2. Chemical Composition of Dune Sand.

Authors	[10]	[50]	[51]	[52]	[53]
SiO_2	74.61	93.56	64.9	64.58	65.63
Al_2O_3	1.35	1	3.0	9.48	12.63
Fe_2O_3	0.86	1	0.7	2.32	3.42
MgO	0.29	-	1.3	2.06	4.52
CaO	17.3	-	14.1	8.62	8.76
Na_2O	-	-	-	2.43	1.83
K_2O	0.47	-	-	1.97	1.87

3. Slump Flow

Concrete workability refers to how easily new concrete may be poured, reinforced and finished, with the fewest uniformity losses possible. Any concrete mixture should be workable enough to be properly poured, solidified and filled into the forms, as well as to encircle any embedded reinforcement or other objects.

Figure 3 shows the slump flow of concrete with the substitution of dune sand. Dawood et al. [40] reported that the slump flow of concrete decreased with the addition of dune sand. The dune sand particle size distribution is mostly responsible for the decline in a slump. It is well-known that when the fines particles content rises, the flowability of concrete declines as well [54]. The lack of chemicals in the concrete resulted in poor workability and a minor slump [55]. As a more natural aggregate was substituted with recycled concrete aggregate (RCA), the densities and slump of fresh concrete reduced. Although the slump was further decreased, the addition of steel fibers increased the densities of both fresh and cured concrete [46]. Increases in dune sand concentration lowered flowability from a high to medium level by 30%, and the addition of steel fibers also dramatically decreased workability [40].

A dramatic decline in a slump was seen in all mixtures, producing concrete with a very low slump when all the fine particles were completely substituted by beach sand (100 percent substitute). When more regular sand is substituted with dune sand, the particle size distribution that results causes the slump to diminish. It is common knowledge that concrete becomes less workable as fine particles are used [56]. The decrease in slump flow due to dune sand may be due to the fact that the mixing water may be found as free layer water, adsorbed layer water and filled water, according to Wang et al. [57]. The many types of water each contribute differently to workability. The solid particles are separated from one another by the free layer of water, which improves workability. The water-adsorbed layer is quite near the solid's surface. This portion of the water will be constrained by the solid particle and unable to flow freely as a result of the solid surface adhering to the water molecule. Therefore, this water has no effect on workability. The filling water does not affect the workability; it only fills the spaces between solid particles. It should be noted that the kind and quantity of the various solid particles determine the types of water that may interchange in the system. For instance, the fine grains will fill the holes of the granular structure by distributing the water in these holes when the right ratio of fine and coarse particles is combined together.

On the other hand, Figure 3 shows that the slump rises as the dune sand content rises, according to Rennani et al. and Leila et al. [48,58]. The spherical form of the dune sand particles, all things being equal, is thought to be responsible for the rise in a slump. This is because spheres travel more easily than angular or awkward-shaped particles because of the ball-bearing effect. The slump did, however, start to subside with dune sand concentrations over 50%. All findings indicate that despite an increase in beach sand content, concrete may still be made to be workable. Additionally, it was found that the amount of dune sands the mixture could contain before slumping down increased with the workability of the mixture.

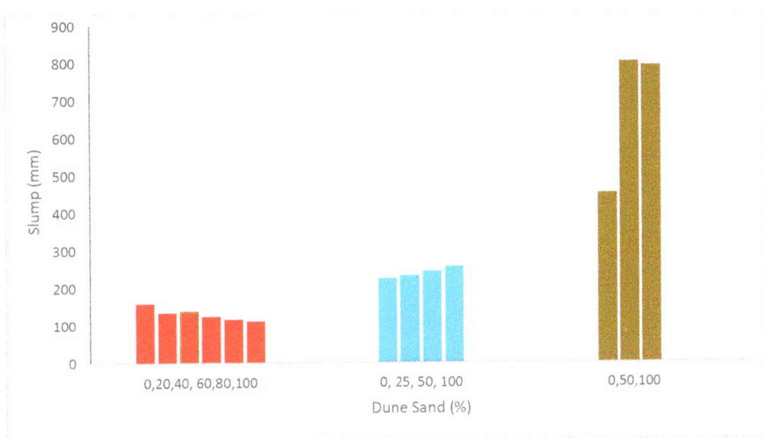

Figure 3. Slump flow: data source [40,48,58].

4. Mechanical Strength

4.1. Compressive Strength

Figure 4 shows concrete compressive capacity with dune sand's substitution. It can be noted that the compressive capacity of concrete was enhanced up to a certain extent by the substitution of dune sand and then decreased. According to certain researchers [56,59], when the dune sand replacement ratio rose, concrete's compressive capacity decreased. Due to the dune sand particles' smooth surfaces and rounded shapes, the binding strength between the cement paste and dune sand may have been reduced [60]. In contrast, several studies [61,62] found that the strength improved to a specific replacement ratio for the dune sand but declined beyond that point. According to Luo et al. [59], high aggregate compactness might cause a rise in strength. According to previous research, dune sand may affect the compactness of aggregates and the strength of the link between dune sand and cement paste. Additionally, the impact of dune sand on each side has a different effect on compressive strength.

It was found that replacing 40% and 50% of the natural dune sand with waste ceramic aggregate resulted in the biggest gain in compressive and flexural capacity when compared to reference mortar. At a 40% and 50% mix ratio of ceramic under various curing conditions, the compressive strength of dune sand ceramic mortar achieves excellent strength [45]. The findings show that as dune sand content rises, concrete strength typically declines. As fine particles have a larger surface area, more grout is required to cover their surface, which results in a loss in strength [56].

The compressive capacity of mortar improves initially with an increase in the dune sand replacement ratio before decreasing and peaking at a 20 percent replacing ratio [47]. According to Appa Rao [11], at a constant water/binder ratio of 0.5, adding silica fume to a mortar up to the point where it replaces 30 percent of the cement results in an increase in compressive strengths [63]. These findings reveal that the development of compressive capacity as a function of time (7, 28 and 90 days) reveals that the compressive capacity is low for all samples during the first seven days but then dramatically rise over the succeeding periods. This is caused by the kinetics of the time-dependent reactions between dune sand powder and portlandite and the hydration of cement [10]. In accordance with the water/binder ratio, Kwan [64] demonstrated that adding silica fume to a mortar with up to 15% of the cement replacement amount results in an increase in compressive strengths after 28 days [65]. Based on our findings, we can conclude that adding the fiber and applying a heat treatment boosted the compressive and flexural capacity. The quantity of dune sand rose, and the compressive strength somewhat improved [66]. Linear connections between compressive capacity and fine dune sand % with respectable correlation coefficients demonstrate that compressive capacity rises as dune sand percentage increases [48]. The findings suggest that when the rate of dune sand replacement increases, the compressive capacity of specimens initially rises and subsequently falls. At curing ages of 7 days and 28 days, respectively, the compressive capacity of the specimen including 50% dune sand improves by 3.87 percent and 10.64 percent compared to the specimen containing 0% dune sand.

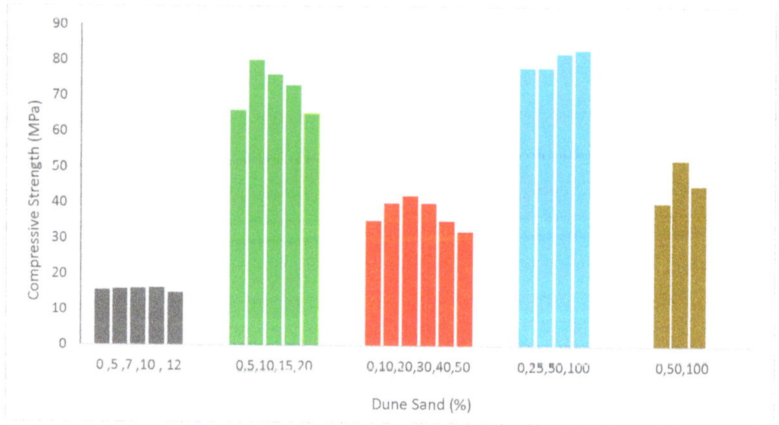

Figure 4. Compressive strength: data source [10,47–49,58].

The compressive capacity increased the percentage of dune sand. However, the highest strength loss was just around 25% [56]. According to a study, concrete's compressive strength declines by 39% when the substitute ratio reaches 100%, but it increases with the addition of single or hybrid fibers. In contrast, sand dunes have a negative impact on compressive capacity, particularly when the proportion increases by more than 40% [40]. The research has shown that dune sand may be utilized as an affordable and easily accessible substitute for sea sand and can therefore assist in stopping the negative impacts on the ecosystem caused by excessive sea sand mining [67].

A study reported no significant changes in the amount of addition or dune sand percent on the setting time of mixes. The pastes' original consistency was dramatically reduced by the addition of dune sand. Due to their higher fineness than that of ordinary Portland cement (OPC), dune sand powder acts as a lubricant by reducing the intergranular vacuum. This cement has a lot of vacuums that need to be filled with water before they can set in the convenience of the mortar. Therefore, in order to decrease the intergranular vacuum and hence the need for water in cement, it is required to enhance the quantities of tiny and big particles. The intergranular vacuums will be filled with the dune sand powder, reducing the need for water [10].

A comparative investigation of compressive capacity with different dune sand dosages at various curing days is shown in Figure 5.

As a reference mix, the compressive capacity of control concrete on day 28 of curing was used to evaluate the compressive strength of different dosages of dune sand at various curing days. For a comparative study, five percent of the recommended dune sand dosage was considered. Compressive strength during seven days of curing is 29% less than the reference compressive strength (concrete at 28 days controls compressive capacity with a 5% replacement of dune sand). The compressive strength with a 5% substitute of dune sand is 21% higher than the reference concrete's compressive strength after 28 days of curing. Compressive capacity is 43% more than the reference concrete at the same dosage of 5% dune sand replacement. Additionally, with a 20 percent replacement of dune sand, the compressive strength at 28 days is the same as the reference concrete's compressive capacity, but by 90 days, it is 14 percent higher than the reference strength.

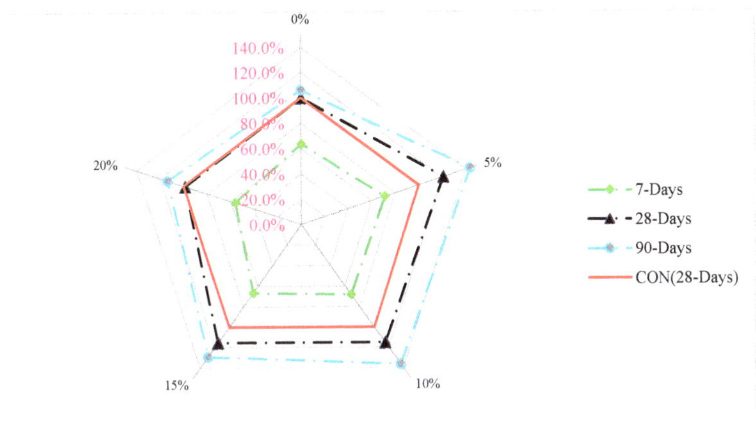

Figure 5. Compressive strength age relation [10].

4.2. Tensile Strength

Figure 6 illustrates the tensile capacity of concrete with the substitution of dune sand. It can be noted that the tensile capacity of concrete reduced with the substitution of dune sand as reported by Abadou et al. [68].

The modulus of elasticity and tensile capacity of concrete is not significantly negatively impacted by increasing the amount of dune sand in concrete [56]. The DS/FA (Dune sand/Fly ash) ratio of 10% was ideal for compressive and tensile capacity under the mixed circumstances of this investigation. The strength differential might be more than 10%, depending on the DS/FA ratio [34]. Since the compressive capacity has greatly increased and the tensile strengths have only marginally diminished, it was possible to create a material that may be considered when constructing pavement constructions by adding dune sand and lime to permeable asphalt [19]. At all evaluated ages up to 44 percent, the recycled aggregate made from concrete waste used to make the dune sand mortar had lower tensile strength than the specimen mortar. When compared to ordinary mortar, the integration of concrete waste aggregates does not seem to have a substantial influence on tensile strength [68]. The tensile and flexural capacity decreased by 29.3 percent and 21.1 percent for splitting and flexural capacity, respectively, and at a slower pace than the compressive strength as the sand dune rate increased. The ratio (40 percent DS without fibers) attained the greatest tensile capacity of 4 MPa (22.3 percent more than the reference samples), according to the tensile capacity data at the age of 28 days.

On the other hand, Figure 6 shows that the tensile capacity of concrete rises as the dune sand content rises, according to Rennani et al. and Leila et al. [48,58]. Similarly, 60 percent dune sand increased tensile strength by 0.91 percent, which is equal to the reference concrete. The mechanical qualities of the hybrid concrete were enhanced by the inclusion of steel fibers, which can make up for much of the lost strength caused by the rise in sand dunes ratios [40]. The tensile strength of the hybrid mixture is 40.2 percent higher than 100 percent dune sand without fibers when dune sand is completely replaced by 100 percent rather than sand with 1 percent fibers [40]. Fiber increased tensile strength due to the prevention of cracks [69]. Even if the cracks appear, the propagation of cracks is restricted by the fibers [70].

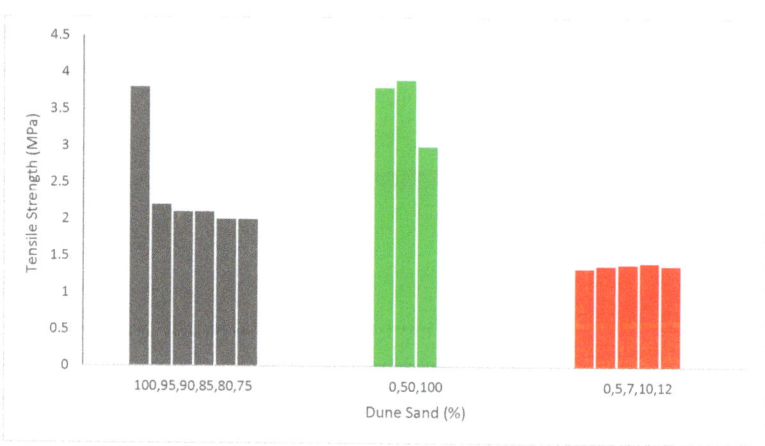

Figure 6. Tensile strength: data source [49,58,68].

4.3. Flexural Strength

Figure 7 reveals the flexural capacity of concrete with the substitution of dune sand. It can be noted that flexural capacity improved by up to 20% with substitution of dune sand. A study reported that the flexural strengths were increased by using waste materials as aggregate. It was found that replacing 40% and 50% of the natural dune sand with waste ceramic aggregate resulted in the biggest gain in compressive and flexural strength when compared to reference mortar [45]. According to the experimental findings, mechanical properties were best when dune sand replacement was 50%. The flexural strength of self-compacting mortar had not been significantly decreased by the replacement of dune sand [54].

The 40% substitution of dune had the maximum flexural strength at seven days, which is 4.3 percent higher than the reference sample. At the same time, the flexural capacity is reduced by 13.7 percent less than the reference specimen when dune sand is substituted 100 percent (completely replaced the sand). The 40 percent dune sand in concrete shows the maximum flexural strength (7.47 MPa) at the age of 28 days, which is 9.85 percent higher than the reference sample. When the dune sand ratio is enhanced, the flexural capacity decreases. At a replacement ratio of 100 percent dune, flexural strength decreased by 21.3 percent when associated to the blank concrete (without dune sand). The rise in fineness of the sand particles in the concrete is the cause of the decreased flexural capacity. However, the flexural capacity of concrete with dune sand substituted was significantly enhanced with the addition of fibers. A concrete mixture of 20% dune sand along with 1% steel fibers achieved the highest flexural capacity of 10.07 MPa at the age of 28 days, which is 45.3 percent higher than the reference concrete [40]. Few researchers considered flexural strength in their studies. Therefore, a more detailed investigation is required.

Flexural and flexure–shear fractures are clearly visible, and the crack patterns of the regular beam resemble those of dune sand beams. It was determined that the presence of the dune may have delayed the initial fracture since it occurred at a force of around 80 KN, as shown in Figure 8, the central flexural zone. Shear fractures began to appear in both shear spans at the supports at a load of around 130 KN, although their dispersion in the shear spans varied. All beams broke in flexure shear at a load of around 280 KN, and the ultimate failure took place near the load point for concrete crushing in the compression zone. Figure 8 shows the failure load for every beam, and it can be inferred that the presence of dune sand may lower the failure load of beams. Although the leading shear crack direction and failure mechanisms varied, the initial fracture, failure load and shear cracks did not.

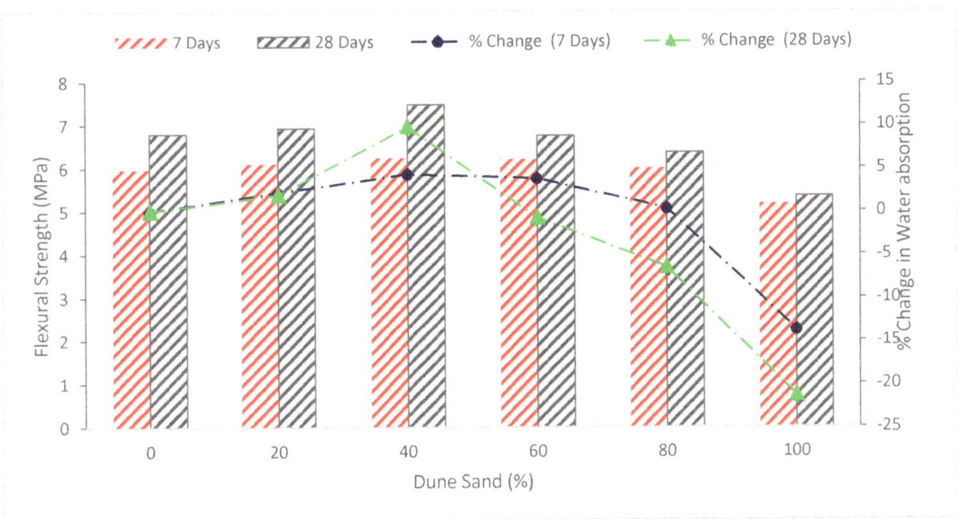

Figure 7. Flexural strength: data source [40].

Figure 8. First crack load and failure load: data source [71].

5. Durability

5.1. Dry Density

The addition of dune sand to the concrete mixture increased the density at replacement ratios of 40 and 60 percent, with the replacement ratio of 40 percent dune sand. The recording density is 2427.35 kg/m^3 with a growth of its quantity 1.88 percent more than the blank concrete, while the replacement ratio is 60 percent, and the increase in density is 0.76%, as associated with the control concrete. Partial substitution of dune with natural sand helped increase the density due to the softness of the dune sand, their spherical form and their overlap between gravel and sand grains, which helped fill all the spaces and holes within the concrete [40]. A study also reported that the filler materials increased the density of concrete due to filling cavities in concrete components, leading to more dense concrete [72]. However, when the percentage of dune sand is raised to 80% and 100%, the density falls, as shown in Figure 9. It might be possible that a higher dose of dune sand reduced the density due to a lack of flowability. The less flowable concrete required more compaction energy as compared to more workable concrete. Therefore, more chance of voids in less workable concrete, which adversely affect the density of concrete. A study also claims that filler material decreased the density of concrete at a higher substitution ratio due to lack of flowability [19].

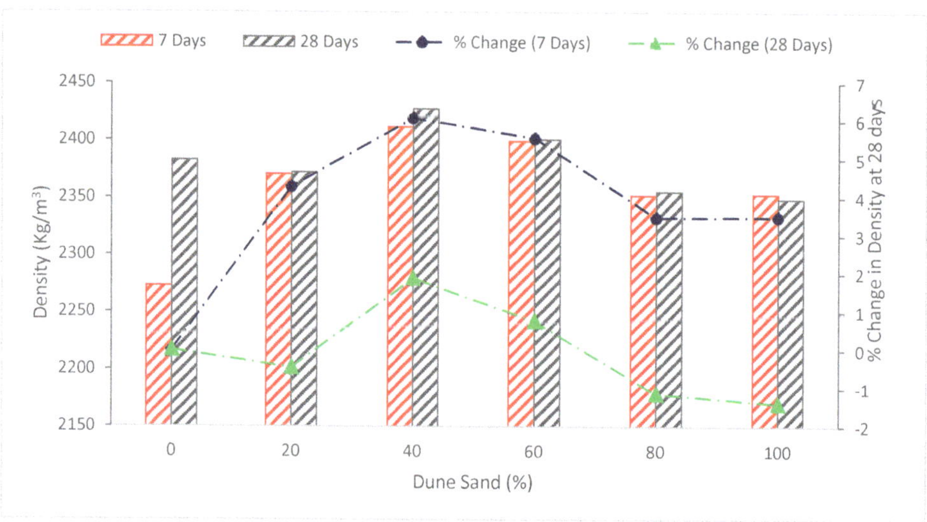

Figure 9. Dry density: data source [40].

5.2. Water Absorption

Figure 10 shows the water absorption of concrete with the substitution of dune sand as fine aggregate at 28 days.

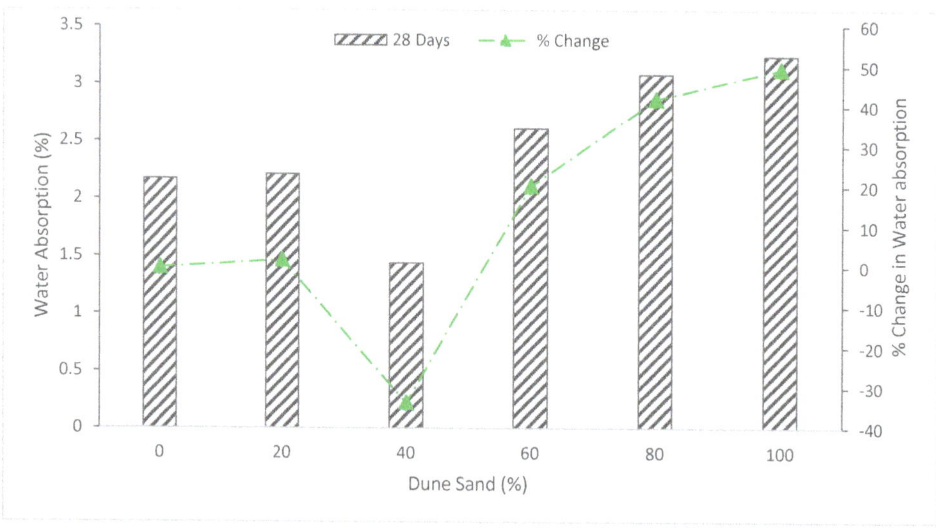

Figure 10. Water absorption: data source [40].

The 40% substitution of dune sand with fine aggregate shows the lowest water absorption rate of 1.44 percent, which is 33.6 percent less than the reference mixture. The decrease in water absorption is due to the micro-filling voids effect of dune sand, which fill the voids, leading to more dense concrete. According to CEB 1989;192, this specimen is regarded as excellent concrete because of its low water absorption. The rate at which water is absorbed increases when the dune sand concentration rises by 60%, 80%, and 100%. The absorption rate at 80% and 100% substitution of dune sand are 41.9 and 49.3 percent, respectively, which is more than the reference mixture, and the concrete quality is medium, according to CEB 1989;192. A similar finding is also reported by other researchers that the quality of concrete decreased at a higher substitution ratio of dune sand [56]. It could be caused by the fineness of the dune sand grains, which absorb more mixing water and have a larger surface area, creating holes or spaces that, in turn, speed up the rate of absorption.

5.3. Ultrasonic Pulse Velocity (UPV)

The findings showed that all concrete mixes had pulse velocities between 4400 and 6100 m/s, which is consistent with excellent and homogenous concrete quality as defined by BS1881, 1983, Part 116. The reference specimen shows a pulse velocity rate of 4894 m/s, and the concrete quality satisfies the requirements of IS code BS1881, 1983. As shown in Figure 11, the pulsing velocity for a concrete mixture of 20% dune sand is 1.98 percent higher than that of the reference specimen (0% dune sand), and it decreases with increasing dune sand content for the ratios 40%, 60%, 80% and 100%. This is consistent with the absorption and density test results, which also show a decrease with increasing dune sand content due to the fineness of the dune [40].

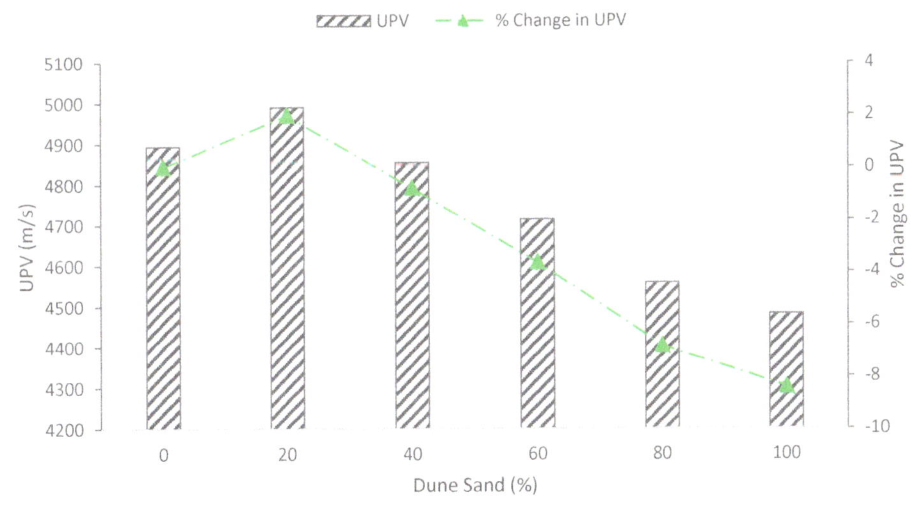

Figure 11. Ultrasonic pulse velocity (UPV): data source [40].

5.4. Sulfate Resistance

5.4.1. Visual Observation

Every month, a comprehensive visual inspection was conducted on the mortar specimens subjected to sulfate assault to assess any evident symptoms of softening, cracking and spalling. Figure 12 depicts typical instances of the harm that sulfate assault on mortar specimens causes after 180 days of immersion in sodium sulfate. It was found that the mortar samples' corners always showed the first signs of degradation, and the broad cracks caused by the expansion strain on the cement matrix were substantial, as shown in

Figure 12. After 180 days of immersion, the dune sand mortar began to visually deteriorate as a result of the mortar structure losing its cohesion. Additionally, a layer of white material was discovered deposited on the mortar faces, which was verified by the XRD investigation to be gypsum and ettringite formation.

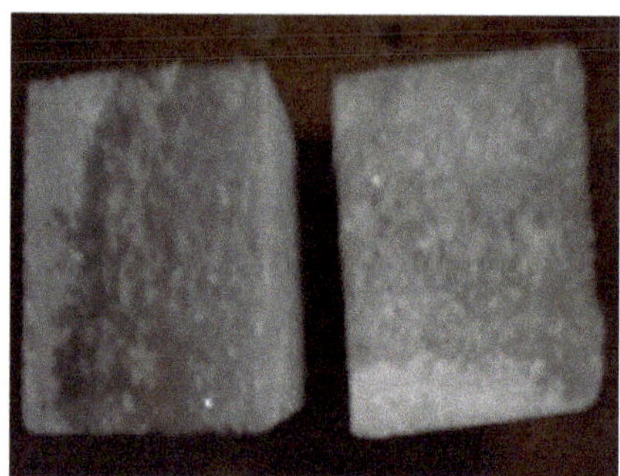

Figure 12. Beam of dune sand mortars immersed in sodium sulfate solution (5%) at 180 days [73].

5.4.2. X-ray Diffraction (XRD)

Figure 13 shows the X-ray Diffraction (XRD) examination of mortar that was exposed to a five percent sodium sulfate solution for 180 days, which shows diffractograms demonstrating the gypsum-specific peaks. On the degraded portions of cube mortar samples that were evaluated for compressive strength, the XRD analysis was performed. It can be seen that some ettringite peaks were found as a result of sulfate assault, together with calcite. Peaks made of quartz and portlandite (made from sand) were visible. The mortars for all showed gypsum and ettringite are two common sulfate attack byproducts that lead to mortar strength loss [74].

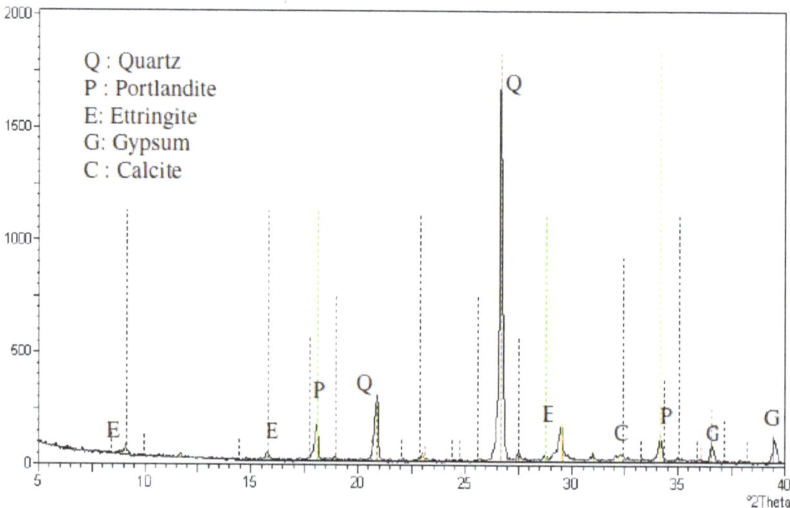

Figure 13. XRD of dune sand mortars immersed in sodium sulfate solution (5%) at 180 days [73].

5.4.3. Strength (Compressive and Flexural)

Strength changes in mortar specimens subjected to sodium sulfate solution (5%) are shown in Figure 14. It can be seen that all mortars demonstrate a strength increase at the beginning of the exposure duration and a subsequent decline in strength development. Strength initially increased in sulfate solution as a result of additional hydration products filling the pores, but strength afterward reduced as a result of microcracking caused by expanding components of sulfate assault. After 180 days of exposure, mortar exhibited a quick loss of strength, and the loss of strength for dune sand mortar was considerable.

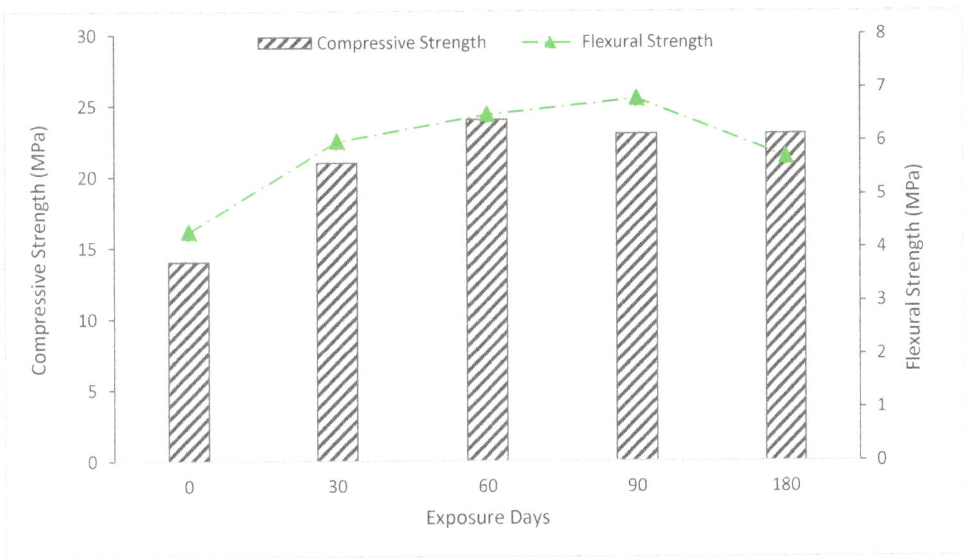

Figure 14. Strength of dune sand mortars immersed in sodium sulfate solution (5%) at 180 days [73].

6. Scanning Electron Microscopy

The hydration process of cementitious materials (ground granulated blast furnace slag) generated using dune sand was explained using the scan electronic microscopy (SEM) method. Figure 15 displays SEM pictures of the microstructure at the 7- and 28-day curing ages. Figure 15 shows that the cementitious material's microstructure is highly thick, particularly after 28 days, which may be explained by the significant decrease in void regions brought on by hydration processes. The existence of different cementation compounds in the dune sand matrix was indicated by several hydration products such as C-S-H gel, Ca (OH)$_2$, and ettringite at the matrix, which also confirmed the development of cementitious compounds. The development of strength is significantly impacted by these hydrated chemicals. At seven days, tabular Ca (OH)$_2$, cotton-shaped C-S-H gels, needle ettringite and their interactions with one another produced a stable paste structure. At 28 days, the pores are completely filled with the C-S-H gel and ettringite, which progressively corrects the structural flaws. Structures become denser with time in comparison to early specimens, which indicates that the specimen's strength grows.

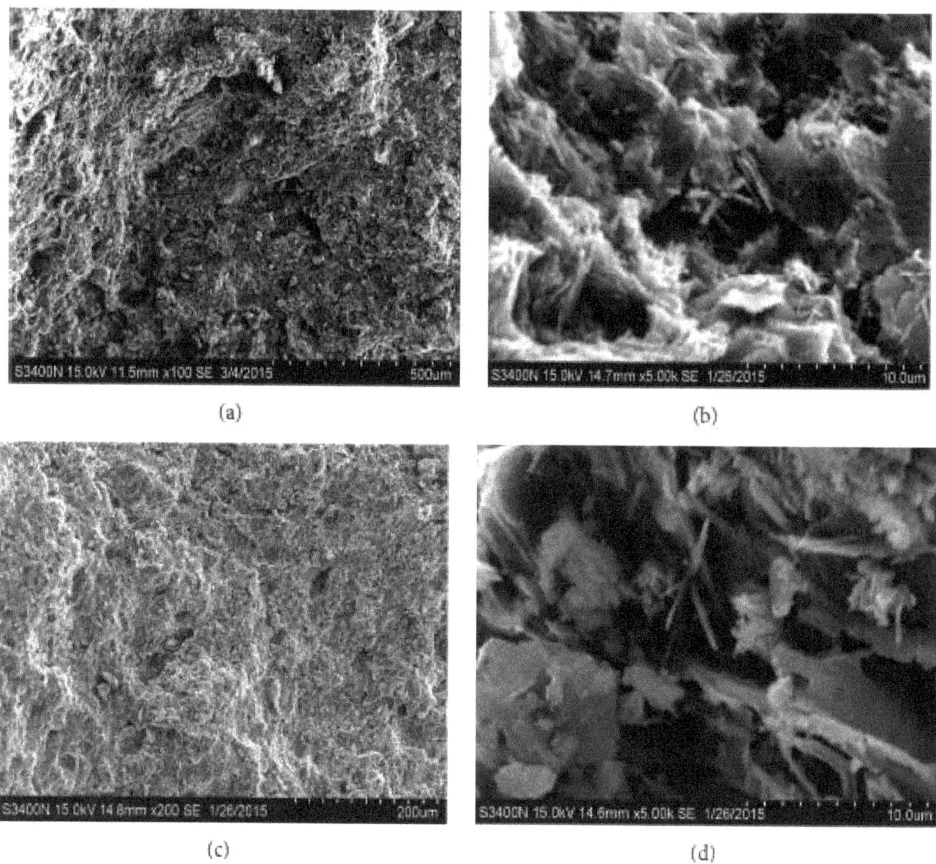

Figure 15. SEM Results (**a**,**b**) 7 days and (**c**,**d**) 28 days [53].

7. Conclusions

The review focused on the use of dune sand as a fine aggregate in the manufacturing of concrete. The physical and chemical composition of dune sand and fresh, strength and durability characteristics of concrete were reviewed and compared. Although using dune sand as a greater or complete replacement for natural sand in the making of concrete has certain negative effects on the performance of the concrete, it maybe used in the creation of concrete to a limited degree. The best substitute dosage for the majority of the attributes examined has been found to be 30 to 40 percent. A detailed conclusion is provided below.

- The physical property of dune depicts that the fineness modulus of dune sand is much lower than river sand. Additionally, a poorly graded and irregular shape adversely affects the flowability of concrete.
- The sum of different chemicals such as silica, iron, lime, alumina and magnesia are greater than 70%. Therefore, it might be possible to use cementitious materials or cement ingredients during the manufacturing of cement.
- Slump decreased with the substitution of dune sand due to its physical nature (rough surface and poorly graded).
- Mechanical strength such as compressive, flexural and tensile capacity is improved to some extent. However, a higher dose or complete substitution adversely affects strength properties. The optimum dose of dune sand varies from 30 to 40%.

- Durability properties such as water absorption, density and sulfate resistance improved with dune sand, but less information is available.

8. Future Studies Recommendation

- The overall studies demonstrate that dune sand has the credibility to be used in concrete. However, the following aspects should be studied before being used in practice.
- The highest strength loss was just around 25%. Therefore, this decline was quite small and can be improved by adding fibers or other pozzolanic materials, such as fly ash and silica fume, waste glass, etc.
- Detailed investigations on durability performance should be explored.
- Dune sand as a cement ingredient in the manufacturing of cement should be explored.
- Treatment of dune sand before use, such as heat or alkaline solution, should be explored.

Funding: This research received no external funding.

Institutional Review Board Statement: Not applicable.

Informed Consent Statement: Not applicable.

Data Availability Statement: All the data available in main text.

Conflicts of Interest: The authors declare no conflict of interest.

References

1. Ahmad, J.; Zaid, O.; Siddique, M.S.; Aslam, F.; Alabduljabbar, H.; Khedher, K.M. Mechanical and durability characteristics of sustainable coconut fibers reinforced concrete with incorporation of marble powder. *Mater. Res. Express* **2021**, *8*, 075505. [CrossRef]
2. Alvee, A.R.; Malinda, R.; Akbar, A.M.; Ashar, R.D.; Rahmawati, C.; Alomayri, T.; Raza, A.; Shaikh, F.U.A. Experimental study of the mechanical properties and microstructure of geopolymer paste containing nano-silica from agricultural waste and crystalline admixtures. *Case Stud. Constr. Mater.* **2022**, *16*, e00792. [CrossRef]
3. Smirnova, O.M.; Menéndez Pidal de Navascués, I.; Mikhailevskii, V.R.; Kolosov, O.I.; Skolota, N.S. Sound-Absorbing Composites with Rubber Crumb from Used Tires. *Appl. Sci.* **2021**, *11*, 7347. [CrossRef]
4. Horňáková, M.; Lehner, P. Analysis of Measured Parameters in Relation to the Amount of Fibre in Lightweight Red Ceramic Waste Aggregate Concrete. *Mathematics* **2022**, *10*, 229. [CrossRef]
5. Althoey, F. Compressive strength reduction of cement pastes exposed to sodium chloride solutions: Secondary ettringite formation. *Constr. Build. Mater.* **2021**, *299*, 123965. [CrossRef]
6. Mesci, B.; Çoruh, S.; Ergun, O.N. Use of selected industrial waste materials in concrete mixture. *Environ. Prog. Sustain. Energy* **2011**, *30*, 368–376. [CrossRef]
7. ZZhu, F.; Wu, X.; Zhou, M.; Sabri, M.M.S.; Huang, J. Intelligent Design of Building Materials: Development of an AI-Based Method for Cement-Slag Concrete Design. *Materials* **2022**, *15*, 3833. [CrossRef] [PubMed]
8. Menéndez, G.; Bonavetti, V.; Irassar, E.F. Strength development of ternary blended cement with limestone filler and blast-furnace slag. *Cem. Concr. Compos.* **2003**, *25*, 61–67. [CrossRef]
9. Bessa, A. Assessing the Contribution of Mineral Additions in Cement Binding Activity in Mortars. In Proceedings of the XXIèmes Rencontres Universitaires de Génie Civil, La Rochelle, France, 2–3 June 2003; Université de Cergy-Pontoise: Cergy, France, 2003.
10. Guettala, S.; Mezghiche, B. Compressive strength and hydration with age of cement pastes containing dune sand powder. *Constr. Build. Mater.* **2011**, *25*, 1263–1269. [CrossRef]
11. Althoey, F.; Farnam, Y. The effect of using supplementary cementitious materials on damage development due to the formation of a chemical phase change in cementitious materials exposed to sodium chloride. *Constr. Build. Mater.* **2019**, *210*, 685–695. [CrossRef]
12. Abdelgader, H.S.; Fediuk, R.S.; Kurpinska, M.; Khatib, J.; Murali, G.; Baranov, A.V.; Timokhin, R.A. Mechanical Properties of Two-Stage Concrete Modified by Silica Fume. *Mag. Civ. Eng.* **2019**, *89*, 26–38.
13. Habert, G. Assessing the Environmental Impact of Conventional and "green" Cement Production. In *Eco-Efficient Construction and Building Materials: Life Cycle Assessment (LCA), Eco-Labelling and Case Studies*; Woodhead: Cambridge, UK, 2013; ISBN 9780857097675.
14. Du, H.; Tan, K.H. Concrete with Recycled Glass as Fine Aggregates. *ACI Mater. J.* **2014**, *111*, 47–57. [CrossRef]
15. Al-Zubaidi, A.B.; Al-Tabbakh, A.A. Recycling Glass Powder and Its Use as Cement Mortar Applications. *Int. J. Sci. Eng. Res.* **2016**, *7*, 555–564.

16. Shekhawat, B.S.; Aggarwal, V. Utilisation of Waste Glass Powder in Concrete-A Literature Review. *Int. J. Innov. Res. Sci. Eng. Technol.* **2007**, *3297*, 2319–8753.

17. Ahmad, J.; Aslam, F.; Martinez-Garcia, R.; De-Prado-Gil, J.; Qaidi, S.M.A.; Brahmia, A. Effects of Waste Glass and Waste Marble on Mechanical and Durability Performance of Concrete. *Sci. Rep.* **2021**, *11*, 21525. [CrossRef]

18. Dolage, D.A.R.; Dias, M.G.S.; Ariyawansa, C.T. Offshore Sand as a Fine Aggregate for Concrete Production. *Br. J. Appl. Sci. Technol.* **2013**, *3*, 813–825. [CrossRef]

19. Ahmad, J.; Martínez-García, R.; De-Prado-Gil, J.; Irshad, K.; El-Shorbagy, M.A.; Fediuk, R.; Vatin, N.I. Concrete with Partial Substitution of Waste Glass and Recycled Concrete Aggregate. *Materials* **2022**, *15*, 430. [CrossRef] [PubMed]

20. Mavroulidou, M.; Lawrence, D. Can Waste Foundry Sand Fully Replace Structural Concrete Sand? *J. Mater. Cycles Waste Manag.* **2019**, *21*, 594–605. [CrossRef]

21. Vigneshpandian, G.V.; Shruthi, E.A.; Venkatasubramanian, C.; Muthu, D. Utilisation of Waste Marble Dust as Fine Aggregate in Concrete. *IOP Conf. Ser. Earth Environ. Sci.* **2017**, *80*, 012007. [CrossRef]

22. Idir, R.; Cyr, M.; Tagnit-Hamou, A. Use of Waste Glass as Powder and Aggregate in Cement Based Materials. In Proceedings of the SBEIDCO—1st International Conference on Sustainable Built Environment Infrastructures in Developing Countries ENSET, Oran, Algeria, 12–14 October 2009; pp. 109–116.

23. Muduli, R.; Mukharjee, B.B. Performance assessment of concrete incorporating recycled coarse aggregates and metakaolin: A systematic approach. *Constr. Build. Mater.* **2020**, *233*, 117223. [CrossRef]

24. Zegardło, B. Heat-Resistant Concretes Containing Waste Carbon Fibers from the Sailing Industry and Recycled Ceramic Aggregates. *Case Stud. Constr. Mater.* **2022**, *16*, e01084. [CrossRef]

25. Sambhaji, Z.K.; Autade, P.B. Effect of Copper Slag as a Fine Aggregate on Properties of Concrete. *Int. Res. J. Eng. Technol.* **2016**, *3*, 410–414.

26. Kumar, B.N.S.; Suhas, R.; Shet, S.U.; Srishaila, J.M. Utilization of Iron Ore Tailings as Replacement to Fine Aggregates in Cement Concrete Pavements. *Int. J. Res. Eng. Technol.* **2014**, *3*, 369–376.

27. Hama, S.M.; Hilal, N.N. Fresh Properties of Concrete Containing Plastic Aggregate. In *Use of Recycled Plastics in Eco-Efficient Concrete*; Elsevier: Amsterdam, The Netherlands, 2019; pp. 85–114.

28. Rajput, S.P.S.; Chauhan, M.S. Suitability of Crushed Stone Dust as Fine Aggregate in Mortars. *Micron (no. 30)* **2014**, *89*, 35–59.

29. Prakash, R.; Raman, S.N.; Subramanian, C.; Divyah, N. Eco-Friendly Fiber-Reinforced Concretes. In *Handbook of Sustainable Concrete and Industrial Waste Management*; Elsevier: Amsterdam, The Netherlands, 2022; pp. 109–145.

30. Jin, B.H.; Song, J.X.; Liu, H.F. Engineering Characteristics of Concrete Made of Desert Sand from Maowusu Sandy Land. In *Applied Mechanics and Materials*; Trans Tech: Zurich, Switzerland, 2012; Volume 174, pp. 604–607.

31. Khay, S.E.E.; Neji, J.; Loulizi, A. Compacted Dune Sand Concrete for Pavement Applications. *Proc. Inst. Civ. Eng.-Constr. Mater.* **2011**, *164*, 87–93. [CrossRef]

32. Westerholm, M.; Lagerblad, B.; Silfwerbrand, J.; Forssberg, E. Influence of fine aggregate characteristics on the rheological properties of mortars. *Cem. Concr. Compos.* **2008**, *30*, 274–282. [CrossRef]

33. Zaitri, R.; Bederina, M.; Bouziani, T.; Makhloufi, Z.; Hadjoudja, M. Development of high performances concrete based on the addition of grinded dune sand and limestone rock using the mixture design modelling approach. *Constr. Build. Mater.* **2014**, *60*, 8–16. [CrossRef]

34. Lee, E.; Ko, J.; Yoo, J.; Park, S.; Nam, J. Effect of Dune Sand on Drying Shrinkage Cracking of Fly Ash Concrete. *Appl. Sci.* **2022**, *12*, 3128. [CrossRef]

35. Wille, K.; Naaman, A.E.; El-Tawil, S.; Parra-Montesinos, G.J. Ultra-high performance concrete and fiber reinforced concrete: Achieving strength and ductility without heat curing. *Mater. Struct.* **2012**, *45*, 309–324. [CrossRef]

36. Ran, Q.; Somasundaran, P.; Miao, C.; Liu, J.; Wu, S.; Shen, J. Adsorption Mechanism of Comb Polymer Dispersants at the Cement/Water Interface. *J. Dispers. Sci. Technol.* **2010**, *31*, 790–798. [CrossRef]

37. Luo, F.J.; He, L.; Pan, Z.; Duan, W.H.; Zhao, X.L.; Collins, F. Effect of very fine particles on workability and strength of concrete made with dune sand. *Constr. Build. Mater.* **2013**, *47*, 131–137. [CrossRef]

38. Belferrag, A.; Kriker, A.; Khenfer, M.E. Improvement of the Compressive Strength of Mortar in the Arid Climates by Valorization of Dune Sand and Pneumatic Waste Metal Fibers. *Constr. Build. Mater.* **2013**, *40*, 847–853. [CrossRef]

39. Krobba, B.; Bouhicha, M.; Zaidi, A.; Lakhdari, M. Formulation of a Repair Mortar Based on Dune Sand and Natural Microfibers. In *Concrete Solutions*; CRC Press: Boca Raton, FL, USA, 2014; pp. 91–95.

40. Dawood, A.O.; Jaber, A.M. Effect of Dune Sand as Sand Replacement on the Mechanical Properties of the Hybrid Fiber Reinforced Concrete. *Civ. Environ. Eng.* **2022**, *18*, 111–136. [CrossRef]

41. Chauvin, J.J. *Les Sables, Guide Technique d'utilisation Routière*; Institut des Sciences et des Techniques de l'Équipement et de l'Environnement pour le Développement: Paris, France, 1987.

42. Wang, W.-H.; Han, L.-H.; Li, W.; Jia, Y.-H. Behavior of concrete-filled steel tubular stub columns and beams using dune sand as part of fine aggregate. *Constr. Build. Mater.* **2014**, *51*, 352–363. [CrossRef]

43. Setra, L. *Guide Technique, Réalisation des Remblais et des Couches de Forme*; LCPC: Paris, France, 2000.

44. Reddy, B.V.; Gupta, A. Influence of sand grading on the characteristics of mortars and soil–cement block masonry. *Constr. Build. Mater.* **2008**, *22*, 1614–1623. [CrossRef]

45. Abadou, Y.; Mitiche-Kettab, R.; Ghrieb, A. Ceramic waste influence on dune sand mortar performance. *Constr. Build. Mater.* **2016**, *125*, 703–713. [CrossRef]

46. Kachouh, N.; El-Hassan, H.; El Maaddawy, T. Effect of steel fibers on the performance of concrete made with recycled concrete aggregates and dune sand. *Constr. Build. Mater.* **2019**, *213*, 348–359. [CrossRef]

47. Liu, Y.; Li, Y.; Jiang, G. Orthogonal experiment on performance of mortar made with dune sand. *Constr. Build. Mater.* **2020**, *264*, 120254. [CrossRef]

48. Rennani, F.Z.; Makani, A.; Agha, N.; Tafraoui, A.; Benmerioul, F.; Zaoiai, S. Mechanical Properties of High-Performance Concrete Made Incorporating Dune Sand as Fine Aggregate. *Rev. Rom. Ing. Civ.* **2020**, *11*, 37–46.

49. Siala, A.; Khay, S.E.E.; Loulizi, A.; Neji, J. Improving the performance of porous asphalt with reclaimed asphalt, dune sand and lime. *Proc. Inst. Civ. Eng.-Constr. Mater.* **2021**, *174*, 214–226. [CrossRef]

50. Smaida, A.; Haddadi, S.; Nechnech, A. Improvement of the mechanical performance of dune sand for using in flexible pavements. *Constr. Build. Mater.* **2019**, *208*, 464–471. [CrossRef]

51. El-Hassan, H.; Hussein, A.; Medljy, J.; El-Maaddawy, T. Performance of Steel Fiber-Reinforced Alkali-Activated Slag-Fly Ash Blended Concrete Incorporating Recycled Concrete Aggregates and Dune Sand. *Buildings* **2021**, *11*, 327. [CrossRef]

52. He, M.; Wang, Y.; Yuan, K.; Sheng, Z.; Qiu, J.; Liu, J.; Wang, J. Synergistic effects of ultrafine particles and graphene oxide on hydration mechanism and mechanical property of dune sand-incorporated cementitious composites. *Constr. Build. Mater.* **2020**, *262*, 120817. [CrossRef]

53. Jiang, C.; Zhou, X.; Tao, G.; Chen, D. Experimental Study on the Performance and Microstructure of Cementitious Materials Made with Dune Sand. *Adv. Mater. Sci. Eng.* **2016**, *2016*, 2158706. [CrossRef]

54. Benabed, B.; Azzouz, L.; Kadri, E.-H.; Kenai, S.; Belaidi, A.S.E. Effect of fine aggregate replacement with desert dune sand on fresh properties and strength of self-compacting mortars. *J. Adhes. Sci. Technol.* **2014**, *28*, 2182–2195. [CrossRef]

55. Zhang, G.; Song, J.; Yang, J.; Liu, X. Performance of mortar and concrete made with a fine aggregate of desert sand. *Build. Environ.* **2006**, *41*, 1478–1481. [CrossRef]

56. Al-Harthy, A.; Halim, M.A.; Taha, R.; Al-Jabri, K. The properties of concrete made with fine dune sand. *Constr. Build. Mater.* **2007**, *21*, 1803–1808. [CrossRef]

57. Wang, A.; Zhang, C.; Sun, W. Fly ash effects: I. The morphological effect of fly ash. *Cem. Concr. Res.* **2003**, *33*, 2023–2029. [CrossRef]

58. Zeghichi, L.; Benghazi, Z.; Baali, L. Comparative Study of Self-Compacting Concrete with Manufactured and Dune Sand. *J. Civ. Eng. Arch.* **2012**, *6*, 1429–1434.

59. Seif, E.; Sedek, E.S. Performance of Cement Mortar Made with Fine Aggregates of Dune Sand, Kharga Oasis, Western Desert, Egypt: An Experimental Study. *Jordan J. Civ. Eng.* **2013**, *7*, 270–284.

60. Mehta, P.K.; Monteiro, P.J.M. *Concrete: Microstructure, Properties, and Materials*; McGraw-Hill Education: New York, NY, USA, 2014; ISBN 0071797874.

61. Rmili, A.; Ben Ouezdou, M.; Added, M.; Ghorbel, E. Incorporation of Crushed Sands and Tunisian Desert Sands in the Composition of Self Compacting Concretes Part II: SCC Fresh and Hardened States Characteristics. *Int. J. Concr. Struct. Mater.* **2009**, *3*, 11–14. [CrossRef]

62. Bouziani, T.; Bederina, M.; Hadjoudja, M. Effect of dune sand on the properties of flowing sand-concrete (FSC). *Int. J. Concr. Struct. Mater.* **2012**, *6*, 59–64. [CrossRef]

63. Rao, G. Development of strength with age of mortars containing silica fume. *Cem. Concr. Res.* **2001**, *31*, 1141–1146. [CrossRef]

64. Kwan, A.K.H. Use of Condensed Silica Fume for Making High-Strength, Self-Consolidating Concrete. *Can. J. Civ. Eng.* **2000**, *27*, 620–627. [CrossRef]

65. Saidi, M.; Safi, B.; Samar, M.; Benmounah, A. Formulation and Physicochemical Characterization of Ultra-High Performance Fiber Concrete Based of Sand Dunes (UHPFC). In Proceedings of the 3rd International Conference on Material Modeling incorporating the 13th European Mechanics of Materials Conference, Warsaw, Poland, 8–11 September 2013.

66. Benmerioul, F.; Tafraoui, A.; Makani, A.; Zaouai, S. A Study on Mechanical Properties of Self-Compacting Concrete Made Utilizing Ground Dune Sand. *Rev. Rom. Mater. J. Mater.* **2017**, *47*, 328–335.

67. Mohammed, M.; Abdelouahed, K.; Allaoua, B. Compressive Strength of Dune Sand Reinforced Concrete. *AIP Conf. Proc.* **2017**, *1814*, 020023.

68. Abadou, Y.; Ghrieb, A.; Bustamante, R. Crushed concrete waste influence on dune sand mortar performance. Contribution to the valorization. *Mater. Today Proc.* **2020**, *33*, 1758–1761. [CrossRef]

69. Ahmad, J.; Manan, A.; Ali, A.; Khan, M.W.; Asim, M.; Zaid, O. A Study on Mechanical and Durability Aspects of Concrete Modified with Steel Fibers (SFs). *Civ. Eng. Arch.* **2020**, *8*, 814–823. [CrossRef]

70. Ahmad, J.; Aslam, F.; Martinez-Garcia, R.; El Ouni, M.H.; Khedher, K.M. Performance of sustainable self-compacting fiber reinforced concrete with substitution of marble waste (MW) and coconut fibers (CFs). *Sci. Rep.* **2021**, *11*, 23184. [CrossRef]

71. Li, Z.; Yang, S.; Luo, Y. Experimental evaluation of the effort of dune sand replacement levels on flexural behaviour of reinforced beam. *J. Asian Arch. Build. Eng.* **2020**, *19*, 480–489. [CrossRef]

72. Ahmad, J.; Aslam, F.; Zaid, O.; Alyousef, R.; Alabduljabbar, H. Mechanical and durability characteristics of sustainable concrete modified with partial substitution of waste foundry sand. *Struct. Concr.* **2021**, *22*, 2775–2790. [CrossRef]

73. Azzouz, L.; Benabed, B.; Belaidi, A.S.E.; Menadi, B. Physical, Mechanical and Durability of Dunes Sand Mortar in The Region of Laghouat–Algeria. 2008. Available online: https://www.thbbakademi.org/wp-content/uploads/2020/12/B2008.012-1.pdf (accessed on 22 June 2022).
74. Al-Dulaijan, S.; Maslehuddin, M.; Al-Zahrani, M.; Sharif, A.; Shameem, M.; Ibrahim, M. Sulfate resistance of plain and blended cements exposed to varying concentrations of sodium sulfate. *Cem. Concr. Compos.* **2003**, *25*, 429–437. [CrossRef]

materials

Article

Overview of Concrete Performance Made with Waste Rubber Tires: A Step toward Sustainable Concrete

Jawad Ahmad [1], **Zhiguang Zhou** [1,*], **Ali Majdi** [2], **Muwaffaq Alqurashi** [3] and **Ahmed Farouk Deifalla** [4]

1 Department of Disaster Mitigation for Structures, Tongji University, Shanghai 200092, China
2 Department of Building and Construction Technologies and Engineering, Al-Mustaqbal University College, Hillah 51001, Iraq
3 Civil Engineering Department, College of Engineering, Taif University, P.O. Box 11099, Taif 21944, Saudi Arabia
4 Structural Engineering Department, Faculty of Engineering and Technology, Future University in Egypt, New Cairo 11845, Egypt
* Correspondence: zgzhou@tongji.edu.cn

Abstract: Utilizing scrap tire rubber by incorporating it into concrete is a valuable option. Many researchers are interested in using rubber tire waste in concrete. The possible uses of rubber tires in concrete, however, are dispersed and unclear. Therefore, a compressive analysis is necessary to identify the benefits and drawbacks of rubber tires for concrete performance. For examination, the important areas of concrete freshness, durability, and strength properties were considered. Additionally, several treatments and a microstructure investigation were included. Although it has much promise, there are certain obstacles that prevent it from being used as an aggregate in large numbers, such as the rubber's weak structural strength and poor binding performance with the cement matrix. Rubber, however, exhibits mechanical strength comparable to reference concrete up to 20%. The evaluation also emphasizes the need for new research to advance rubberized concrete for future generations.

Keywords: waste tires; concrete; aggregate; compressive strength; treatments; durability

Citation: Ahmad, J.; Zhou, Z.; Majdi, A.; Alqurashi, M.; Deifalla, A.F. Overview of Concrete Performance Made with Waste Rubber Tires: A Step toward Sustainable Concrete. *Materials* **2022**, *15*, 5518. https://doi.org/10.3390/ma15165518

Academic Editors: Stefano Guarino and Flaviana Tagliaferri

Received: 8 July 2022
Accepted: 5 August 2022
Published: 11 August 2022

Publisher's Note: MDPI stays neutral with regard to jurisdictional claims in published maps and institutional affiliations.

1. Introduction

Due to the new infrastructure being built, a significant quantity of waste concrete is generated from destroyed buildings every year. Construction and demolition trash is generated annually in China and the EU at 450 million tons and 200 million tons, respectively. About half of the rubber in a tire is made of natural rubber, butadiene rubber, and styrene-butadiene rubber. The other parts include carbon black, metal, textile, zinc oxide, sulfur, and additives [1]. Tires represent environmental dangers owing to their composition, which makes them exceedingly durable, non-biodegradable, and a fire hazard, as well as a breeding ground for rats, mice, and mosquitoes [2]. The proper disposal of used tires has grown to be a significant environmental issue [3]. Each year, the EU member states produce more than three million tons of scrap tires [4], and a stock of 600 hundred tons is present. Due to the desire to create a green environment, keep it clean, and minimize carbon emissions, many western nations, including Canada, are having trouble maintaining and reusing structural waste [5]. Figure 1 shows the recycling percentages of different waste.

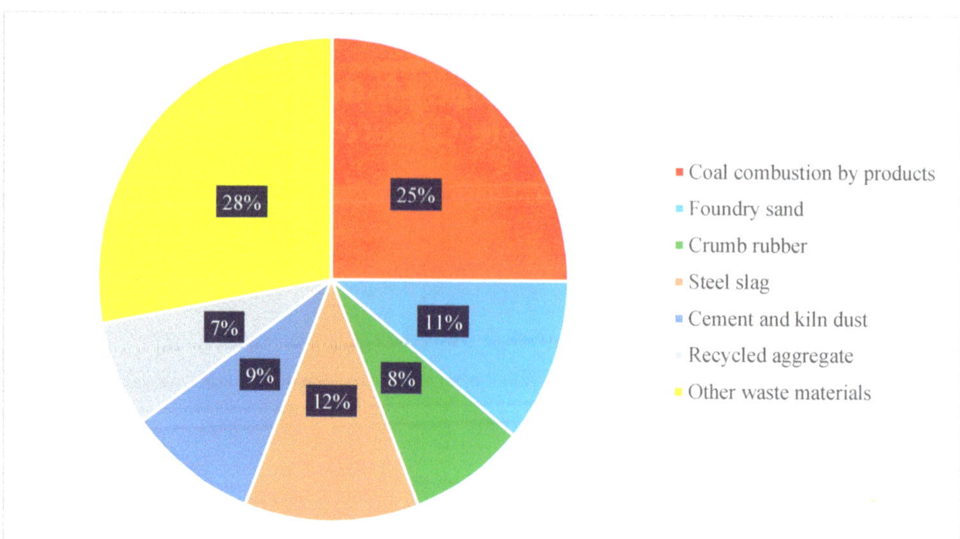

Figure 1. Utilization of different waste materials: data source [6].

In other words, as seen in Figure 2, these tires from motor vehicles will produce a significant number of waste tires and seriously harm the environment. A total of 290 million tires are produced annually in the United States alone, in addition to the 275 million tires that are already in storage [7]. Therefore, the need for a recycling concept for such used tires has increased.

Figure 2. Waste rubber tires [8].

Recycling used car tires as substitute aggregates to create new-class concrete is a creative solution with favorable effects on the environment, the economy, and performance. Numerous current findings examine the improvement in the compatibility of crumb rubber particles when utilized as a sand substitute. Shredded and/or crumb rubber particles have been extensively explored as a replacement for concrete aggregate [9]. A study concluded that styrene rubber materials in construction buildings improves the quality of concrete for sustainability and durability of the structures [10].

In addition to landfilling, there are other options, such as energy recovery, which is often carried out in cement kilns [11], and pyrolysis of tire rubber to produce carbon black [12]. However, because of the challenges in marketing low-quality pyrolysis end products, the ultimate solution is really not commercially feasible. The recycling of rubber and steel fibers is possible via the process of shredding used tires, which is typically done after an electromagnetic separation.

Waste tire features that, if managed improperly, might endanger the environment can be utilized to the building industry's benefit. Numerous research has been done on concrete that substitutes natural aggregate with scrap rubber from old tires in varying percentages [13]. Figure 3 shows the manufacturing process of waste rubber tires for concrete. The qualities of the concrete produced are dramatically changed when used as a partial replacement for natural coarse and/or fine aggregate. According to studies, adding rubber to concrete regularly lowers the material's compressive, flexure, and elasticity when compared to normal concrete [14,15]. The weak bond between the rubber surface and paste is primarily caused by the hydrophobic nature of rubber particles and their extreme external irregularity [16]. Based on a microstructural analysis, Taha et al. [17] found that a significant drop in capacity may be related to the behavior of tire rubber particles as a soft aggregate. Concrete's loss of strength when using rubber as an aggregate has an impact on the rubber's content, particle size, and characteristics, as well as the mix's parameters and ratios [18].

Figure 3. Flow chart of waste rubber tires from waste into concrete [19].

Concrete strength qualities are impacted using scrap rubber tires as aggregates. Some of its qualities, including toughness, impact resistance, energy absorption, and sound and temperature isolation, are improved. However, it lessens various other strength characteristics, including workability, split tensile strength, and compressive strength [20]. It is advised to utilize rubber in concrete as foundation pads for rotating equipment and railway stations, as vibration-dampers, or anywhere blast or impact resistance is needed [21]. Additionally, it may be utilized as a shock absorber in structures made to withstand seismic waves [22].

In order to explore sound absorption and the ultrasonic modulus of tire rubber concrete, the research used the ultrasonic echo method to conduct an ultrasonic analysis. The authors came to the conclusion that rubberized concrete is a powerful sound and shaking energy absorber [23]. According to reports, rubber concrete might be utilized to make buildings' seismic shock-wave absorbers and sound barriers for highway construction [24].

The research looked at the possibility of using scrap rubber from the car sector as fine aggregates in the cementitious matrix to create lightweight building materials. The outcomes of the tests showed that the existence of rubber has a tendency to prevent water propagation and minimize water absorption, providing a superior defense against corrosion for the steel reinforcement [25].

Aiello and Leuzzi [26] substituted 0%, 15%, 30%, 50%, and 75% of the sand with the same volume of rubber when adding rubber as a substitute for sand. They observed that because of the reduction in specific gravity, both the density and compressive strength drastically decreased. Atahan and Yücel [27] employed rubber to substitute fine aggregate in a volume ratio of 0%, 20%, 40%, 60%, 80%, and 100%. They observed a steady decline in the samples' compressive strength until 100% of the fine aggregate was replaced with rubber, at which point 93% of its strength was lost. Additionally, they noted a considerable decrease in the elastic modulus, which reached a 96% loss for 100% replacement of the aggregate. Furthermore, Batayneh et al. [28] replaced the fine aggregate up to 100% by volume with six different mixes including rubber. Their findings demonstrated that rubberized concrete still has enough strength to be employed as lightweight concrete despite the reduction in compressive strength. They advised using these sorts of mixtures for partition walls, traffic barriers, pavement, and walkways in this respect [29].

Brief literature shows that several researchers focus to utilized rubber tire waste in concrete. However, the potential application of rubber tires in concrete is scattered and not clear. No one can easily judge the suitability of waste rubber tires in concrete. Therefore, a compressive review is required to summarize the positive and negative impact of rubber tires as aggregate on concrete performance. The important properties of fresh concrete properties (slump, fresh density, and air content), mechanical strength (compressive, tensile, and flexural strength), and durability (permeability, water absorption, and chloride ions penetration) were considered for analysis. Furthermore, different treatments, microstructure study (scan electronic microscopy), and application of rubberized concrete were also included. Finally, the review also highlights the future research aspects for future generations to further improvement in rubberized concrete.

2. Fresh Concrete

2.1. Slump Flow

Figure 4 and Table 1 show the flowability of concrete with the replacement of rubber as aggregate in concrete. The flowability of concrete decreased with the addition of rubber as aggregate.

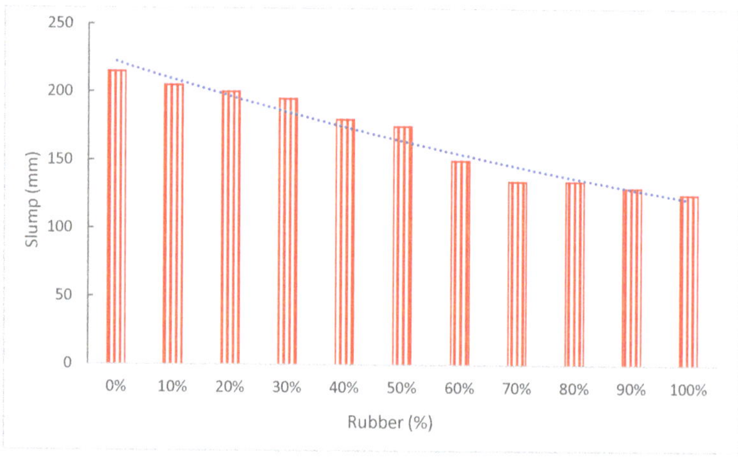

Figure 4. Slump flow: data source [30].

Table 1. Summary of slump flow.

Reference	Rubber Tire	Slump (mm)
[26]	0%, 25%, 50%, and 75%	180, 220, 215, and 215
[31]	0%, 5%, 10%, and 15%	80, 75, 64, and 55
[32]	0%, 5%, 10%, 15%, 20%, and 25%	0, 0, 7, 20, 55, and 87
[30]	0%, 10%, 20%, 30%, 40%, 50%, 60%, 70%, 80%, 90%, and 100%	215, 205, 200, 195, 180, 175, 150, 135, 135, 130, and 125
[33]	0%, 5%, 10%, and 15%	74.50, 74.00, 72.50, and 70.00
[34]	0%, 6%, 12%, 18%, 24%, 30%, 36%, 42%, 48%, 54%, and 60%	140, 138, 139, 130, 110, 70, 75, 22, 20, 10, and 0

In addition, mixes created with fine crumb rubber were better to work with than those prepared with coarse tire chips or a combination of tire chips and crumb rubber, according to research. The slump was also shown to diminish as rubber proportion improved [35]. The rubber's uneven surface roughness caused greater interparticle friction, which led to an increase in the admixture dose [36]. The drop almost remained constant until 12% replacement after having drastically decreased from 140 mm to 110 mm at 24% replacement [37]. The use of rubber aggregates is also believed to reduce slump due to their irregular forms and sharp edges [38]. Similarly, when natural coarse aggregates are partially substituted with rubber aggregates, the slump is reduced as a result of the shape of the rubber aggregate particles [39]. The decrease in flowability of concrete with the incorporation of rubber is mainly due to the rough and angular surface of rubber particles, as shown in Figure 5.

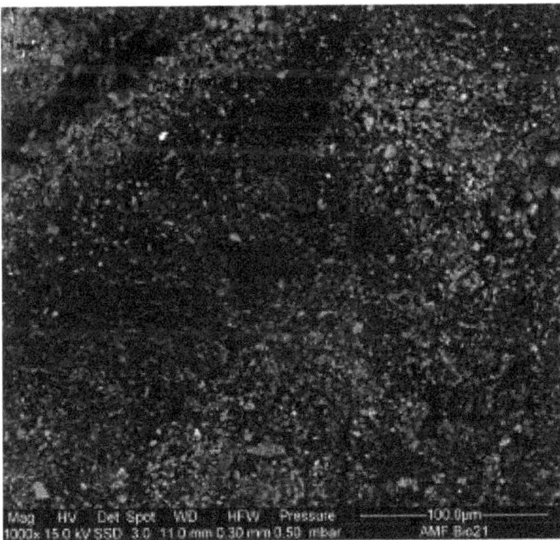

Figure 5. SEM of rubber particle [40]: used as per Elsevier permission.

Khatib and Bayomy [41] claim that adding more rubber to concrete reduces both the slump and unit weight. At a 5% crumb rubber content, the slump on rubber without fibers and rubber with fibers fell by 6.67% and 12.50%, respectively. The slump then gradually decreases for rubberized concrete with a crumb rubber content of 10% or 15%. Rubber without fibers and rubber with fibers slumped the least at a 20% crumb rubber concentration, measuring 2 cm and 2.5 cm, respectively. This represents a decrease of 71.43% and 64.29% in comparison to control mixes [42].

The fresh mix performance was completely different for materials with a higher natural aggregate to rubber replacement ratio. As additional rubber was included, the droop became worse. However, the nonwetting rubber particles improved the flow of newly mixed concrete [43]. The slump value reduces as recycled concrete aggregate, rubber, and fiber replacement levels rise. The impact of fiber on slump value is greater than that of rubber and recycled concrete aggregate. [44]. The friction force between the components of concrete was enhanced because of the smaller width and longer length. The workability of rubberized concrete demonstrates an increase in a slump as the overall aggregate content rises [32]. Due to their smaller surface area, less paste was needed to coat them. Research, however, asserts that the slump became worse as the rubber dose was increased. The rubber's nonwetting particles were what made the flow of newly mixed concrete better [45]. Rubber particles may be added to concrete to lower the slump value. As the replacement level increases, hardened rubberized lightweight aggregate concrete loses dry unit weight. This loss of weight causes larger holes in the concrete, which absorb more water and leave no or less free water available for flowability.

Concrete with a larger proportion of rubber particles was less workable. This is mostly because of rubber particles' greater water absorption rates, which cause the mixture to have less free water overall and, as a consequence, have lower workability. Different sizes of rubber show a decline in flow with a rise in tire particle replacements [40].

According to the experimental findings, rubberized concrete absorbed more water than regular concrete, and the amount of rubber in the concrete enhanced the water absorption. Therefore, water, chloride, and chemical assaults on rubberized concrete are more likely to occur [46]. Due to its weak bonding with wet cement paste, the capillary absorption of concrete rises with an increase in the concentration of rubber particles. Since all of the coarse rubber particles have larger absorption rates than the fine rubbers, the size of the replacement aggregate is significant to the increase in water absorption through capillarity [47]. Furthermore, since recycling involves grinding, the surface of the finer waste tire particles is rougher than that of the coarse waste tire. The rubber surfaces of tire crumbs and shreds may be seen in SEM pictures to have rough and jagged surfaces. It is obvious that tire crumbs have more jagged edges than tire shreds [40]. According to research, this roughness increases the frictional resistance to concrete's flowability, causing mixes with tire crumb replacement to slump less [48]. When compared to concrete without rubber, even though the flow value reduced as the rubber percentage improved, the mixture still provided a workable mix up to a 20% rubber content [42].

2.2. Fresh Density and Air Content

Air content and fresh concrete density are correlated. Fresh concrete's density will decline as the amount of air in the mixture rises. Concrete's properties might alter when different materials are used. It can be noted that the density of various materials differs in the new concrete density test. Figure 6 displays the fresh density of concrete with the incorporation of rubber. It can be noted that rubber decreased the fresh density of concrete. Rubber naturally had a lower specific gravity than the fine aggregate. Because fresh concrete has a low specific gravity, investigations show that its density declined as crumb rubber concentration rose [49]. Research suggested that the density of rubberized concrete improved with the size of the rubber particles, with the lowest density being 1900 kg/m^3 and the greatest being 2240 kg/m^3. This is due to the fact that the tiny rubber filled up any gaps between the small concrete particles and served as a fine aggregate, increasing the density [50]. According to research, the spaces around the rubber cause the concrete's density to decrease as its rubber concentration increases [51]. Gesog Lu [52] found that mixing rubber with concrete resulted in lighter-weight concrete. The densities of the rubberized concrete were 2–11% lighter than those of the reference sample. Due to the increased densification of the concrete structure, Pelisser et al. [53] indicated that although the density of rubberized concrete was 13% less dense than regular concrete, only a 9% loss was seen when silica fume was added to it. Torgal et al. [54] replaced all of the mineral aggregates with crumb rubber and tire chips, replacing plenty of the coarse aggregate. It was noted that the density with the substitute of the sand was decreased by 34%, while the density of the concrete with the replacement of the coarse aggregate was reduced by 45%. The combination resulted in a 33% overall drop in density.

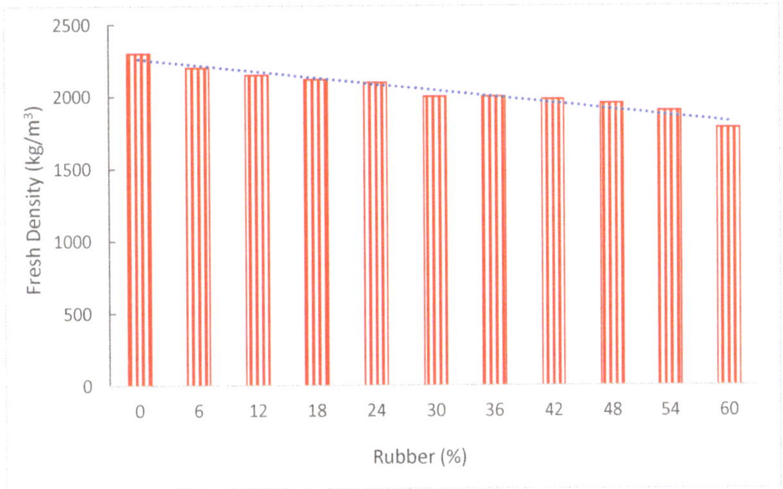

Figure 6. Fresh density of concrete: data source [34].

Figure 7 shows the air content of concrete with the substantiation of rubber as aggregate. The air content improved as the substitution ratio of rubber enhanced.

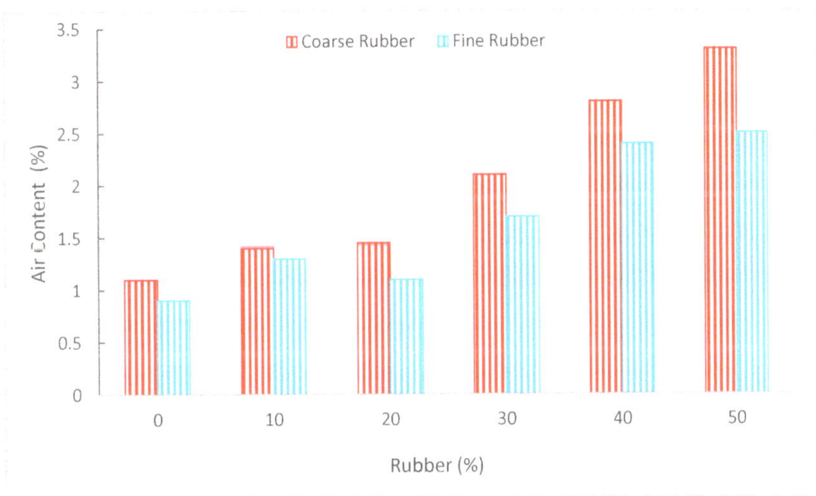

Figure 7. Air content: data source [55].

However, it can be observed that rubber as coarse aggregate increased air content more than the fine aggregate. When crumb rubber was added to the concrete, Kardos et al. [56] discovered an increase in air content. Due to the large specific area of the fine crumb rubber, as the crumb rubber content grew, the air content of the fresh concrete increased. As a result, increasing crumb rubber (CR) will cause more air to get trapped in the concrete, as indicated by [57]. Due to the rubbers' non-polar nature, water will be repelled, and air will be readily trapped in the concrete [49]. Richardson et al. [57] compared the air entrainment percentages for conventional concrete, which was 1.9%, and less than 0.5 mm crumb rubber, which was 3.3%. Compared to the sample with less than 0.5 mm crumb rubber, ordinary concrete contained 74% more air than that sample. This is a noteworthy change because a 3% air entrainment is sufficient to provide freeze/thaw protection, especially because the mixture also contains evenly distributed crumb rubber particles that will allow for pressure absorption and have a particle size that is suitable for effective freeze/thaw protection. According to research by Al-Akhras et al. [58], the amount of air in the concrete improves with the size of the tire particle. According to Benazzouk et al. [59], a larger rubber volume ratio results in more air content. The usage of tire rubber ash in concrete may reduce the air content, according to Akhras et al. [60]. With a growth in tire rubber ash content, the mortar's air content was reduced. In the mortar containing 10% tire rubber ash, the air content dropped from 2.6% in the reference sample to 1.5%.

3. Mechanical Strength

3.1. Compressive Strength (CS)

Figure 8 displays the compressive strength (CS) of concrete with the substitution of rubber as aggregate. It can be noted that CS decreased as the substitution ratio of rubber increased. The qualities of the concrete are dramatically changed when rubber is used as a partial replacement for natural coarse and/or fine aggregate in concrete. According to studies, adding rubber to concrete regularly lowers the material CS and elasticity when compared to normal concrete [14]. Although it is often believed that rubberized concrete has limited mechanical strength, Issa and Salem found that it had high CS [61]. Youssf et al. found that the usage of 10%, 20%, 30%, 40%, and 50% volume rubber/sand substitution decreased the concrete's CS by 21.3%, 37.9%, 54.3%, 62.5%, and 66.4% at 28 days, respectively, compared to reference concrete [16]. The impact of rubber-based aggregate particle size on rubberized concrete CS. According to the authors, crumb rubber fraction 2/4 lowered CS more than fraction 4/6 [13]. For all samples of recycled rubber mortar, Guelmine et al. [62] noted a rise in the damage factor of both the compressive and flexural capacity, which rose with a rise in the utilized raised temperature.

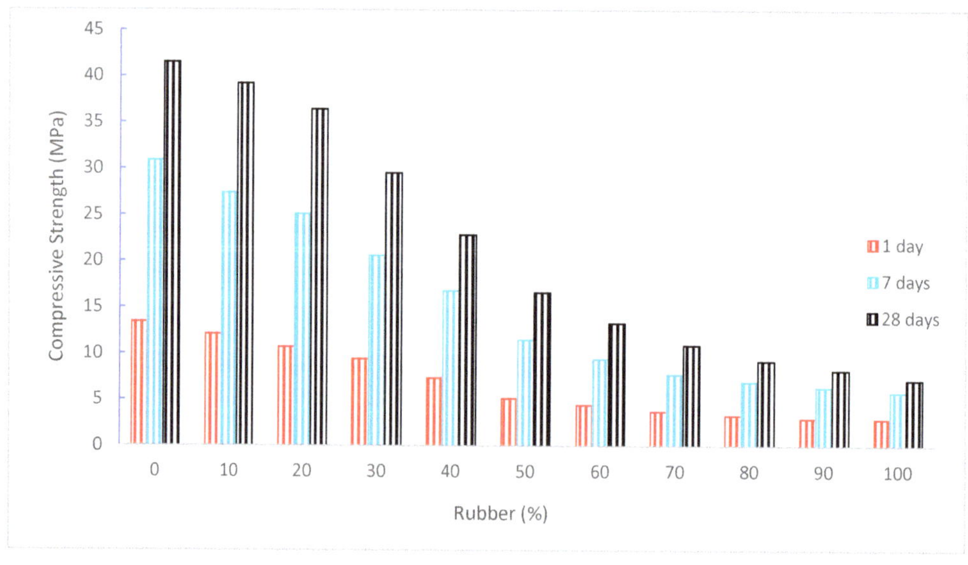

Figure 8. Compressive strength: data source [30].

Depending on the dose of rubber-based aggregate, the strength of concrete containing that aggregate was lowered. The lowest compressive and capacity were seen when rubber fine particles replaced fine natural aggregate by 30% [45]. The scientists also noted that the addition of rubber tire aggregate to concrete results in a reduction in compressive and flexural capacity [63]. Concrete CS is decreased by 10% to 23% when rubber particles are substituted for 7.5% and 10% of the typical concrete aggregate. When rubber is substituted by 7.5% to 10%, the tensile capacity decreases by 30% to 60% [64]. Concrete CS was reduced by 90% when chipped rubber was replaced by 100%. However, when crumb rubber was utilized in lieu of sand in concrete at a 100% replacement rate, the strength was reduced by 80% [65]. When crumb rubber was utilized to substitute fine aggregate, there was a reduction in 28-day CS of 15%, 25%, and 67% for replacement levels of 25%, 50%, 75%, and 100%. Similar to chipped rubber, replacing coarse aggregates with them causes a 40%, 48%, 73%, and 78% drop in 28-day CS for replacement levels of 25%, 50%, 75%, and 100%, respectively [17].

Obinna Onuaguluchi et al. [66] discovered a considerable increase in the combination of coated rubber crumb and silica fume, however, in terms of compressive and tensile capacity. At 5% and 10% sand substitute, this combination performed better than the reference mixture due to the interaction between silica fume and limestone powder. Improvements in the grading of the coarse and fine aggregates may have contributed to a similar small increase in CS of the sample containing 5% chipped rubber [67]. The NaOH-treated rubber particles exhibit improved cement paste cohesiveness, suggesting that the method improved the flexural strength but that the CS decreased by 33% [68]. When tire rubber ash substituted the fine aggregates up to 10%, Smadi [60] noted an improvement in CS. When the tire rubber ash content was 2.5%, 5%, 7.5%, and 10%, respectively, the enhancement in CS of mortar after 90 days was 14%, 21%, 29%, and 45%. According to Feng Lie et al. [69], the CS of rubberized concrete diminishes as the quantity of rubber is enhanced. The CS of rubberized concrete was reduced by larger rubber particles. As a result, rubberized concrete has a greater CS the finer the rubber particles are created.

When 5% of the concrete was replaced, the presence of crumb rubber caused a 63.64% strength reduction relative to the goal strength. CS decreased by 52.73% when 2% of the material was replaced with rubber [70]. Although employing rubber instead of natural aggregates in concrete is a workable solution for recycling scrap rubber tires, earlier research advised against using rubberized concrete for primary structural elements due to its weak CS [71]. According to Meherier [36], concrete CS significantly decreased when rubber content exceeded 20%, although it displayed greater strain capacity compared to control concrete specimens [44]. Comparatively, to control concrete, CS decreases when rubber is added to concrete mixtures; however, it rises as the amount of fiber content increases. The best CS is shown by mix (30% recycled aggregate concrete and 2% fiber content without rubber), which is 26.9% more than the reference sample at 28 days [44]. When coarse aggregates were completely replaced, CS decreased by a maximum of 85%, and when fine aggregates were completely substituted with rubber particles, CS decreased by a maximum of 65% [72]. Due to rubber's lower elastic modulus compared to natural aggregates and its poor adherence to cement paste, both the CS and elastic modulus decrease. For the finer rubber replacement, the strength loss is less noticeable [40]. The effects of adding waste rubber to composite Portland concrete were explored by Albano et al. [73]. Flowability, density, compressive capacity, and tensile capacity were shown to decline as rubber concentration and particle size rose. The concrete CS and static elastic modulus were all negatively impacted by the use of tire rubber. Higher rubber contents and larger rubber sizes tended to cause a more significant reduction in these strength values [74].

The causes of the rubberized concrete's declining flexural and CS as per past study [67] are, (a) The cement paste, including rubber particles, would encircle the aggregate. Without the rubber, this cement mix would be considerably softer. Due to the fast formation of fractures surrounding the rubber particles during loading, the specimens fail quite quickly. (b) Compared to cement paste and natural aggregate, there would not be a suitable binding between rubber and mortar. As a result of the applied stresses' uneven distribution, fractures may result. (c) The physical and mechanical characteristics of the component materials affect the CS. Rubber will weaken the materials if it replaces any of them in whole or in part. (d) Rubber has a propensity to rise upward during vibration due to its low specific gravity and lack of adhesion to other materials in concrete. This results in a larger rubber concentration at the top layer. Reduced strengths result from a sample of concrete that is so non-homogeneous. Table 2 shows the summary of CS of rubberized concrete as per past literature.

Table 2. Summary of compressive strength.

Reference	Rubber Tire	Compression Strength (MPa)
[45]	0%, 10%, 20%, and 30%	61.5, 28, 11, and 3.5
[63]	0%, 25%, 50%, 75%, and 100%	31.9, 19.6, 13.8, 9.9, and 7.5
[70]	1%, 2%, and 5%	20, 15, and 12
[74]	0%, 5%, 10%, 15%, 20%, 25%, and 30%	54, 50, 45, 40, 35, 36, and 30
[75]	0%, 5%, 7.5%, 10%, 12.5%, 15%, 17.5%, and 20%	71.0, 70.5, 68.8, 66.5, 61.3, 54.8, 47.5, 37.3, and 30.3
[67]	0%, 5%, 7.5%, and 10%	32, 35, 30, and 25
[26]	0%, 25%, 50%, and 75%	45.80, 23.90, 20.87, and 17.42
[31]	0%, 5%, 10%, and 15%	27, 21, 17, and 12
[30]	0%, 10%, 20%, 30%, 40%, 50%, 60%, 70%, 80%, 90%, and 100%	41.5, 39.2, 36.4, 29.5, 22.8, 16.6, 13.3, 10.9, 9.20, 8.20, and 7.10
[35]	0%, 5%, 10%, 15%, and 20%	65, 60, 50, 40, and 35
[76]	0%, 5%, 10%, and 15%	55, 45, 35, and 25
[39]	0%, 8%, 10%, 20%, and 30%	38, 32, 27, 15, and 13
[33]	0%, 5%, 10%, and 15%	78.05, 68.12, 59.94, and 55.15
[77]	0%, 5%, 7.5%, and 10%	71, 68.8, 66.5, and 61.3
[78]	0%, 5%, 10%, 15%, 20%, 25%, 30%, and 35%	40, 44, 45, 38, 37, 37, 36, and 34
[79]	0%, 10%, 20%, and 30%	37.4, 40, 28.3, and 24.8
[80]	0%, 20%, 40%, 60%, 80%, and 100%	25.3, 18.9, 12.2, 8.0, 4.4, and 2.5
[81]	0%, 5%, 10%, 20%, and 30%	64, 46, 34, 14, and 10

Figure 9 depicts a relative analysis in which several mixes with increasing percentages of rubber are evaluated using the 28-day CS of the control mix as the reference mix. At a 10% rubber replacement, the concrete strength (CS) is 6% less than the reference concrete's CS after 28 days of curing.

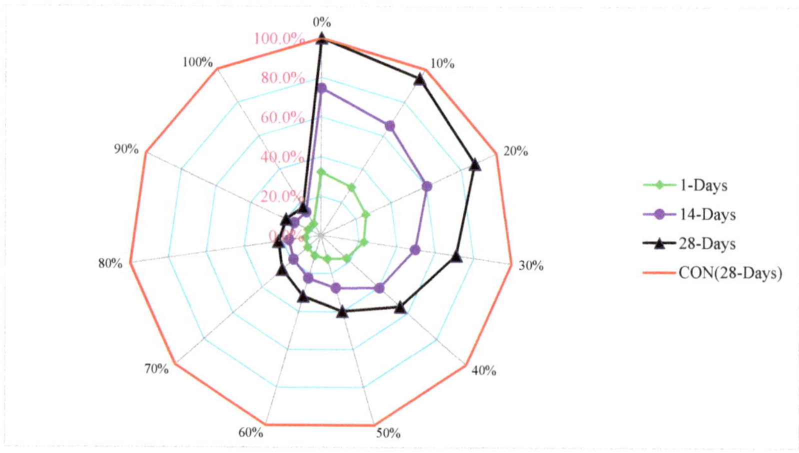

Figure 9. Relative compressive strength.

At a 20% rubber replacement, CS is just 12% less than the benchmark concrete CS. Rubber may be used up to 20% in concrete without having a materially detrimental impact on strength since the CS of concrete with a replacement of rubber is almost equivalent to the CS of reference concrete. However, when the percentages climbed (beyond 20%), a significant decline in CS was seen. Rubber replacement, even at 50%, shows CS 60% lower than reference concrete. Therefore, it is suggested to use rubber up to 20% in concrete. For a higher dose of rubber, treatment of rubber particles should be applied, or it can be used in precast non-load bearing building walls and precast roofs for green buildings.

3.2. Tensile Strength (TS)

Figure 10 shows the tensile strength (TS) of concrete with the substitution of rubber as aggregate. It can be noted that TS decreased as the substitution ratio of rubber increased, similar to the CS of concrete.

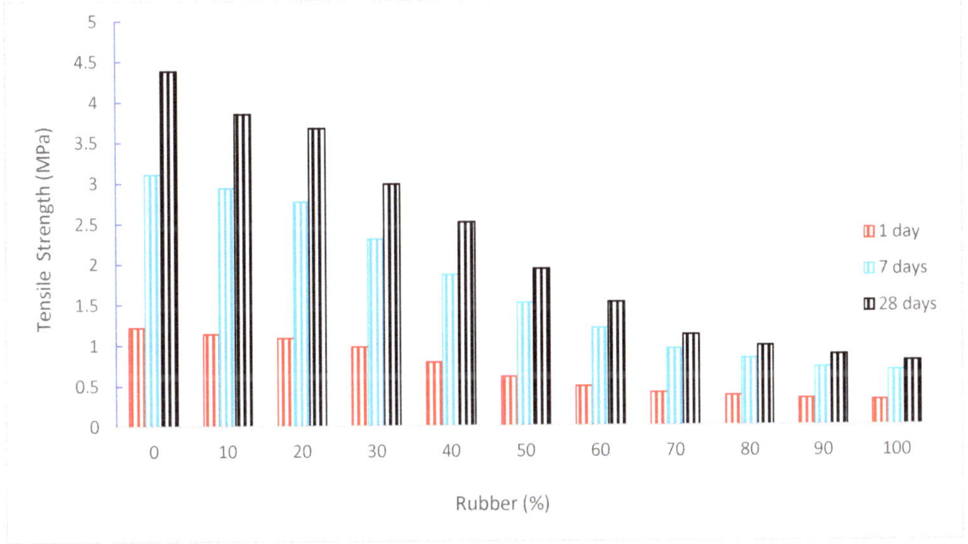

Figure 10. Tensile strength: data source [30].

Concrete's compressive strength (CS) is decreased by 10% to 23% when rubber particles are substituted for 7.5% and 10% of the typical concrete's aggregate. When rubber is substituted by 7.5% to 10%, the tensile strength (TS) of concrete decreases by 30% to 60% [64], which indicate that TS is more adversely affected as compared to CS. Concrete with a 10% replacement of shredded rubber shows a significant drop in TS of up to 76.59%, while concrete with a 5% replacement of shredded rubber shows a reduction of 55.42%. When 1% of shredded tire rubber was added to the overall aggregate composition, a small decrease of 7.62% was seen [70]. Research showed that when the rubber % rose, the values of splitting TS and FS dropped. To obtain results equivalent to the control concrete specimens, the research advised not exceeding a 10% rubber replacement [44]. In the case of rubber, concrete's splitting tensile strength diminishes as the rubber level rises. Tensile strength (TS) values for 10% recycled concrete aggregate (RCA) are 3.16 MPa, 2.82 MPa, and 2.79 MPa for 0%, 5%, and 10% replacement levels of CR, respectively, when the fiber content is 1%. [44]. Studies have shown that adding fibers to concrete mixtures improves TS and ductility while also lowering the incidence of spalling [82]. As the proportion of rubber substitution in concrete grew, the TS of the concrete decreased. Concrete made using chipped rubber instead of aggregates has a lower TS than concrete made with rubber powder as a cement substitute [83]. However, Farhan et al. [84] found that because of the fragility of these added particles, the TS of rubber is decreased as a result of the addition of rubber particles in cement-stabilized aggregate mixture. However, a rise in post-peak behavior was seen, which led to an improvement in the modified mixes' toughness. Additionally, the original combinations' high stiffness was reduced after being partially replaced with rubber particles.

To investigate the strength and toughness characteristics of rubberized concrete mixes, Eldin and Senouci carried out tests. Their findings showed that when the coarse aggregate was completely replaced with chipped tire rubber, the compressive strength decreased by around 85%, while the splitting TS decreased by about 50% [67]. With a higher proportion of rubber substitution in concrete, the TS of the material decreased [85]. Although concrete is a desired and practical material for construction, it has drawbacks, including poor TS, low ductility, low energy absorption, and limited capacity for contraction and shrinkage [30]. In mixes including rubber particles, significant reductions in compressive strength, flexural strength, and breaking TS were observed. When the replacement ratio is 50%, the largest strength loss occurs. It lessens various other mechanical qualities, including slump flow, TS, and CS [11]

Despite losing some of its CS and TS, Deepak and Naidu [86] found that rubber increases the fire resistance of concrete. Due to the rubber aggregate's poor interfacial connection with the cement paste, the CS and TS of concrete also decrease as rubber aggregate content rise [87]. The findings of the TS test showed that increasing the sand replacement with rubber causes the TS to drop [33]. The results showed that when coarse aggregate was fully replaced by an equal volume of chip rubber, compressive strength decreased by 85%, and TS decreased by 15%, but when fine aggregates were replaced with crumb, CS decreased by up to 65% and TS decreased by 50% [83]. Concrete loses some of its TS when rubber is replaced. TS decreases by around 2–12% when shredded rubber replaces 5–10% of the coarse aggregate [88]. Concrete containing coarse rubber aggregate has a higher strength than concrete containing finer rubber aggregate in the same percentage, as discovered by Mavroulidou and Figueiredo [89]. The control combination yielded the greatest TS and FS, and when the crumb rubber content rose, a systematic decline in strengths was seen [79]. Nearly all replacement levels of treated rubberized concrete are found to have greater FS and TS than standard conventional concretes [90]. According to research, there is an 85% drop in compressive strength and a 50% reduction in TS in rubberized concrete with tire articles and crumb rubber of diameters 36, 24, and 18 mm. However, there is a significant energy absorption [91]. Rubberized concrete's TS declines, yet the strain at failure rises in line with it [92]. The TS of rubberized concrete that included tire chips and crumb rubber as replacements for aggregates with diameters of 38, 24, and 19 mm was reduced by 50% in tests, but it also absorbed the greatest energy under tensile loading [91]. Table 3 shows the summary of TS of concrete as per past studies.

Table 3. Summary of tensile strength (TS).

Reference	Rubber Tire	Tensile Strength (MPa)
[76]	0%, 5%, 10%, and 15%	4.2, 4.0, 3.5, and 3.0
[70]	1%, 2%, and 5%	2.7, 2.0, and 0.8
[74]	0%, 5%, 10%, 15%, 20%, 25%, and 30%	3.2, 2.7, 2.6, 2.5, 2.3, 2.2, and 2.1
[67]	0%, 5%, 7.5%, and 10%	3.0, 2.0, 1.6, and 1.4
[31]	0%, 5%, 10%, and 15%	3.02, 2.50, 2.33, and 2.04
[30]	0%, 10%, 20%, 30%, 40%, 50%, 60%, 70%, 80%, 90%, and 100%	4.38, 3.85, 3.67, 2.98, 2.51, 1.93, 1.52, 1.12, 0.98, 0.87, and 0.79
[35]	0%, 5%, 10%, 15%, and 20%	4.7, 4.5, 4.3, 4.0, and 3.7
[33]	0%, 5%, 10%, and 15%	4.90, 4.82, 4.63, and 4.20
[77]	0%, 5%, 7.5%, and 10%	3.0, 3.0, 1.5, and 1.4

Table 3. *Cont.*

Reference	Rubber Tire	Tensile Strength (MPa)
[79]	0%, 10%, 20% and 30%	3.6, 2.3, 2.7, and 2.3
[80]	0%, 20%, 40%, 60%, 80%, and 100%	2.8, 1.8, 1.4, 0.9, 0.5, and 0.2
[81]	0%, 5%, 10%, 20%, and 30%	3.48, 3.68, 3.08, 1.83, and 1.70

The concrete TS shows the same pattern as the concrete CS. Therefore, a substantial link between the TS and CS of concrete. Figure 11 shows linear regression analysis between CS and TS of concrete. A regression line that has an R^2 value greater than 90% seems to be straight. Therefore, the equation shown in Figure 11 can be used to predict the TS from the CS of concrete.

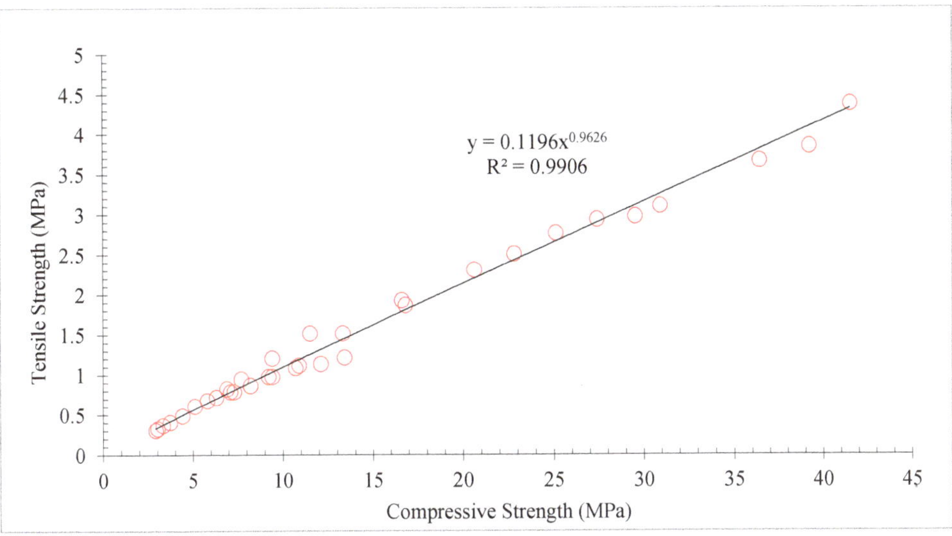

Figure 11. Correlation between compressive and tensile strength: data source [30].

3.3. Flexural Strength (FS)

Figure 12 shows the flexural strength (FS) of concrete with the substitution of rubber as aggregate. It can be noted that CS decreased as the substitution ratio of rubber increased in the same pattern as CS of concrete. According to studies, adding rubber to concrete regularly lowers the material CS, FS, and elasticity when compared to normal concrete [14]. Hora and Reiterman found that elastic modulus and FS decreased very little after 200 cycles of freezing and thawing [93]. With increases in the tire chip contents, FS was found to be significantly less affected than compressive strength. The specimens of FS lost up to 35% of their flexural capacity [63].

FS was found to improve by 12% when shredded rubber was covered with a NaOH solution that was the opposite of what the authors had previously found [70]. A study also discovered that when the crumb rubber (CR) % rose, the values of TS and FS declined. To obtain results equivalent to the control concrete specimens, the research advised not exceeding a 10% CR replacement [48]. FS is seen to rise with rising fiber levels but decreases with increasing recycled concrete aggregate (RCA) and CR levels. The greatest value, which is 8.6% greater than the control concrete mixture, comes from the mix (30% RCA and 2% fiber content without any CR) [44]

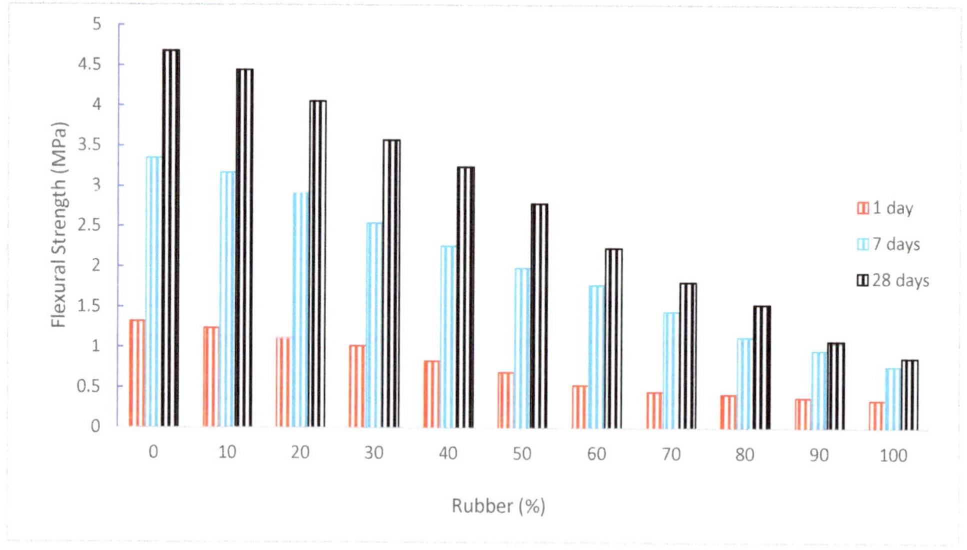

Figure 12. Flexural strength: data source [30].

The findings of compressive and flexural tests showed that substituting coarse aggregate rather than fine aggregate resulted in a greater loss of mechanical characteristics of rubber concrete. By replacing the coarse aggregate with rubber shreds, the post-cracking behavior of rubberized concrete was favorably impacted, displaying excellent energy absorption and ductility indices in the range seen for fibrous concrete [26]. Concrete's FS was decreased when rubber was swapped out for gravel or cement. Replacement of coarse aggregates resulted in a decrease of roughly 37%, while replacement of cement resulted in a reduction of 29% [67].

The specimens containing tire rubber fibers up to 20% exhibited higher FS than the control specimens [94]. Concrete's FS decreased for both classes when rubber was substituted for cement or aggregate, although the rate of decline was different [75]. A wider group of studies would enable the optimization of waste tire rubber particle volume percentages and size distribution in concrete to enhance post-cracking behavior while preventing the deterioration of CS and FS [26]. Ganesan et al. [95] confirmed that an increase in crumb rubber % resulted in an increase in the FS of concrete. A small number of investigations, meanwhile, indicated that the FS decreased as the replacement amount of rubber components rose.

In mixes including rubber particles, significant reductions in FS and breaking tensile strength were observed. When the replacement ratio is 50%, the largest strength loss occurs [30]. The investigation of mortars containing TRC led to reductions in FS and a rise in the likelihood of accumulative plastic fractures [96]. The FS decreases as the percentage of rubber used to replace sand increases. For instance, when no finer is employed, a 15% substitution of sand with TRC results in a 17% reduction in FS. However, when fractures emerge, fiber addition may increase FS [33]. The findings showed that although CS decreased by 33%, FS improved as a consequence of this technique. In research, rubber particles were substituted for sand in varying proportions (10% to 100%). The investigation's findings showed that when the amount of rubber particles increases, mechanical strengths such as CS, TS, and FS decrease. [30]. According to research by Da Silva et al. [97], rubber content in concrete paving blocks ranged from 10% to 50%. The amount of rubber present reduced as the rubber content rose. This decrease for concrete with 50% rubber-containing particles was 32%.

Five indicated rubber contents with volume percentages ranging from 10% to 50% were evaluated by Benazzouk et al. [25]. A rubber volume ratio between 20% and 30% increased the FS despite the fact that the compressive strength was significantly reduced [25]. The results of numerous early studies showed that adding rubber crumbs to concrete improves impact resistance but decreases compressive and FS. There is widespread consensus that the dramatic reduction in mechanical characteristics is brought on by the addition of more rubber [41]. The control combination yielded the greatest TS and FS when the crumb rubber content rose; a systematic decline in strengths was seen [79]. Nearly all replacement levels of treated rubberized concrete are found to have greater FS and TS than standard conventional concretes [90]. Table 4 shows the summary of FS of rubberized concrete as per past literature.

Table 4. Summary of flexural strength (FS).

Reference	Rubber Tire	Flexure Strength (MPa)
[45]	0%, 10%, 20%, and 30%	6.8, 5.7, 3.1, and 1.5
[63]	0%, 25%, 50%, 75%, and 100%	3.8, 3.5, 3.1, 2.8, and 2.4
[70]	1%, 2%, and 5%	3.0, 3.0, and 4.2
[75]	0%, 5%, 7.5%, 10%, 12.5%, 15%, 17.5%, and 20%	7.2, 7.3, 6.9, 6.9, 6.6, 6.1, 5.7, 5.7, and 5.5
[67]	0%, 5%, 7.5%, and 10%	5.3, 5.2, 3.8, and 3.4
[26]	0%, 25%, 50%, and 75%	3.52, 2.93, 2.52, and 2.52
[31]	0%, 5%, 10%, and 15%	4.77, 5.97, 4.32, and 3.87
[32]	0%, 5%, 10%, 15%, 20%, and 25%	3.9, 3.8, 3.6, 3.3, 3.1, and 2.7
[30]	0%, 10%, 20%, 30%, 40%, 50%, 60%, 70%, 80%, 90%, and 100%	4.68, 4.44, 4.05, 3.57, 3.24, 2.79, 2.23, 1.81, 1.53, 1.08, and 0.87
[35]	0%, 5%, 10%, 15%, and 20%	7.0, 6.5, 6.0, 6.0, and 5.5
[76]	0%, 5%, 10%, and 15%	8.4, 8.0, 6.0, and 5.0
[33]	0%, 5%, 10%, and 15%	8.45, 8.03, 7.48, and 6.98
[77]	0%, 5%, 7.5%, and 10%	7.2, 6.9, 6.9, and 6.6
[78]	0%, 5%, 10%, 15%, 20%, 25%, 30%, and 35%	5.5, 6.0, 5.4, 5.1, 5.0, 4.5, 4.4, and 4.1
[80]	0%, 20%, 40%, 60%, 80%, and 100%	3.6, 2.5, 2.0, 1.3, 0.77, and 0.64
[81]	0%, 5%, 10%, 20%, and 30%	0.25, 0.32, 0.41, 0.25, and 0.19

The concrete FS shows the same pattern as the concrete CS. Therefore, a substantial link between the FS and CS of concrete. Figure 13 shows the linear regression analysis between CS and FS of concrete. A regression line that has an R^2 value greater than 90% seems to be straight. Therefore, the equation shown in Figure 13 can be used to predict the FS from the CS of concrete.

3.4. Failure Modes

The concrete cylinder's failure pattern during a compressive strength test is seen in Figure 14. Most of the concrete cylinder's failures were of the cone and shear types. Contrary to the control batch's brittle collapse, the concrete cylinders containing rubber and fiber showed a gradual and progressive failure. In contrast to sand, rubber is softer than that material, making it more pliable under compression. Fiber is effective in preventing cracking because it creates a strong bridging effect between aggregate and cement paste and maintains the aggregate particle's cohesiveness.

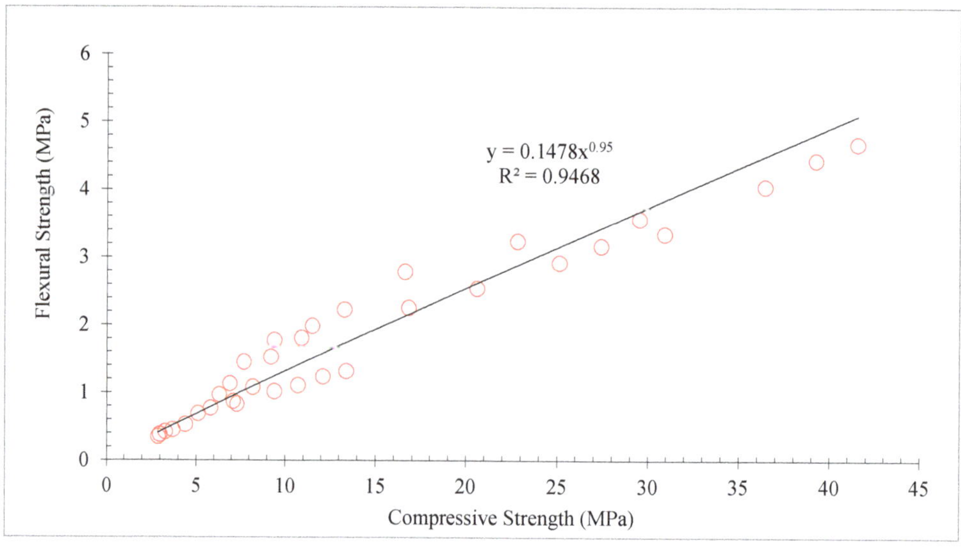

Figure 13. Correlation between compressive and flexure strength: data source [30].

Figure 14. Cylinder failure under compression: (**a**) reference and (**b**) combination of RCA, rubber, and fibers [44]: used as per Elsevier permissions.

The rubberized fiber-based concrete cylinders exhibited uneven failure planes because the fiber keeps the aggregate and cement paste together to prevent the fracture, but the control mix and the mixes with simply RCA replacement broke along well-defined failure planes, as seen in Figure 15.

The failure pattern of the concrete beam during an FS test is seen in Figure 16. In the control and just recycled concrete aggregate (RCA) replacement specimens, the load-deflection curve quickly plummeted and split into two halves. When the maximum stress was applied, the beams with fiber, especially the combinations with 2% fiber, did not completely fracture as they did with other specimens. As a result of their more ductile behavior, the collapse occurred gradually. In summary, fiber-containing concrete exhibits higher hardness ratings and is more ductile when compared to control concrete mixtures. Later, as shown in Figure 16, those beams are physically cracked to investigate the failure surfaces. On each side of the crack near the point of collapse, it is evident that the fiber builds a bridge within the concrete matrix.

Figure 15. Splitting cylinder failure: (**a**) reference and (**b**) combination of RCA, rubber, and fibers [44]: used as per Elsevier permissions.

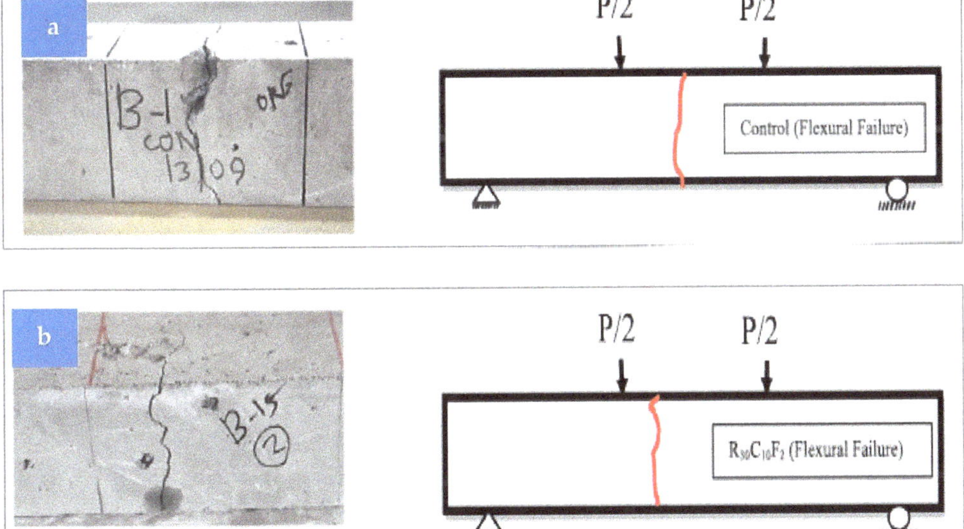

Figure 16. Beam failure: (**a**) reference and (**b**) combination of RCA, rubber, and fibers [44]: used as per Elsevier permissions.

4. Enhancing Properties of Rubber

4.1. Water Washing

The rubber is cleaned with water to remove additives, organics, contaminants, and dirt that were placed on its surface during manufacture. According to certain studies, rubberized concrete made with rubber that has been cleaned with water is a little bit stronger than the rubberized concrete used as a control. The cement's compressive strength rose by 15% [98]. When compared to a 24 h soak, a 2 h soak of crumb rubber in tap water yields far superior results for the rubber mortar's strength and bonding abilities. In addition to enhancing the interfacial transition zone between the rubber particles and the cement paste, it also exhibits a higher compressive strength after 28 days when compared to crumb rubber mortar that had been soaked for 24 h [99].

The strength of rubberized cement cannot be increased only by water washing. Water cleaning, on the other hand, may effectively remove rubber contaminants that are soluble in water before adding the combination since it is the most cost-effective and ecologically beneficial way. It shows a positive development in the rubber exterior's hydrophilicity [100]. There is little information accessible in this area.

4.2. Silane Coating Agents (SCA)

The only chemically altered rubber surface is the latex treatment. This procedure involves soaking rubber in a solution containing CS2, KMnO4, acid, SCA, acetone, $Ca(OH)_2$, and NaOH, followed by UV exposure and partial oxidation. Chemical treatment cleans the oil from the rubber surface, gets rid of the dust and grime, and makes the rubber more hydrophilic and uneven. By modifying the rubber with NaOH liquor, rubberized cement characteristics are most often improved. This is due to the improved wear resistance, fracture energy, and FS shown by rubber that has been treated with NaOH liquid. In addition, the research found [53] that the combination of silica fume (SF) and NaOH treatment might produce an interfacial transition zone (ITZ) that is more stable. The enhanced water affinity of rubber following the surface treatment may have contributed to the small rise in the density of rubber-modified cement composites after the silane coupling agent was applied to the rubber particles' surfaces. Another explanation might be that the air gaps surrounding the rubber particles decreased as a consequence of the chemical link that formed between the rubber particles and cement hydration products. The increase in density for cement composites incorporating silane-treated rubber varied from 1.8% to 3.6% for the 5–25% rubber concentrations employed in the research. The improvement in compressive strength of rubber-modified cement composites was more significantly impacted by the silane coupling agent treatment than was the increase in density. The compressive strength of cement composites made with silane-treated rubber was 24%, 9%, 18%, 14%, and 22% higher than that of the paste containing as-received rubber, respectively, at rubber contents of 5%, 10%, 15%, 20%, and 25%. The chemical link created by the silane coupling agent between rubber particles and cement paste was primarily responsible for the increase in compressive strength. Figure 17 depicts the silane coupling agent's reaction process. Methoxy groups and reactive vinyl or epoxy group (X) are found in silane coupling agents (OR). The methoxy group becomes the hydroxyl group by hydrolysis (OH). The OH groups are further chemically or physically attached to an inorganic substance via dehydration condensation or a hydrogen bond (cement paste). An organic substance is chemically joined to the X group (rubber). The chemical bonds created by the silane coupling agent will need more energy to break, resulting in enhanced compressive strength of the composites containing silane-treated rubber.

According to research [101], treating rubber with NaOH liquid reduces the mixture's plummet by 25%. Additionally, NaOH treatment may help make rubberized cement more durable. A reduction in the mixture's resistivity and an increase in the adherence of the rubber to the cement were also effects of the alteration of the NaOH solution [102]. The specimens with reduced electrical resistance have enhanced long-term durability. By promoting adhesion at the interface, the SCA functions as an aggregate to improve adhesion between the concrete matrix and rubber. It causes the cement matrix and rubber to physically and chemically mix to form a firmly linked structure by acting on the inorganic/organic interface section. According to research [103], the rubber's tensile stiffness and strength are enhanced after SCA processing. According to a study [104], the hydrolysate of SCA may react with concrete paste to strengthen their bond and noticeably improve the microstructure. The SCA treatment, which results in the FS and CS of rubberized mortar, may significantly enhance the mechanical performance of the rubberized cement. When rubber was treated by Guo et al. [102] utilizing SCAs, the bond between the rubber aggregates and the cement paste was enhanced.

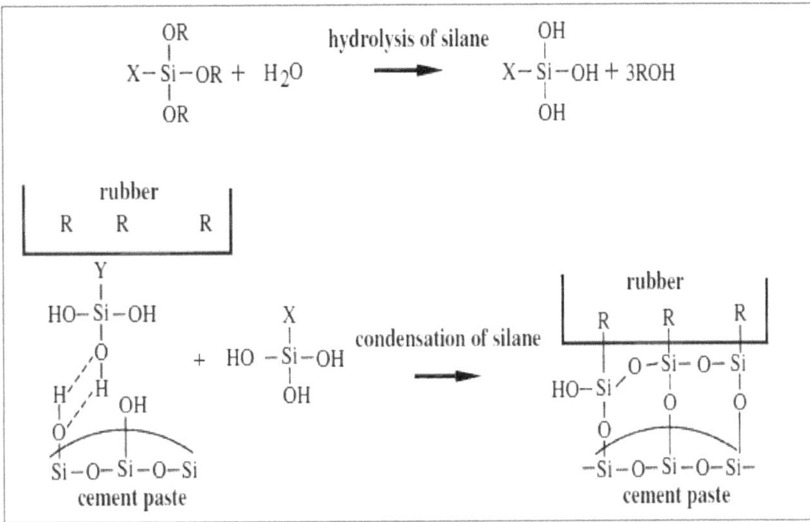

Figure 17. Mechanism of silane coating agents [105]: used as per Elsevier permissions.

4.3. Cement Coating

The density and compressive strength of rubber-modified cement composites were further improved by a cement coating created using a silane coupling agent around rubber particles. The 30% cement coating was more successful in improving density than the 60% cement coating. The compressive capacity of concrete modified with rubber was greatly enhanced when the silane coupling agent was combined with cement coating as opposed to the silane coupling agent alone. At the rubber contents utilized in the investigation, the silane coupling agent surface treatment increased compressive strength by 9% to 24%, but the 30 weight % cement coating increased compressive strength by 27% to 110%. The increase in compressive strength was significantly greater for the 60 weight % cement coatings, ranging from 53% to 168%. The rubber-modified cement composite was able to preserve a considerably greater percentage of the compressive capacity of the reference pastes without rubber, thanks to the rubber's considerable rise in the compressive capacity as a result of its two-staged surface treatment. For instance, the composite comprising 60% cement-coated rubber could sustain 94% of the strength of the reference paste at a rubber level of 5%. The compressive capacity formed with 60 weight % cement-coated rubber was greater than 50% from the control paste, even at a rubber content of 15%.

The hard shell that formed around rubber particles because of cement hydration, as shown in Figure 18, enhanced the stiffness compatibility between rubber and cement paste, which was the cause of the strength gain. There were some visible patches without a cement coat. It can be noted that 30% of cement-coated rubber particles looked to be darker as compared to 60%. This clarified why the 60% cement coating improved the compressive capacity of rubberized concrete more than the 30 weight % coating. According to research, rubberized cement's resistance might be strengthened by using tiny rubber particles that were only rinsed in water, improving the cement's strength by 16%. Additionally, a technique of reformatory water soaking has been found to control rubber's hydrophilicity. Prior to mixing, the rubber is submerged in water for 24 h. Silica fumes (SF), mortar, cement paste (LP), and limestone powder are examples of pre-cementitious materials that are often employed. By covering the rubber with cementitious materials, the automatic properties of various rubberized cement may be effectively enhanced. A concrete matrix, cement coating, or rubber particles treated with silane coupling agents (SCAs) are more tightly coupled than untreated ones, according to the splitting tensile strength test. The coated rubber retained cement hydration products on its outside, while the uncoated rubber was left exposed. Typically, rubber processing entails air-seasoning the outside and coating the rubber with cementitious materials [106].

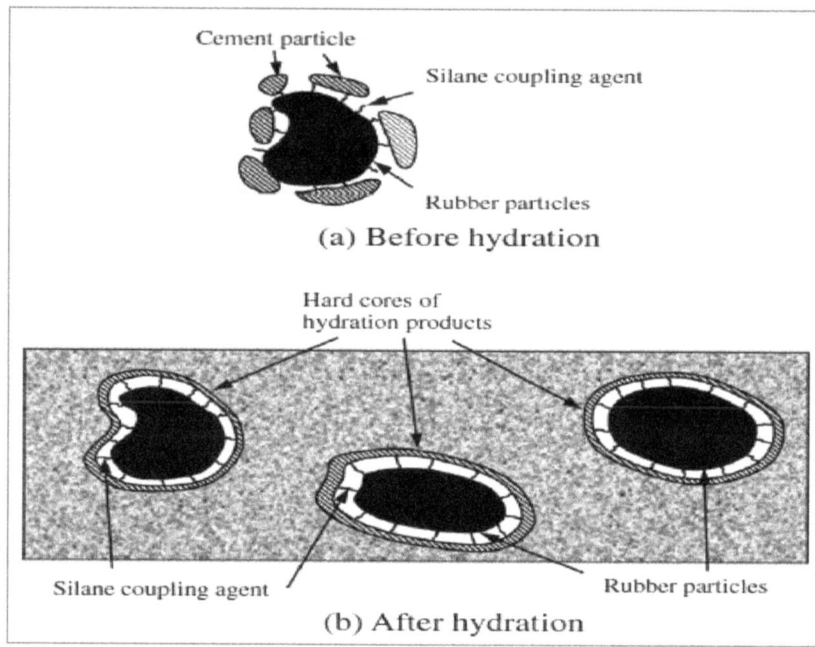

Figure 18. Mechanism of Cement Coating [105]: Used as per Elsevier permissions.

Rubber's elastic modulus and interfacial transition zone (ITZ) adhesion were both enhanced by the procedure of coating it with cementitious ingredients. As a result, it was very successful in enhancing the strength and durability characteristics of rubberized concrete. Additionally, the basic materials employed in this process have a variety of uses. They thus hold promise for industrial and large manufacture of rubberized cement for structural purposes [100]. Ordinary rubberized concrete (ORC) gets a coating treatment to increase bending resistance.

5. Durability

5.1. Permeability and Water Absorption

Figure 19 shows the water permeability of concrete with the substitution of rubber as aggregate. It can be noted that the permeability of concrete increased as the substitution ratio of rubber increased. However, the permeability of Ca(ClO)$_2$-treated concrete is comparable to reference concrete. In concrete mixes, replacing rubber increases the depth of water permeability. In comparison to the second combination, the first mixture exhibits a greater rise in water permeability depth. Mixtures with substitutions of 5% and 7.5% rubber are categorized as having low permeability, while mixtures with replacements of 10% tire rubber are categorized as having medium permeability. In concrete mixes, replacing rubber improved water penetrability depth and enhanced water absorption when coarse aggregate was replaced but decreased water absorption when cement was replaced [67].

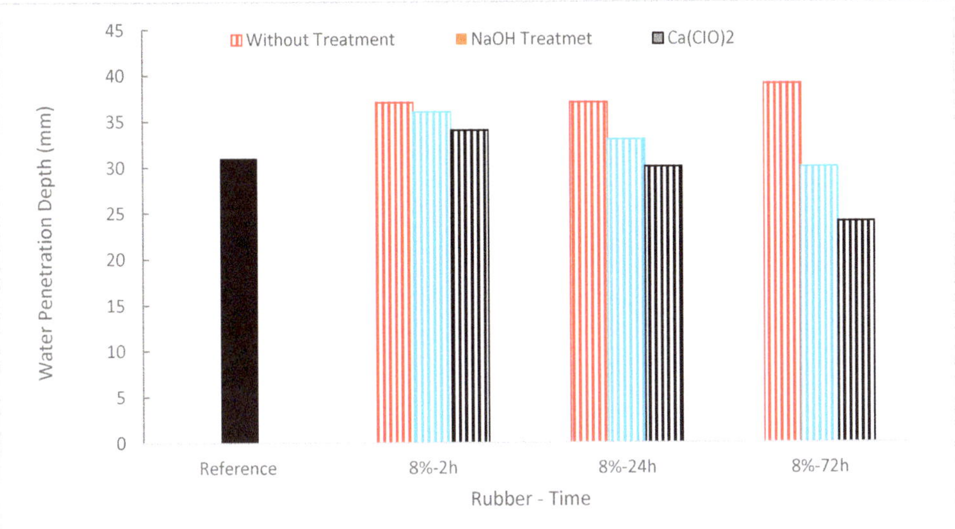

Figure 19. Water Penetration Depth: Data Source [39].

The results indicate that the $Ca(ClO)_2$ treatment of rubber aggregates is the most successful method for lowering the permeability of concrete mixes. Additionally, the highest outcomes were seen during a 72 h therapy period. While the aggregates' permeability was likewise decreased after being treated with NaOH, the improvement was noticeably less than after being treated with $Ca(ClO)_2$. This decrease is also due to better rubber aggregate/paste bonding, which reduces the ITZ's porosity [39]. For the samples containing waste rubber, a greater penetration depth was seen. This can be due to a decreased binding between rubber aggregate and cement paste and a greater water-cement ratio [67]. Additionally, the permeability of concrete has a considerable impact on its longevity. The water absorption test gauges the concrete's capacity to transfer fluids [107]. The results of the tests showed that the existence of rubber tends to prevent water transmission and minimize water absorption, providing superior defense against corrosion for steel reinforcement [25].

The initial combination samples that were examined for water absorption looked to have fractured during oven drying, which led to noticeably higher values. The weaker link between the cement pastes and the bigger rubber particles (as opposed to the powder rubber in the second batch) may be the cause of this breaking. This rise is brought about by the substitution of rubber for sand, which has various forms and structures and develops some porosity, enhancing water absorption. Conversely, improving the quantity of fiber in concrete decreases water absorption [33].

The authors draw the conclusion that the water absorption is higher than it was for the control combination. This is because the connection between the cement pastes and big rubber particles has decreased. In contrast, when the proportion of replacement is raised, the water absorption of the second combination containing powdered tire rubber decreases. It seems that filling cavities with powdered rubber has decreased the porosity of the concrete, which has decreased water absorption in this combination [67]. By preventing water from spreading, rubber minimizes water absorption in concrete and helps to better protect the steel reinforcement against water [25].

5.2. Chloride Ion Penetration

One significant unrecognized risk to the safety of buildings is the durability of concrete in the sea environment. The impact of dry-wet alternation substantially speeds up the diffusion of corrosive ions, and the tidal range region develops the maximum severe area for concrete structure corrosion. Among them, chloride ion diffusion is one of the primary causes of the durability of concrete [108].

Figure 20 shows that adding rubber to concrete significantly lowers the absorption of free chloride ions in various depths of concrete and that the capability of concrete to resist chloride ion erosion is in the order of RC-2 group > RC-1 group > RC-3 group > OC group, indicating that adding the right quantity of rubber to concrete can enhanced concrete's resistance to chloride ion destruction and decrease the interruption of chloride ions. It is recommended that the rubber percentage be 10% if only to lower the free chloride ion absorption at various depths of the concrete [109].

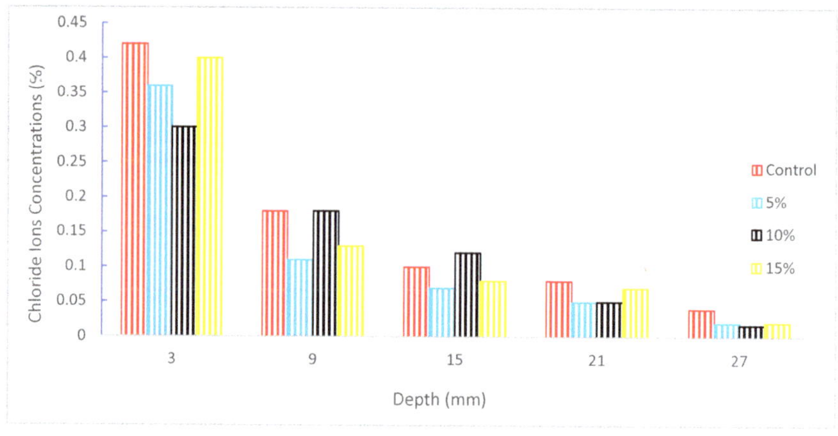

Figure 20. Chloride ion penetration: data source [109].

According to Oikonomou and Mavridou [110], as the quantity of rubber in mortar grew, the chloride ion diffusion dropped. When compared to the reference mix, there was a decrease of 14.22% in the 2.5% rubber mix and a reduction of 35.85% in the 15% rubber mix. Comparing concrete with 12.5% tire rubber to control mix concrete, the mixture with the bitumen emulsion showed a decrease in chloride ion penetration of up to 55.89%. On rubberized concrete, Bravo and Brito [111] tested for chlorine migration. For 5–15 replacement with tire rubber, an increase in chloride diffusion coefficient was seen. The chloride diffusion coefficient rises as rubber aggregate size increases. Concrete with tire aggregates ground mechanically provided greater resistance to chloride penetration than concrete with tire aggregates ground cryogenically. Chloride penetration was reduced when the curing time was extended.

Rubber may prevent concrete fractures from forming and can lower the peak value of steel corrosion, according to research by Jian Liang et al. [112]. Han Zhu [113] investigated the resistance of rubber concrete to chloride ion penetration under various conservational temperatures and discovered that rubber can decrease reinforcement corrosion and that the durability of rubber concrete is altered with various environmental temperatures. The chloride ion transport and erosion process of rubber concrete were thoroughly investigated and estimated by Han Qinghua et al. [114]. Using both macroscopic and microscopic simulations. The findings demonstrate that rubber has the capacity to significantly lower the chloride ion diffusion coefficient and increase the robustness of concrete structures.

According to research [115], replacing some of the coarse and fine aggregates with crumb rubber and rubber chips, respectively, gradually increased the diffusion of chloride ions. When silica fume was put into the concrete mix, it was discovered that the penetration was reduced. The increased resistance was attributed to the silica fume's ability to fill spaces in cement pastes and transition zones between aggregate and paste. The addition of silica fume causes a decrease in the permeability of chloride ions. The cause was linked to the mortar's decreased calcium hydroxide content [116].

6. Elevated Temperature

It was nearly impossible to see any difference when samples of rubberized concrete were exposed to temperatures of 150 °C and 200 °C. Figure 21 compares samples of material containing 30% rubber aggregate after being heated to 300 and 400 degrees Fahrenheit to a sample that was not heated. The reference white-grayish color of the sample shifted to a light brown and black hue. It resulted from the acceleration of rubber particle dissolution at higher temperatures, which began at about 300 °C. The subsequent burning of the rubber produced a black color.

Figure 21. Effect of elevated temperature on rubberized concrete: (**A**) unheated, (**B**) 300 °C, and (**C**) 400 °C [45]: used as per Elsevier permission.

All samples had an increase in mass loss as the heating temperature rose. The early mass loss was caused by the evaporation of capillary and gel water from the cement matrix, followed by the escape of absorbed and interlayer water, as demonstrated by the thermogravimetric measurement of rubber particles, which displayed that rubber disintegration begins quickly on the attainment of a temperature of 300 °C. The delivery of chemically bonded water, which is a component of cement hydration results and is particularly resistant to evaporation, may be responsible for the mass loss at higher temperatures [117]. The CSH phase's dehydration process occurred at temperatures as high as 300 °C [118]. Ettringite and mono sulfate phases are also dehydrated between 110 and 156 °C, according to Liu et al. [119]. According to other scientists, the mass loss of all samples was comparable up to 300 [62]. At 400 °C, the mass loss for samples including rubber increased significantly, but the control sample only had a little rise of around 2.6%. The reference sample revealed the same mass loss result (2.6%) for 400 °C as Medine et al. [120]. It is clear that at 400 °C, the mass loss improved as the concrete's rubber-based aggregate content rose.

When related to the strength characteristics found for materials held at laboratory temperature, a study [45] found that both the compressive and tensile capacity increased at a temperature of 150 °C. Gupta et al. [117] showed a similar rise in compressive capacity up to 150 °C and suggested that this rise may be related to a decrease in calcium hydroxide and un-hydrated area fraction, which is advantageous for the development of the compact concrete microstructure. On the other hand, Guelmine et al. [62] found that the damage factor rose with an increase in the used raised temperature of recycled rubber mortar, affecting both the CS and FS. For samples subjected to temperatures of 200 °C, 300 °C, and 400 °C, respectively, this concrete performance was seen. Damage factors between control concrete and rubberized concrete with varying amounts of rubber were mostly convergent up to 300 °C. A significant increase in the damage factor was seen during exposure to 400 °C, and this rise varied depending on the amount of rubber present. Typically, a steep slope of the damage factor function was seen for mixtures with significant rubber content. This conclusion is consistent with findings published, for instance, by Thomas and Gupta [121]. For control concrete, the major source of the damage processes is the temperature gradient that is applied to the samples, which results in water evaporation and the breakdown of the CSH, ettringite, and mono sulfate phases [122]. The burning of the rubber-based aggregate, which produced gaps and hence amplified its permeability, was specifically to blame for the extra damage to rubberized concrete at 400 °C.

7. Microstructure Analysis

Scan electronic microscopy (SEM) and digital microscopic pictures of concretes with rubber aggregates treated for 72 h with water, NaOH, and $Ca(ClO)_2$ were collected for microstructural investigations are shown in Figure 22. Interfacial regions were photographed. When rubber aggregates were treated with water, distinct grooves with a gap width of around 7 mm were seen at the interfacial transition zone (ITZ). After being exposed to NaOH for 72 h, the rubber aggregate particles narrowed to 2.4–3 mm. There are no gaps at the interface after a 72 h $Ca(ClO)_2$ treatment. This is explained by stronger aggregate/matrix bonding brought on by the rough tire surface and decreased porosity brought on by potential Friedel's salt production. This is consistent with the strength data, which demonstrated that treating aggregates with $Ca(ClO)_2$ for 72 h almost totally neutralized the strength decline shown for rubber particles that were not treated.

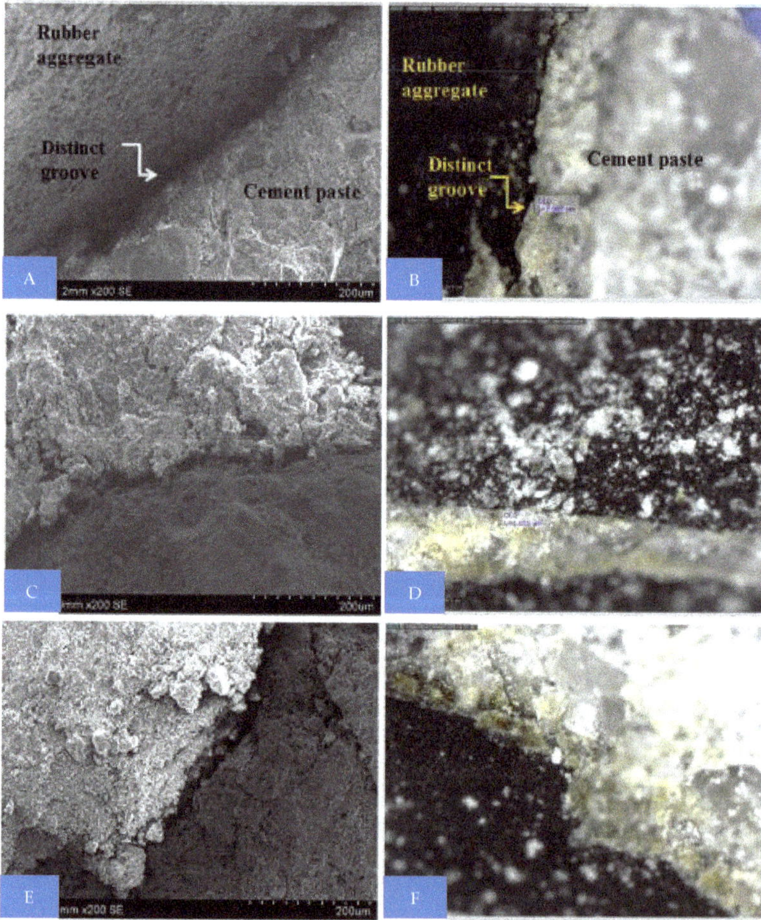

Figure 22. SEM results: (**A,B**) water, (**C,D**) NaOH treated, and (**E,F**) Ca(CIO$_2$) treated [39].

According to Pelisser et al. [53], treatment with NaOH and the addition of silica fume (SF) may produce an ITZ that is denser than concrete with untreated rubber. There is a noticeable reduction in the porosity of the rubber–cement matrix contact. The compressive strength of ORC after 28 days is just 14% lower than that of the reference concrete without rubber. The rubberized concrete mixture also includes natural zeolite to help rubber react with the NaOH solution treatment [31]. The rubber's surface may be modified using NaOH solution to strengthen the connection between it and the cement. The mechanical qualities of rubberized concrete gradually enhance as a result of the synergistic interaction between natural zeolite and NaOH treatment.

When rubber is treated with a NaOH solution, the ITZ porosity of rubberized concrete is reduced, and the rubber–cement adhesion is increased, which lowers the mixture's resistivity [123]. In order to improve adhesion between the two materials, SCA may be used as an aggregate or as an adhesion promoter at the interface between the rubber and cement matrix. It works to join the rubber and cement matrix chemically and physically into a tightly bound structure by acting on the organic/inorganic interface [102].

8. Applications of Rubberized Concrete

When compared to traditional concrete, rubberized concrete is more resilient to pressure, impact, and temperature. It is also cheaper and more cost-effective. It has been noted that the compressive and tensile strengths of rubber-modified concrete (RMC) are quite low. However, they have increased acid resistance, minimal shrinkage, strong impact resistance, suitable water resistance with low absorption, and great sound and thermal insulation. In comparison to a standard concrete mix, studies reveal that CRC (crumb rubber concrete) specimens stayed intact after failure (did not shatter). For construction that has to have strong impact resistance qualities, this behavior could be advantageous. Concrete samples aggregated with thick rubber were notably noticeable to have rubberized concrete with improved impact resistance.

Additionally, the special properties of rubberized concrete will find new applications in building structures for use as earthquake shock-wave absorbers, sound barriers, and highway construction as shock absorbers. It lessens plastic shrinkage, cracking, and concrete's susceptibility to disastrous collapse.

Presently, precast sidewalk panels, non-load-bearing building walls, and precast roofs for green buildings all employ concrete that has had scrap tires added to it [124]. It is often used for construction-related tasks such as constructing ramps that are skid-resistant and constructing walkways and courts for enjoyment. These concretes are anticipated to be applied in architectural applications such as nailing concrete, where high strength is not required, wall panels that need low unit weight, construction elements, and impact-prone Jersey barriers, and railroads to secure rails to the ground, thanks to this new property [22].

Waste tire-modified concrete mixes might provide a feasible replacement to the standard weight concrete since rubberized concrete can also be utilized in non-load-bearing elements such as lightweight concrete walls, building facades, or other light architectural units [41]. Wherever cement-stabilized aggregate bases are required, especially below flexible pavements, rubberized mixes may be employed. The other practical uses may also be poured in bigger sheets than regular concrete and are ideal for usage in regions that often freeze and thaw.

9. Conclusions

The practice of rubber tires in concrete undoubtedly offers advantages, and the building sector as a whole cannot ignore this trend. For instance, rubber tires may be utilized as a substitute for natural aggregates. More than 100 modern and historical pieces of literature were evaluated to study the impact of rubber tires on the strength characteristics, freshness, and durability of rubberized concrete. The following were the key findings.

- Increase in rubber concentration. Rubberized concrete loses workability. However, it may be enhanced by adding admixtures such as plasticizers or other filler ingredients;

- The lower specific gravity and tendency to absorb air of rubber, rubberized concrete density reduces significantly when rubber content is increased. Rubberized concrete is hence advantageous for lightweight buildings;
- Concrete's mechanical strength may generally be decreased by adding rubber, and this tendency becomes worse as rubber content rises. Due to the poor adherence of rubber with cement paste, a broad and porous weak interfacial transition zone (ITZ) was seen in rubberized concrete. The detrimental effects of rubber on the strength qualities of regular concrete may be lessened if the bond is strengthened at ITZ by any practical and affordable techniques. As a result, the construction industry would be able to employ rubberized concrete efficiently in a variety of concrete buildings;
- The decline in flexural capacity was lower than the decline in compressive capacity;
- The majority of studies feel that rubberized concrete with NaOH treatment has improved mechanical qualities. Other studies, however, asserted that the strength characteristics of rubberized concrete that has been treated with NaOH solution remain unchanged or even improve. The inconsistent findings might have been caused by varying rubber particle sizes, rubber suppliers, solution concentrations, and processing times;
- The silane coating agents (SCA) process transforms the rubber's hydrophobic surface into a hydrophilic one and creates a chemical link between it and the cement matrix, enhancing the rubberized concrete's mechanical characteristics and durability.

10. Recommendations

- Rubberized concrete performs badly at the moment. Pozzolanic filler additives could make it perform better. However, more detailed research is required before it may be used in a practical setting;
- The microstructure of rubberized concrete should be properly studied;
- Steel reinforcing bars' corrosion behavior in rubberized concrete is recommended to be explored;
- Rubber surface treatment raises the price of utilizing rubber as a concrete aggregate. The cost of rubber surface modification should be investigated to evaluate its cost-effectiveness and identify the cheapest and most effective approach, which is crucial for more field applications;
- The thermal properties of rubberized concrete should be explored in more detail;
- The dry shrinkage and freeze–thaw action of rubberized concrete should be studied in detail.

Author Contributions: Conceptualization, J.A. and Z.Z.; Formal analysis, Z.Z., A.M., and M.A.; Funding acquisition, M.A.; Methodology, J.A.; Project administration, A.F.D.; Software, J.A. and A.M.; Supervision, M.A. and A.F.D.; Writing—original draft, J.A.; Writing—review and editing, Z.Z., A.M., and A.F.D. All authors have read and agreed to the published version of the manuscript.

Funding: This research was funded by Taif University Researchers Supporting Project number (TURSP-2020/324).

Institutional Review Board Statement: Not applicable.

Informed Consent Statement: Not applicable.

Data Availability Statement: All the data are available in main text.

Acknowledgement: The authors would like to acknowledge Taif University Researchers Supporting Project number (TURSP-2020/324), Taif University, Taif, Saudi Arabia.

Conflicts of Interest: The authors declare no conflict of interest.

References

1. Danon, B.; Görgens, J. Determining Rubber Composition of Waste Tyres Using Devolatilisation Kinetics. *Thermochim. Acta* **2015**, *621*, 56–60. [CrossRef]
2. Di Mundo, R.; Petrella, A.; Notarnicola, M. Surface and Bulk Hydrophobic Cement Composites by Tyre Rubber Addition. *Constr. Build. Mater.* **2018**, *172*, 176–184. [CrossRef]
3. Smirnova, O.M.; de Navascués, I.M.P.; Mikhailevskii, V.R.; Kolosov, O.I.; Skolota, N.S. Sound-Absorbing Composites with Rubber Crumb from Used Tires. *Appl. Sci.* **2021**, *11*, 7347. [CrossRef]
4. Fenner, R.A.; Clarke, K. Environmental and structural implications for the re-use of tyres in fluvial and marine-construction projects. *Water Environ. J.* **2003**, *17*, 99–105. [CrossRef]
5. Shahria Alam, M.; Slater, E.; Muntasir Billah, A.H.M. Green Concrete Made with RCA and FRP Scrap Aggregate: Fresh and Hardened Properties. *J. Mater. Civ. Eng.* **2013**, *25*, 1783–1794. [CrossRef]
6. Kaliyavaradhan, S.K.; Ling, T.-C.; Guo, M.-Z.; Mo, K.H. Waste Resources Recycling in Controlled Low-Strength Material (CLSM): A Critical Review on Plastic Properties. *J. Environ. Manage.* **2019**, *241*, 383–396. [CrossRef]
7. Rubber Manufactures Association. *US Scrap Tire Markets 2003 Edition*; Rubber Manufactures Association: Washington, DC, USA, 2004.
8. Alfayez, S.A.; Suleiman, A.R.; Nehdi, M.L. Recycling Tire Rubber in Asphalt Pavements: State of the Art. *Sustainability* **2020**, *12*, 9076. [CrossRef]
9. Najim, K.B.; Hall, M.R. A Review of the Fresh/Hardened Properties and Applications for Plain-(PRC) and Self-Compacting Rubberised Concrete (SCRC). *Constr. Build. Mater.* **2010**, *24*, 2043–2051. [CrossRef]
10. Arbili, M.M.; Ghaffoori, F.K.; Awlla, H.A.; Alzeebaree, R.; Ibrahim, T.K. Utilization of Styrene-Butadiene Rubber (SBR) Polymer Replacement of Fine Aggregate in Concrete. In *Proceedings of the IOP Conference Series: Earth and Environmental Science*; IOP Publishing: Bristol, UK, 2021; Volume 856, p. 12032.
11. Siddique, R.; Naik, T.R. Properties of Concrete Containing Scrap-Tire Rubber–an Overview. *Waste Manag.* **2004**, *24*, 563–569. [CrossRef]
12. Neocleous, K.; Pilakoutas, H.; Waldron, P. From Used Tires to Concrete Fiber Reinforcement. In Proceedings of the 2nd International FIB Congress, Naples, Italy, 5–8 June 2006.
13. Girskas, G.; Nagrockienė, D. Crushed Rubber Waste Impact of Concrete Basic Properties. *Constr. Build. Mater.* **2017**, *140*, 36–42. [CrossRef]
14. Snelson, D.G.; Kinuthia, J.M.; Davies, P.A.; Chang, S.-R. Sustainable Construction: Composite Use of Tyres and Ash in Concrete. *Waste Manag.* **2009**, *29*, 360–367. [CrossRef] [PubMed]
15. Thomas, B.S.; Gupta, R.C.; Panicker, V.J. Recycling of Waste Tire Rubber as Aggregate in Concrete: Durability-Related Performance. *J. Clean. Prod.* **2016**, *112*, 504–513. [CrossRef]
16. Youssf, O.; Hassanli, R.; Mills, J.E. Mechanical Performance of FRP-Confined and Unconfined Crumb Rubber Concrete Containing High Rubber Content. *J. Build. Eng.* **2017**, *11*, 115–126. [CrossRef]
17. Reda Taha, M.M.; El-Dieb, A.S.; Abd El-Wahab, M.A.; Abdel-Hameed, M.E. Mechanical, Fracture, and Microstructural Investigations of Rubber Concrete. *J. Mater. Civ. Eng.* **2008**, *20*, 640–649. [CrossRef]
18. Benazzouk, A.; Mezreb, K.; Doyen, G.; Goullieux, A.; Quéneudec, M. Effect of Rubber Aggregates on the Physico-Mechanical Behaviour of Cement–Rubber Composites-Influence of the Alveolar Texture of Rubber Aggregates. *Cem. Concr. Compos.* **2003**, *25*, 711–720. [CrossRef]
19. Mhaya, A.M.; Baghban, M.H.; Faridmehr, I.; Huseien, G.F.; Abidin, A.R.Z.; Ismail, M. Performance Evaluation of Modified Rubberized Concrete Exposed to Aggressive Environments. *Materials* **2021**, *14*, 1900. [CrossRef]
20. Mohammed, B.S.; Hossain, K.M.A.; Swee, J.T.E.; Wong, G.; Abdullahi, M. Properties of Crumb Rubber Hollow Concrete Block. *J. Clean. Prod.* **2012**, *23*, 57–67. [CrossRef]
21. Fattuhi, N.I.; Clark, L.A. Cement-Based Materials Containing Shredded Scrap Truck Tyre Rubber. *Constr. Build. Mater.* **1996**, *10*, 229–236. [CrossRef]
22. Topcu, I.B. The Properties of Rubberized Concretes. *Cem. Concr. Res.* **1995**, *25*, 304–310. [CrossRef]
23. Khaloo, A.R.; Dehestani, M.; Rahmatabadi, P. Mechanical Properties of Concrete Containing a High Volume of Tire–Rubber Particles. *Waste Manag.* **2008**, *28*, 2472–2482. [CrossRef]
24. Avcular, N. Analysis of Rubberized Concrete as a Composite Material. *Cem. Concr. Res.* **1997**, *27*, 1135–1139.
25. Benazzouk, A.; Douzane, O.; Langlet, T.; Mezreb, K.; Roucoult, J.M.; Quéneudec, M. Physico-Mechanical Properties and Water Absorption of Cement Composite Containing Shredded Rubber Wastes. *Cem. Concr. Compos.* **2007**, *29*, 732–740. [CrossRef]
26. Aiello, M.A.; Leuzzi, F. Waste Tyre Rubberized Concrete: Properties at Fresh and Hardened State. *Waste Manag.* **2010**, *30*, 1696–1704. [CrossRef] [PubMed]
27. Atahan, A.O.; Yücel, A.Ö. Crumb Rubber in Concrete: Static and Dynamic Evaluation. *Constr. Build. Mater.* **2012**, *36*, 617–622. [CrossRef]
28. Batayneh, M.K.; Marie, I.; Asi, I. Promoting the Use of Crumb Rubber Concrete in Developing Countries. *Waste Manag.* **2008**, *28*, 2171–2176. [CrossRef]
29. Al-Shathr, B.S.; Gorgis, I.N.; Motlog, R.F. Effect of Using Plastic and Rubber Wastes as Fine Aggregate on Some Properties of Cement Mortar. *Eng. Tech. J.* **2016**, *3*, 1688–1699.

30. Lv, J.; Zhou, T.; Du, Q.; Wu, H. Effects of Rubber Particles on Mechanical Properties of Lightweight Aggregate Concrete. *Constr. Build. Mater.* **2015**, *91*, 145–149. [CrossRef]
31. Jokar, F.; Khorram, M.; Karimi, G.; Hataf, N. Experimental Investigation of Mechanical Properties of Crumbed Rubber Concrete Containing Natural Zeolite. *Constr. Build. Mater.* **2019**, *208*, 651–658. [CrossRef]
32. Banerjee, S.; Mandal, A.; Rooby, J. Studies on Mechanical Properties of Tyre Rubber Concrete. *SSRG Int. J. Civ. Eng* **2016**, *3*, 18–21. [CrossRef]
33. Hesami, S.; Hikouei, I.S.; Emadi, S.A.A. Mechanical Behavior of Self-Compacting Concrete Pavements Incorporating Recycled Tire Rubber Crumb and Reinforced with Polypropylene Fiber. *J. Clean. Prod.* **2016**, *133*, 228–234. [CrossRef]
34. Abdullah, W.A.; Muhammad, M.A.; Abdulkadir, M.R. Experimental Investigation of Some Mechanical Properties of Rubberized Concrete with Highest Possible Rubber Content. *J. Duhok Univ.* **2020**, *23*, 509–522. [CrossRef]
35. Mishra, M.; Panda, K.C. Influence of Rubber on Mechanical Properties of Conventional and Self Compacting Concrete. In *Advances in Structural Engineering*; Springer: Berlin/Heidelberg, Germany, 2015; pp. 1785–1794.
36. Choudhary, S.; Chaudhary, S.; Jain, A.; Gupta, R. Assessment of Effect of Rubber Tyre Fiber on Functionally Graded Concrete. *Mater. Today Proc.* **2020**, *28*, 1496–1502. [CrossRef]
37. AbdulKadir, M.R. Effect of High Temperature on Mechanical Properties of Rubberized Concrete Using Recycled Tire Rubber as Fine Aggregate Replacement. *Eng. Technol. J.* **2018**, *36*, 906–913.
38. Aslani, F.; Ma, G.; Wan, D.L.Y.; Muselin, G. Development of High-Performance Self-Compacting Concrete Using Waste Recycled Concrete Aggregates and Rubber Granules. *J. Clean. Prod.* **2018**, *182*, 553–566. [CrossRef]
39. Khern, Y.C.; Paul, S.C.; Kong, S.Y.; Babafemi, A.J.; Anggraini, V.; Miah, M.J.; Šavija, B. Impact of Chemically Treated Waste Rubber Tire Aggregates on Mechanical, Durability and Thermal Properties of Concrete. *Front. Mater.* **2020**, *7*, 90. [CrossRef]
40. Karunarathna, S.; Linforth, S.; Kashani, A.; Liu, X.; Ngo, T. Effect of Recycled Rubber Aggregate Size on Fracture and Other Mechanical Properties of Structural Concrete. *J. Clean. Prod.* **2021**, *314*, 128230. [CrossRef]
41. Khatib, Z.K.; Bayomy, F.M. Rubberized Portland Cement Concrete. *J. Mater. Civ. Eng.* **1999**, *11*, 206–213. [CrossRef]
42. Nur, O.F.; Albarqi, K.; Melinda, A.P.; Al Jauhari, Z. The Effect of Waste Tyre Rubber on Mechanical Properties of Normal Concrete and Fly Ash Concrete. *GEOMATE J.* **2021**, *20*, 55–61.
43. *EN, B.S. 12350-2*; Testing Fresh Concrete, Part 2: Slump-Test. European Committee for Standardization: Brussels, Belgium, 2009.
44. Hossain, F.M.Z.; Shahjalal, M.; Islam, K.; Tiznobaik, M.; Alam, M.S. Mechanical Properties of Recycled Aggregate Concrete Containing Crumb Rubber and Polypropylene Fiber. *Constr. Build. Mater.* **2019**, *225*, 983–996. [CrossRef]
45. Záleská, M.; Pavlík, Z.; Čítek, D.; Jankovský, O.; Pavlíková, M. Eco-Friendly Concrete with Scrap-Tyre-Rubber-Based Aggregate—Properties and Thermal Stability. *Constr. Build. Mater.* **2019**, *225*, 709–722. [CrossRef]
46. Pham, T.M.; Elchalakani, M.; Hao, H.; Lai, J.; Ameduri, S.; Tran, T.M. Durability Characteristics of Lightweight Rubberized Concrete. *Constr. Build. Mater.* **2019**, *224*, 584–599. [CrossRef]
47. Topçu, İ.B.; Unverdi, A. Scrap Tires/Crumb Rubber. In *Waste and Supplementary Cementitious Materials in Concrete*; Elsevier: Amsterdam, The Netherlands, 2018; pp. 51–77.
48. Su, H.; Yang, J.; Ling, T.-C.; Ghataora, G.S.; Dirar, S. Properties of Concrete Prepared with Waste Tyre Rubber Particles of Uniform and Varying Sizes. *J. Clean. Prod.* **2015**, *91*, 288–296. [CrossRef]
49. Grinys, A.; Balamurugan, M.; Augonis, A.; Ivanauskas, E. Mechanical Properties and Durability of Rubberized and Glass Powder Modified Rubberized Concrete for Whitetopping Structures. *Materials* **2021**, *14*, 2321. [CrossRef]
50. Gesoğlu, M.; Güneyisi, E.; Khoshnaw, G.; İpek, S. Investigating Properties of Pervious Concretes Containing Waste Tire Rubbers. *Constr. Build. Mater.* **2014**, *63*, 206–213. [CrossRef]
51. Gisbert, A.N.; Borrell, J.M.G.; García, F.P.; Sanchis, E.J.; Amorós, J.E.C.; Alcaraz, J.S.; Vicente, F.S. Analysis Behaviour of Static and Dynamic Properties of Ethylene-Propylene-Diene-Methylene Crumb Rubber Mortar. *Constr. Build. Mater.* **2014**, *50*, 671–682. [CrossRef]
52. Gesoğlu, M.; Güneyisi, E.; Khoshnaw, G.; İpek, S. Abrasion and Freezing–Thawing Resistance of Pervious Concretes Containing Waste Rubbers. *Constr. Build. Mater.* **2014**, *73*, 19–24. [CrossRef]
53. Pelisser, F.; Zavarise, N.; Longo, T.A.; Bernardin, A.M. Concrete Made with Recycled Tire Rubber: Effect of Alkaline Activation and Silica Fume Addition. *J. Clean. Prod.* **2011**, *19*, 757–763. [CrossRef]
54. Pacheco-Torgal, F.; Ding, Y.; Jalali, S. Properties and Durability of Concrete Containing Polymeric Wastes (Tyre Rubber and Polyethylene Terephthalate Bottles): An Overview. *Constr. Build. Mater.* **2012**, *30*, 714–724. [CrossRef]
55. Topçu, I.B.; Uygunoglu, T. Sustainability of Waste Rubber in Construction. In *Sustainability of Construction Materials*, 2nd ed.; Elsevier: Amsterdam, The Netherlands, 2016; pp. 599–626.
56. Kardos, A.J.; Durham, S.A. Strength, Durability, and Environmental Properties of Concrete Utilizing Recycled Tire Particles for Pavement Applications. *Constr. Build. Mater.* **2015**, *98*, 832–845. [CrossRef]
57. Richardson, A.; Coventry, K.; Edmondson, V.; Dias, E. Crumb Rubber Used in Concrete to Provide Freeze–Thaw Protection (Optimal Particle Size). *J. Clean. Prod.* **2016**, *112*, 599–606. [CrossRef]
58. Li, G.; Garrick, G.; Eggers, J.; Abadie, C.; Stubblefield, M.A.; Pang, S.-S. Waste Tire Fiber Modified Concrete. *Compos. Part B Eng.* **2004**, *35*, 305–312. [CrossRef]
59. Benazzouk, A.; Douzane, O.; Mezreb, K.; Laidoudi, B.; Quéneudec, M. Thermal Conductivity of Cement Composites Containing Rubber Waste Particles: Experimental Study and Modelling. *Constr. Build. Mater.* **2008**, *22*, 573–579. [CrossRef]

60. Al-Akhras, N.M.; Smadi, M.M. Properties of Tire Rubber Ash Mortar. *Cem. Concr. Compos.* **2004**, *26*, 821–826. [CrossRef]
61. Issa, C.A.; Salem, G. Utilization of Recycled Crumb Rubber as Fine Aggregates in Concrete Mix Design. *Constr. Build. Mater.* **2013**, *42*, 48–52. [CrossRef]
62. Guelmine, L.; Hadjab, H.; Benazzouk, A. Effect of Elevated Temperatures on Physical and Mechanical Properties of Recycled Rubber Mortar. *Constr. Build. Mater.* **2016**, *126*, 77–85. [CrossRef]
63. Toutanji, H.A. The Use of Rubber Tire Particles in Concrete to Replace Mineral Aggregates. *Cem. Concr. Compos.* **1996**, *18*, 135–139. [CrossRef]
64. Khorrami, M.; Vafai, A.; Khalilitabas, A.A.; Desai, C.S.; Ardakani, M.H. Experimental Lnvestigation on Mechanical Characteristics and Environmental Effects on Rubber Concrete. *Int. J. Concr. Struct. Mater.* **2010**, *4*, 17–23. [CrossRef]
65. Topçu, İ.B.; Demir, A. Durability of Rubberized Mortar and Concrete. *J. Mater. Civ. Eng.* **2007**, *19*, 173–178. [CrossRef]
66. Onuaguluchi, O.; Panesar, D.K. Hardened Properties of Concrete Mixtures Containing Pre-Coated Crumb Rubber and Silica Fume. *J. Clean. Prod.* **2014**, *82*, 125–131. [CrossRef]
67. Ganjian, E.; Khorami, M.; Maghsoudi, A.A. Scrap-Tyre-Rubber Replacement for Aggregate and Filler in Concrete. *Constr. Build. Mater.* **2009**, *23*, 1828–1836. [CrossRef]
68. Segre, N.; Joekes, I. Use of Tire Rubber Particles as Addition to Cement Paste. *Cem. Concr. Res.* **2000**, *30*, 1421–1425. [CrossRef]
69. Liu, F.; Chen, G.; Li, L.; Guo, Y. Study of Impact Performance of Rubber Reinforced Concrete. *Constr. Build. Mater.* **2012**, *36*, 604–616. [CrossRef]
70. Deshpande, N.; Kulkarni, S.S.; Pawar, T.; Gunde, V. Experimental Investigation on Strength Characteristics of Concrete Using Tyre Rubber as Aggregates in Concrete. *Int. J. Appl. Eng. Res. Dev* **2014**, *4*, 97–108.
71. Ling, T.-C.; Nor, H.M.; Hainin, M.R. Properties of Crumb Rubber Concrete Paving Blocks with SBR Latex. *Road Mater. Pavement Des.* **2009**, *10*, 213–222. [CrossRef]
72. Eldin, N.N.; Senouci, A.B. Measurement and Prediction of the Strength of Rubberized Concrete. *Cem. Concr. Compos.* **1994**, *16*, 287–298. [CrossRef]
73. Albano, C.; Camacho, N.; Reyes, J.; Feliu, J.L.; Hernández, M. Influence of Scrap Rubber Addition to Portland I Concrete Composites: Destructive and Non-Destructive Testing. *Compos. Struct.* **2005**, *71*, 439–446. [CrossRef]
74. Gesoglu, M.; Güneyisi, E.; Hansu, O.; İpek, S.; Asaad, D.S. Influence of Waste Rubber Utilization on the Fracture and Steel–Concrete Bond Strength Properties of Concrete. *Constr. Build. Mater.* **2015**, *101*, 1113–1121. [CrossRef]
75. Sofi, A. Effect of Waste Tyre Rubber on Mechanical and Durability Properties of Concrete–A Review. *Ain Shams Eng. J.* **2018**, *9*, 2691–2700. [CrossRef]
76. Najim, K.B.; Hall, M.R. Mechanical and Dynamic Properties of Self-Compacting Crumb Rubber Modified Concrete. *Constr. Build. Mater.* **2012**, *27*, 521–530. [CrossRef]
77. Valente, M.; Sibai, A. Rubber/Crete: Mechanical Properties of Scrap to Reuse Tire-Derived Rubber in Concrete: A Review. *J. Appl. Biomater. Funct. Mater.* **2019**, *17*, 2280800019835486. [CrossRef]
78. Fakhri, M. The Effect of Waste Rubber Particles and Silica Fume on the Mechanical Properties of Roller Compacted Concrete Pavement. *J. Clean. Prod.* **2016**, *129*, 521–530. [CrossRef]
79. Aly, A.M.; El-Feky, M.S.; Kohail, M.; Nasr, E.-S.A.R. Performance of Geopolymer Concrete Containing Recycled Rubber. *Constr. Build. Mater.* **2019**, *207*, 136–144. [CrossRef]
80. Alam, I.; Mahmood, U.A.; Khattak, N. Use of Rubber as Aggregate in Concrete: A Review. *Int. J. Adv. Struct. Geotech. Eng.* **2015**, *4*, 92–96.
81. Grinys, A.; Sivilevičius, H.; Daukšys, M. Tyre Rubber Additive Effect on Concrete Mixture Strength. *J. Civ. Eng. Manag.* **2012**, *18*, 393–401. [CrossRef]
82. Najimi, M.; Farahani, F.M.; Pourkhorshidi, A.R. Effects of Polypropylene Fibers on Physical and Mechanical Properties of Concretes. In Proceedings of the Third International Conference on Concrete and Development, Tehran, Iran, 27 April 2009; pp. 1073–1081.
83. Senouci, E.A.B. Rubber-Tire Particles as Concrete Aggregates. *ASCE J. Mater. Civ. Eng.* **1993**, *5*, 478–496.
84. Farhan, A.H.; Dawson, A.R.; Thom, N.H. Characterization of Rubberized Cement Bound Aggregate Mixtures Using Indirect Tensile Testing and Fractal Analysis. *Constr. Build. Mater.* **2016**, *105*, 94–102. [CrossRef]
85. Nawy, E.G. *Concrete Construction Engineering Handbook*; CRC Press: Boca Raton, FL, USA, 2008; ISBN 0429127243.
86. Deepak, W.S.; Naidu, G.T. Effect on Compressive Strength of Concrete Using Sea Sand as a Partial Replacement for Fine Aggregate. *Int. J. Res. Eng. Technol.* **2015**, *4*, 180–183.
87. Eldin, N.N.; Senouci, A.B. Observations on Rubberized Concrete Behavior. *Cem. Concr. Aggregates* **1993**, *15*, 74–84. [CrossRef]
88. Khorami, M.; Ganjian, E.; Vafaii, A. Mechanical Properties of Concrete with Waste Tire Rubbers as Coarse Aggregates. In Proceedings of the Special Sections on International conference on Sustainable Construction Materials and Technologies, Coventry, UK, 11–13 June 2007; pp. 85–90.
89. Thiruppathi, R. Discarded Tyre Rubber as Concrete Aggregate: A Possible Outlet for Used Tyres. In Proceedings of the 2013 International Conference on Current Trends in Engineering and Technology (ICCTET), Coimbatore, India, 3 July 2013; pp. 202–207.
90. Tarry, S.R. Effect of Partial Replacement of Coarse Aggregates in Concrete by Untreated and Treated Tyre Rubber Aggregates. *Int. J. Adv. Sci. Res.* **2018**, *3*, 65–69.

91. KIST, B.; Kigali, R. A Review on Construction Technologies That Enables Environmental Protection: Rubberized Concrete. *Am. J. Eng. Appl. Sci* **2008**, *1*, 40–44.

92. Kaloush, K.E.; Way, G.B.; Zhu, H. Properties of Crumb Rubber Concrete. *Transp. Res. Rec.* **2005**, *1914*, 8–14. [CrossRef]

93. Hora, M.; Reiterman, P. Assessment of the Air-Entraining Effect of Rubber Powder and Its Influence on the Frost Resistance of Concrete. *Rev. Rom. Mater. J. Mater.* **2016**, *46*, 327–333.

94. Yilmaz, A.; Degirmenci, N. Possibility of Using Waste Tire Rubber and Fly Ash with Portland Cement as Construction Materials. *Waste Manag.* **2009**, *29*, 1541–1546. [CrossRef] [PubMed]

95. Ganesan, N.; Raj, J.B.; Shashikala, A.P. Flexural Fatigue Behavior of Self Compacting Rubberized Concrete. *Constr. Build. Mater.* **2013**, *44*, 7–14. [CrossRef]

96. Meddah, A.; Beddar, M.; Bali, A. Use of Shredded Rubber Tire Aggregates for Roller Compacted Concrete Pavement. *J. Clean. Prod.* **2014**, *72*, 187–192. [CrossRef]

97. da Silva, F.M.; Barbosa, L.A.G.; Lintz, R.C.C.; Jacintho, A.E.P.G.A. Investigation on the Properties of Concrete Tactile Paving Blocks Made with Recycled Tire Rubber. *Constr. Build. Mater.* **2015**, *91*, 71–79. [CrossRef]

98. Raffoul, S.; Garcia, R.; Pilakoutas, K.; Guadagnini, M.; Medina, N.F. Optimisation of Rubberised Concrete with High Rubber Content: An Experimental Investigation. *Constr. Build. Mater.* **2016**, *124*, 391–404. [CrossRef]

99. Roychand, R.; Gravina, R.J.; Zhuge, Y.; Ma, X.; Mills, J.E.; Youssf, O. Practical Rubber Pre-Treatment Approch for Concrete Use—an Experimental Study. *J. Compos. Sci.* **2021**, *5*, 143. [CrossRef]

100. Jiang, Z.; Zhang, X.; Zhang, Y.; Yu, L.; Hu, X.; Zhou, X.; Zhang, Y. Surface Treatment of Rubberized Waste Reinforced Concrete. *Front. Built Environ.* **2021**, 57. [CrossRef]

101. Youssf, O.; ElGawady, M.A.; Mills, J.E.; Ma, X. An Experimental Investigation of Crumb Rubber Concrete Confined by Fibre Reinforced Polymer Tubes. *Constr. Build. Mater.* **2014**, *53*, 522–532. [CrossRef]

102. Guo, S.; Dai, Q.; Si, R.; Sun, X.; Lu, C. Evaluation of Properties and Performance of Rubber-Modified Concrete for Recycling of Waste Scrap Tire. *J. Clean. Prod.* **2017**, *148*, 681–689. [CrossRef]

103. Colom, X.; Canavate, J.; Carrillo, F.; Velasco, J.I.; Pagès, P.; Mujal, R.; Nogués, F. Structural and Mechanical Studies on Modified Reused Tyres Composites. *Eur. Polym. J.* **2006**, *42*, 2369–2378. [CrossRef]

104. Stewart, A.; Schlosser, B.; Douglas, E.P. Surface Modification of Cured Cement Pastes by Silane Coupling Agents. *ACS Appl. Mater. Interfaces* **2013**, *5*, 1218–1225. [CrossRef] [PubMed]

105. Huang, B.; Shu, X.; Cao, J. A Two-Staged Surface Treatment to Improve Properties of Rubber Modified Cement Composites. *Constr. Build. Mater.* **2013**, *40*, 270–274. [CrossRef]

106. Kashani, A.; Ngo, T.D.; Hemachandra, P.; Hajimohammadi, A. Effects of Surface Treatments of Recycled Tyre Crumb on Cement-Rubber Bonding in Concrete Composite Foam. *Constr. Build. Mater.* **2018**, *171*, 467–473. [CrossRef]

107. Ahmad, J.; Zaid, O.; Siddique, M.S.; Aslam, F.; Alabduljabbar, H.; Khedher, K.M. Mechanical and Durability Characteristics of Sustainable Coconut Fibers Reinforced Concrete with Incorporation of Marble Powder. *Mater. Res. Express* **2021**, *8*, 075505. [CrossRef]

108. Jiang, L.; Li, C.; Zhu, C.; Song, Z.; Chu, H. The Effect of Tensile Fatigue on Chloride Ion Diffusion in Concrete. *Constr. Build. Mater.* **2017**, *151*, 119–126. [CrossRef]

109. Su, D.; Pang, J.; Huang, X. Experimental Study on the Influence of Rubber Content on Chloride Salt Corrosion Resistance Performance of Concrete. *Materials* **2021**, *14*, 4706. [CrossRef]

110. Oikonomou, N.; Mavridou, S. Improvement of Chloride Ion Penetration Resistance in Cement Mortars Modified with Rubber from Worn Automobile Tires. *Cem. Concr. Compos.* **2009**, *31*, 403–407. [CrossRef]

111. Bravo, M.; de Brito, J. Concrete Made with Used Tyre Aggregate: Durability-Related Performance. *J. Clean. Prod.* **2012**, *25*, 42–50. [CrossRef]

112. Liang, J.; Zhu, H.; Chen, L.; Han, X.; Guo, Q.; Gao, Y.; Liu, C. Rebar Corrosion Investigation in Rubber Aggregate Concrete via the Chloride Electro-Accelerated Test. *Materials* **2019**, *12*, 862. [CrossRef] [PubMed]

113. Zhu, H.; Liang, J.; Xu, J.; Bo, M.; Li, J.; Tang, B. Research on Anti-Chloride Ion Penetration Property of Crumb Rubber Concrete at Different Ambient Temperatures. *Constr. Build. Mater.* **2018**, *189*, 42–53. [CrossRef]

114. Han, Q.; Wang, N.; Zhang, J.; Yu, J.; Hou, D.; Dong, B. Experimental and Computational Study on Chloride Ion Transport and Corrosion Inhibition Mechanism of Rubber Concrete. *Constr. Build. Mater.* **2021**, *268*, 121105. [CrossRef]

115. Gesoğlu, M.; Güneyisi, E. Strength Development and Chloride Penetration in Rubberized Concretes with and without Silica Fume. *Mater. Struct.* **2007**, *40*, 953–964. [CrossRef]

116. Torii, K.; Kawamura, M. Pore Structure and Chloride Ion Permeability of Mortars Containing Silica Fume. *Cem. Concr. Compos.* **1994**, *16*, 279–286. [CrossRef]

117. Gupta, T.; Siddique, S.; Sharma, R.K.; Chaudhary, S. Effect of Elevated Temperature and Cooling Regimes on Mechanical and Durability Properties of Concrete Containing Waste Rubber Fiber. *Constr. Build. Mater.* **2017**, *137*, 35–45. [CrossRef]

118. Ma, Q.; Guo, R.; Zhao, Z.; Lin, Z.; He, K. Mechanical Properties of Concrete at High Temperature—A Review. *Constr. Build. Mater.* **2015**, *93*, 371–383. [CrossRef]

119. Liu, S.; Wang, L.; Gao, Y.; Yu, B.; Bai, Y. Comparing Study on Hydration Properties of Various Cementitious Systems. *J. Therm. Anal. Calorim.* **2014**, *118*, 1483–1492. [CrossRef]

120. Medine, M.; Trouzine, H.; De Aguiar, J.B.; Asroun, A. Durability Properties of Five Years Aged Lightweight Concretes Containing Rubber Aggregates. *Period. Polytech. Civ. Eng.* **2018**, *62*, 386–397. [CrossRef]
121. Ocholi, A.; Ejeh, S.P.; Yinka, S.M. An Investigation into the Thermal Performance of Rubber-Concrete. *Acad. J. Interdiscip. Stud.* **2014**, *3*, 29.
122. Collier, N.C. Transition and Decomposition Temperatures of Cement Phases–a Collection of Thermal Analysis Data. *Ceramics-Silikaty* **2016**, *60*, 338–343. [CrossRef]
123. Si, R.; Guo, S.; Dai, Q. Durability Performance of Rubberized Mortar and Concrete with NaOH-Solution Treated Rubber Particles. *Constr. Build. Mater.* **2017**, *153*, 496–505. [CrossRef]
124. Tomosawa, F.; Noguchi, T.; Tamura, M. The Way Concrete Recycling Should Be. *J. Adv. Concr. Technol.* **2005**, *3*, 3–16. [CrossRef]

Review

Concrete Made with Partially Substitutions of Copper Slag (CPS): State of the Art Review

Jawad Ahmad [1,*], Ali Majdi [2], Ahmed Farouk Deifalla [3,*], Haytham F. Isleem [4] and Cut Rahmawati [5]

1 Department of Civil Engineering, Military College of Engineering, Risalpur, Sub Campus of Natioanl University of Sciences and Technology, Islamabad 44000, Pakistan

2 Department of Building and Construction Technologies and Engineering, Al-Mustaqbal University College, Hillah 51001, Iraq; alimajdi@mustaqbal-college.edu.iq

3 Structural Engineering Department, Faculty of Engineering and Technology, Future University in Egypt, New Cairo 11845, Egypt

4 Department of Construction Management, Qujing Normal University, Qujing 655011, China; hathamisleem@mail.qjnu.edu.cn

5 Department of Civil Engineering, Universitas Abulyatama, Aceh Besar 23372, Indonesia; cutrahmawati@abulyatama.ac.id

* Correspondence: jawadcivil13@scetwah.edu.pk (J.A.); ahmed.deifalla@fue.edu.eg (A.F.D.)

Abstract: Copper slag (CPS) is a large amount of waste material produced during the manufacture of copper. The disposal of this waste material becomes a problem for environmental concerns. Therefore, it is necessary to explore feasible alternate disposal options. They may also be utilized in concrete manufacturing to cut down on the usage of cement and natural aggregates. A lot of researchers focus on utilizing CPS in concrete, either as a cement replacement or as a filler material. This article aims to summarize the literature already carried out on CPS in conventional concrete to identify the influence of CPS on the fresh, hardened and durability performance of cement concrete. Results indicate that CPS improved the strength and durability performance of concrete but simultaneously decreased the slump value of concrete. Furthermore, an increase in the durability performance of concrete was also observed with CPS. However, the higher dose results declined in mechanical and durability aspects owing to a scarcity of flowability. Therefore, it is suggested to use the optimum dose of CPS. However, a different researcher recommends a different optimum dose ranging from 50 to 60% by weight of fine aggregate depending on the source of CPS. The review also recommends future researcher guidelines on CPS in concrete.

Keywords: copper slag; mechanical strength; flowability; chemical composition and microstructure analysis

Citation: Ahmad, J.; Majdi, A.; Deifalla, A.F.; Isleem, H.F.; Rahmawati, C. Concrete Made with Partially Substitutions of Copper Slag (CPS): State of the Art Review. *Materials* **2022**, *15*, 5196. https://doi.org/10.3390/ma15155196

Academic Editors: Stefano Guarino and Flaviana Tagliaferri

Received: 9 June 2022
Accepted: 3 July 2022
Published: 27 July 2022

Publisher's Note: MDPI stays neutral with regard to jurisdictional claims in published maps and institutional affiliations.

1. Introduction

A widely utilized raw material in construction is concrete, which is the basis for all construction and development initiatives around the globe, serving as the base for all buildings and infrastructure [1–3]. The environmental impact of concrete's primary ingredients changes depending on the kind of concrete and the amount of cement applied. Because concrete is utilized in such huge amounts across the globe, it raises several questions about its long-term viability [4]. An increase in the amount of riverbed sand and gravel, which are used as concrete components, is raising significant worry among environmentalists. The increased removal of natural sand from riverbeds has come from the extensive usage of concrete, which has occurred because of the rapid urbanization and industrialization of the world's population. Enhancement of riverbed distance, a decrease in the water table, revelation of bridge substructures, the most significant influence on rivers, deltas, coastal and marine ecologies, land loss as a result of the river or coastal erosion and a reduction in the quantity of deposit sources are just a few of the negative consequences of sedimentation [5]. Moreover, owing to limits on removal of sand from rivers, the construction industry's viability has been seriously threatened, resulting in a significant increase in sand charges [6,7]. Fine aggregates in concrete may be made from a variety of industrial wastes [8–10].

The term "sustainable building" refers to management that is accountable for providing a favorable environment that considers ecological and resource development [9–13]. Concrete is rapidly becoming a crucial building materials across the globe, due to its low superior performance. However, manufacturing cement has an impact on ecological systems [14–16]. Cement production, which is a major constituent in concrete, is a considerable source of greenhouse gas discharges CO_2 [17–19]. Currently, the globe generates about 3.6 billion metric tons of material every year [20]. The amount is predicted to reach more than 5 billion metric tons by 2030 [17,21]. Despite the fact that each country's situation varies, over half of the world's ordinary Portland cement (OPC) generates 11 billion metric tons of concrete each year, with the balance being utilized in projects [22]. To minimize CO_2 emissions, waste materials should be used instead of cement in concrete.

The industrial sector has seen significant expansion, resulting in a vast number of by-products whose dumping has come to be a serious problem, since it impacts the ecosystem's [23–25]. The use of such relevant by-products in the building sector, particularly in concrete manufacturing, will help to reduce environmental stress. Several studies have already demonstrated that using industrial waste such as fly ash [25], rice husk ash [26], bagasse ash [27], silica fume [28], blast furnace slag [29], copper slag [30], waste glass [31] and waste marble [32], etc., were considered to be advantageous. Similar copper slags are also valuable options to be used as concrete ingredients.

Copper slag (CPS) is a metallurgical waste product that is created through the matte smelting of copper metal. CPS is an industrial waste substance that is formed as a by-product of the copper production process. It is a smooth and glassy by-product of the matte smelting and refining phases involved in the pyrometallurgical removal of copper. Copper is extracted and purified from copper oxide ores using aqueous (water-based) solutions at room temperature, often in three steps: heap leaching, solvent extraction and electrowinning.

1. Heap leaching is the method of extracting metals from chemical solutions by allowing them to percolate. Low-grade ore that would otherwise not be economically sent through a milling process is often utilized in heap leaching. The crushed ore is placed into a heap on top of an impermeable layer, on a little slope, after mining, shipping and crushing to a constant gravel or golf ball size. The copper from the ore is dispersed in the leaching agent (diluted sulfuric acid), which is sprayed via sprinklers on top of the heap pile and allowed to flow down into the heap. A small pool is used to collect the copper sulfate and sulfuric acid "pregnant" leach solution that results. Currently, concentrations of the copper complex range from 60 to 70 percent.

2. The second stage is solvent extraction, which involves stirring and allowing two immiscible (non-mixing) liquids to separate, causing the copper to transfer from one liquid to the other. A solvent is aggressively combined with the pregnant leach solution. The copper migrates into the solvent from the leach solution. The two liquids are then allowed to separate depending on solubility, with the contaminants staying in the leach solution while copper remain in solution in the solvent. The remaining leach solution is then recycled by adding more acid and returning it to the heap leaching sprinklers.

3. The last stage is an electrolysis process known as electrowinning. An inert anode (positive electrode) and the copper solution from the prior phase, which functions as an electrolyte, are both contacted by an electrical current. Next, 99.99 percent pure copper is deposited onto a cathode (negative electrode) as positively charged copper ions (referred to as cations) emerge from solution. The manufacture process of copper slag in the industry is displayed in Figure 1. CPS is often a dark black color, as seen in Figure 2.

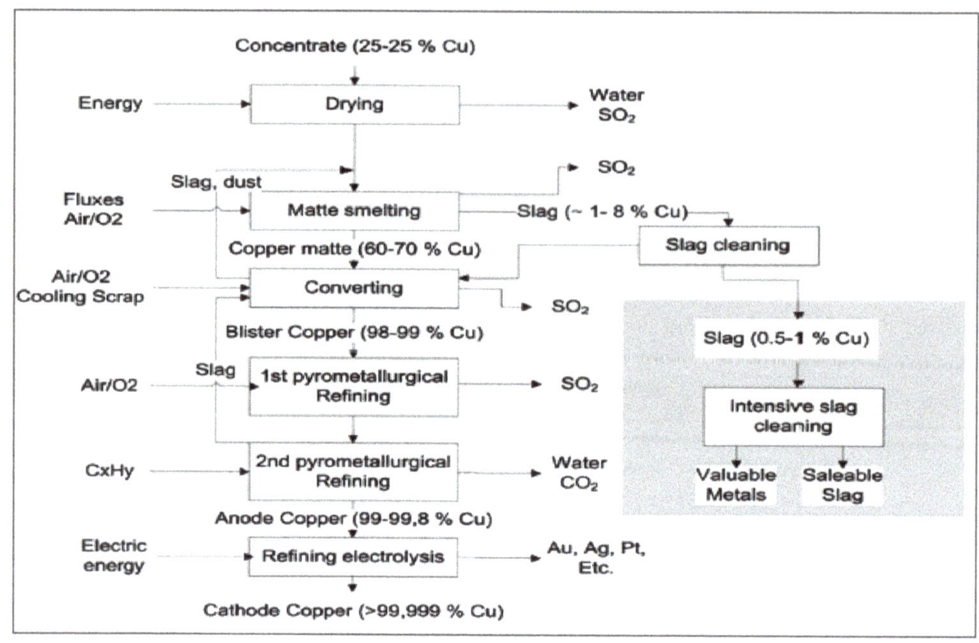

Figure 1. Flow Chart of CPS Extraction [33].

Figure 2. Cooling of CPS in air [33,34].

It has been calculated that the copper industry generates about 24.6 million tons of slag per year across the globe. Even though CPS is frequently employed in the sandblasting business and the manufacture of abrasive tools, the rest is placed without any further usage or recovery possible. CPS contains mechanical and chemical properties that allow it to be utilized in concrete as a substitute for cement or sand, depending on the application. For example, copper slag possesses a variety of mechanical features that make it a desirable choice for aggregate application, including great soundness characteristics, outstanding abrasion resistance and excellent stability [35]. Because it takes 2.2 million tons of CPS to generate each year [35], environmental protection organizations and governments are concerned about the usage and disposal of this CPS waste.

Following the escalation of the issue, various investigations have shown a wide range of potential reuse and recycling options for this particular material. Among the potential options for metal recovery from slags containing significant levels of metallic elements are different techniques such as electric arc furnace melting, leaching and flotation (as well as other methods). However, since the metals present in copper slag are often found in trace levels, their recovery may not be economically feasible in most cases. Instead, various applications for copper slag were investigated and eventually accepted [35]. An alternate conceivable use for this material might be as an aggregate and perhaps as a cement substitute in the manufacturing of concrete, based on its physical and chemical qualities.

Additionally, copper slag has pozzolanic capabilities due to its low CaO level, as well as the presence of other oxides such as Al_2O_3, SiO_2, and Fe_2O_3. Pozzolans are classified as siliceous or siliceous and aluminous minerals that have no or little cementitious properties; however, when finely ground, they chemically react with calcium hydroxide (CH) in the presence of water to produce compounds with cementitious characteristics (calcium silicate hydrates). It is possible to reduce disposal costs by using CPS in concrete as binder or filler. This may also assist with conserving the environment by lowering the amount of waste produced. Because of the massive amount of CPS being generated, environmental pollution has an adverse effect on the growth of the nation. The effective and environmentally friendly use of CPS, as well as the promotion of green building, are significant concerns of this assessment.

Although some studies have discovered that copper slag possesses pozzolanic activity, the activity is quite low, limiting the use of copper slag as a mineral additive in concrete [36,37].

The silica modulus (ratio of the actual amount of lime in raw meal/clinker to the theoretical lime required by the major oxides (SiO_2, Al_2O_3 and Fe_2O_3) in the clinker) of the activator increases the hydration degree of quick-cooled copper slag and the polymerization of amorphous hydration products. In quickly cooled copper slag, instead of high Fe particles, high Si particles provide primary activities. Alkali-activated quick cooled copper slag mortars exhibit reasonably high compressive strengths, particularly when the silica modulus is greater than one, indicating that alkali-activated copper slag for building may be feasible in the future [38]. As the silica modulus decreases, the intensity of the reaction rises. However, silica modulus reductions have a detrimental effect on the mechanical characteristics of the resulting concrete. The ideal silica modulus requirement is determined by the raw sample's mineralogical makeup. The silica modulus of the used alkaline solution determines the rate of partial dissolution of the semi-crystalline and crystalline mineral phases found in natural pozzolan [39].

The dissolution of soluble components in raw materials, the accumulation of soluble components and the creation of oligomers, and the polymerization of oligomers and the precipitation of hydration products are the three steps of the early hydration process of alkali-activated material [40]. Some active components in copper slag, such as Si, Al and Ca, are thought to dilute when exposed to high OH- concentrations. The concentration of ions then rises, and oligomers form. The hydration products then precipitate, linking the unreacted copper slag particles [41]. A study [42] concluded that the microstructure is densified by adding CPS up to 60% because of its pozzolanic action.

2. Physical and Chemical Compositions of CPS

It is possible to determine the applicability and ability of using industrial wastes in concrete based on their physical properties, which include specific gravity, absorption coefficients, grain size, fineness modulus, moisture content, bulk density specific surface area and unit weight. According to previous investigations, the physical parameters of CPS are listed in Table 1. CPS has specific gravity ranging from 2.4 to 3.5, which is somewhat higher than the specific gravity of aggregate (2.4 to 3.5). CPS has an absorption capability of 0.36 percent according to published data, as shown in Table 1. Because the absorption capacity of CPS is lower than that of fine aggregate, the flowability of concrete will increase. It is composed mostly of particles with uniform, angular shapes, with the majority of the particles measuring between 4.75 and 0.075 mm in size. [43]. However, a study suggested that the particle size of copper slag should be below 10 mm [44]. Copper slag has a density ranging between 3.16 and 3.87 kg/m^3, which fluctuates depending on the quantity of iron included in it.

Table 1. Physical Aspects of CPS.

Authors	Manjunatha et al. [45]	Jabri et al. [46]	Mavroulidou et al. [47]	Raju et al. [30]	Maharishi et al. [33]
Specific gravity	3.51	2.4	-	3.52	3.30
Water absorption (%)	0.36	-	0.11	-	0.36
Fineness modulus	3.11	-	2.97	3.68	3.18
Moisture content (%)	-	-	-	-	-
Density (kg/m^3)	-	-	3.73	-	-
Specific surface area, (m^2/kg)	-	-	-	-	-
Initial setting (min)	-	250	-	-	-
Fineness (cm^2/g)	-	1261	-	-	-

The scan electronic microscopy (SEM) of CPS is shown in Figure 3. The particles have an uneven morphology, rough and irregular in shape, as may be expected. Furthermore, it is obvious that CPS has a reasonably smooth surface, which is responsible for the greater workability of new concrete using CPS as a partial fine aggregate when compared to a mix made with 100 percent natural sand [48].

Figure 3. SEM of CPS [49].

Four to five percent alumina, four to six percent calcium oxide, thirty-five percent to thirty-seven percent iron, thirty percent to thirty-four percent silica, and one percent copper are the primary elements of CPS [50]. Other slag aggregates, such as electric furnace ferronickel slags, have a chemical composition that is predominantly composed of SiO$_2$, MgO and Fe$_2$O$_3$ as the primary constituents [51]. Iron slag is mostly composed of the elements SiO$_2$, Al$_2$O$_3$, CaO and MgO, which account for 95 percent of the total composition.

It also contains manganese, iron and sulfur compounds, as well as tiny quantities of numerous other elements [52]. According to previous research, the chemical makeup of CPS is shown in Table 2.

Table 2. Chemical compounds of CPS.

Authors	Raju et al. [30]	Najimi et al. [53]	Jabri et al. [54]	Chithra et al. [55]	Rajasekar et al. [48]
SiO_2	25.84	9.57	33.05	25.84	27
Al_2O_3	0.22	4.43	2.79	0.22	3.0
Fe_2O_3	68.29	57.42	53.45	68.29	0.60
MgO	-	1.56	1.56	-	4.0
CaO	0.15	22.5	6.06	0.15	63
Na_2O	0.58	1.47	0.28	0.58	-
K_2O	0.23	-	0.61	0.28	1.3

X-ray diffraction (XRD) analysis revealed that the mineralogical components included in this slag are pyroxene ($CaZnSi_2O_6$), fayalite $S(iO_4Fe_2)$, anorthite ($CaAl_2Si_2O_8$), quartz (SiO_2) and magnetite (Fe_3O_4) [53]. Figure 4 shows the results of the XRD analysis (Fe_3O_4). The amorphous nature of the SiO_2 found in CPS has a significant impact on concrete, from the initial hydration to the ultimate strength [56]. According to ASTM [57], it is possible to employ pozzolanic materials that have accumulated more than 70% of a chemical (silica dioxide, calcium oxide, aluminum oxide, magnesium oxide, sodium oxide and iron oxide). From the Table 2, CPS has a greater than 70% accumulation of the mentioned chemical, making it suitable for use as a cementitious material.

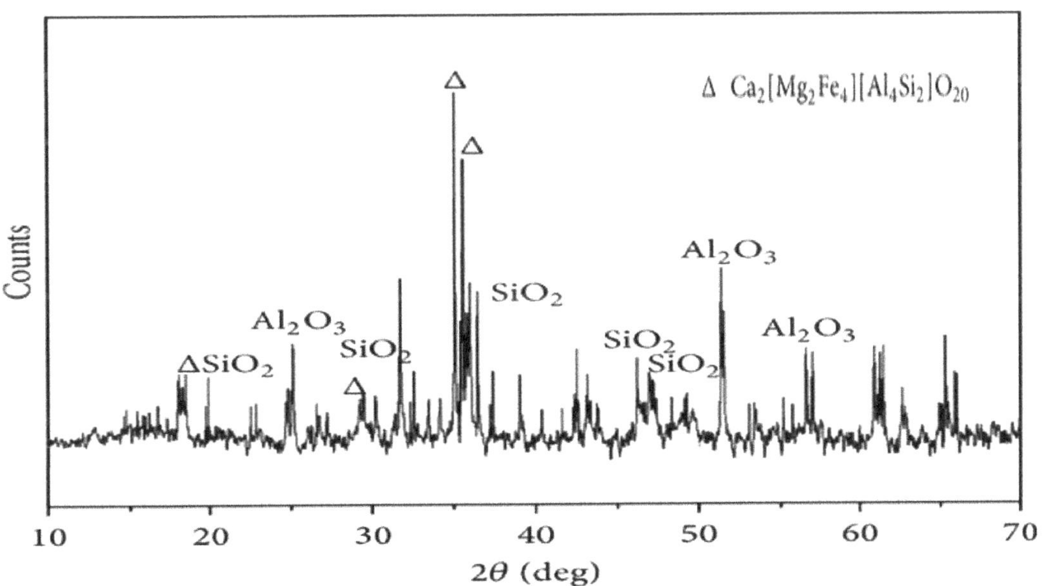

Figure 4. XRD test results of CPS [58]: Used as per Elsevier permission.

3. Fresh Properties

3.1. Workability of Concrete

Figure 5 shows the outcomes of the flowability of concrete for each CPS proportion ratio for each of the two distinct waters to binder ratios (w/c). Concrete made with w/c 0.55 possess high flowability (100–175 mm) to have a maximum slump value of 175 mm, and therefore higher workable concrete. However, the slumps of copper slag mixes were less than control concrete for a 0.45 w/c and high percentages of copper slag are substituted. In such a case, concrete has very little workability, which could be unsatisfactory for several useful purposes. Furthermore, lower workability also adversely affects strength properties.

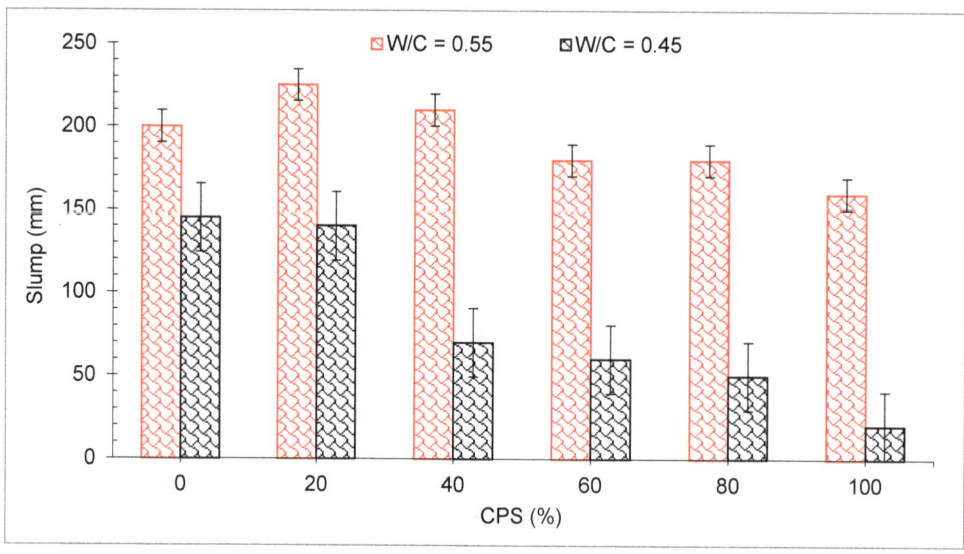

Figure 5. Workability of concrete [47].

It was hypothesized that the CPS would experience more slump collapse (more workable concrete) when compared with natural aggregate due to lower water absorption. However, the findings proved that the supposition is wrong. Slumps were probably influenced to some degree due to the angular shape of CPS, which enhances the friction between concrete components and ultimately reduced the flowability of concrete. In addition, a researcher discovered that when sand was substituted with CPS, the workability of the concrete decreased, contrary to what was expected [59]. Workable mixtures may be created for all sand substitute ratios provided the water is properly managed. CPS particles have a glassy and smooth surface, which might be a contributing factor to the enhanced slump value of concrete mixtures [49]. Furthermore, the SEM picture (Figure 3) of CS revealed that CPS is composed mostly of spherical particles which allow for more efficient movement of ingredients. There was also a modest delay in the setting time mixtures when CPS was added to the mix, as shown in Figure 6, which might be related to the existence of heavy metals in the CPS that delayed the hydration of the cement during the setting process [60].

According to one investigation, the recorded slump was 150 mm when CPS was utilized to substitute 100 percent of the copper. According to the authors, the low water absorption features of CPS and its glassy surface when compared to aggregate caused in a noteworthy increase in the flowability of concrete due to a significant rise in the amount of free water remaining after the absorption and hydration processes were completed. The improvement in flowability of concrete may be made with the same amount of sand substituted, and these mixes may have better flowability, as well as higher strength and durability than standard high-performance concrete (HPC) mixes. Furthermore, it was highlighted that mixes containing significant levels of CPS exhibit bleeding and segregation which might have a harmful influence on the strength of concrete [61].

3.2. Setting Times of Mortar

Figure 6 depicts the start and final setting times for the cement that will be blended with different percentages of CPS. It can be noted that the initial and final setting time of 10 percent substation of CPS is much less than the cement. It is well known that CPS reacting passively with cement hydrates results in the creation of calcium silicate hydrates CSH gel and calcium sulfate, and this reaction helps to construct the structure of cement pastes during the hydration process and hence shortening the setting time of the cement. As the percentage of CPS replacement is increased to 20 percent and 30 percent, reduced setting times are observed when compared with the percentages of 0 and 10 percent of cement paste mixed with CPS, respectively, because of the larger need for water with a higher concentration of CS, which results in a denser structure and hence, a shorter setting time. In addition, according to one research study, slag will aid in the setting of cement paste by lowering the induction time [62]. With the smaller particles and greater slag dose, the initial setting time was reduced, and the induction time was shortened. However, because of the reduction in the particle size of CPS, the research discovered a much greater delay in the setting time [63]. The delay in setting time with CPS substitution may be due to pozzolanic reaction, as the pozzolanic reaction continues gradually, as associated with hydration of cement.

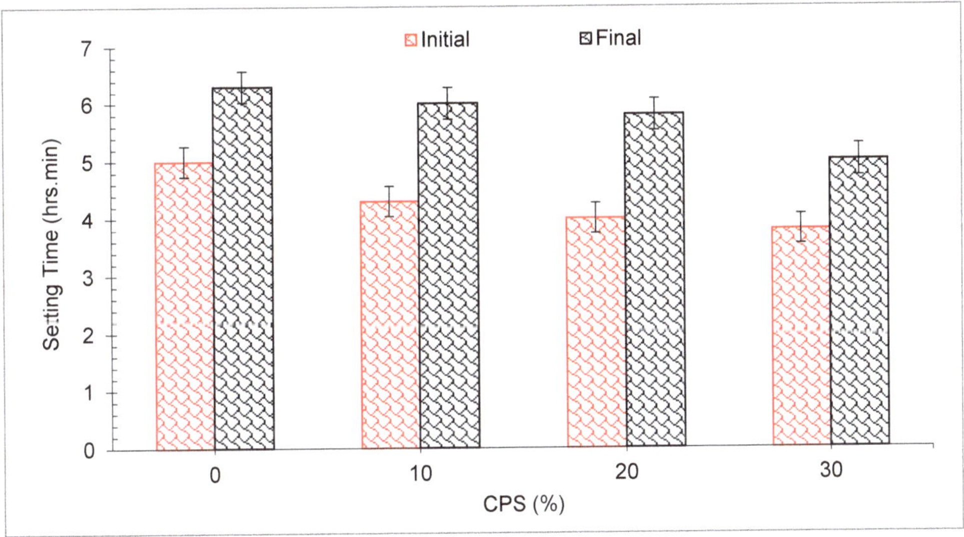

Figure 6. Setting time of mortar with CPS [64].

4. Mechanical Strength

4.1. Compressive Strength (CMS) of Concrete

The compressive strength (CMS) outcomes at each curing interval are shown in Figure 7, with variable percentages of CPS substation instead of natural sand. It is seen that as the proportion of CPS replacement ratio by weight of river sand rises, in an increase in CMPS is seen up to 60% substitute of aggregate with CPS. However, a minor drop in CMS is seen for the mixes containing 80 and 100 percent CPS, respectively, when compared with the blank mix (control mix). The maximum CMS at 7, 14, 28 and 90 days curing was 11.64 percent, 16.60 percent, 6.89 percent, and 9.66 percent in comparison with blank concrete at 7, 14, 28, and 90 days, respectively. The use of small particles of CPS in concrete improves the packing density and makes it less permeable. Additionally, CPS particles have angular edges which aid in the enhancement of matrix cohesion by increasing the surface area of the particle [54]. Additionally, according to research, the use of sharp-edged waste copper slag particles increases the bonding of concrete with its other component elements. As a result, improvement in strength is seen until the mix contains 60 percent CPS. Through compressive strength tests, it has been shown that CPS may be partially substituted with fine aggregate up to 60% for M40 grade concrete without impacting the strength qualities of concrete [65]. The CMS of the mortar improves gradually when the curing period is increased with the addition of mineral additives. When slag is used in place of cement, the concentration of the hydration component $Ca(OH)_2$ decreases, while small particles of mineral mixture fill the spaces between cement particles, making the cement mortar denser and strengthening the interfacial area [66]. According to the findings of the study, the CMS of concrete is equivalent to the control concrete up to 40 percent substitution of CPS [33]. Under sulfate exposure circumstances, again, mass is found in CPS concrete, causing in a reduction in CMS [67]. The majority of the studies have observed a continuous rise in CMS, with an enhancement in CPS content up to a 50 percent proportion of sand [61,67]. However, the optimum dose of CPS varies due to different sources of CPS. Table 3 displays the summary of CMS with partial substitution of CPS in concrete, as per past studies.

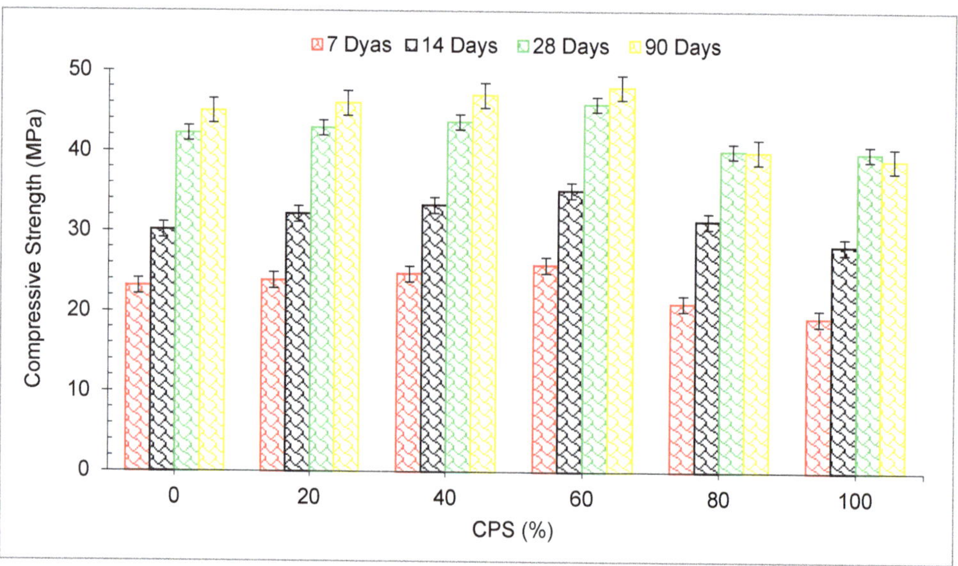

Figure 7. Compressive strength of concrete [45].

Figure 8 displays the strength age relation of CMS with different proportions of CPS at various periods of curing. CMS (28 days) of the blank mix (control) is chosen as benchmark strength (reference strength). From Figure 8, the maximum CMS is achieved at 60% substitution of CPS. CMS at 7 days and 14 days is 39% and 17% lower than reference compressive strength at 60% substitution of CPS. However, CMS of concrete at 28 and 90 days is 9% and 14% greater than the reference CMS at 60% substation of CPS. It can be concluded that the CPS up to 60% can be used in concrete without any harmful effect on the compressive strength.

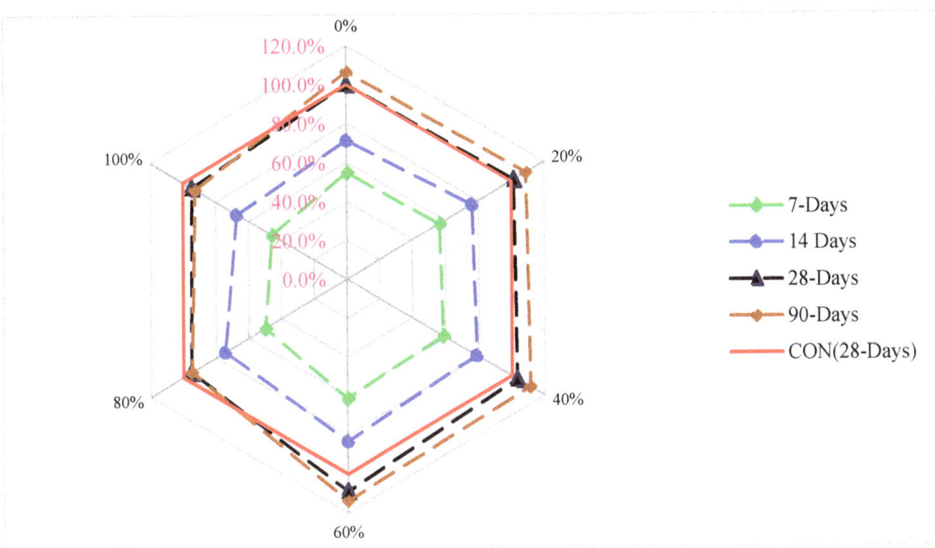

Figure 8. Relative compressive strength of concrete: data source [45].

4.2. Split Tensile Strength (TS)

Figure 9 represents the split tensile strength (TS) test results of all mixes (0 to 100 CPS substitution) at each curing period, respectively. In the same way, as the partial replacement of CPS by river sand grows, the TS improves as the partial replacement of CPS by river sand increases, up to a maximum of 60 percent substitution of CPS instead of natural river sand in comparison to mixing. Following that, a minor drop in TS is seen for the mixes with 80 percent and 100 percent replacement of CPS, respectively as compared to the control mix. When compared to the control mix, the 60 percent substation of CPS achieved maximum TS of 11.75 percent, 11.05 percent, 4.69 percent and 6.53 percent at 7-, 14-, 28-, and 90-day curing periods, respectively, compared to the control concrete TS. The filling effects of tiny particles of discarded CPS are responsible for the improvement in TS. Furthermore, according to one research study, the compressive, tensile and flexural strengths of concrete were equivalent to those of the control mix when up to 50% CPS replacement by weight of sand was used, but the TS started to drop when the CPS concentration increased further, up to 80% [54,63,68,69].

Numerous studies have indicated that the CMS and TS of concrete specimens created using CPS as fine and coarse particles are higher than those of regular concrete [61]. As a consequence of using CPS aggregate rather than natural aggregate, the CMS of the concrete improved by about 10–15 percent after 28 days and the TS improved by approximately 10–18 percent after 28 days [61]. The addition of CPS enhanced the tensile property of the TS by up to 40 percent replacement. Following that, it gradually decreased, but did not fall below the maximum TS obtained with fly ash [70]. In most cases, the use of CPS as a partial replacement in concrete produced with 100 percent cement resulted in an increase in TS, except for concrete containing 100 percent CPS [49]. According to the findings of one investigation, concrete made with CPS as a partial substitute had greater TS up to 60%. A further increase in the usage of CPS over this replacement ratio resulted in a 7–10 percent decrease in the TS [71]. Furthermore, Table 3 displays the summary of TS of concrete with various percentages of CPS.

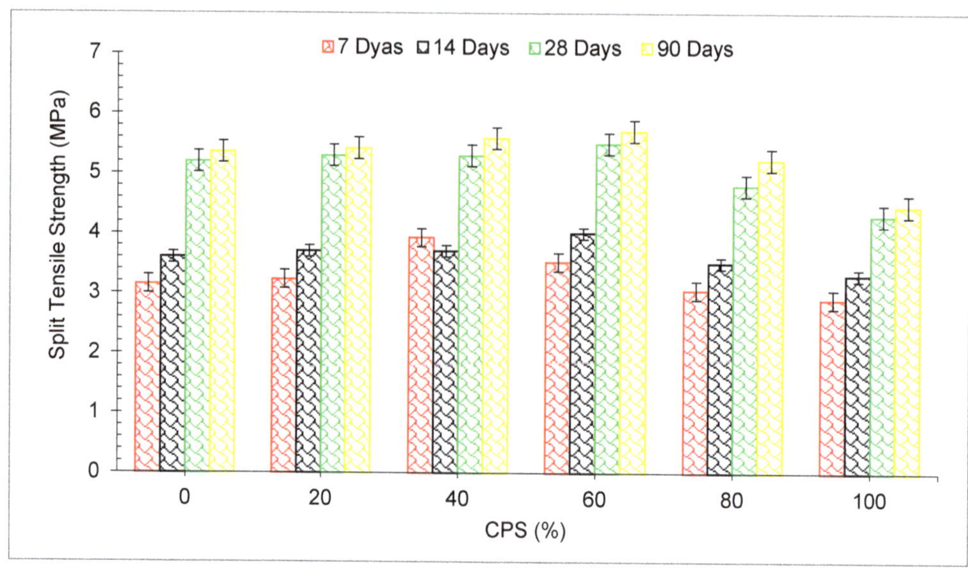

Figure 9. Split tensile strength of concrete [45].

Figure 10 shows the TS age relation with distinct percentages of CPS at various periods of curing. TS (28 days) of the blank mix (control) is taken as a reference mix. It can be noted that the maximum TS is achieved at 60% substitution of CPS instead of natural river sand. TS at 7 days and 14 days is 33% and 23% lower than reference TS at 60% replacement of CPS instead of natural river sand. However, TS of concrete at 28 and 90 days of curing is 6% and 10% more than the reference TS at 60% substation of CPS. However, Wang et al. reported a 40% optimum dose [44]. At a higher dose (80 and 100%), split tensile is lower than the reference TS, even at 90 days of curing.

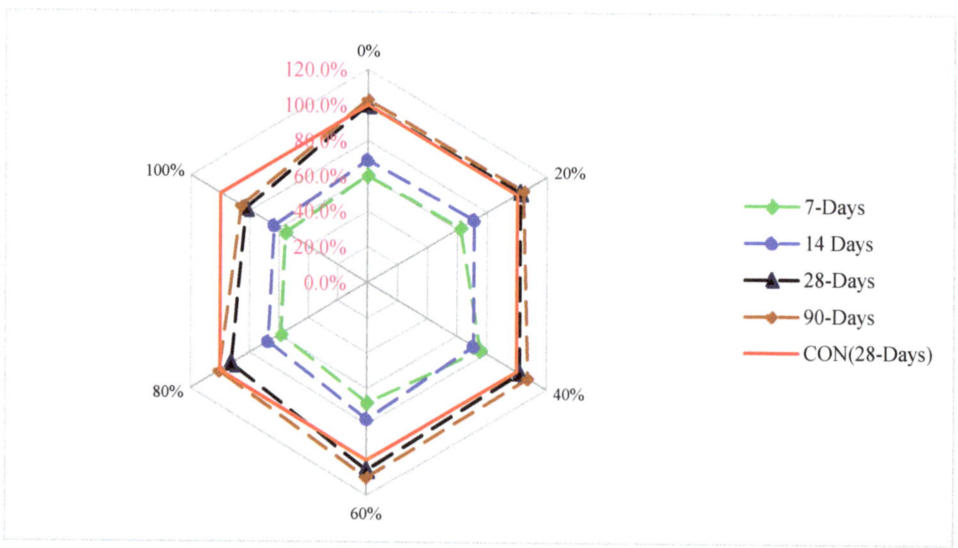

Figure 10. Relative split tensile strength of concrete: data source [45].

Figure 11 shows the correlation between compressive strength (CMS) and split tensile strength (TS) with various proportions of CPS at different days of curing. TS depends on the CMS of concrete. TS is about 10 to 50% of CMS of concrete. It can be noted that the trendline between CMS and TS seems to be straight. Therefore, a strong correlation exists between CMS and TS with an R^2 value greater than 90%.

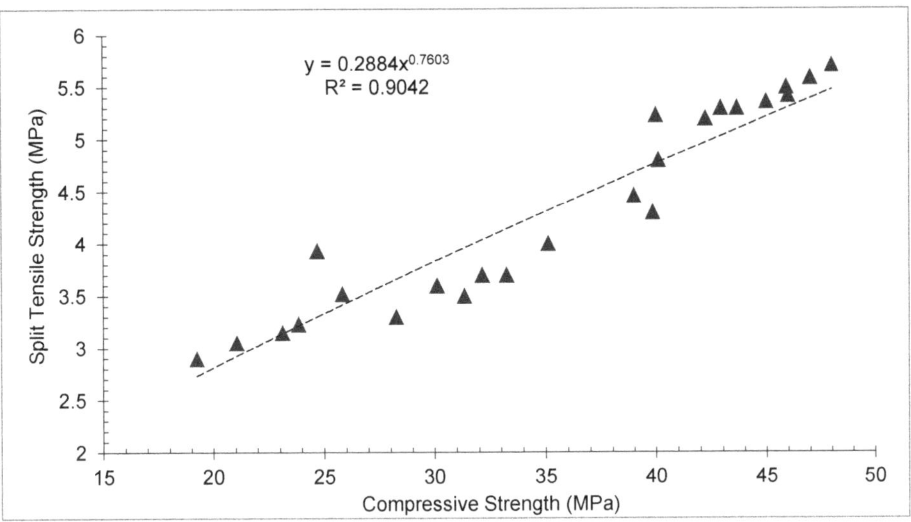

Figure 11. Correlation between CMS and TS: data source [45].

4.3. Flexural Strength (FS)

The flexural strength (FS) of concrete for all mixtures at each curing time is shown in Figure 12. It can be shown that as the percentage of CPS replaced by river sand rose, an increase in FS was noted up to a mix of 60% proportion of CPS when compared with a mixture of 0% CPS (control). The CMS and TS tests both indicated an increasing trend in progress over time. Further increases in the substitution ratio of CPS resulted in a minor reduction in FS for the mix, and 80 and 100% substation of CPS when compared with the control mix. Comparing the FS of the mix (60% substitution of CPS) to the control mix, it can be shown that the maximum FS of the mix (60% substitution of CPS) was 13.38 percent, 11.08 percent, 8.65 percent and 12.61 percent after 7, 14, 28 and 90 days of curing time, respectively. The use of tiny particles of CPS improves the interlocking ability of component materials, which results in an increase in FS and stiffness of concrete [72].

Analysis of FS of the mortar containing 10 percent CPS reveals that FS increased by 12.9 percent more than the control mixture. According to some theories, this improvement may be attributed to the increase in the compactness and durability matrix of the CPS mortar. Despite this, the findings of the 20 and 30 percent CPS tests indicate that the gain in FS is much more significant. The improvement in FS of the 20 percent CS mortar was about 18.2 percent higher than the FS of the control mortar. The concrete containing 30 percent CPS resulted in the greatest improvement in FS, which was 38.7 percent more than the control mortar. This substantial increase in FS was most likely due to the increased compaction and uniformity of the distribution of CPS in mortar mixes, which has resulted in a successful improvement in FS [73]. The FS of concrete was equivalent to that of the control mix when up to 50 percent CPS replacement for sand was used, but it declined when the CPS content of the concrete increased further [54]. The FS of concrete was improved when 40 percent CPS was used, but the FS reduced when the amount of CPS used exceeded 40 percent. The lowest FS of 6.16 MPa was achieved when 100 percent CPS was used, and this was achieved by adding 1 percent nano silica to the 100 percent copper slag mix.

The FS of concrete improved from 6.16 to 6.49 MPa throughout the testing process [74]. Furthermore, Table 3 displays the summary FS of concrete with different percentages of CPS.

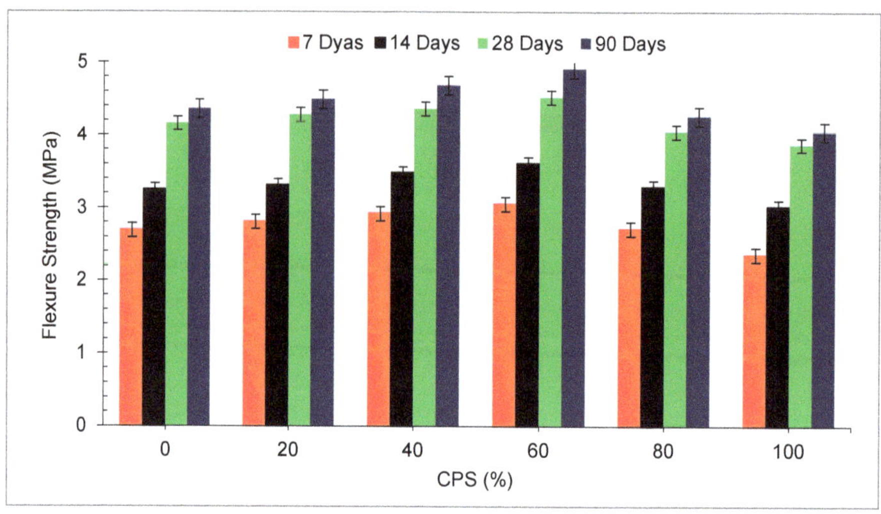

Figure 12. Flexural strength of concrete: data source [45].

Figure 13 reveals the relative flexural strength (FS) of concrete with various percentages of CPS at different periods of curing. FS (28 days) of the blank mix (control) is taken as the reference mix. It can be noted that maximum flexural strength is attained at 60% replacement of CPS instead of natural river sand. FS at 7 days and 14 days is 36% and 13% lower than reference FS at 60% substitution of CPS instead of natural river sand. However, the FS of concrete at 28 and 90 days is 9% and 18% higher than the reference FS at 60% substation of CPS. At a higher dose (80 and 100%), FS is lower than the reference tensile strength even at 90 days. Therefore, it is recommended that CPS is used up to 60% substation instead of the natural river without any negative effect on the flexural strength of concrete.

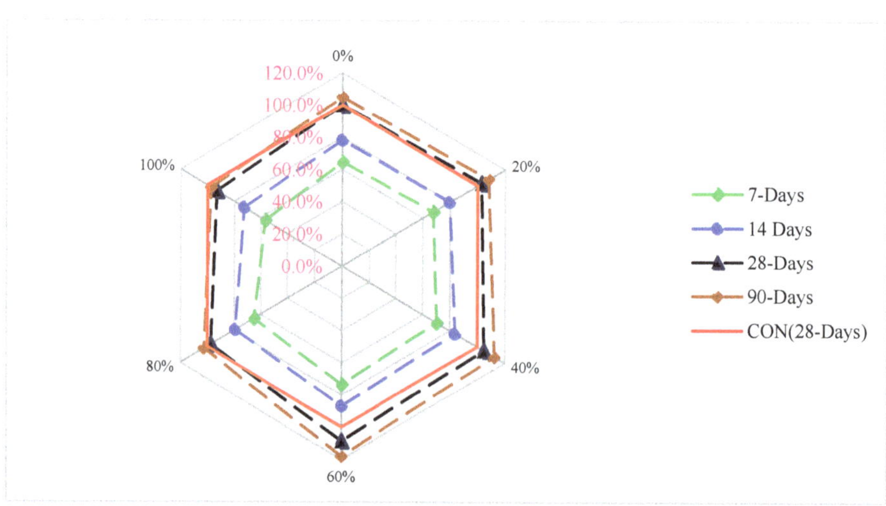

Figure 13. Relative flexural strength of concrete: data source [45].

Figure 14 reveals the correlation between the compressive and flexural capacity of concrete with distinct percentages of CPS at various days of curing. *FS* is almost 10% to 20% of *CMS* varying on the mix design of concrete. It can be noticed that the trendline among CMS and FS seems to be straight. Therefore, a strong correlation has occurred among CMS and FS with an R^2 value more than 90%.

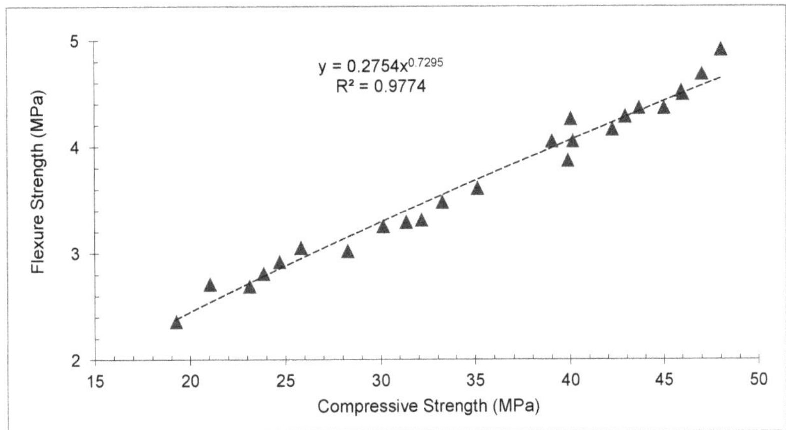

Figure 14. Correlation between compressive and flexural strength: data source [45].

Table 3. Summary of slump and mechanical strength of concrete with copper slag (CPS).

Reference	Percentage of (CPS)	Slump (mm)	Compression Strength (MPa)				Flexural Strength (MPa)		Split Tensile Strength (MPa)			
	W/C=0.55		7D		28D				7D		28D	
	0%	200	23.9		34.6				2.4		2.7	
	20%	225	29.0		38.6				2.5		2.8	
[47]	40%	210	25.7		33.2		-		2.3		2.7	
	60%	180	25.8		34.0				2.4		2.8	
	80%	180	24.5		32.8				2.3		2.6	
	100%	160	23.6		30.7				-		-	
			7D	28D	56D	90D	28D		28D			
	0%		30	44	43	45	7		3.5			
	20%		34	45	45	47	7		3.8			
[30]	40%	-	35	47	50	53	7.5		4			
	60%		33	45	47	50	7		4			
	80%		33	46	48	51	6.5		3.9			
	100%		34	46	46	50	6.5		3.9			
			7D	28D		90D	28D	90D	28D		90D	
	0%		29.7	40.0		44.0	3.28	3.35	3.74		4.05	
[53]	5%	-	27.5	37.5		43.1	3.09	3.30	3.72		4.02	
	10%		25.2	36.0		41.7	3.02	3.17	3.71		4.02	
	15%		23.5	35.2		39.5	2.98	3.12	3.67		3.98	
	0%CS	65	7D	28D	56D	90D	28D		28D			
	10%CPS+90%S	80	23.3	24.6	25.3	27	7.7		3			
	20%CPS+80%S	80	29	31	34.7	36	7.2		3.5			
[54]	40%CPS+60%S	110	30.6	39.8	40	42	6.5		3.8			
	50%CPS+50%S	130	30	42.7	44.5	50.3	7.3		4.1			
	60%CPS+40%S	165	28	39.2	42	47.8	6.3		3.6			
	80%CPS+20%S	190	26.8	35	40.1	44.8	7.2		3.6			
	100%CPS	200	23.3	26.1	32	35.5	5.9		3.4			

Table 3. *Cont.*

Reference [45]

Reference	Percentage of (CPS)	Slump (mm)	Compression Strength (MPa)				Flexural Strength (MPa)			Split Tensile Strength (MPa)		
			7D	14D	28D	90D	14D	28D	90D	14D	28D	90D
[45]	0%	65	23.12	30.12	42.21	45	3.25	4.16	4.36	3.6	5.3	5.36
	20%	70	23.85	32.15	42.90	46	3.31	4.28	4.49	3.7	5.3	5.42
	40%	72	24.69	33.26	43.65	47	3.48	4.36	4.68	3.7	5.3	5.59
	60%	75	25.81	35.12	45.92	48	3.61	4.52	4.91	4.0	5.5	5.71
	80%	82	21.05	31.34	40.12	40	3.29	4.05	4.26	3.5	4.8	5.23
	100%	80	19.25	28.26	39.86	39	3.02	3.87	4.05	3.3	4.3	4.69

Reference [61]

Reference	Percentage of (CPS)	Slump (mm)	Compression Strength (MPa)		Flexural Strength (MPa)	Split Tensile Strength (MPa)
			7D	28D	28D	28D
[61]	0%CS	28	76.9	93.9	14.6	5.4
	10%CPS+90%S	28	79.6	99.8	13	5.2
	20%CPS+80%S	50	74.5	95.3	12.4	6.2
	40%CPS+60%S	85	76.4	95.2	12.5	6.1
	50%CPS+50%S	115	77.8	96.8	12.9	6.1
	60%CPS+40%S	128	69.0	83.0	11.1	4.8
	80%CPS+20%S	143	63.8	83.6	10.3	4.7
	100%CPS	150	63.4	82.0	10.1	4.4

Reference [49]

Reference	Percentage of (CPS)	Slump (mm)	Compression Strength (MPa)		Flexural Strength (MPa)	Split Tensile Strength (MPa)	
			14D	28D		14D	28D
[49]	0%	52	32.1	35.7		3.77	3.9
	25%	57	37.1	38.5	-	4.3	4.36
	50%	63	38.2	39.9		4.37	4.43
	75%	68	31.8	34.1		4.24	4.29
	100%	74	28.8	30.0		4.13	4.18

Reference [74]

Reference	Percentage of (CPS)	Slump (mm)	Compression Strength (MPa)	Flexural Strength (MPa)	Split Tensile Strength (MPa)
			28D	28D	
[74]	0%		93	6.2	
	20%		97	6.5	
	40%	-	100	7.1	-
	60%		95	6.9	
	80%		91	6.4	
	100%		87	6.1	

Reference [75]

Reference	Percentage of (CPS)	Slump (mm)	Compression Strength (MPa)		Flexural Strength (MPa)		Split Tensile Strength (MPa)	
			7D	28D	7D	28D	7D	28D
[75]	0%		35	40	4.0	5.0	2.5	3.0
	5%	-	32	43	4.1	5.1	2.6	3.3
	10%		30	41	4.0	5.0	2.5	3.2
	15%		28	32	3.2	4.0	1.7	2.5

Reference [76]

Reference	Percentage of (CPS)	Slump (mm)	Compression Strength (MPa)		Flexural Strength (MPa)		Split Tensile Strength (MPa)	
			28D	56D	28D	56D	28D	56D
[76]	0%		45	49	3.6	4.0	3.6	4.0
	5%		50	55	3.5	4.5	3.5	4.5
	10%		47	51	3.3	4.4	3.3	4.4
	15%	-	45	49	3.1	4.2	3.1	4.2
	20%		40	47	3.0	4.0	3.0	4.0
	25%		38	44	2.9	3.9	2.9	3.9
	30%		36	42	2.6	3.6	2.6	3.6

Reference [77]

Reference	Percentage of (CPS)	Slump (mm)	Compression Strength (MPa)			Flexural Strength (MPa)	Split Tensile Strength (MPa)		
			7D	14D	28D		7D	14D	28D
[77]	0%		17.03	21.66	29.25		1.82	2.03	2.73
	10%		18.74	23.70	29.85		2.12	2.19	2.95
	20%		20.22	25.22	32.07		2.21	2.31	3.09
	30%	-	23.11	27.33	37.55	-	2.22	2.38	3.42
	40%		24.66	28.59	39.48		2.38	2.5	3.49
	50%		20.96	25.9	33.03		2.05	2.26	2.48
	60%		16.48	20.45	28.66		1.98	2.12	2.33

Table 3. *Cont.*

Reference	Percentage of (CPS)	Slump (mm)	Compression Strength (MPa)				Flexural Strength (MPa)		Split Tensile Strength (MPa)			
			28D	56D					28D	56D		
[65]	0%		30	31					10.66	11.72		
	20%		35	37					9.94	10.59		
	40%	-	36	38			-		10.81	10.53		
	60%		39	38					10.43	11.14		
	80%		42	42					11.07	11.70		
	100%		36	36					12.18	1257		
			7D	28D					7D	28D		
[33]	0%		24	33					2.8	3.3		
	20%		26	28					3.0	3.4		
	40%	-	31	37			-		3.3	3.5		
	60%		26	31					3.0	3.2		
	80%		25	28					2.9	3.1		
	100%		20	21					2.5	2.6		
			28D									
[78]	0%	29	29.19									
	10%	34	31.56									
	20%	43	34.59									
	30%	46	41.7									
	40%	51	38.74			-				-		
	50%	55	42.22									
	60%	57	34.81									
	70%	62	32.74									
	80%	66	31.7									
	90%	69	30.15									
	100%	78	30									
			7D	28D			7D	28D				
[64]	0%		40	45			8.0	9.0		-		
	10%	-	42	47			10.0	10.0				
	20%		43	49			11.0	10.0				
	30%		45	50			11.5	12.5				
			7D	28D			28D					
[79]	0%		42	62			3.5			-		
	20%		50	62			3.4					
	40%	-	52	70			3.5					
	60%		50	68			3.3					
	80%		40	60			3.2					
	100%		30	50			2					
			7D	14D	28D	56D	28D		7D	14D	28D	56D
[80]	0%		17.70	25.20	37.35	41.15	25.41		4.02	5.20	6.91	8.54
	15%		22.66	30.04	39.90	44.20	25.87		4.39	5.69	7.90	9.14
	30%	-	25.90	32.40	43.94	50.39	6.16		5.21	7.14	9.57	
	45%		22.04	29.90	39.79	45.84	5.77		4.38	6.64	8.13	10.1
	60%		18.13	26.83	35.14	42.65	5.19		4.20	5.77	7.91	9.86

D—days, CPS—copper slag, S—sand.

5. Durability

5.1. Water Absorption and Voids

The average water absorption of the mix with CPS substation is shown in Figure 15. It can be noted that, except for the 20 percent CPS mix having w/c = 0.55, higher water absorption was exhibited compared with blank concrete. The w/c = 0.45 mixes showed less water absorption when compared to the corresponding control mix, except for the two mixes with the greatest CPS proportions, i.e., 80 and 100% which had the maximum water absorption. However, all the mixes had low water absorption rates of less than three percent. This was within predicted limits for high durability in terms of liquid infiltration in concrete, which was observed by the researcher [81]. The rise in water absorption, which was detected mostly in the $w/c = 0.55$ mixes, was caused by an excess of water, i.e., CPS grains showed less water absorption than sand, which might have resulted in a larger porosity of the mixture [47]. According to the findings of the research, up to 40% replacement of sand by CPS results in a general decrease in surface water absorption, after which the water absorption swiftly rises as the amount of CPS increases. The research indicates that the replacement of CPS for 40 percent of the cement resulted in a decrease in surface water absorption [61]. Similar trends were found with the rate of copper slag substitution, which increased up to 40% while the rate of surface water absorption reduced [82]. Water absorption of concrete with filler materials reduced because of the micro filling of spaces in the concrete, which led to a more compact mass and hence reduced the water absorption. However, due to the lack of flowability of filler materials, a larger dosage of filler materials might result in increased water absorption [7].

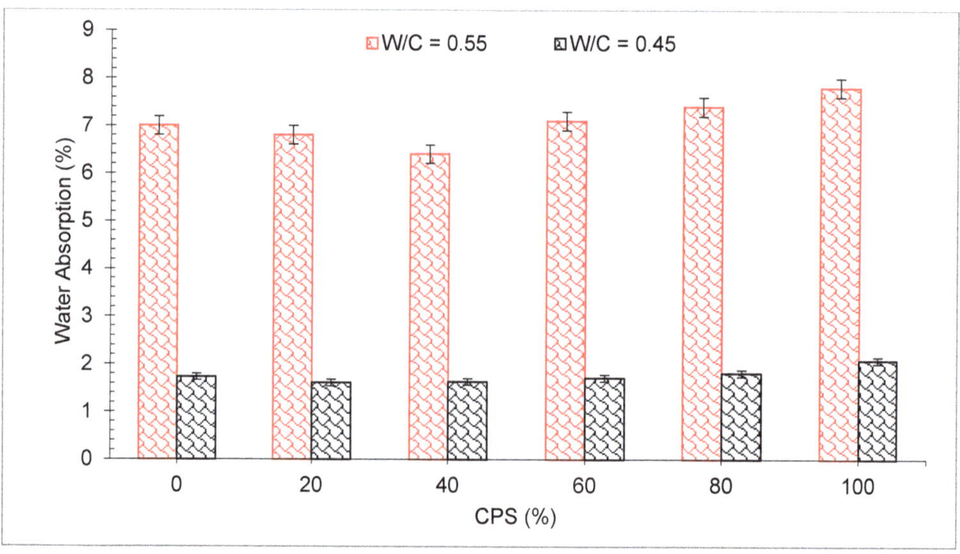

Figure 15. Water absorption of concrete: data source [47].

The percentage of voids in concrete mixes containing CPS is shown in Figure 16. It has been concluded that the voids contained in concrete mixtures follow a pattern similar to that of water absorption. The proportion of voids may decrease by 40% when fine aggregates are replaced with CPS. The percentage of voids in mix containing CPS 40% was determined to be 6.56 percent, which is the least void-containing of all the mixes. The presence of voids increased when CPS was substituted for more than 40% of the cement. Among all concrete mixes tested, the proportion of voids in CPS 100 percent concrete was the highest (7.82 percent). It was found that the 100% CPS mix had an even larger proportion of voids than the control mix. The settlement of CPS as a consequence of its heavy weight in contrast with natural causes water to rise to the surface, causing in cavities and a permeable microstructure on the surface [67]. The findings suggest that the substitution of 40% CPS as a partial alternative for natural sand will result in concrete that poses a comparable challenge to water absorption and cavities.

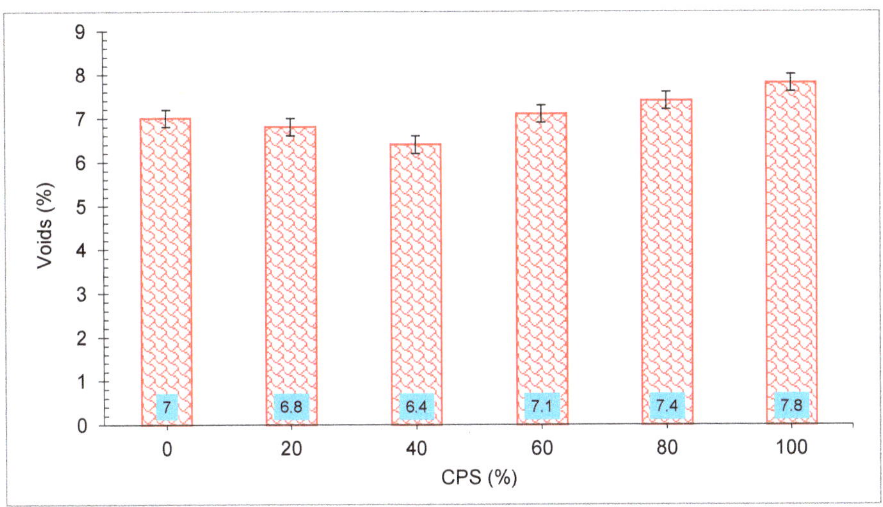

Figure 16. Voids in concrete: data source [33].

5.2. Accelerated Corrosion Testing Results

The results of the faster corrosion assessment in terms of mass failure of the entrenched reinforcement are depicted in Figure 17. Both control mixes (with w/c ratios of 0.45 and 0.55, respectively) showed more corrosion (as seen by the larger mass loss of the bars) than the concrete made with CPS.

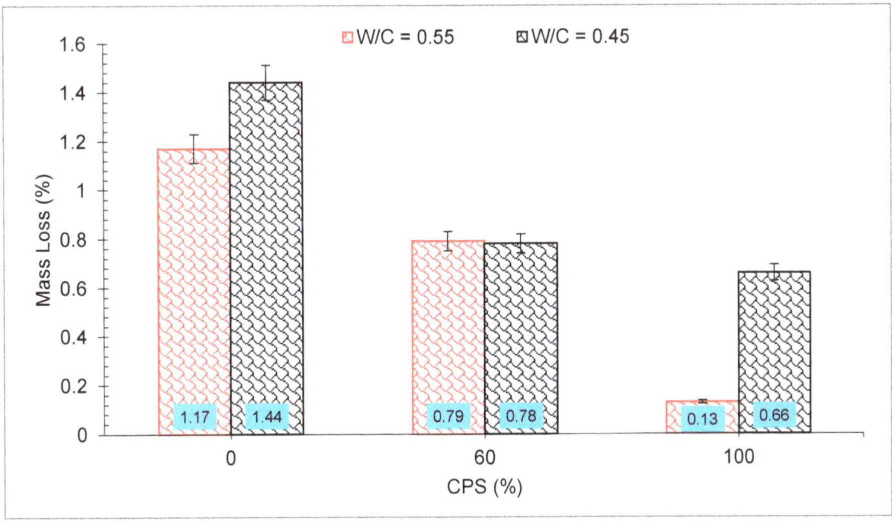

Figure 17. Corrosion test results: data source [47].

The 0.45 control mix had the greatest amount of corrosion, as well as several surface fractures near the reinforcing bar, which were the most severe. However, the $w/c = 0.55$ mix with 100 percent CPS demonstrated only a minor loss in mass compared to the other mixtures. Utilizing CPS as a substitute for sand, the researchers discovered that the corrosion resistance of the resultant mix was greater than that of mix made with natural sand. However, reverse results were examined by Brindha et al. [83], who demonstrated that for CPS utilized for up to 50% sand substitute concentrations, further raising CPS proportions resulted in a modest rise in the rust ratio when associated with blank concrete

5.3. Acid Resistance

The concrete including various amounts of CPS percentages was preserved in water for 28 days before being subjected to the sulfuric acid mixture for 56 days and examined for corrosion. The compressive capacity and mass of concrete sample were measured before and after contact with acid to determine the severity of the acid assault on the specimens. When exposed to a sulfuric acid solution for 56 days, the mass of concrete specimens varied, as shown in Figure 18. The results of the tests indicated that after 56 days of exposure to the sulfuric acid solution, all of the specimens lost weight. Weight loss could be minimized by up to 40% by increasing the percentage of CPS replacement. The CPS 40% mix had the smallest change in mass of all the mixes, with a change of just 5.63 percent. When comparing the drop in mass after 56 days of exposure for CPS 40% mix and the control mix of 0% CPS, it was found that the former was around four percent less. An excessive increase in the CPS fraction over 40% of sand has a negative effect on the ability to withstand the acid assault. There was a significant difference in mass between the 100% CPS concrete mix and the other mixtures tested. The change in mass for the 100% CPS concrete mix after 56 days of exposure was considerably larger than the change in the mass of the control mix.

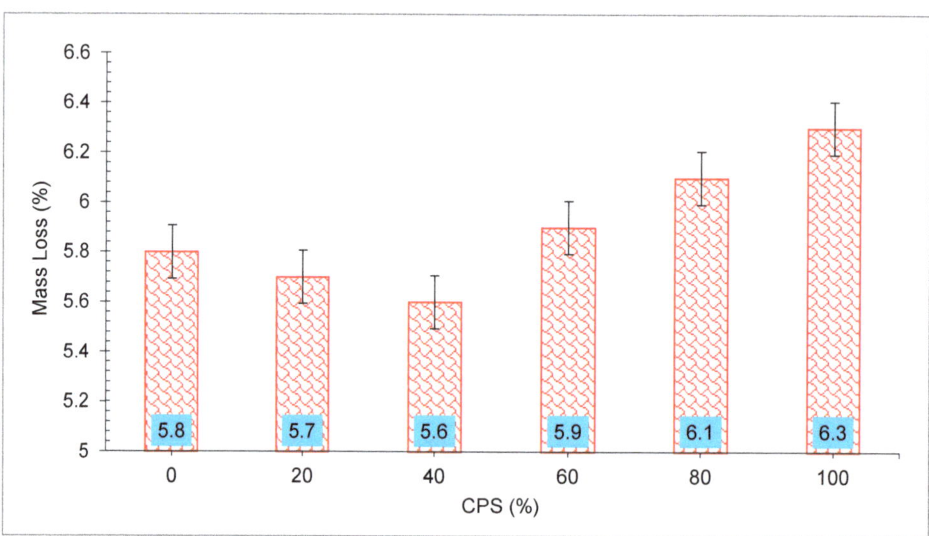

Figure 18. Acid resistance of concrete: data source [33].

5.4. Electrical Resistivity Results

The electrical resistance is defined as the voltage age ratio multiplied by the electrical current that runs across a sample (voltage to current ratio). It is also referred to as material resistance to electric current in certain circles. Understanding the strength of the flow of the electrical current is important, since it may aid in predicting the probability of reinforcing corrosion occurring. The fluctuation in electrical resistance with hardening time and the percentage of CPS is shown in Figure 19. The electrical resistance of CPS cement mortars rose substantially over the duration of time. Resistance to 30% CPS mortars was much greater than that of 20% CPS mortars. The resistance of the control mortar did not grow with time, especially over a considerable length of time, and it remained lower than the resistance of the CPS mortars. The electrolytic current that passes through the wetting cement mortar is responsible for the wetness. The electrical resistivity may be used to determine porosity and permeability in a more indirect manner [84]. Mortar resistivity assessments are affected by a variety of parameters, including the makeup of the binder phase, the constitution of the liquid phase and the connectivity of the pore system [85].

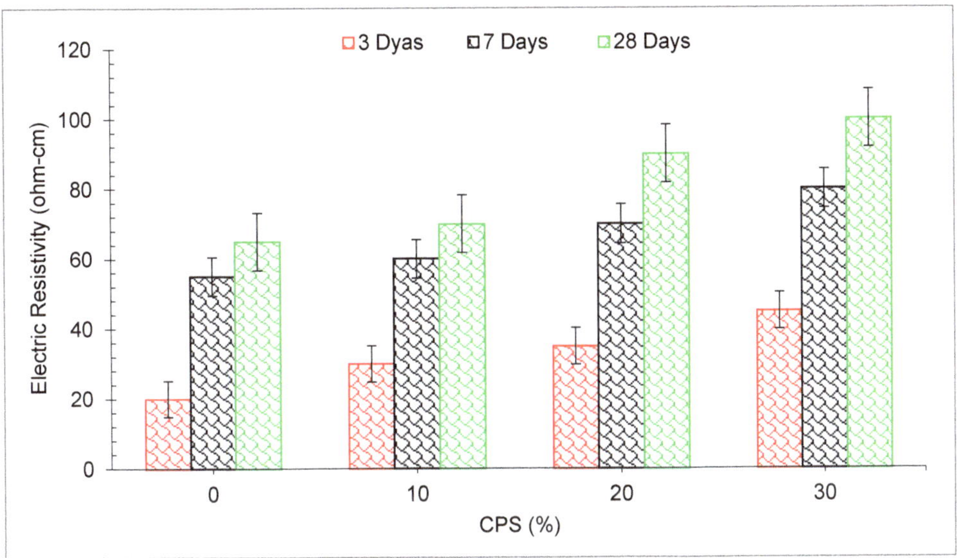

Figure 19. Electric resistivity of concrete: data source [64].

5.5. Sulfate Resistance

Extremely serious environmental deteriorations brought on by sulfate assault have an impact on the long-term durability of concrete buildings. Structures made of concrete, such as foundations, bridges, piers, concrete pipelines, etc., experience expansion and cracking as a result of sulfate assault, which worsens the condition. If the sulfate ions originate from sea water, ground water or soil, they will be present in the solution together with other ions including magnesium, sodium, calcium and potassium [86]. The findings demonstrate that adding CPS as an additional cementitious material to concrete increased its acidic resistance [87]. For improved resistance against concrete buildings vulnerable to sulfate assault, this research recommends using CPS as an alternative to fine aggregates with mineral admixtures [88].

In another investigation, iron slag was used as a replacement for fine aggregates in the SCC, and it was shown that specimens attacked by sulfate showed no mass loss despite developing white deposits after 91 days. After 28 days of sulfate exposure, it was discovered that each degree of iron slag replacement had a lower-than-5% effect on the compressive strength [89]. Sharma and Khan [67] stated that despite the improvement in the workability and compressive strength, the sulfate attack reduced as fine aggregate replaced up to 20% by CS. Najimi et al. [90] evaluated the durability of CS contained concrete exposed to sulfate attack. They claimed that use of CS as cement at 5%, 10% and 15%, caused deteriorative sulfate expansions decreased by 57.4%, 63.4% and 64.7%, respectively, compared with the control mixture. According to Gevaudan et al. [91] the addition of copper and cobalt to alkali-activated cement would lessen the rate of calcium sulfate generation and, as a consequence, would result in less permeability and corrosion when exposed to acidic environments. By forming a passivation and protective barrier against acid attack, copper and cobalt ions boost this cement's acid resistance and decrease deterioration.

6. Scanning Electron Microscopy (SEM)

Figure 20A,B depict the SEM of concrete with full replacement of sand with CPS. It can be noted that the complete replacement of sand with CPS causes additional water to become stagnant in the concrete, resulting in an increase in the number of cavities and vessel channels in the finished product. The creation of these voids and capillary channels has an influence on the interlocking connection among the cement and the aggregates, causing in a loss of capacity, while the durability of the concrete will be affected because of the poor connection of cement paste with aggregate. However, the performance of concrete with 100% CPS can be improved with the supplement of secondary cementitious or filler materials. Combining pozzolanic reaction and filling voids of mineral admixture improved the performance of concrete [92,93]. Research was carried out using nanosilica with 100 percent CPS as fine aggregate in concrete [74]. Nano silica particles have a filler effect which causes an extremely dense structure to form because of their presence. The addition of nanosilica to concrete helps to prevent segregation and bleeding while also improving the cohesiveness of the concrete. A little amount of nanosilica is added to the cementitious matrix to lower its viscosity, offset the detrimental effects of trapped air and reduce the permeability of the cured concrete. From Figure 20C, it is noted that nanosilica (as pozzolanic material [94]) reacts with calcium hydroxide, and the reaction starts the creation of the secondary C-SH gel. This secondary C-S-H gel fills all of the pores in the solid state, making the concrete compact and improving its load capacity attributes.

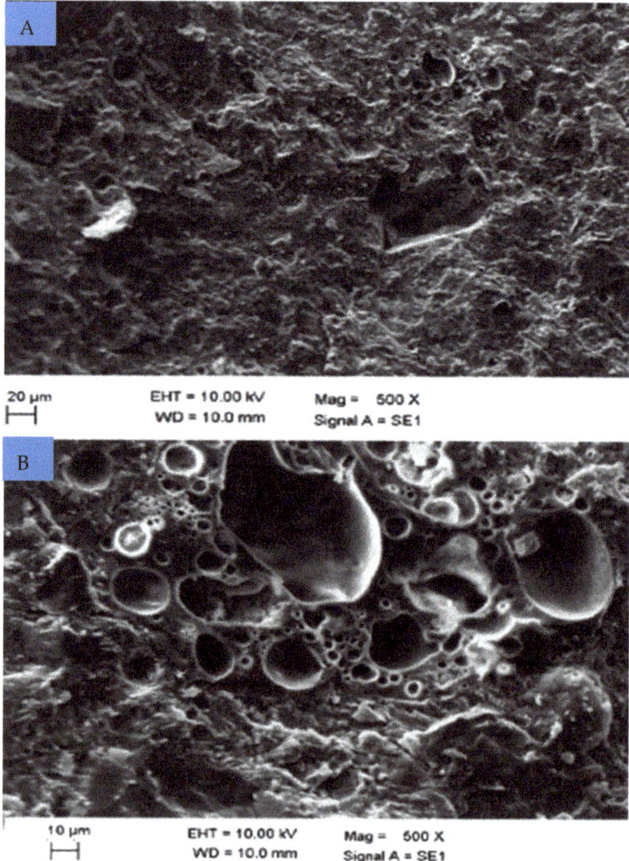

Figure 20. *Cont.*

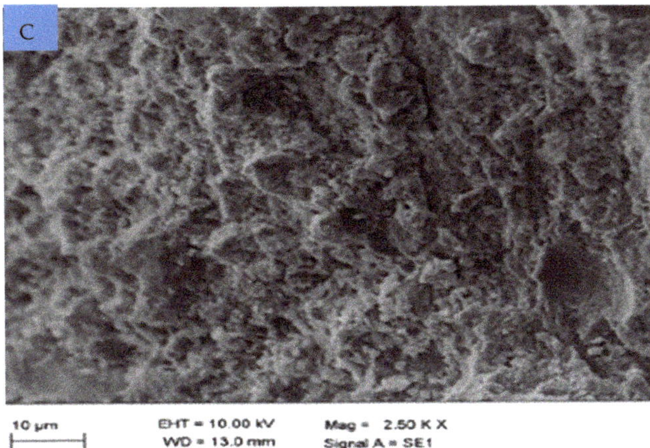

Figure 20. SEM Results, (**A**) 100% of CPS at 20 µm, (**B**) 100% of CPS at 10 µm and (**C**) 100% of CPS with 2% Nano Silica at 10 µm [74].

7. X-Ray Diffraction (XRD)

X-ray diffraction (XRD) patterns after 7, 28 and 90 days of curing and the blended cement pastes containing 30 percent copper slag are shown in Figure 4. To explore how the hydration products of blended cement vary with curing time, the spectra in the range of 2θ between 10 and 50 degrees are stacked. Similar groups of diffraction peaks may be seen in the mixed cements of various ages. It is obvious that the portlandite (CH), calcite ($CaCO_3$), larnite (C_2S), and ettringite are the major minerals in the paste samples [95].

Portlandite is produced when cement hydrates, and it is the main crystalline mineral in the pastes. A study also claimed that CH forms weak pockets which adversely affect the strength of concrete [93]. Furthermore, CH is chemically active, reacting other chemicals and resulting deterioration of the concrete structure [96]. The carbonation of CH during the setting and hardening of the pastes is to blame for the presence of calcite [97]. Due to its poor reactivity, C_2S is present as a non-hydrated cement component [98]. In addition, it is challenging to use XRD to identify calcium silicate hydrates (C-S-H) gel, which is often classified as an amorphous phase. C-S-H gel improved the binding properties of cement paste, which results in more mechanical strength durability.

It is generally known that the reactive element in pozzolanic materials, amorphous silica, may interact with CH to create more hydration products. The consumption of CH in cement pastes is often used to measure the intensity of pozzolanic reaction. In mixed cement pastes, the CH diffraction peaks are located at 18.02 and 34.05 degree. As indicated in Figure 21, the peak CH intensity at 18.02 has been compared.

The findings indicate that this peak intensity gradually and noticeably decreased throughout the course of the curing process, with a residual amount still present in the pastes after 90 days. This suggests that the pozzolanic reaction results from the inclusion of UGCS and would continue even after 90 days of cure time. In the part that follows, a quantitative study will be carried out while taking the impact of $CaCO_3$ into account to precisely quantify the level of pozzolanic reaction [97].

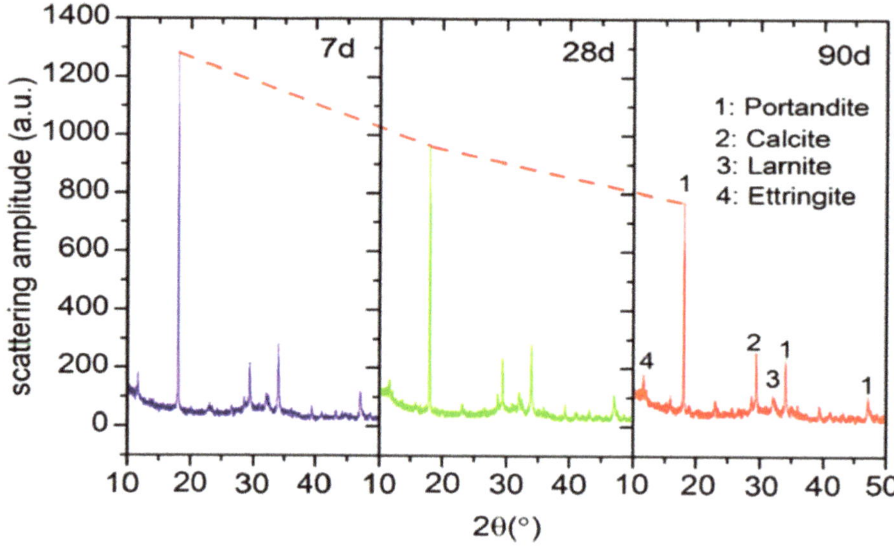

Figure 21. XRD of cement paste with 30% copper slag [99].

8. Hazards and Safety of Copper Slag

Copper slags are used in the production of cement, aggregates, landfill, ballast, abrasives, roofing granules, glass, tiles and bituminous pavements, among other things. Copper slag has lately been studied in terms of its properties and applications [35]. The biggest worry in the large-scale usage of copper slags, however, is the fear of environmental danger due to heavy metal leaching from the slag and its long-term stability under harsh environmental conditions. The chemical compositions of the slags reveal extremely low heavy metal concentration, and the leach test findings demonstrate that simulated leachate in an aggressive laboratory test removes very little of any of these metals. The quantities eliminated are substantially below the regulated limits set by the US (and possibly other national) drinking water quality standards [100]. A similar study also claims that the leach test conservatively predicts that the slags will not release enough As, Cd, Cr, Pb, or Se to damage groundwater [100].

However, The Cu and Pb concentrations exceed the authorized limits (1.0 and 0.3 mg/L, respectively) at the 30 percent replacement level of CPS, whereas the other elements (Zn, Mn, Ni, Cr and As) are within the limits. Cr has the lowest concentration (it is close to the detection limit of 0.003 mg/L). When the CPS content is raised to 50 wt.%, the excess quantities of Cu and Pb concentrations rise as well compared with the 30 wt.% content instances. Excess concentrations of the metals Zn, Ni and As are also found. The findings clearly reveal that employing CPS (both air cooled and water cooled) as a partial cement replacement offers a significant danger of heavy metal leaching into the environment (namely, Cu, Pb, Zn, Ni and As), particularly at high replacement levels. Other CPS applications, such as aggregates in concrete and raw materials for alkali activation, may also provide a risk of heavy metal leaching [101].

The results of the leaching procedure (TCLP), acid leaching and repeated extraction tests performed on a large number of slag samples of varied compositions produced from the use of numerous copper concentrate show that heavy metals have little leachability and that long-term stability is assured even in harsh environments. Leaching experiments on mechanically activated samples provide an indication of the heavy metals' resistance to leaching even after weathering. The heavy metals contained in the slag are stable, according to numerous extractions leaching experiments, and are unlikely to dissolve considerably even after repeated leaching in an acid rain environment. The greatest concentration of all components is significantly below the USEPA 40CFR Part 261 permitted limits [102]. A study [102] also observed that the heavy metals included in the slag are relatively stable and have low leachability, according to the TCLP, repeated extraction process tests and sulfuric acid leaching findings. The TCLP test results are far below the USEPA's 40CFR Part 261 standards. The heavy metals found in the slag are very stable, according to several extraction leaching experiments, and are unlikely to dissolve considerably even in acid rain in a natural setting. The greatest concentration of elements recovered by the multiple extraction technique is less than the USEPA 40CFR Part 261 mandated limits for the elements covered by this standard.

It is indicated that the slag is safe for use in a broad range of applications, including Portland cement, building materials such as tiles and bituminous pavement projects. The slag samples are non-toxic and pose no risk to the environment.

9. Conclusions

The focus on green construction has resulted in an ongoing search for alternative materials to be employed in concrete construction. In this review, a complete parametric analysis was carried out to determine the impacts of CPS on physical and chemical properties, concrete qualities both fresh and hardened and concrete's long-term durability performance. Consequently, the findings are given below.

- The physical property of CPS shows that the particle nature of CPS is rough and angular which adversely affects the flowability of concrete.
- The chemical composition of CPS ensures that it can be used as binding material.
- The slump value of concrete was reduced with the replacement of CPS due to angular and rough surface texture.
- The setting time increased with CPS as the pozzolanic reaction proceeded slowly.
- CPS up to 60% can be used without any harmful impact on the mechanical strength of concrete. The improvement in compressive, split tensile strength and flexure at 28 days was 9%, 6% and 9% higher than control concrete, respectively.
- The higher dose of CPS (80 and 90%) resulted in a decline in the mechanical strength of concrete due to the absence of flowability.
- A good correlation was observed between two specified strengths with an R^2 value greater than 90%.
- The durability performance of concrete, such as water absorption and voids, corrosion resistance, acid resistance and electric resistivity increased with CPS.
- SEM results reveal that the performance of concrete with CPS can be improved with the addition of secondary cementitious materials.

The overall studies demonstrate that the CPS has the credibility to be utilized partially in concrete, either as a binding material or as a sand. The optimal percentages are an essential parameter for good strength. Different researchers recommend a different optimum value of CPS due to a change of source. The typical range of optimum value of CPS is from 50 to 60% by weight of fine aggregate. Furthermore, less information is available on dry shrinkage and creep properties of concrete with CPS. No or little information is available about the alkali silica reaction (ASR) connected with the CPS substitution. Lastly, although CPS can be utilized in concrete and the mechanical capacity can be enhanced, concrete is still low in tension. Therefore, further research was suggested to enhance the ductility of concrete with the supplement of various kinds of fibers. A study [103] concluded that the optimum mix substituted silica fume and copper slag for 7% and 20% of the cement, respectively, to provide a workable, resilient, cost-effective and durable mix design. However, there is less information about the economic benefits of CPS, and detailed investigation is required.

Author Contributions: Writing—original draft preparation, J.A.; Conceptualization, J.A., A.M. and A.F.D.; methodology, J.A.; validation, A.F.D., A.M. and C.R.; formal analysis, C.R.; investigation, A.M.; resources, H.F.I.; writing—original draft preparation, J.A.; writing—review and editing, A.F.D. and A.M.; visualization, C.R. and H.F.I.; project administration, J.A. and H.F.I.; funding acquisition, A.F.D. and H.F.I. All authors have read and agreed to the published version of the manuscript.

Funding: This paper is funded by the deanship of King Khalid University under grant number RGP. RGP.2/104/43.

Institutional Review Board Statement: Not applicable.

Informed Consent Statement: Not applicable.

Data Availability Statement: All the data are available in manuscript.

Acknowledgments: The authors extend their appreciation to the Deanship of Scientific Research at King Khalid University for funding this work through Large Groups Project under grant number RGP.2/104/43.

Conflicts of Interest: The authors have no conflict interest.

References

1. Ahmad, J.; Zaid, O.; Siddique, M.S.; Aslam, F.; Alabduljabbar, H.; Khedher, K.M. Mechanical and Durability Characteristics of Sustainable Coconut Fibers Reinforced Concrete with Incorporation of Marble Powder. *Mater. Res. Express* **2021**, *8*, 075505. [CrossRef]
2. Taskin, A.; Fediuk, R.; Grebenyuk, I.; Elkin, O.; Kholodov, A. Effective Cement Binders on Fly and Slag Waste from Heat Power Industry of the Primorsky Krai, Russian Federation. *Int. J. Sci. Technol. Res.* **2020**, *9*, 3509–3512.
3. Althoey, F.; Farnam, Y. The Effect of Using Supplementary Cementitious Materials on Damage Development Due to the Formation of a Chemical Phase Change in Cementitious Materials Exposed to Sodium Chloride. *Constr. Build. Mater.* **2019**, *210*, 685–695. [CrossRef]
4. Singh, N.; Gupta, A.; Haque, M.M. A Review on the Influence of Copper Slag as a Natural Fine Aggregate Replacement on the Mechanical Properties of Concrete. *Mater. Today Proc.* **2022**, *62*, 3624–3637. [CrossRef]
5. Ahmad, J.; Aslam, F.; Martinez-Garcia, R.; De-Prado-Gil, J.; Qaidi, S.M.A.; Brahmia, A. Effects of Waste Glass and Waste Marble on Mechanical and Durability Performance of Concrete. *Sci. Rep.* **2021**, *11*, 21525. [CrossRef] [PubMed]
6. Dolage, D.A.R.; Dias, M.G.S.; Ariyawansa, C.T. Offshore Sand as a Fine Aggregate for Concrete Production. *Br. J. Appl. Sci. Technol.* **2013**, *3*, 813–825. [CrossRef]
7. Ahmad, J.; Aslam, F.; Zaid, O.; Alyousef, R.; Alabduljabbar, H. Mechanical and Durability Characteristics of Sustainable Concrete Modified with Partial Substitution of Waste Foundry Sand. *Struct. Concr.* **2021**, *22*, 2775–2790. [CrossRef]
8. Amin, S.K.; Allam, M.E.; Garas, G.L.; Ezz, H. A Study of the Chemical Effect of Marble and Granite Slurry on Green Mortar Compressive Strength. *Bull. Natl. Res. Cent.* **2020**, *44*, 19. [CrossRef]
9. Abdallah, S.; Fan, M. Characteristics of Concrete with Waste Glass as Fine Aggregate Replacement. *Int. J. Eng. Tech. Res.* **2014**, *2*, 11–17.
10. Corinaldesi, V.; Gnappi, G.; Moriconi, G.; Montenero, A. Reuse of Ground Waste Glass as Aggregate for Mortars. *Waste Manag.* **2005**, *25*, 197–201. [CrossRef]
11. Du Plessis, C. A Strategic Framework for Sustainable Construction in Developing Countries. *Constr. Manag. Econ.* **2007**, *25*, 67–76. [CrossRef]
12. Kansal, K.G.R. Waste Glass Powder as a Partial Replacement of PPC. *Int. J. Sci. Res.* **2016**, *5*, 1414–1418.
13. Oh, D.-Y.; Noguchi, T.; Kitagaki, R.; Park, W.-J. CO_2 Emission Reduction by Reuse of Building Material Waste in the Japanese Cement Industry. *Renew. Sustain. Energy Rev.* **2014**, *38*, 796–810. [CrossRef]
14. Naik, T.R. Sustainability of Concrete Construction. *Pract. Period. Struct. Des. Constr.* **2008**, *13*, 98–103. [CrossRef]
15. Vigneshpandian, G.V.; Shruthi, E.A.; Venkatasubramanian, C.; Muthu, D. Utilisation of Waste Marble Dust as Fine Aggregate in Concrete. *IOP Conf. Ser. Earth Environ. Sci.* **2017**, *80*, 012007. [CrossRef]
16. Xiao, R.; Huang, B.; Zhou, H.; Ma, Y.; Jiang, X. A State-of-the-Art Review of Crushed Urban Waste Glass Used in OPC and AAMs (Geopolymer): Progress and Challenges. *Clean. Mater.* **2022**, *4*, 100083. [CrossRef]
17. Imbabi, M.S.; Carrigan, C.; McKenna, S. Trends and Developments in Green Cement and Concrete Technology. *Int. J. Sustain. Built Environ.* **2012**, *1*, 194–216. [CrossRef]
18. Singh Shekhawat, B.; Aggarwal, V. Utilisation of Waste Glass Powder in Concrete-A Literature Review. *Int. J. Innov. Res. Sci. Eng. Technol.* **2007**, *3297*, 2319–8753.
19. Council, W.B. World Business Council for Sustainable Development. *Cem. Sustain. Initiat. Cem. Ind. Energy CO_2 Perform. Get. Numbers Right* **2009**. [CrossRef]
20. Humphreys, D. Long-Run Availability of Mineral Commodities. *Miner. Econ.* **2013**, *26*, 1–11. [CrossRef]

21. Müller, N.; Harnisch, J. *How to Turn Around the Trend of Cement Related Emissions in the Developing World*; WWF—Lafarge Conservation Partnership: Gland, Switzerland, 2008.
22. Smith, R.A.; Kersey, J.R.; Griffiths, P.J. The Construction Industry Mass Balance: Resource Use, Wastes and Emissions. *Construction* **2002**, *4*, 680.
23. Cui, X.; Sun, S.; Han, B.; Yu, X.; Ouyang, J.; Zeng, S.; Ou, J.; Khushnood, R.A.; Nawaz, A.A.; Siddique, R.; et al. Evaluation of the Reinforcement Efficiency of Low-Cost Graphite Nanomaterials in High-Performance Concrete. *Constr. Build. Mater.* **2018**, *47*, 3875–3882. [CrossRef]
24. Lee, H.; Hanif, A.; Usman, M.; Sim, J.; Oh, H. Performance Evaluation of Concrete Incorporating Glass Powder and Glass Sludge Wastes as Supplementary Cementing Material. *J. Clean. Prod.* **2018**, *170*, 683–693. [CrossRef]
25. Fediuk, R.S.; Yushin, A.M. The Use of Fly Ash the Thermal Power Plants in the Construction. *Proc. IOP Conf. Ser. Mater. Sci. Eng.* **2015**, *93*, 12070. [CrossRef]
26. Handayani, L.; Aprilia, S.; Rahmawati, C.; Al Bakri, A.M.M.; Aziz, I.H.; Azimi, E.A. Synthesis of Sodium Silicate from Rice Husk Ash as an Activator to Produce Epoxy-Geopolymer Cement. *Proc. J. Phys. Conf. Ser.* **2021**, *1845*, 12072. [CrossRef]
27. Srinivas, D.; Suresh, N.; Lakshmi, N.H. Experimental Investigation on Bagasse Ash Based Geopolymer Concrete Subjected to Elevated Tempreture. *Proc. IOP Conf. Ser. Earth Environ. Sci.* **2021**, *796*, 12028. [CrossRef]
28. Abdelgader, H.; Fediuk, R.; Kurpińska, M.; Elkhatib, J.; Murali, G.; Baranov, A.V.; Timokhin, R.A. Mechanical Properties of Two-Stage Concrete Modified by Silica Fume. *Mag. Civ. Eng.* **2019**, *89*, 26–38.
29. Suda, V.B.R.; Rao, P.S. Experimental Investigation on Optimum Usage of Micro Silica and GGBS for the Strength Characteristics of Concrete. *Mater. Today Proc.* **2020**, *27*, 805–811. [CrossRef]
30. Raju, S.; Dharmar, B. Mechanical Properties of Concrete with Copper Slag and Fly Ash by DT and NDT. *Period. Polytech. Civ. Eng.* **2016**, *60*, 313–322. [CrossRef]
31. Nassar, R.-U.-D.; Soroushian, P. Green and Durable Mortar Produced with Milled Waste Glass. *Mag. Concr. Res.* **2012**, *64*, 605–615. [CrossRef]
32. Belouadah, M.; Rahmouni, Z.E.A.; Tebbal, N. Experimental Characterization of Ordinary Concretes Obtained by Adding Construction Waste (Glass, Marble). *Procedia Comput. Sci.* **2019**, *158*, 153–162. [CrossRef]
33. Maharishi, A.; Singh, S.P.; Gupta, L.K. Strength and Durability Studies on Slag Cement Concrete Made with Copper Slag as Fine Aggregates. *Mater. Today Proc.* **2021**, *38*, 2639–2648. [CrossRef]
34. Sharifi, Y.; Afshoon, I.; Asad-Abadi, S.; Aslani, F. Environmental Protection by Using Waste Copper Slag as a Coarse Aggregate in Self-Compacting Concrete. *J. Environ. Manage.* **2020**, *271*, 111013. [CrossRef]
35. Gorai, B.; Jana, R.K. Characteristics and Utilisation of Copper Slag—A Review. *Resour. Conserv. Recycl.* **2003**, *39*, 299–313. [CrossRef]
36. Liu, J.; Guo, R.; Shi, P.; Huang, L. Hydration Mechanisms of Composite Binders Containing Copper Slag at Different Temperatures. *J. Therm. Anal. Calorim.* **2019**, *137*, 1919–1928. [CrossRef]
37. Murari, K.; Siddique, R.; Jain, K.K. Use of Waste Copper Slag, a Sustainable Material. *J. Mater. Cycles Waste Manag.* **2015**, *17*, 13–26. [CrossRef]
38. Liu, J.; Guo, R. Hydration Properties of Alkali-Activated Quick Cooled Copper Slag and Slow Cooled Copper Slag. *J. Therm. Anal. Calorim.* **2020**, *139*, 3383–3394. [CrossRef]
39. Firdous, R.; Stephan, D. Effect of Silica Modulus on the Geopolymerization Activity of Natural Pozzolans. *Constr. Build. Mater.* **2019**, *219*, 31–43. [CrossRef]
40. Wang, S.-D.; Scrivener, K.L. Hydration Products of Alkali Activated Slag Cement. *Cem. Concr. Res.* **1995**, *25*, 561–571. [CrossRef]
41. Brough, A.R.; Atkinson, A. Sodium Silicate-Based, Alkali-Activated Slag Mortars: Part I. Strength, Hydration and Microstructure. *Cem. Concr. Res.* **2002**, *32*, 865–879. [CrossRef]
42. Chakrawarthi, V.; Avudaiappan, S.; Amran, M.; Dharmar, B.; Raj Jesuarulraj, L.; Fediuk, R.; Aepuru, R.; Vatin, N.I.; Saavedra Flores, E. Impact Resistance of Polypropylene Fibre-Reinforced Alkali–Activated Copper Slag Concrete. *Materials* **2021**, *14*, 7735. [CrossRef]
43. Wang, G.C. *The Utilization of Slag in Civil Infrastructure Construction*; Woodhead Publishing: Sawston, UK, 2016; ISBN 0081003978.
44. Wang, R.; Shi, Q.; Li, Y.; Cao, Z.; Si, Z. A Critical Review on the Use of Copper Slag (CS) as a Substitute Constituent in Concrete. *Constr. Build. Mater.* **2021**, *292*, 123371. [CrossRef]
45. Manjunatha, M.; Reshma, T.V.; Balaji, K.; Bharath, A.; Tangadagi, R.B. The Sustainable Use of Waste Copper Slag in Concrete: An Experimental Research. *Mater. Today Proc.* **2021**, *47*, 3645–3653. [CrossRef]
46. Al-Jabri, K.S.; Taha, R.A.; Al-Hashmi, A.; Al-Harthy, A.S. Effect of Copper Slag and Cement By-Pass Dust Addition on Mechanical Properties of Concrete. *Constr. Build. Mater.* **2006**, *20*, 322–331. [CrossRef]
47. Mavroulidou, M. Mechanical Properties and Durability of Concrete with Water Cooled Copper Slag Aggregate. *Waste Biomass Valorization* **2017**, *8*, 1841–1854. [CrossRef]
48. Rajasekar, A.; Arunachalam, K.; Kottaisamy, M. Assessment of Strength and Durability Characteristics of Copper Slag Incorporated Ultra High Strength Concrete. *J. Clean. Prod.* **2019**, *208*, 402–414. [CrossRef]
49. Esfahani, S.M.R.A.; Zareei, S.A.; Madhkhan, M.; Ameri, F.; Rashidiani, J.; Taheri, R.A. Mechanical and Gamma-Ray Shielding Properties and Environmental Benefits of Concrete Incorporating GGBFS and Copper Slag. *J. Build. Eng.* **2021**, *33*, 101615. [CrossRef]

50. Mirhosseini, S.R.; Fadaee, M.; Tabatabaei, R.; Fadaee, M.J. Mechanical Properties of Concrete with Sarcheshmeh Mineral Complex Copper Slag as a Part of Cementitious Materials. *Constr. Build. Mater.* **2017**, *134*, 44–49. [CrossRef]
51. Saha, A.K.; Khan, M.N.N.; Sarker, P.K. Value Added Utilization of By-Product Electric Furnace Ferronickel Slag as Construction Materials: A Review. *Resour. Conserv. Recycl.* **2018**, *134*, 10–24. [CrossRef]
52. Singh, G.; Siddique, R. Strength Properties and Micro-Structural Analysis of Self-Compacting Concrete Made with Iron Slag as Partial Replacement of Fine Aggregates. *Constr. Build. Mater.* **2016**, *127*, 144–152. [CrossRef]
53. Najimi, M.; Pourkhorshidi, A.R. Properties of Concrete Containing Copper Slag Waste. *Mag. Concr. Res.* **2011**, *63*, 605–615. [CrossRef]
54. Al-Jabri, K.S.; Al-Saidy, A.H.; Taha, R. Effect of Copper Slag as a Fine Aggregate on the Properties of Cement Mortars and Concrete. *Constr. Build. Mater.* **2011**, *25*, 933–938. [CrossRef]
55. Chithra, S.; Kumar, S.R.R.S.; Chinnaraju, K. The Effect of Colloidal Nano-Silica on Workability, Mechanical and Durability Properties of High Performance Concrete with Copper Slag as Partial Fine Aggregate. *Constr. Build. Mater.* **2016**, *113*, 794–804. [CrossRef]
56. Kubissa, W.; Jaskulski, R.; Gil, D.; Wilińska, I. Holistic Analysis of Waste Copper Slag Based Concrete by Means of EIPI Method. *Buildings* **2019**, *10*, 1. [CrossRef]
57. *ASTM ASTM D6868 Standard*; Specification for Biodegradable Plastics Used as Coatings on Paper and Other Compostable Substrates. ASTM International: West Conshohocken, PA, USA, 2017.
58. Mantry, S.; Jha, B.B.; Satapathy, A. Evaluation and Characterization of Plasma Sprayed Cu Slag-Al Composite Coatings on Metal Substrates. *J. Coat.* **2013**, *2013*, 1–7. [CrossRef]
59. De Schepper, M.; Verlé, P.; Van Driessche, I.; De Belie, N. Use of Secondary Slags in Completely Recyclable Concrete. *J. Mater. Civ. Eng.* **2015**, *27*, 4014177. [CrossRef]
60. Gupta, N.; Siddique, R. Strength and Micro-Structural Properties of Self-Compacting Concrete Incorporating Copper Slag. *Constr. Build. Mater.* **2019**, *224*, 894–908. [CrossRef]
61. Al-Jabri, K.S.; Hisada, M.; Al-Oraimi, S.K.; Al-Saidy, A.H. Copper Slag as Sand Replacement for High Performance Concrete. *Cem. Concr. Compos.* **2009**, *31*, 483–488. [CrossRef]
62. Ting, L.; Qiang, W.; Shiyu, Z. Effects of Ultra-Fine Ground Granulated Blast-Furnace Slag on Initial Setting Time, Fluidity and Rheological Properties of Cement Pastes. *Powder Technol.* **2019**, *345*, 54–63. [CrossRef]
63. Ayano, T.; Sakata, K. Durability of Concrete with Copper Slag Fine Aggregate. In Proceedings of the 5th International Can-MET/ACI Conference on Durability of Concrete 2000, Barcelona, Spain, 4–9 June 2000; American Concrete Institute: Farmington Hills, MI, USA, 2000; pp. 141–157.
64. Gopalakrishnan, R.; Nithiyanantham, S. Microstructural, Mechanical, and Electrical Properties of Copper Slag Admixtured Cement Mortar. *J. Build. Eng.* **2020**, *31*, 101375. [CrossRef]
65. Siddique, R.; Singh, M.; Jain, M. Recycling Copper Slag in Steel Fibre Concrete for Sustainable Construction. *J. Clean. Prod.* **2020**, *271*, 122559. [CrossRef]
66. Kwon, Y.-H.; Kang, S.-H.; Hong, S.-G.; Moon, J. Acceleration of Intended Pozzolanic Reaction under Initial Thermal Treatment for Developing Cementless Fly Ash Based Mortar. *Materials* **2017**, *10*, 225. [CrossRef] [PubMed]
67. Sharma, R.; Khan, R.A. Durability Assessment of Self Compacting Concrete Incorporating Copper Slag as Fine Aggregates. *Constr. Build. Mater.* **2017**, *155*, 617–629. [CrossRef]
68. Caliskan, S.; Behnood, A. Recycling Copper Slag as Coarse Aggregate: Hardened Properties of Concrete. In Proceedings of the 7th International Conference on Concrete Technology in Developing Countries, Kuala Lumpur, Malaysia, 5–8 October 2004; pp. 91–98.
69. Khanzadi, M.; Behnood, A. Mechanical Properties of High-Strength Concrete Incorporating Copper Slag as Coarse Aggregate. *Constr. Build. Mater.* **2009**, *23*, 2183–2188. [CrossRef]
70. Sambangi, A.; Arunakanthi, E. Fresh and Mechanical Properties of SCC with Fly Ash and Copper Slag as Mineral Admixtures. *Mater. Today Proc.* **2021**, *45*, 6687–6693. [CrossRef]
71. Sharma, R.; Khan, R.A. Sustainable Use of Copper Slag in Self Compacting Concrete Containing Supplementary Cementitious Materials. *J. Clean. Prod.* **2017**, *151*, 179–192. [CrossRef]
72. Tangadagi, R.B.; Manjunatha, M.; Bharath, A.; Preethi, S. Utilization of Steel Slag as an Eco-Friendly Material in Concrete for Construction. *J. Green Eng.* **2020**, *10*, 2408–2419.
73. Li, Q.; Zhang, L.; Gao, X.; Zhang, J. Effect of Pulverized Fuel Ash, Ground Granulated Blast-Furnace Slag and CO2 Curing on Performance of Magnesium Oxysulfate Cement. *Constr. Build. Mater.* **2020**, *230*, 116990. [CrossRef]
74. Babu, K.M.; Ravitheja, A. Effect of Copper Slag as Fine Aggregate Replacement in High Strength Concrete. *Mater. Today Proc.* **2019**, *19*, 409–414. [CrossRef]
75. Junwei, S.; Shenglei, F.; Xiong, R.; Ouyang, Y.; Qingli, Z.; Jielu, Z.H.U.; Zhang, C. Mechanical Properties, Pozzolanic Activity and Volume Stability of Copper Slag-Filled Cementitious Materials. *Mater. Sci.* **2020**, *26*, 218–224.
76. Afshoon, I.; Sharifi, Y. Use of Copper Slag Microparticles in Self-Consolidating Concrete. *ACI Mater. J.* **2017**, *114*, 691–699. [CrossRef]
77. Jayapal Naganur, C. Effect of Copper Slag as a Partial Replacement of Fine Aggregate on the Properties of Cement Concrete. *Int. J. Res.* **2014**, *1*, 8.

78. Sambhaji, Z.K.; Autade, P.B. Effect of Copper Slag as a Fine Aggregate on Properties of Concrete. *Int. Res. J. Eng. Techn.* **2016**, *3*, 410–414.
79. Li, X.M.; Zou, S.H.; Zhao, R.G.; Deng, L.Y.; Ji, N.; Ren, C.L. Preparation and Mechanical Properties of Steel Fiber Reinforced High Performance Concrete with Copper Slag as Fine Aggregate. *Proc. IOP Conf. Ser. Mater. Sci. Eng.* **2019**, *531*, 12037. [CrossRef]
80. Tiwary, A. Effect of Copper Slag and Fly Ash on Mechanical Properties of Concrete. *Int. J. Civ. Eng. Technol.* **2018**, *9*, 354–362.
81. Neville, A.M. *Properties of Concrete*, 4th ed.; Longman: London, UK, 1995.
82. Velumani, M.; Nirmalkumar, K. Durability and Characteristics of Copper Slag as Fine Aggregate and Fly Ash as Cement in Concrete. In Proceedings of the Second International Conference on Current Trends in Engineering and Technology-ICCTET, Coimbatore, India, 8 July 2014; IEEE: New York, NY, USA, 2014; pp. 222–227.
83. Brindha, D.; Baskaran, T.; Nagan, S. Assessment of Corrosion and Durability Characteristics of Copper Slag Admixed Concrete. *Int. J. Civ. Struct. Eng.* **2010**, *1*, 192.
84. Hassaan, M.Y.; El Desoky, M.M.; Salem, S.M.; Yousif, A.A. Some Physical Properties of Anhydrous and Hydrated Brownmillerite Doped with NaF. *Cem. Concr. Res.* **2003**, *33*, 697–702. [CrossRef]
85. Kurda, R.; de Brito, J.; Silvestre, J.D. Water Absorption and Electrical Resistivity of Concrete with Recycled Concrete Aggregates and Fly Ash. *Cem. Concr. Compos.* **2019**, *95*, 169–182. [CrossRef]
86. Al-Akhras, N.M. Durability of Metakaolin Concrete to Sulfate Attack. *Cem. Concr. Res.* **2006**, *36*, 1727–1734. [CrossRef]
87. Sharifi, Y.; Afshoon, I.; Nematollahzade, M.; Ghasemi, M.; Momeni, M.-A. Effect of Copper Slag on the Resistance Characteristics of SCC Exposed to the Acidic Environment. *Asian J. Civ. Eng.* **2020**, *21*, 597–609. [CrossRef]
88. Sharma, R.; Khan, R.A. Sulfate Resistance of Self Compacting Concrete Incorporating Copper Slag as Fine Aggregates with Mineral Admixtures. *Constr. Build. Mater.* **2021**, *287*, 122985. [CrossRef]
89. Singh, G.; Siddique, R. Effect of Iron Slag as Partial Replacement of Fine Aggregates on the Durability Characteristics of Self-Compacting Concrete. *Constr. Build. Mater.* **2016**, *128*, 88–95. [CrossRef]
90. Najimi, M.; Sobhani, J.; Pourkhorshidi, A.R. Durability of Copper Slag Contained Concrete Exposed to Sulfate Attack. *Constr. Build. Mater.* **2011**, *25*, 1895–1905. [CrossRef]
91. Gevaudan, J.P.; Caicedo-Ramirez, A.; Hernandez, M.T.; Srubar III, W. V Copper and Cobalt Improve the Acid Resistance of Alkali-Activated Cements. *Cem. Concr. Res.* **2019**, *115*, 327–338. [CrossRef]
92. Ahmad, J.; Tufail, R.F.; Aslam, F.; Mosavi, A.; Alyousef, R.; Faisal Javed, M.; Zaid, O.; Khan Niazi, M.S. A Step towards Sustainable Self-Compacting Concrete by Using Partial Substitution of Wheat Straw Ash and Bentonite Clay Instead of Cement. *Sustainability* **2021**, *13*, 824. [CrossRef]
93. Ahmad, J.; Aslam, F.; Martinez-Garcia, R.; El Ouni, M.H.; Khedher, K.M. Performance of Sustainable Self-Compacting Fiber Reinforced Concrete with Substitution of Marble Waste (MW) and Coconut Fibers (CFs). *Sci. Rep.* **2021**, *11*, 1–22. [CrossRef]
94. Alvee, A.R.; Malinda, R.; Akbar, A.M.; Ashar, R.D.; Rahmawati, C.; Alomayri, T.; Raza, A.; Shaikh, F.U.A. Experimental Study of the Mechanical Properties and Microstructure of Geopolymer Paste Containing Nano-Silica from Agricultural Waste and Crystalline Admixtures. *Case Stud. Constr. Mater.* **2022**, *16*, e00792. [CrossRef]
95. Chen, Q.; Zhang, Q.; Qi, C.; Fourie, A.; Xiao, C. Recycling Phosphogypsum and Construction Demolition Waste for Cemented Paste Backfill and Its Environmental Impact. *J. Clean. Prod.* **2018**, *186*, 418–429. [CrossRef]
96. Ahmad, J.; Zaid, O.; Shahzaib, M.; Abdullah, M.U.; Ullah, A.; Ullah, R. Mechanical Properties of Sustainable Concrete Modified by Adding Marble Slurry as Cement Substitution. *AIMS Mater. Sci.* **2021**, *8*, 343–358. [CrossRef]
97. Papadakis, V.G.; Vayenas, C.G.; Fardis, M.N. A Reaction Engineering Approach to the Problem of Concrete Carbonation. *AIChE J.* **1989**, *35*, 1639–1650. [CrossRef]
98. El-Didamony, H.; Sharara, A.M.; Helmy, I.M.; Abd El-Aleem, S. Hydration Characteristics of β-C2S in the Presence of Some Accelerators. *Cem. Concr. Res.* **1996**, *26*, 1179–1187. [CrossRef]
99. Feng, Y.; Zhang, Q.; Chen, Q.; Wang, D.; Guo, H.; Liu, L.; Yang, Q. Hydration and Strength Development in Blended Cement with Ultrafine Granulated Copper Slag. *PLoS ONE* **2019**, *14*, e0215677. [CrossRef] [PubMed]
100. Alter, H. The Composition and Environmental Hazard of Copper Slags in the Context of the Basel Convention. *Resour. Conserv. Recycl.* **2005**, *43*, 353–360. [CrossRef]
101. Wang, D.; Wang, Q.; Huang, Z. Reuse of Copper Slag as a Supplementary Cementitious Material: Reactivity and Safety. *Resour. Conserv. Recycl.* **2020**, *162*, 105037. [CrossRef]
102. Shanmuganathan, P.; Lakshmipathiraj, P.; Srikanth, S.; Nachiappan, A.L.; Sumathy, A. Toxicity Characterization and Long-Term Stability Studies on Copper Slag from the ISASMELT Process. *Resour. Conserv. Recycl.* **2008**, *52*, 601–611. [CrossRef]
103. Shirdam, R.; Amini, M.; Bakhshi, N. Investigating the Effects of Copper Slag and Silica Fume on Durability, Strength, and Workability of Concrete. *Int. J. Environ. Res.* **2019**, *13*, 909–924. [CrossRef]

Review

Progressing towards Sustainable Machining of Steels: A Detailed Review

Kashif Ishfaq [1,*], Irfan Anjum [1], Catalin Iulian Pruncu [2,*], Muhammad Amjad [3], M. Saravana Kumar [4] and Muhammad Asad Maqsood [1]

1   Department of Industrial and Manufacturing Engineering, University of Engineering & Technology, Lahore 548900, Pakistan; irfanhanjum@gmail.com (I.A.); 2016im11@student.uet.edu.pk (M.A.M.)
2   Design, Manufacturing & Engineering Management, University of Strathclyde, Glasgow G1 1XJ, Scotland, UK
3   Department of Mechanical, Mechatronics and Manufacturing Engineering, University of Engineering & Technology, Lahore 548900, Pakistan; amjad9002@uet.edu.pk
4   Department of Production Engineering, National Institute of Technology, Tiruchirappalli 620015, Tamil Nadu, India; saravana312@gmail.com
*   Correspondence: kashif.ishfaq@uet.edu.pk (K.I.); Catalin.pruncu@strath.ac.uk (C.I.P.)

Abstract: Machining operations are very common for the production of auto parts, i.e., connecting rods, crankshafts, etc. In machining, the use of cutting oil is very necessary, but it leads to higher machining costs and environmental problems. About 17% of the cost of any product is associated with cutting fluid, and about 80% of skin diseases are due to mist and fumes generated by cutting oils. Environmental legislation and operators' safety demand the minimal use of cutting fluid and proper disposal of used cutting oil. The disposal cost is huge, about two times higher than the machining cost. To improve occupational health and safety and the reduction of product costs, companies are moving towards sustainable manufacturing. Therefore, this review article emphasizes the sustainable machining aspects of steel by employing techniques that require the minimal use of cutting oils, i.e., minimum quantity lubrication, and other efficient techniques like cryogenic cooling, dry cutting, solid lubricants, air/vapor/gas cooling, and cryogenic treatment. Cryogenic treatment on tools and the use of vegetable oils or biodegradable oils instead of mineral oils are used as primary techniques to enhance the overall part quality, which leads to longer tool life with no negative impacts on the environment. To further help the manufacturing community in progressing towards industry 4.0 and obtaining net-zero emissions, in this paper, we present a comprehensive review of the recent, state of the art sustainable techniques used for machining steel materials/components by which the industry can massively improve their product quality and production.

Keywords: sustainable manufacturing; minimum quantity lubrication; cryogenic machining; solid lubricants; vegetable oils; steels

check for **updates**

Citation: Ishfaq, K.; Anjum, I.; Pruncu, C.I.; Amjad, M.; Kumar, M.S.; Maqsood, M.A. Progressing towards Sustainable Machining of Steels: A Detailed Review. *Materials* **2021**, *14*, 5162. https://doi.org/10.3390/ma14185162

Academic Editors: Stefano Guarino and Flaviana Tagliaferri

Received: 31 July 2021
Accepted: 3 September 2021
Published: 8 September 2021

Publisher's Note: MDPI stays neutral with regard to jurisdictional claims in published maps and institutional affiliations.

1. Introduction

Manufacturing products while conserving natural resources and causing no negative environmental impacts is called sustainable manufacturing. Manufacturing industries create products for fulfilling human needs; however, this includes the consumption of huge amounts of raw resources and the generation of wastes which are increasing day by day and can be very detrimental for our environment.

The following three stages of product waste are primary factors for waste generation and the degradation of the environment:

- In the manufacturing processes;
- During usage of the product;
- At the end of the life of the product.

The production of metals triggers the consumption of natural resources and has created a harmful effect on humankind. To avoid using resources needed by future generations, it is necessary to use fewer natural resources and reduce the negative environmental impact caused by manufacturing systems. That is why industries are now moving towards sustainable manufacturing. Early ideas about sustainable manufacturing first appeared in the 1970s and 1980s [1–5].

1.1. Manufacturing Industries and Sustainable Manufacturing

It is well noted that machining is widely used to produce automotive parts within the manufacturing industry sector. In all machining operations, cutting fluids play a vital role in reducing the machining cost by increasing tool life. It was observed that 7–17% of the cost incurred in the machining of a part is associated with using cutting fluids. Further, the tooling cost is about 2–4%, so it is necessary to improve the whole process. In addition, the use of cutting fluids causes health diseases like skin problems, allergies, eyes problems, and cancer in workers. Here, skin problem is about 80% [6]. Lawal et al. [7] also witnessed that major skin problems, about 80% in quantity, are due to cutting fluids. They also proposed that vegetable oil-based and metal working liquids have been proven to be environmentally sustainable in the dielectric regime.

Strict environmental regulations demand that cutting oil used during machining processes should be recycled or disposed of in such a way that it will not spoil the environment and will be harmless for all interested parties. These fluids are extremely costly to dispose of or store. The cost is about double the machining cost depending on the cutting fluid which is being used. Mineral oils used as cutting fluids are difficult to dispose of into the environment without any prior treatment [8–10].

1.2. Need for Sustainable Manufacturing

Jordi Oliver Solà, Chief Executive Officer (CEO) of a circular economy consulting group, has demonstrated the importance of sustainability, not only from an ethical or environmental point of view, but also that it is needed for markets to be competitive and important for the survival of any sector [11]. Therefore, to save natural resources, sustainable manufacturing is very important [12,13].

The need for sustainable manufacturing techniques is also depicted by the three pillars of sustainable manufacturing, as shown below in Figure 1. One of these is the need for improvement from an economic, social, and environmental point of view. It brings balance between social, economic, and environmental aspects [14,15]. This technique mainly deals with the minimal usage of cutting fluids. It does not mean just stoping the supply of cutting oils to make the environment better. Cutting oils serve many purposes like lubrication and temperature reduction in the cutting zones.

Figure 1. Three Pillars of Sustainability, reprinted with permission from ref. [16]. Copyright 2017 BSP books Pvt Ltd.

Concerns about environmental impacts, climate change, occupational health and safety, and machining costs have forced companies to move towards sustainable techniques. As per the investigation of Jayal et al. [17], the selection of sustainability aspects occurs mainly because of factors like the increase in diseases in shop floor workers, inflexibility in government plans, and when targeted to minimize the cost of production. So, sustainable machining is highly recommended where traditional cutting methods became null. Currently, advanced technologies like cryogenic cooling, nano cutting fluids, dry cutting, and minimum quantity lubrication (MQL), etc. are being used [18].

Several studies have been published that emphasize the importance of sustainability. For example, Zein et al. [19] presented certain resources which are associated with the manufacturing technologies, including production tools and methods that directly correlate with economic impact. They outlined that the sustainability of a firm can be affected by manufacturing approaches. Jayal et al. [17] established a case study on machining techniques by improving the model at the process, product, and system-level for sustainable manufacturing. Sarkis [20] found a relationship between environmental concerns and manufacturing activities. The study concluded that sustainable machining not only deals with environmental initiatives but also included techniques that empower benefits for humanity. Lu et al. [21] developed a metric to ensure the sustainability of the manufacturing process. They also established the interrelationship between the elements of metrics and studied the potential impact. The metrics broadly covered the given elements, i.e., social, environmental, and economic. Jawahir and Dillon [22] and Hegab et al. [23] highlighted six of the most important factors that alter the paradigm of sustainability in manufacturing processes. The factors included cost, energy, the safety of workers, personal health, environmental impact, and waste. The researchers said that out of the six aforementioned elements, waste, cost, and energy can be more easily computed than the rest of the elements. Waas et al. [24] made a framework for the sustainability of manufacturing sectors by taking the hierarchy of social needs and combinations. Then they used the Delphi technique to propose the metrics for each category. Some researchers suggested rules to achieve sustainability in the manufacturing firms, as demonstrated by Lovin et al. [25]. The rules are (1) minimal usage of energy and material, (2) usage of cleaner production, recycling and conversion techniques to reusable substance, (3) adoption of a solution based system (i.e., supply chain structure) rather than a proactive business model, and (4) reinvestment in natural substitutions that are available for distinct materials, such as investing in renewable resources instead of non-renewable substances. From the machining perspective, Diaz Elsayed et al. [26] discussed a detailed study about the combination of green and lean in the automotive organization. The purpose of their research was to determine the effect of green-lean in the manufacturing sector. They concluded that grouping of green-lean proved an effective way of improving manufacturing firms in terms of waste reduction, less resource utilization, and energy consumption. Thus, they stated that the use of green-lean is a sustainable manufacturing approach for different enterprises. Abdul Rashid et al. [27] also investigated environmental performance by employing sustainable manufacturing techniques. They proposed that the main environmental initiatives are entirely based on manufacturing practices. In the same vein, Rusinko [28] established the relationship between manufacturing activities and their results. The results revealed that manufacturing cost is decreased by preventing waste and unnecessary substances. According to Gimenez et al. [29], organizations should improve their environmental, social, and economic behavior to get a sustainable approved system. The aforementioned literature divulged the importance of sustainability in manufacturing or business firms. Therefore, the current study was conducted to scrutinize a systematic review of the sustainable machining of steel, as it is being used in scattered application areas including aerospace, automotive, nuclear power plants, and medical equipment, etc.

There are numerous benefits, i.e., financial, environmental, and safety, which are related to the three pillars as discussed before. The need to turn towards sustainable manufacturing is due to many reasons like occupational health-related problems, environmental regulations, and unsafe or polluted environments for workers, but the largest is the waste cost in using too much cutting fluid in metal cutting industries. In such industries, the costs associated with purchasing, maintaining, the makeup of cutting fluid, cutting oil, and system cleaning are more prominent.

The consideration of the following points allows companies to improve these three pillars:

- Efficient resource utilization (Energy, Material, Water, Labor, etc.);
- Improvement in the application of metalworking fluids;
- Adopting other sustainable manufacturing techniques;
- Lean Implementation;
- Improvement in the working environment by applying best machining practices;
- Most important, training to all employees related to sustainable machining.

Figure 2 shows the basic objectives by which pressure was built on manufactures to change their way of working. It is clearly depicted that there is a need to change the whole scenario of conventional working in manufacturing industries to improve socially, economically, and environmentally. Technology revolution should be introduced in manufacturing industries to lower the cost per piece of product. Whereas Figure 3 shows the breakdown of the product cost, including cooling and lubricating costs, which are about 18%.

Figure 2. Pressure to change the paradigms of the manufacturing industry, reprinted with permission from ref. [30]. Copyright 2010 Elsevier Ltd.

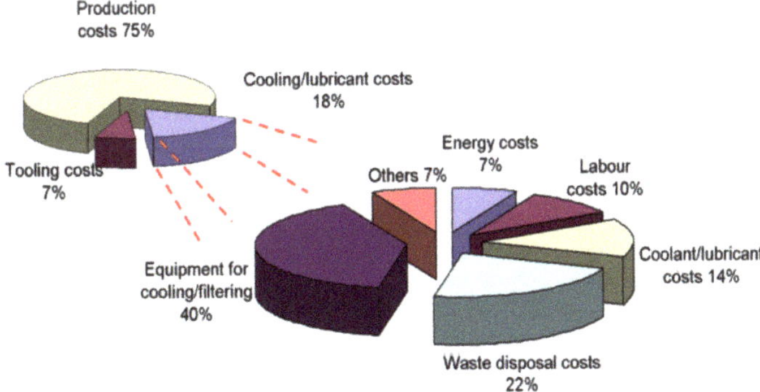

Figure 3. Cooling and lubricating costs incorporated in the automotive sector, reprinted with permission from ref. [31]. Copyright 2010 Elsevier Ltd.

Manufacturing companies can improve their costs and tackle environmental issues by implementing sustainable principles. Implementation can be done by analyzing the current situation of the process or system in any industry. There is a need to adopt alternate technologies for redesigning systems for the effective realization of these principles in factories [30]. The key methods that provide a direct path to create a cleaner manufacturing sector are depicted in Figure 4.

Figure 4. Implementation of Clean Manufacturing process, reprinted with permission from ref. [31]. Copyright 2010 Elsevier Ltd.

Further, some key characteristics of sustainable machining are presented in Figure 5, which clearly shows that this technique justified three pillars of sustainable manufacturing. Figure 6 shows the evolution over time of sustainable manufacturing, which depicts the critical importance of embedding sustainable manufacturing by 2025. It is assumed that the industries will work on 6-reduction (6R) elements rather than 3-reduction (3R) entities which are used in the actual green manufacturing model [32].

Figure 5. Characteristics of Sustainable machining, reprinted with permission from ref. [17]. Copyright 2010 CIRP, Published by Elsevier Ltd.

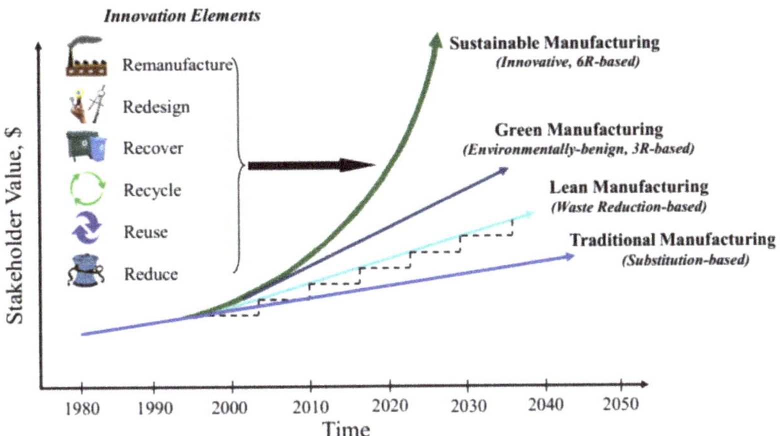

Figure 6. Evolution with the time of sustainable manufacturing, reprinted with permission from ref. [32]. Copyright 2010 CIRP, Published by Elsevier Ltd.

As per the given literature, it has been found that certain areas need to be reviewed. For example, material wastage and the amount of disposed-off material during the process may cause the cost of machining to be high. Moreover, suspended particles enter the environment and damage the quality of the air. The disturbing air quality index influences human life, and thus sustainability is compromised. Therefore, this review has been developed to understand the conditions/parameters of different sustainable machining techniques which alter the cost, pollute the environment, and decrease the overall productivity. For this context, a PRISMA approach has been adopted to study the variants of sustainable manufacturing processes as far as steel material is considered. The current review is restricted to sustainable techniques used for the machining of steel in order to ensure cost-effective, environmentally stable, and eco-friendly processes.

Even though this review provides a comprehensive discussion on sustainable manufacturing techniques, other elements may be added, like frostbite hazards in cryogenic machining and initial setup cost, which is difficult to afford by any local industry. Therefore, a comprehensive investigation is required to mitigate the aforesaid issues of cryogenic machining by ensuring a controlled temperature environment. The mathematical modeling of the sustainable cutting mechanisms with respect to the cutting of steel is still an area that needs special focus.

This paper is presented in the following order: (i) A brief introduction with mechanical properties of some steel grades is proposed in Section 2. (ii) The comprehensive methodology is given in Section 3. (iii) Different sustainable techniques employed in a couple of manufacturing sectors are demonstrated in Section 4. It constitutes different subsections; each outlines the discussion, significance, advantages, disadvantages, and limitations of each sustainable technique separately. (iv) The detailed discussion about the present work, along with comparisons between each sustainable technique, has been granted in Section 5. (v) Section 6 illustrates the multiple challenges faced with the implementation of sustainable manufacturing. (vi) Fundamental issues associated with additive manufactured steel has been given in Section 7. (vii) The findings are summarized in Section 8. (viii) Finally, future implications have been revealed in Section 9.

2. Steels' Classification, Properties, Machining Difficulties, and Sustainability Requirements in Steels' Machining

An alloy of iron (Fe) with minimal carbon content is referred to as steel. Carbon (C), generally up to 1.5%, is present in steel [33]. As per the literature, about 1808 million tons of steel were produced in 2018 worldwide. This is depicted in Figure 7, along with the emissions of CO_2 [34]. If the machinability perspective of steel is under consideration, then up to 29% of steel is employed in machining, as given by Diva Metal Ltd. Company [35]. Figure 8 represents the division of steel in different applications. Steel exists in the form of different variants like structural steel, heat resistant steel, and tool steel, etc. Another important class of steel is named Alloy Steel, a standard form that constitutes various elements (i.e., nickel, magnesium, copper, titanium, vanadium, silicon, boron, and manganese, etc.) in different proportions that range from 1.0% to 50% by weight. Alloy steel can be categorized into low alloy steel (LAS) and high alloy steel (HAS). Usually, the phrase "alloy steel" is related to LAS. Nickel (Ni) is a prime element in LAS that has the ability to increase the strength and ductility of different engineering applications, including jet engines, spacecraft, and nuclear reactors. Interestingly, Ni also amplifies the characteristics of ferrite steel, such as stability at low-temperature toughness, which allows them to be used in cryogenic applications [36]. For instance, steel with 9% Ni can be employed for liquefied natural gas (LNG) handling and storage purposes. Moreover, it assists in nitriding, carburizing, and tool steel due to tremendous properties like good strength, high hardness, superior toughness, the ability to withstand elevated temperatures, excellent wear, and corrosion resistance. The other combination, such as an alloy of Fe with C, is known as the simplest alloy. The ferromagnetic feature of Fe permits the use of the simplest steel in magnetic applications like electric motors, transformers, etc. [37]. The details about some key classes of steel are presented in the forthcoming sections.

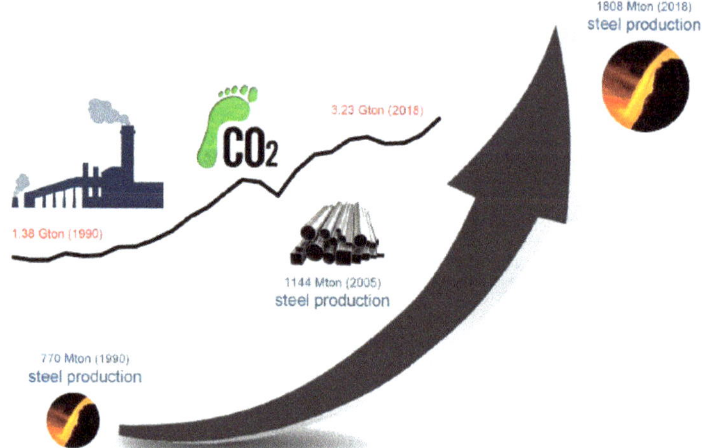

Figure 7. Worldwide steel production in 2018, along with the CO_2 emission, reprinted with permission from ref. [34]. Copyright 2019 Elsevier Ltd.

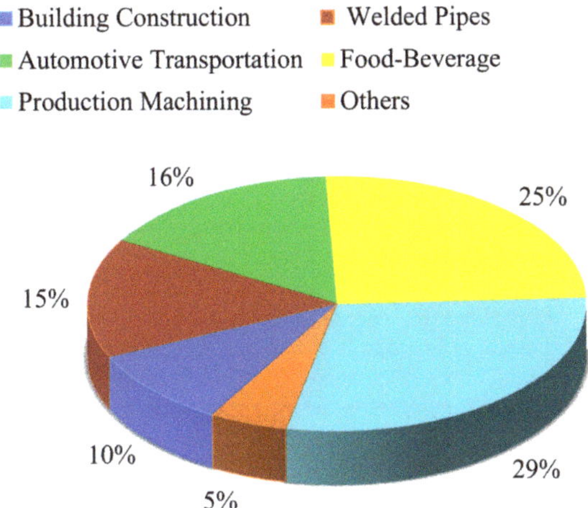

Figure 8. Division of steel usage in different applications.

2.1. Structural Steel (SS)

SS is a commonly used building material in the construction industry. The performance of SS is now predictable and depends on standards recognized by the American Institute of Steel Construction (AISC), which elaborate shapes, sizes, elemental composition, as well as mechanical attributes. SS is 100% recyclable and has proven to be one of the most reprocessed materials in the world [38]. The fundamental classification of structural steel is:

i. Carbon-manganese steel;
ii. High strength low alloy (HSLA) steel;
iii. High strength quenched and tempered alloy steels.

From the above-mentioned classes of structural steel, HSLA is important because it provides good mechanical properties, high resistance to rust, and high weldability with a carbon percentage between 0.05–0.25%. The other benefits of HSLA steel are: (i) light in weight, (ii) good strength to wear ratio, (iii) and control over internal and external stresses. However, HSLA has limitations in terms of acquiring more power (>25–30%) as compared to carbon steel. Additionally, such steel has sensitivity in directional properties [39]. The chemical composition of HSLA-80 steel is presented in Table 1, whereas the mechanical properties are listed in Table 2.

Table 1. Chemical composition of HSLA Steel, reprinted with permission from ref. [39]. Copyright 2018 Elsevier Ltd.

Elements	Cu	Ni	Cr	Mn	Si	Mo	C	Nb	S	P
wt. %	1.25	0.83	0.78	0.54	0.37	0.19	0.07	0.03	0.024	0.022

Table 2. Mechanical characteristics of HSLA steel, reprinted with permission from ref. [39]. Copyright 2018 Elsevier Ltd.

Properties	Units	Values
Yield Stress	(MPa)	450 ± 32
Ultimate strength	(MPa)	778 ± 17
Elastic Modulus	(GPa)	203 ± 5
Total Strain	%	21 ± 2

2.2. Heat Resistant Steel

Heat resistant steels (HRS) are a unique class of steel alloys that can easily be operated at temperatures as high as 750 °C. To attain their specific properties, all HRS are composed of numerous elements, two of which are considered basic elements, i.e., Chromium (Cr) and Nickel (Ni). Cr is preferred for corrosion resistance, while Ni is useful to obtain high strength and ductility. The other elements (aluminum, cobalt, manganese, niobium, copper, zirconium, and phosphorous, etc.) are added to achieve high-temperature properties along with good weldability. Based on chemical stability, high strength, and superb corrosion resistance, HRS are divided into three types; (1) Low alloy steels, (2) Martensitic steels, (3) Austenitic steels [40].

Low allow steels are extensively used in pressure-based applications like steam boilers and thermal power plants due to unique characteristics such as mechanical strength, great toughness, and sufficient rust resistance ability. Such steel alloys are mostly preferred in thick components such as headers, pipes, and control valves. The different grades of low alloy steel have applications in distinct areas. Grade 11 (1CrMoV) and 22 (2.25Cr1Mo) are used in power developing industries. The mechanical properties of these low alloy steels are tabulated in Table 3. In the same way, Fe-0.1C-xMn and Fe-0.1C-xNi, where x = 1.5, 3% by mass, are special kinds of low alloy steels used for cryogenic treatments to decrease the corrosion property as well as improve the microstructure. The chemical composition of all the said low alloy steels are mentioned in Table 4.

Table 3. Mechanical features of low alloy steels (grade 11 & 22), reprinted with permission from ref. [41]. Copyright 2007 Elsevier B.V.

Alloys	Yield Stress (MPa)	Ultimate Tensile Stress (MPa)	Elongation in %
1CrMoV	205	415	30
2.25Cr1Mo	205	415	30

Table 4. Chemical composition of different grades of low alloy steels, reprinted with permission from refs. [41,42]. Copyright 2015 Elsevier Ltd.

Alloys	C	Si	Mn	P	S	Ni	Fe
Fe-0.1C-1.5Mn	0.10	0.01	1.48	0.001	0.002	0.01	Balance
Fe-0.1C-3Mn	0.10	0.02	2.96	0.001	0.002	0.01	Balance
Fe-0.1C-1.5Ni	0.09	0.01	0.02	0.001	0.002	1.58	Balance
Fe-0.1C-3Ni	0.09	0.01	0.02	0.001	0.002	3.16	Balance
1CrMoV	0.15	0.50	0.60	0.025	0.025	0.03	Balance
2.25Cr1Mo	0.082	0.23	0.41	0.051	0.0054	0.03	Balance

Martensitic heat resistant steels contain medium and high chromium contents of about 5–9% Cr and 12%, respectively. They have been fabricated for power plant materials where the temperature is significantly higher, such as 650 °C. The high percentage of Cr in such steel alloys enhances the creep strength and corrosion resistance of the materials because of a low coefficient of thermal expansion and high thermal conductivities in contrast to austenitic steels. Moreover, the emission of hazardous fumes and gases and the efficiency of the power plant are also raised due to the application of Cr. 9Cr and 12Cr, a special series of martensitic alloy steels [40]. The elemental composition of some martensitic alloys is given in Table 5, and mechanical properties are elaborated in Table 6.

Table 5. Elemental composition of some martensitic alloys, reprinted with permission from refs. [43–45]. Copyright 2020 International Atomic Energy Agency (IAEA).

Alloys	Cr	Mo	V	Nb	C	Mn	Cu	Si	N	Ni	P	S	W
9Cr-1Mo	8.55	0.88	0.21	0.08	0.1	0.51	0.18	0.32	0.035	0.15	0.012	0.005	-
9Cr-1MoVNb	8.44	0.89	0.24	0.08	0.086	0.37	0.03	0.16	0.054	0.11	0.012	0.003	-
9Cr-1MoVNb-2Ni	8.57	0.98	0.22	0.066	0.064	-	-	-	0.053	2.17	-	-	0.01
12Cr-1MoVW	11.99	0.93	0.27	0.018	0.21	-	-	-	0.020	0.43	-	-	0.54

Table 6. Mechanical characteristics of martensitic alloys, reprinted with permission from refs. [43–45]. Copyright 2020 International Atomic Energy Agency (IAEA).

Alloys	Yield Strength (MPa)	Ultimate Tensile Stress (MPa)	Elongation in %
9Cr-1Mo	533	683	26.0
9Cr-1MoVNb	547	697	11.9
9Cr-1MoVNb-2Ni	148	171	19.1
12Cr-1MoVW	110	142	19.6

Austenitic steel is also called austenitic stainless steel, with a Cr percentage of about 13% by weight at room temperature. The high cost associated with these alloys is due to the high percentage of supplementary elements compared to other steel alloys. The applications of austenitic alloys are limited to those conditions where chances of corrosion are substantial, such as in boiler tubes. They have similar properties to martensitic steels, except high thermal loading can lead to wear and tear over the surface. The FeCrNi is the most commonly used austenitic steel. However, AISI (American Iron and Steel Institute) 302, 304, 321, 347, 316, 309, ASS304L, ASS316L, and other alloys are also employed in different application sectors. The elemental composition of few austenitic alloys is provided in Table 7.

Table 7. Elemental composition of some austenitic alloys, reprinted with permission from refs. [36,46]. Copyright 2018 Elsevier Ltd.

Alloys	Fe	C	Mn	Si	Mo	Co	Cr	Cu	Ni	Others
ASS304L	70.78	0.025	1.140	0.410	0.360	0.210	18.40	0.180	8.190	0.305
ASS316L	67.69	0.018	1.28	0.38	2.42	0.21	16.63	0.21	10.85	0.312
AISI 304	Balance	0.06	3.97	0.49	0.008	0.11	17.61	1.17	8.85	0.076
AISI 201	Balance	0.04	7.38	0.588	0.008	0.072	17.40	2.17	3.13	0.292

2.3. Tool Steel

Tool steels are alloy steels that are appropriate for the manufacture of tools due to their excellent properties like high hardness, low deformation, minimal abrasion, and no wear and tear, even at elevated temperatures. Apart from the mentioned properties, these steels have a high magnitude of tensile and compressive yield strength which tends to minimize the plastic deformations at the stress concentration points in the tooling [47]. There are different variants of tool steels, including cold working, hot working, high speed, vibration resistance, water hardening, and some unusual purposes. The selection of this group is based on cost, temperature, surface hardness, ductility, and toughness values. In severe circumstances, carbide tool steels are utilized. They have applications in cutting, drawing dies, pressing, cold extrusion dies, broaches, thread rolling, forming rolls, and coining of materials. Another important application of tool steel is in the injection molding process, where durability plays an integral role. The common scale of tool steel grade is AISI-SAE. The chemical composition of some tool steel is given in Table 8.

Table 8. Chemical composition of some tool steel alloys, reprinted with permission from ref. [48]. Copyright 2018 MDPI Metals.

Alloys	C	Si	Mn	P	S	Ni	Cr	Mo	Cu	V	W
AISI D2	1.56	0.24	0.25	0.025	0.001	0.175	11.31	0.83	0.14	0.25	-
AISI M4	1.33	0.33	0.26	0.03	0.03	0.3	4.25	4.88	0.25	4.12	5.88
HWS	1.08	1.38	0.34	-	-	-	7.80	1.86	-	2.66	1.73

Different machining methods have been practiced in past investigations on the steel material, including milling, drilling, broaching, grinding, planing, and turning, etc. However, some methods induced complications while machining steel materials. For instance, Nagy et al. [49] said that machining (turning operation) of super duplex stainless steel is highly challenging when cutting tool inserts made up of PVD coating are used. The difficulties may be due to continuous and long chips formation, which is often problematic in the context of chip handling. Furthermore, long chips rolled on the part, and accordingly, stimulates surface imperfections. Eventually, surface quality is compromised. A similar problem has engaged with the austenitic steel. Sunil Magadum et al. [50] claimed that high strength, greater toughness, large fatigue, and corrosion resistivity are the prime reasons behind the poor machinability of steel. All the stated factors cause build-up-edge, irregular electrode wear, early tool failure during cryogenic machining of SS304 steel. Ingle et al. [51] proposed that there are certain grades of steel that have a machinability rating of 40%. Those grades belong to the austenitic steel such as 302B, 309, 309S, 330, 384, and 314. The authors demonstrated that a rating of less than 100% refers to the difficulty of machining alloys. As long as a rating is going down, then the difficulty level raises accordingly. The issues attributed to the aforesaid grades of austenitic steel are characterized by high ductility, toughness, prolong work-hardening, and less thermal conductivity. The machinability issues of martensitic steel grades (414, 422, 431, 440A, 440B, and 440C) has also been discussed for the austenitic steel grades.

Steel materials are the most used materials in designing and manufacturing automotive components and in several other industrial sectors. The growth of the manufacturing industry, together with the need for cleaner production, makes the integration of sustainable techniques necessary. To help the manufacturing sector and research community find the best option to meet these goals, we present in this work a detailed review of major sustainable techniques used in the manufacturing of steel materials. Further, the details presented in this review can act as a guide in selecting the best solution to be integrated towards achieving net-zero emissions in their manufacturing process.

Numerous studies have been presented on the various grades of steel. Laleh et al. [52] demonstrated the unexpected behavior of LPBF 316L in the context of erosion and corrosion. They proposed that lower erosion and corrosion resistance of the selected austenitic stainless steel is due to its minimum repassivation through traditional techniques. Thompson [53] contrasted HSLA-80 steel with two alternatives of HSLA, i.e., HSLA-80/100 & HSLA-100, considering yield strength, fracture, and results of Charpy impact test. They found that outcomes of yield strength are enough to study the microstructure, as long as strength and toughness are concerned. Durmusoglu et al. [54] joined the HSLA-80 steel by employing gas metal arc welding based on the high strength of weld metal followed by the heat-affected zone (HAZ) and target metal. Furthermore, the author detected that martensite needle-like sand is looked up in the HAZ, whereas the weld metal has residual austenite. Rajbongshi et al. [55] analyzed the effect of the surface topology of AISI D2 steel at the flank side using texturing and non-texturing coated carbide tools. Two responses (flank wear and surface integrity) were evaluated against three factors, i.e., speed, feed, and depth of cut. The results predicted that texturing tools yield minimal flank wear and less surface roughness (SR). Rath et al. [56] investigated the effect of dry machining on the newly developed grade AISI D3 steel using a mixed ceramic insert (Al_2O_3 + TiCN). Three control parameters (cutting speed, feed rate, and depth of cut) were used to evaluate the influence on cutting forces, SR, electrode wear, and chip thickness. They revealed that feed rate is the most dominant factor, which alters the magnitude of all the defined responses magnificently. Kajendirakumar et al. [57] also conducted a study on AISI D3 steel. They optimized the process parameters via the electric discharge machining (EDM) technique by utilizing grey relational analysis. Material removal rate (MRR) and SR were taken as output responses. They said that optimum parameters were achieved at low pulse on time, high pulse off time, and a large value of current. Guo et al. [58] studied the microstructure and characteristics of heat resistant steel (2.25Cr1Mo0.25V) using the Wire-Arc AM (WAAM) process. They claimed that subtract after processing through the WAAM technique exhibit high quality, excellent metallurgical features, and defect-free surface. Baddoo [59] has proposed a review article about the challenges, applications, and opportunities of stainless steel in the construction sector. The author stated that stainless steel had been proven to be a good alternative in construction sectors because of its good mechanical strength and high ductility. However, these are also fundamental requirements of any architectural applications. Ramana et al. [60] depicted the influence of powder (Nickel) contained EDM on MRR, tool wear rate (TWR) using die steel material against copper electrode. They estimated that nickel in dielectric fluid substantially improves both the said output when pulse-on/off time and current are considered as input variables.

3. Methodology

This module describes a detailed methodology for the sustainable machining of steel that has undergone a comprehensive review procedure. A PRISMA (Preferred Reporting Items for Systematic Reviews and Meta-Analysis) approach was adopted, as displayed in Figure 9, to study the multiple intents of sustainable techniques [61]. The different aspects of sustainable machining techniques for the steel material comprise processing, benefits, drawbacks, and limitations. Afterward, the three pillars of sustainability, such as social-environment -economic, are critically reviewed and highlighted during steel machining to find out the research gaps and future implications. From this perspective, different literature has been studied from various Journals, including Science Direct, Tandfonline, MDPI, Springer, Hindawi, Wiley, Web of Science, etc. The iterative forward and backward strategy was practiced in the identification process to collect the explicit information using the Keywords, Sustainable manufacturing, MQL, Cryogenic machining, Solid lubricants, Vegetable oils, and Steels.

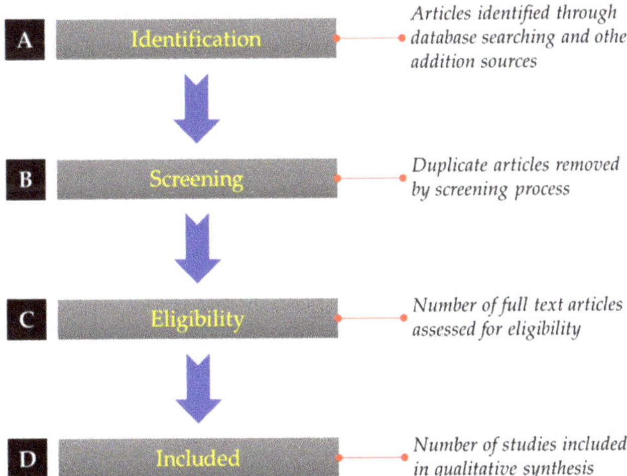

Figure 9. A PRISMA Methodology, reprinted with permission from ref. [61]. Copyright 2009 BMJ Publishing Group Ltd.

For the acquisition of Journal articles, books, reports, and web pages, the string of sustainable machining was utilized in each of the databases' searches. Then a screening operation was performed to find out the future implications in sustainability machining of steel by appraising the existing issues and tentative solutions in a contextual manner. Each of the content of the research articles has been extensively examined while taking the sustainability viewpoint of steel into account. Based on the following established criteria, a large amount of information taken from published literature was systematically organized for assessing future research possibilities:

- Studies belonging to human health, environmental and economic impact on the machining of steel;
- Investigations related to the mechanical and chemical characterization of the machining under special cutting oils/fluids;
- Articles linked with the MQL machining attributes of steel material plus the cryogenic treatment cutting effect on the suitability of steel;
- Content affiliated to the behavior of dry machining of steel.

The references have been cited within a broad time span from 1982 to 2021. Out of the complete list of references, about 43% of articles have been selected from the last six years (2015–2021). The fundamental information based on the sustainable machining techniques for steel was collected and organized, then sub-categorized as per the importance in the respective studies. A comprehensive revision of the research records was developed to examine the sustainability aspect of steel, keeping an eye on its machining attributes. In the present study, challenges to the sustainable machining of steel were also described, and the discussion of this research is summarized in the Conclusion. Finally, future directions and research limitations have been consolidated using the identified knowledge about the sustainable machining aspect of steel.

4. Sustainable Techniques

Different sustainable techniques, i.e., cryogenic cooling, MQL, solid lubricants, and other techniques which are being used in the auto industry that fulfill the overall objectives of this review, are depicted in Figure 10 [8]. The techniques mentioned in Figure 10 have certain benefits, as portrayed in Figure 11.

Figure 10. Sustainable Techniques, reprinted with permission from ref. [8]. Copyright 2015 Elsevier Ltd.

Figure 11. Benefits associated with sustainable machining.

4.1. Cryogenic Cooling

In cryogenic cooling, low temperature (below $-150\ ^\circ$C) materials and medium are used for cooling purposes. Liquid nitrogen, whose boiling point is ($-195.82\ ^\circ$C) and frozen carbon dioxide, whose sublimation point is ($-78.5\ ^\circ$C), are two common media used in this process. Nitrogen is employed to cool down the temperature in the cutting zone because of exothermic conditions. The large amount of heat that is generated during machining causes tool failure and tends to alter the mechanical properties of the specimen. Therefore, to minimize the detrimental effects due to heat and elevated temperature, nitrogen is used, which decreases wear and tear as well as improves the build-up edge [62]. Cryogenic is an eco-friendly technique that shows better results at higher cutting speeds. It is best to control machining temperature along with enhanced tool life [63]. The schematic diagram of the cryogenic cooling setup is represented in Figure 12.

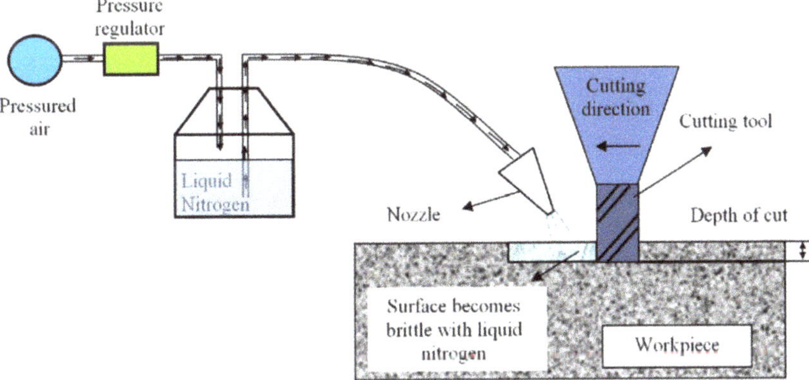

Figure 12. Diagram of cryogenic cooling technique setup, reprinted with permission from ref. [64]. Copyright 2010 Elsevier Ltd.

It was noted during the comparison of dry cutting, MQL, and cryogenic machining that the cryogenic technique is better in increasing tool life with the reduction of cutting temperature. With this product, life improved due to better surface quality [65].

The use of liquid nitrogen in hard turning caused the improvement in cutting speed, and higher productivity and greater tool life were achieved. All of the surface finishes also improved as it causes a decrease in machined surface temperature. Also, it is good for the environment and has no toxic properties [66].

Figure 13 shows the environmental impact of different cooling techniques in the machining of AISI 304. Wet cooling has a tremendous impact on the environment, like ozone depletion, etc. Cryo MQL-CO_2 is best found in all these.

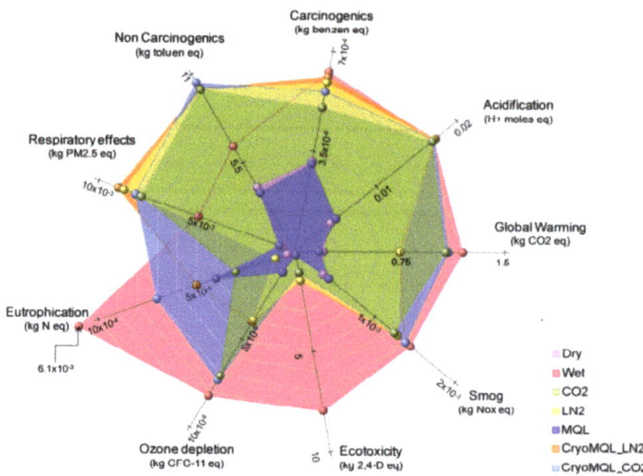

Figure 13. The impact on the environment by different cooling techniques, reprinted with permission from ref. [67]. Copyright 2016 Elsevier Ltd.

It was observed that cryogenic machining, which is suitable for environmental impact, may also have other benefits in terms of lesser tool life and low power consumption as compared to dry cutting. Figure 14 shows a graphical representation of tool life in different cooling techniques, which clearly depicts that tool life is longer in CryoMQL-CO_2 as compared to other techniques [67]. In the milling of hardened AISI D3 steel, the effect of cryogenic cooling (liquid nitrogen) was noted for tool life, surface roughness, and cutting forces. Cutting forces were reduced by 20% to 27%, and surface roughness was decreased up to 16 to 29% due to less cutting temperature at the tool chip interface. Tool life was increased up to 26% to 35% as compared to dry cutting conditions [68].

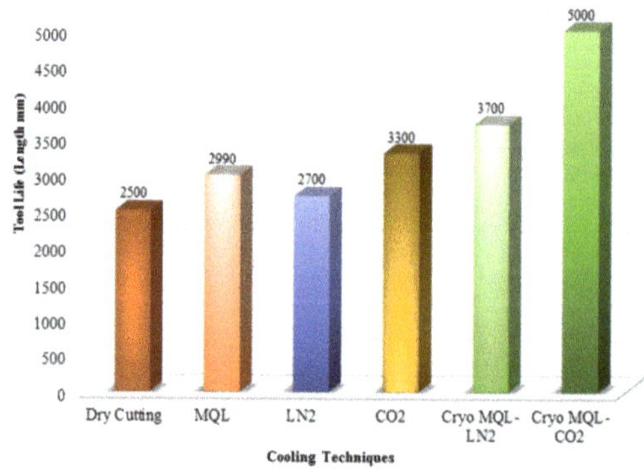

Figure 14. Comparison of Tool life between different cooling techniques, reprinted with permission from ref. [67]. Copyright 2016 Elsevier Ltd.

The cryo-cooling process consists of many input variables: cooling rate, soaking time and temperature, tempering temperature, and its required time [69]. Gill et al. [70] flourished that three of above parameters (cooling rate, soaking time, and soaking temperature) have been extensively increased the tool life upto 98% by compromising the mechanical characteristics of it. Stratton [71] put forward that cooling rate must be low enough to avoid cracking and deforming in the material. Molinari et al. [72] reported about soaking time that must be less than 35 h. It also stated that tool fracture mainly because of insufficient cooling rate, so the optimum value for cooling rate should be near to 30 °C/h. Barron [73] had observed the effect of soaking temperature (189.15 K and 77.15 K) on the wear resistance property of M2 Steel. Besides, many researchers witnessed that increase in hardness, toughness, improving stability and resistance to corrosion is enhanced the tool life [74,75]. Dhar and Kamruzzaman [76] have compared the dry, wet and cryogenic techniques for AISI-4037 Steel. They concluded that cryogenic has been proved as sustainable method followed by dry and wet method in terms of reduction in heat upto 673.15 K. SR is another important criterion to check whether the machining is sustainable or not. Rotella et al. [77] carried out machining under dry, wet and cryogenic condition on Ti-6Al-4V. They noted that cryogenic machining is more prominent than dry and wet machining in term of getting high surface integrity. They also summarized that cryogenic machining has been proved as effective at high feed rates. Kumar and Dhananchezian [64] also demonstrated the similar consideration about SR in cryogenic machining of Ti-6Al-4V. The 35% improvement in SR magnitude has been observed in comparison to dry and wet processing.

In the turning of 17-4 PH SS, different cooling techniques were used like cryogenic, MQL, and wet and dry turning. Different depth of cut (DOC) was used to check the optimum conditions for each technique. It was noted that the cryogenic technique was best in terms of cutting zone temperature decrement, improved surface integrity, and less tool wear. Chip thickness was also less, and also this technique was environmentally friendly. Figure 15 shows the surface morphology obtained after applying different cooling techniques. The surface was smoother in cryogenic as compared to dry machining [78].

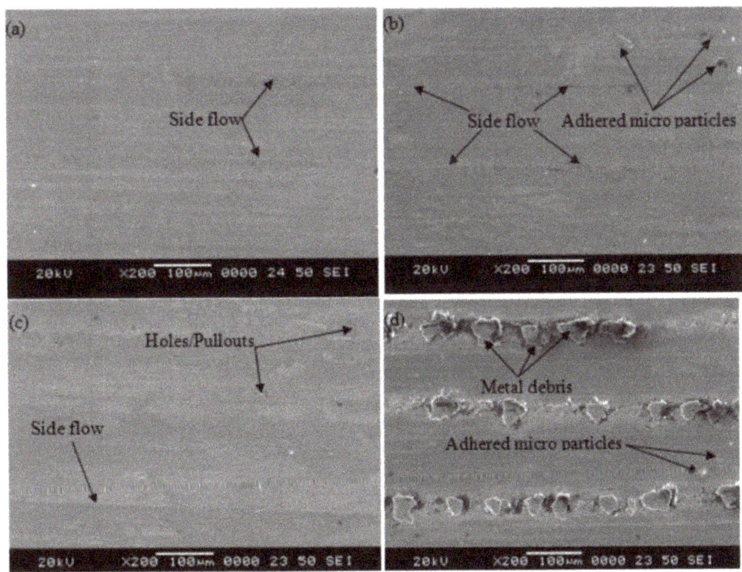

Figure 15. Images of surface morphology under different cooling environments. (**a**) cryogenic, (**b**) wet. (**c**) MQL, (**d**) dry, reprinted with permission from ref. [78]. Copyright 2018 CIRP.

Figure 16 shows the cutting temperature according to the depth of cut increment, which is lower in cryogenic machining than dry, wet, and MQL machining. In machining AISI 52100 Bearing steel, the effect of cryogenic coolant compared to dry cutting on surface integrity was observed. It was noted that residual stresses and white layer formation were less. This layer is non-recommended because it causes fatigue of the product and affects its life. It became evident that it enhances the surface integrity of hard components in many aspects [79]. In hard turning of 17-4 PH stainless steel, the effect of cryogenic machining was found to be positive. It reduced the cutting temperature by using liquid nitrogen as a cooling medium, and it is eco-friendly. This method can be effectively used in any type of hard material [80].

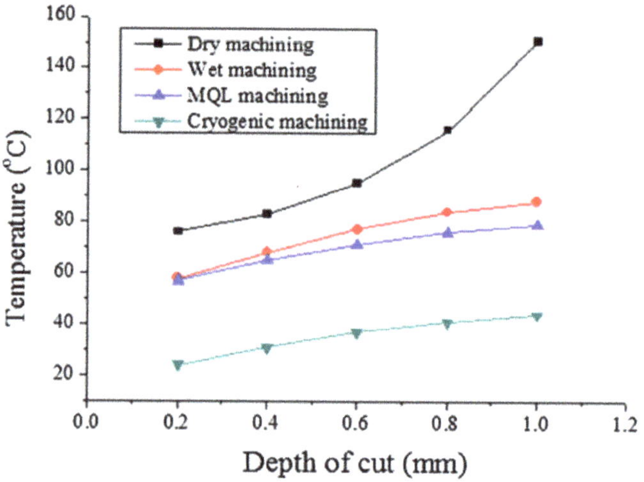

Figure 16. Effect of DOC on cutting temperature under different cooling techniques, reprinted with permission from ref. [78]. Copyright 2018 CIRP.

Nitrogen is most commonly used as it is a safe, noncombustible, noncorrosive gas. The air we breathe has 78% nitrogen gas in it. Liquid nitrogen has the property of easy evaporation, so when it is used in cryogenic machining, it evaporates quickly, and no wastes remain on surfaces, tools, and machines, etc. It contributes to cost savings by avoiding disposal costs [81]. Currently, cryogenic turning is being used to achieve deformation-induced surface hardening. For such purposes, the powerful coolant CO_2-snow is used due to its good wetting behavior [82].

In hard turning of ASP23 steel, CO_2 cryogenic media was used with two types of inserts: one negative and one positive. Tool life was increased in the negative insert up to 19.96%, but in the positive insert, the value of improvement rose to 69.5%. The white layer was also checked. In the negative insert during CO_2 cryogenic machining, it produced a minimal thickness of 2 micrometers. In the positive insert, this layer was not produced. In Figure 17, the microstructure of the material in which machining is done with negative insert using both techniques: dry turning and CO_2 machining [83], is shown. In Figure 18, the microstructure is presented in which machining is done with positive insert using dry turning and CO_2 machining. The white layer is not produced, which indicates good structure.

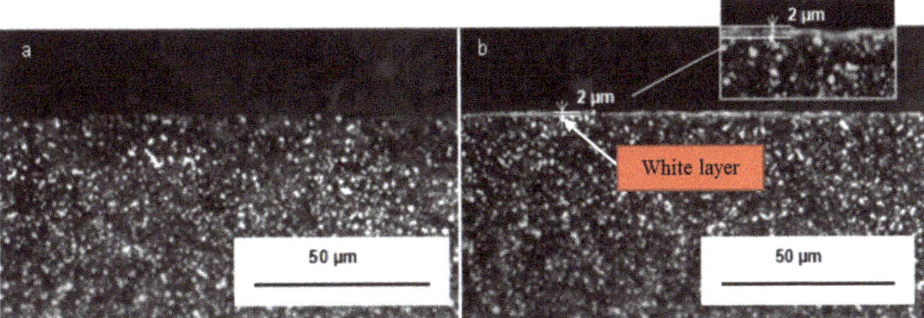

Figure 17. After turning with VNGA160408 (Negative). (**a**) Dry machining; (**b**) CO_2 machining, reprinted with permission from ref. [83]. Copyright 2015 Elsevier Ltd.

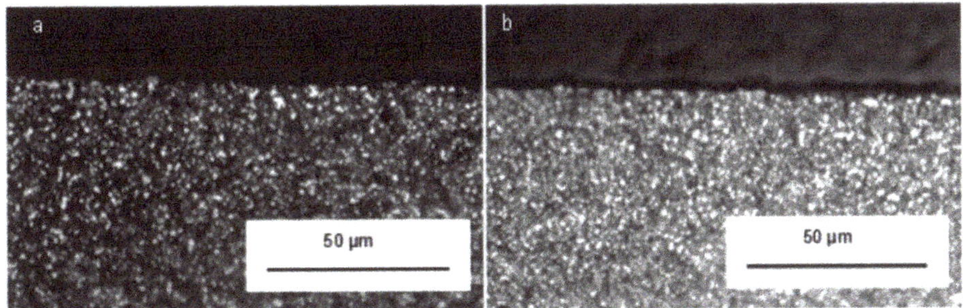

Figure 18. Microstructure after turning with VCGW160408 (positive). (**a**) Dry machining; (**b**) CO_2 machining, reprinted with permission from ref. [83]. Copyright 2015 Elsevier Ltd.

In hard turning of AISI 420 steel, the effect of cryogenic cooling was noted compared to nano fluids. It was noted that tool life at a cutting speed of 75 m/min was increased by approximately 29%. This effect was increased as the speed was increased. Also, the temperature is reduced as compared to nano fluids. Chip morphology was better than nano fluids. It was noted that tool wear was also less [84]. In the machining of AISI 4340, it was found that cutting powers are reduced in cryogenic (LN2) cooling as compared to other water-based cutting fluids. Material removal rate (MRR) was increased with a decrement in surface roughness, which was 0.97 micrometers in cryogenic cooling [85]. The comparison is shown between conventional machining and cryogenic machining. In Figure 19, a conventional machining setup is shown in which cooling and lubricants are required, and waste is generated.

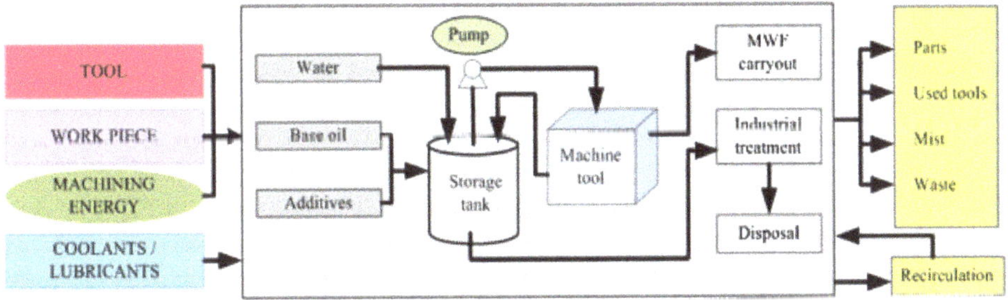

Figure 19. Machining, reprinted from ref. [86]. Copyright 2014 Elsevier Ltd.

In Figure 20, a cryogenic setup is shown. Unlike conventional machining, there is no need for lubricants, and no waste is generated, which is better for the environment and saves on the cost of the product. To observe the cryogenic effect in hard turning of AISI 4340, a setup was done on the shop floor of the CNC turning center. By this process, surface roughness was achieved up to 0.4 micrometers. Tool life of order was achieved 34 min. Cutting forces were reduced by 18%, and power consumption was decreased by about 320 W.

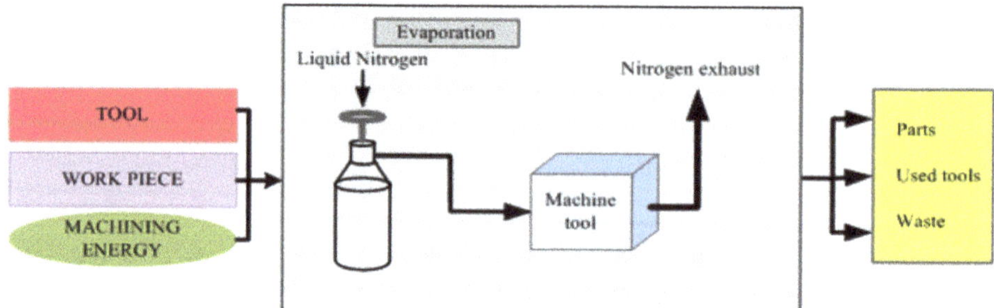

Figure 20. Cryogenic Machining, reprinted from ref. [86]. Copyright 2014 Elsevier Ltd.

In Figure 21, an SEM image was captured to check the flank and rake area of the cutting insert after machining. The insert was chipped off when flood cooling was used while in cryogenic machining abrasion type phenomenon observed at flank face [87].

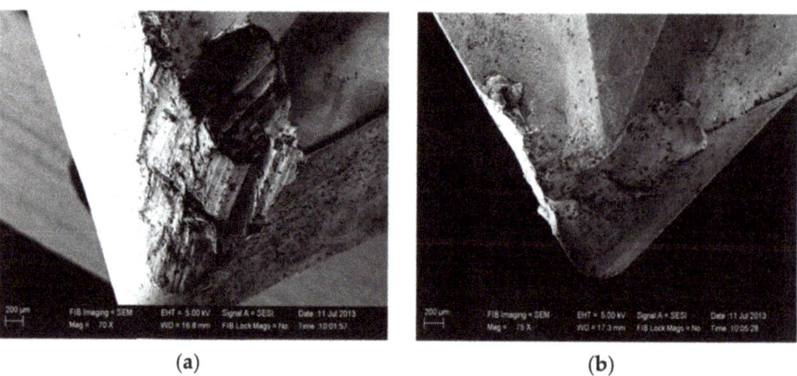

(a)   (b)

Figure 21. Images of (**a**) wet cooling insert (**b**) cryogenic machining insert, reprinted with permission from ref. [87]. Copyright 2017 Elsevier Ltd.

It was noted that power consumption in terms of electricity creates about 99% environmental impacts, which need to be minimized. It was done by choosing the optimal cutting conditions in terms of CO_2 emission, which leads to better environmental impacts [88].

The effects of dry, MQL, flood, and cryogenic machining were observed during turning of 15-5 PH SS, and it was noted that cryogenic machining performed well in terms of tool life which was about 44% of flood and 68% of MQL cooling technique. Surface roughness was better in cryogenic and flood cooling as compared to MQL and dry cutting. In Figure 22, SEM images show the smoothness in the wear pattern of the flank face of the tool in cryogenic as compared to other techniques [89].

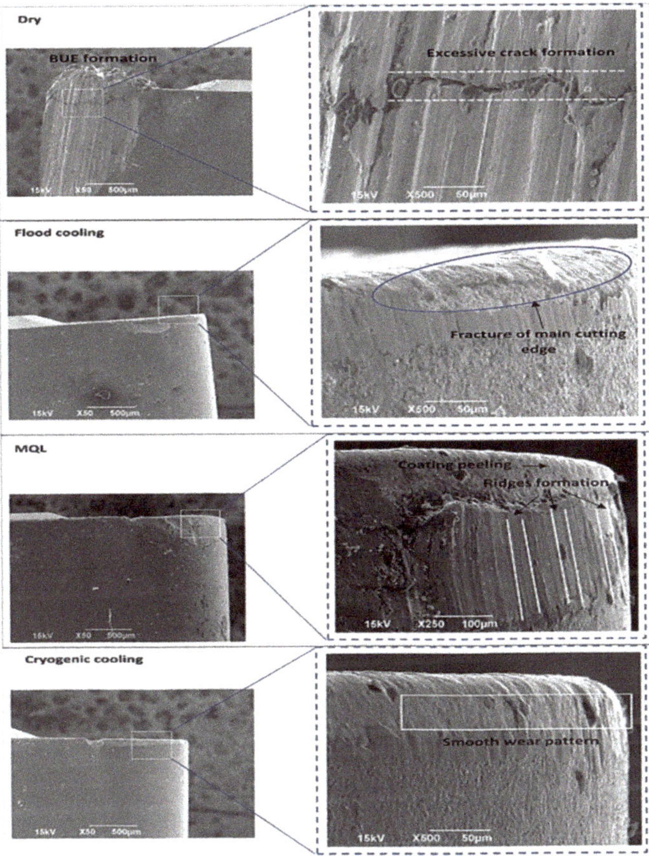

Figure 22. Micrographs of the flank face of cutting inserts under different cutting environments, reprinted with permission from ref. [89]. Copyright 2020 Elsevier Ltd.

The growth of global production and the increase of cutting fluids application has caused intensive research concerning economic and environmental aspects of systems for cooling/lubricating the cutting zone. Thus, recently several cooling/lubrication techniques were developed in order to achieve sustainable manufacturing by reducing or eliminating cutting fluids. Currently, the most widely used cooling/lubricating techniques with a low negative effect on the environment and human operator's health are dry cutting, cryogenic cooling, and minimum quantity lubrication (MQL), etc. [90].

LN2 was found to be good in milling of P20 hardened steel as compared to dry and flood machining. Tool wear was less, about 15%, compared to dry machining, while about 5% compared to wet cooling. Also, it was noted that due to temperature reduction, chip curl was less, which leads to good surface morphology [91]. In the machining of AISI D6 tool steel, a comparison was made between LN2 machining, dry, and wet machining. LN2 was good in surface integrity, and tool life was good, but the production cost for cryogenic setup was more compared to dry machining. This cost varied as the flow rate of LN2 increased [92]. In the milling of AISI D2, the impact of cryogenic cooling was noted compared to a dry and wet cutting environment. Cutting zone temperature was reduced up to 44% by dry and about 36% by wet machining, while cutting forces were reduced by about 10% by dry and about 29% by wet machining [93].

During the study, a comparison was done in the machining of normalized and hardened bearing steel AISI 52100. The response of cryogenic and conventional turning techniques like dry and flood cooling was checked in terms of tool life, surface finish, and productivity. Productivity was higher in cryogenic cooling, and tool life was about 315% in normalized while 15% in the hardened workpiece compared to other techniques. No white layer was formed in cryogenic that are not recommended for machining part. Table 9 shows the MRR for both techniques, and it can be seen that it is about 23% more in cryogenic [94].

Table 9. Optimum productivity comparison through MRR, reprinted with permission from ref. [94]. Copyright 2012 Elsevier B.V.

Material	Conventional Turing (mm^3/min)	Cryogenic Turning (mm^3/min)
Normalized AISI 52100	61,600 (Flood)	75,600
Hardened AISI 52100	5606 (Dry)	6300

During machining of duplex stainless steel, a comparison was conducted between cryogenic cooling and dry cutting. The tool which was used in the machining was coated carbide. Reduction of cutting zone temperature was observed in the case of cryogenic by 53–58%. Required cutting forces were decreased by 30–43%; also, it was noted that surface finish was improved by 18% to 23%. These results were in comparison with dry cutting. Figure 23 shows the cutting temperature for cryogenic and dry machining, which is less in cryogenic machining [95].

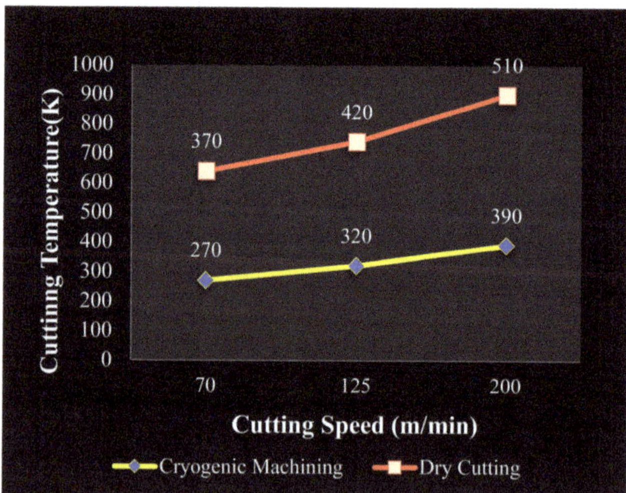

Figure 23. Effect of two cooling techniques on cutting Temperature, reprinted with permission from ref. [95]. Copyright 2018 Elsevier Ltd.

In the hard turning of AISI 52100 bearing steel, the impact of Cryo MQL with two different media (LN_2 and CO_2) was evaluated against conventional and dry turning. Machining was done with two different inserts: one was conventional cubic boron nitride (CBN), and the other was a wiper geometry insert. Less flank wear and crater wear were observed using MQL + CO_2. It was due to the combined effect of minimum quantity lubrication with cryogenic cooling. The surface finish was better, and this technique was found to be eco-friendly. Figure 24 shows the wear pattern of the flank face, which is more in dry cutting, while in Cryo MQL + CO_2, better wear performance was observed, especially by using the wiper geometry insert [96]. Cryogenic machining has major benefits in the sense of environment and product quality, but some limitations like lack of lubrication and chip cleaning. Also, a drawback is the coldness effect for the operator due to high cooling generation during this process [97].

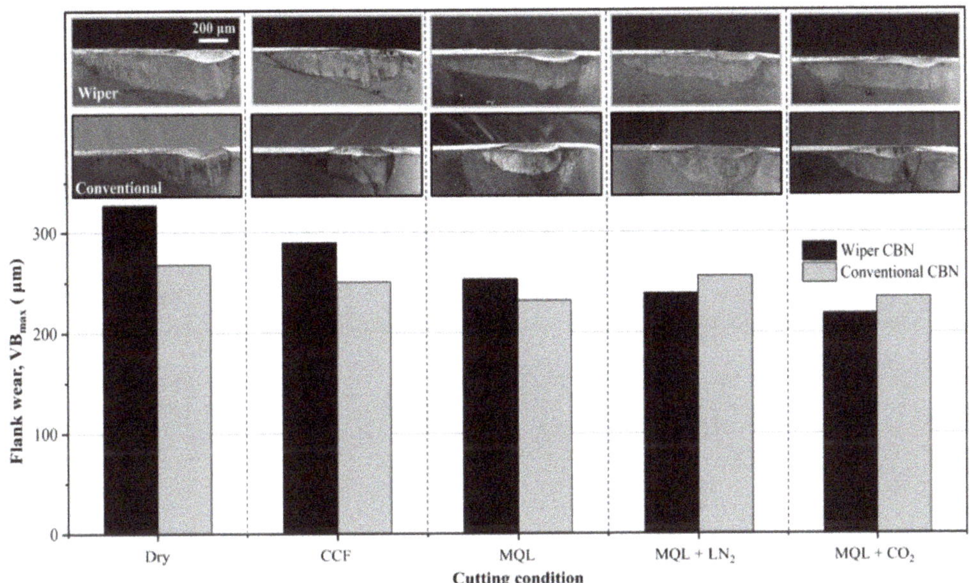

Figure 24. Wear values and SEM images of conventional and wiper CBN inserts under different cutting conditions, reprinted with permission from ref. [96]. Copyright 2020. The Society of Manufacturing Engineers. Published by Elsevier Ltd.

In a nutshell, the cryogenic cooling technique assists us in minimizing chip adherence on a tool. Its benefits include reduction of wear and tear, increase in tool life, improved surface finish, and a decrease in the coefficient of friction. Although some literature has stated that the cryogenic method is beneficial in all aspects, as mentioned earlier, it has certain limitations, as ascertained by Tushar and Suprabhat in their work [98]. The drawbacks are: (1) cryogenic process demands extra control and monitoring over cooling process, (2) a large amount of machining cost belong to process, so any failure during operational hours lead to high maintenance expenses, (3) liquid nitrogen cannot be reused, (4) it is not acceptable for heat treatment processes, (5) and cryogenic fluid, when operated at low temperature, becomes reactive; therefore, it damages the workpiece by directly contacting it.

4.2. Minimum Quantity Lubrication

To avoid using a large amount of cutting fluids, a technique called minimum quantity lubrication, or near dry machining [99], is used in which cutting fluid is supplied at the rate of 100 mL/h. Lawal et al. [100] demonstrated that MQL is a highly competitive approach for a sustainable environment. They explained that minimum usage of cutting fluid in MQL reduces environmental and occupational health hazards. It is well known that metal cutting fluids cause environmental problems. In this case, the amount of cutting oils is greatly reduced which also reduces the environment problem. It was also pointed out that the use of vegetable oils improves the performance of the MQL process, especially in the machining of hard materials, by using water soluble oil in the presence of nano particles. There was no toxic effect generated by using this process which leads to sustainable machining process [101].

Normally, machining is done in dry mode, but the problem which we face is shorter tool life, and sometimes, surface integrity suffers. On the other hand, flooded type coolant application has a higher cost. So, a tradeoff is required in the form of minimal application of lubricants (MQL) which will serve both purposes. In a comparison of wet and MQL, it was found that MQL had better results in tool wear, tool vibration, surface roughness, cutting forces, and cutting temperature during hard turning. About 1.3%, 6.7%, and 8.6% reduction were observed in surface roughness, tool wear, and tool vibration, respectively. Tool wear was less observed in the minimal cutting application as compared to others. Figure 25 shows the surface morphology of three types of cutting techniques in which hard turning with minimal fluid (HTMF) produced the smoothest surface [102].

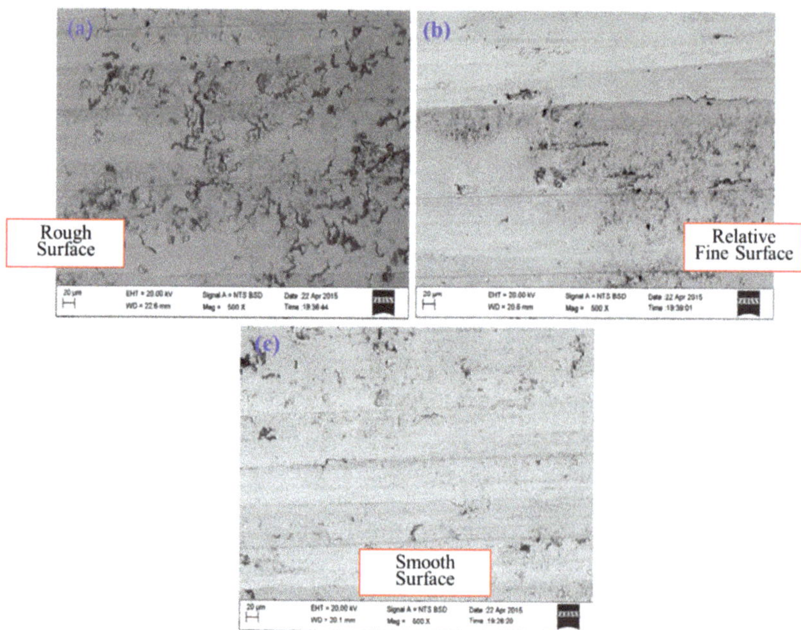

Figure 25. Morphology SEM images; (**a**) By dry turning, (**b**) conventional wet turning, (c) and HTMF application, adapted from ref. [102].

To avoid cutting fluids, dry cutting can be adopted, but this results in shorter tool life at higher cutting parameters, so near dry machining is recommended. Using cutting oil at optimal speeds serves both economic and environmental [103] purposes. In turning of AISI H13 hardened steel under the MQL method, it was noted that the surface finish was improved. This method also has the benefit of being environmentally friendly due to the minimal use of aerosols and cutting oils [104]. An experiment was conducted on heat-treated AISI 4340 steel with a hardness of 52–54 HRC in MQL and dry turning conditions using different bio-cutting oils. It was observed that surface roughness improved as compared to dry turning. At higher cutting speeds, more than 240 m/min, sudden tool failure was observed under MQL conditions [62].

In the machining of AISI 1045, it was found that the cutting temperature and cutting forces were reduced by 10–30% and 5–28%, respectively, in MQL compared to dry machining (see details in Figure 26). This reduction of temperature leads to better tool life and contributes to sustainable manufacturing [105].

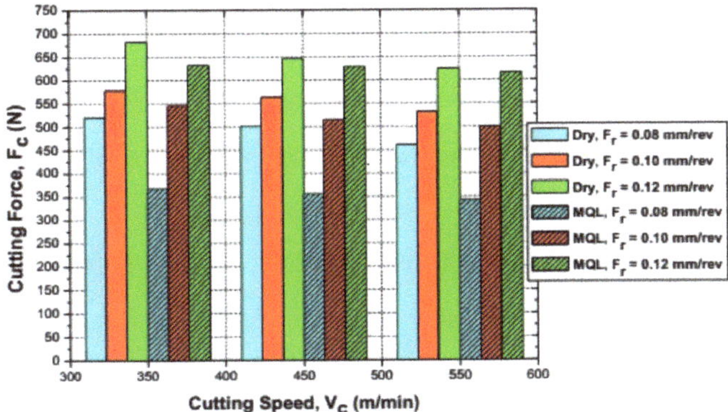

Figure 26. Effect on cutting forces by using Dry and MQL, reprinted with permission from ref. [105]. Copyright 2015 Elsevier B.V.

In turning of AISI D2 steel, the effect of the eco-friendly MQL system was observed compared to dry machining in terms of tool life, tool wear, and surface finish. Reduction of about 100 °C was noted in cutting zone temperature, and surface finish was improved up to 91% compared with dry machining. Tool wear was less and tool life was increased about 267% in chemical vapor deposition (CVD) coated tools [106].

In a study [107] to check the sustainability and effectiveness of different cooling and lubrication techniques, it was found that MQL nanofluids and cryogenic were the best techniques in terms of keeping a balance between the sustainable environment and not compromising machinability efficiency.

MQL technique is an efficient process when we compare it with wet machining. About 15% was saved using this technique. It was noted that it has a better effect in the form of a good surface finish and longer tool life compared to dry machining. When we used biodegradable oils, the effectiveness of this technique increased towards the sustainable point of view. Cutting temperature was reduced by about 50%, which reduced the cutting forces as well [108].

Table 10 shows the different cost estimations of different techniques used in machining. MQL was found better in terms of initial setup and tool cost. Cleaning and disposal costs are comparable with other techniques [109].

Table 10. Qualitative Cost Estimation data for different cooling/lubricating techniques, reprinted with permission from ref. [109]. Copyright 2017 Elsevier Ltd. The symbols used to depict Very low (*), Low (**), Medium (***), High (****), Very High (*****).

Sr. No.	Type	Raw Material Cost	Fluid Consumption	Tool Cost	Equipment Costs	Cleaning Costs	Disposal Cost
1	Cutting fluid	**	*****	**	****	*****	*****
2	Dry Machining	*	*	*****	*	*	***
3	MQL	**	**	**	***	**	**
4	Cryogenic Cooling	***	***	***	*****	*	*
5	Gaseous Cooling	***	***	*****	****	*	*
6	Sustainable Cutting fluid	***	****	**	****	****	***
7	Solid lubricant	****	***	***	***	***	****
8	Nanofluids	*****	****	***	****	****	*****

In an experiment performed on a transmission housing using MQL rather than wet machining, about 15% in savings were achieved. It was noted that due to the reduction of wastewater, it is a sustainable process. One problem is in MQL is the cleaning of chips during machining, especially of hard materials. Figure 27 shows the cost comparison of two types of machining processes, MQL, and wet machining. Operation and maintenance costs are less using MQL. Equipment costs are also less, and the overall cost is about 78% than in wet machining [110].

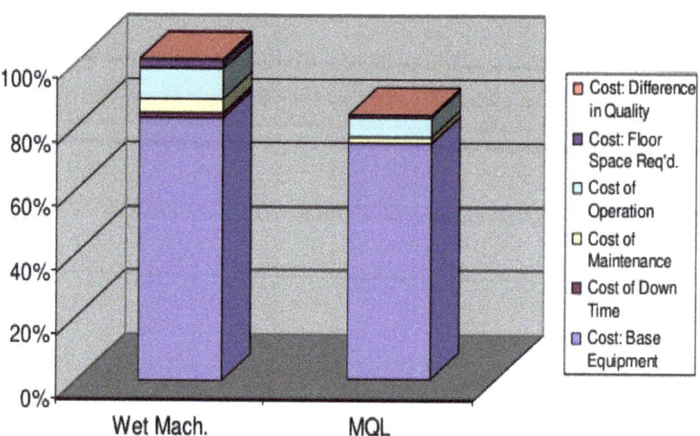

Figure 27. Up to 10 years life cycle of two types of machining techniques, reprinted with permission from ref. [83]. Copyright 2015 Elsevier Ltd.

MQL can be applied in two types of application methods. Different types of MQL systems are shown in Figure 28. In the external application, a compressed air and oil mixture is fed through an external nozzle to the cutting area from a chamber. There are two types of this system. One has an ejector nozzle in which air and oil are supplied separately to the ejector, and mixing is done after the nozzle. In conventional mixing, it is done before the feeding at the cutting zone. In internal application, the mixture is sent through the spindle and tool to the cutting area of the part [111].

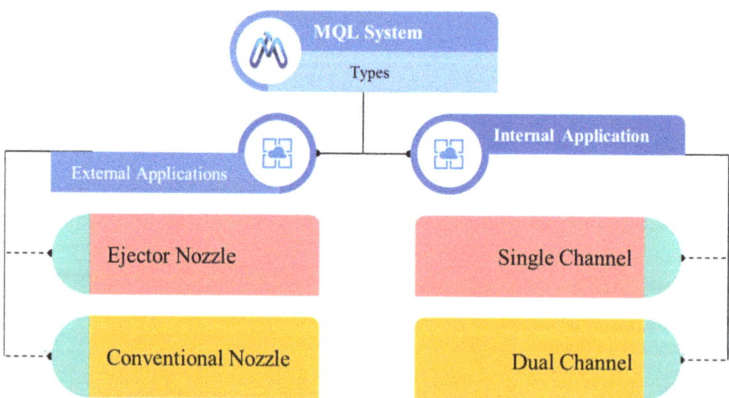

Figure 28. Different types of MQL systems, reprinted from ref. [108]. Copyright 2020 Elsevier Ltd.

Using cutting fluids at a very large scale in machining creates many environmental problems, so it is necessary to adopt a strategy that minimizes the use of these oils but serves the purpose of machining. Also, governments have imposed restrictions on the disposal of such fluids as these cause damage to natural resources. To avoid environmental, regulatory, and health-related problems, MQL is a better technique that serves most of the purposes and also reduces costs [112].

In the context of industry 4.0, sustainable manufacturing is very important. Research was conducted to check the sustainable aspects of MQL on the machining of difficult-to-cut materials, and it concluded that MQL is a tradeoff between flood type and dry cutting. It has more advantages for the environment and is more cost-effective than other techniques. Skin problems created by metalworking fluid (MWF) were reduced by using MQL [113].

In the machining of a mold of tile industry, the impact of sustainable machining was observed. MQL technique was used for such purpose, and it was noted that by using optimal cutting parameters, a major improvement was achieved in the context of a safe cutting environment. There was an approximate 67% reduction in kg CO_2, and about 3357 liters of water were saved. Costs were reduced by about 60% [114].

Four types of cooling techniques (dry, MQL, flood, and solid lubricants with compressed air) were investigated in the machining of AISI 1060 in terms of temperature and surface roughness. In all these, MQL was found to be the best from a sustainable point of view. This technique is responsible for lower manufacturing costs and fewer occupational health and safety problems [115]. Due to sustainability, some properties possessed by MQL are high lubricity, high stability and should be biodegradable. Low consumption of oil is very common in these [116,117].

In addition to the above literature, some studies have also been carried out under vegetable oil mixed MQL conditions. Khan et al. [118] machined low alloy steel of grade AISI 9310 using vegetable oil emulsion. They studied the effect of the MQL process on SR, cutting temperature, chip development, and electrode erosion in different cutting environments. They proposed that surface roughness and tool tip wear were extensively reduced under the MQL environment, and flank wear promisingly improved when machining was treated in vegetable oil. Likewise, some investigations are compiled based on conventional machining in MQL conditions. For instance, Braga et al. [119] compared the results of two scenarios; one in MQL state and the second in the mixture of Al-Si (7% Si) alloy. They conducted a drilling process in both conditions and then measured the potency of each. The results yielded the same SR values in both of the aforementioned drilling conditions, which generally confirms the sustainability of vegetable oil-based machining. In another work, Kishawy et al. [120] used Al alloy (Al-356) to examine the effect of high-speed face milling under dry, wet, and MQL setups at various cutting conditions such as speed of cutting up to 5225 m/min. They illustrated that high cutting forces were noted in the case of dry cutting while fewer cutting forces were observed in wet machining. Whereas in MQL, intermediate cutting forces were marked.

MQL has some disadvantages: (1) removal of chips from the machining zone is not carried out properly, (2) MQL permits corrosion in the work parts or in the chips, (3) there must be great care taken in nozzle adjustment, as it should be more than 1 or 2 inches from the tool, (4) MQL is limited to chip heat removal only, it does not cool down the workpiece and tool, (5) mist creation is also one of the major drawbacks of MQL [121].

4.3. Dry Cutting

Sustainable manufacturing refers to the use of all available natural resources which reduce environmental pollution. Machining is one of these which is very much energy-intensive, and we have to bring improvements to reduce energy consumption. Using cutting fluids is very common for this process but has tremendous impacts on the environment as there are certain disposal costs associated with it. There is no convenient method to dispose of it after proper treating due to which skin diseases are common. To eliminate these problems, dry machining is used in which the need for cutting oils is eliminated. Dixit et al. [90] reported that the use of dry machining significantly minimized air and water pollution. They called dry cutting an eco-friendly process. Eco-friendly refers to such techniques in which detrimental wastes and tiny particles are excluded. Schultheiss et al. [122] urged that dry machining can be performed without any fluid; therefore, for sustainability, it is more appreciated to engage dry processing than the traditional machining approach where usually dielectric fluid is used to compensate for the generated heat while operating on the work part. Dry cutting is more suitable for low-strength materials, and using coated tools is recommended, which can reduce heat generation [123–126]. In the machining of 15-5 PHSS hardened steel, an experiment was done to compare different cooling techniques like dry, wet, and cryogenic cooling. Sustainability assessments were conducted, and it was noted that dry cutting is more optimal towards sustainable assessment indicators, but when we talk about the combined effects of productivity and environment, cryogenic machining is good [127].

In the machining of stainless steel under dry cutting conditions, the effect of feed rate, cutting speed, and depth of cut was observed. The main objective was to reduce the energy cost and machining cost, which is the ultimate objective of sustainable machining. It was concluded that at a higher feed rate and cutting speed with a lower depth of cut, the energy consumption was reduced by 33.46% with a 17.81% reduction in machining cost [128]. Dry cutting is more useful in the context of the environment as there is no need to dispose of the water and metalworking cutting fluids. Cost is also saved, but the problem is high cutting zone temperature and shorter tool life if the cutting parameters are high. Surface quality is better than wet cutting [129].

Although dry cutting is good to retain the sustainability factor while machining, it also has certain disadvantages. For example, Chetan et al. [130] and Rotella et al. [77] outlined that adhesion between the electrode and chips takes place in specific tool and workpiece materials. The said issue led to the reduction of the material erosion rate, and thus the quality of the machined surface is compromised. Moreover, there is the chance of a greater heat-affected zone over the surface, which decreases the strength and durability of the workpiece. Therefore, sustainability plays a prime role in the metal processing areas.

4.4. Cryogenic Treated Tools

Tool life is very important to increase the productivity of any machining industry. It is necessary to use tools that have a long tool life without the use of cutting oils for environmental protection. What are the requirements for sustainable machining? Cutting tools without any treatment wear very rapidly due to heat generation on the cutting zone. Cryogenic treatment is done on cutting tools to compensate for this. Cryogenic treatment is an add-on process that is required to improve tool life. The ultimate goal is to improve the performance, which cuts down the machining cost. It is a subzero heat treatment process that affects the entire cross-section area of cutting tools. Life enhancement of tools is accomplished by microstructure changes of the tool during cryogenic treatment. Two types of treatments are used; one is shallow, and the other is deep cryogenic treatment. Shallow treatment: $-80\,^\circ$C to $-145\,^\circ$C. Deep cryogenic treatment: $-145\,^\circ$C or below. It was noted that the performance of deep cryogenic treatment is more effective than shallow treatment [131].

Cryogenic treatment is an advanced process for increasing tool life, reducing wear resistance, improving the strength and microstructure of the tool [132–134]. With the help of cryogenic treatment on the tool, productivity in terms of tool durability is escalated satisfactorily. Much past literature based on cryogenic treatment has been enlisted. For example, Ramji et al. [135] studied the effect of drilling processes on non-treated and cryogenically treated tools, and a combination of cryogenically treated and heat-treated carbide tipped drills on thrust, SR, and torque of drilled holes in diverse cutting conditions. They concluded that cutting forces, thrust, and torque were reduced when cryogenic treated and a combination of heat-treated carbide insert was used. Gill et al. [136] evaluated the effect of cryogenic treatment of tools on cooling rate. They demonstrated that when cooling and heating are performed at different rates (say 0.5 °C/min and 1 °C/min), then the wear resistance of the tool and micro-cracks on the surface was improved, respectively. Another study conducted by Silva et al. [74] reported the impact of cryogenic treatment of M2 HSS tools and said that 65–34.3% improvement was observed in the reduction of tool fracture while drilling on steel. Cryogenic treatment has numerous benefits in traditional machining, including milling, drilling, and turning. It has been extensively used outside the conventional machining zone for microstructure analysis and wear resistance tests for increasing tool life [136–140].

Furthermore, an experiment was done to check the impact of cryogenic treatment on Tungsten carbide inserts. It was noted that the inserts' life was increased up to 36% with deep cryogenic treatment compared to non-treated inserts. Cutting forces were lesser, and performance was more consistent. Tool life was about 56% in deep cryogenic treatment than by non-treated insert at cutting speed of 110 m/min [92,93].

Cryogenic treatment has many benefits due to its enhancement of cutting tool properties by changing the austenite phase to the marten site phase by heat treatment. By doing this, the hardness and toughness of cutting tools improved [141–143]. In the machining of PHSS, cryogenic treat inserts were used. Due to lesser flank wear, tool life was improved as compared to non-treated tools. These tools resulted in lesser cutting forces, enhanced surface finish with longer tool life [144]. In the machining of 15-5 PHSS cryo- treated inserts were used, and it was noted that cutting forces were reduced, and due to high hardness and strength, the wear of the tool was less as compared to conventional types of tools [145].

Deep cryogenic treatment in hard turning of AISI D2 steel with ceramic cutting tools improves the surface roughness by 32.97%, and improvement in tool life was observed 21.79% [146]. In the turning of C 45 steel, the impact of cryo-treated tungsten carbide inserts was noted compared to non-treated inserts. Treated inserts were found best in machinability and long tool life. Tool tip temperature was decreased due to higher thermal conductivity by cryogenic treatment. This treatment is limited to smooth turning [147]. Contrarily, in cryogenic treatment of cutting tools, machinability increases, and due to good thermal conductivity, cutting temperature decreased. These types of tools are not preferable for interrupted cutting due to breakage problems. This statement indirectly limits the use of cryogenic treatment of tools.

4.5. Solid Lubricants

In the solid lubricant-assisted machining of hardened steel, it was found that this technique is suitable for an ecofriendly environment with less cost of production and helps in the reduction of waste as well as occupational health and safety. It was noted that as demand for sustainable machining is increasing day by day, so solid lubricant assisted machining is emerging as a sustainable alternative machining process [148].

It was noted in a review that the performance of solid lubricants at higher cutting parameters is high, which leads to enhanced productivity. Also, it was observed that there is no negative impact while using these, but the issue is selecting the right type of solid lubricant [136].

In the turning of hardened steel, the effect of solid lubricants was noted, and it was discovered that Molybdenum disulfide is better than graphite. It was observed that solid lubricants are better than dry or wet turning in terms of improved surface finish and from an environmental point of view. The good lubricating effect of these solid lubricants caused the reduction of cutting zone temperature and tool wear. This is becoming a good alternative to dry and wet turning [149].

In the turning of AISI 1040 steel, the impact of solid lubricants (MoS_2) was noted in terms of toxic effect, surface finish, and machinability efficiency. It was concluded that the surface finish was improved by 5% to 30%. The chip thickness ratio was reduced. The friction was reduced in this process, so the material removal rate was high, which leads to high productivity. Also, not using cutting fluids leads to better environmental impact [150].

In machining, the effect of SAE 40 oil with different percentages of graphite and boric acid was studied. It resulted that the boric acid (20%) in SAE 40 oil was performing well. The surface finish was improved, and less tool wear and lesser cutting forces were observed to boric acid lubricious film formation, which lessens the friction forces and cutting temperature. Figure 29 shows the impact of boric acid and graphite on cutting temperature compared to dry and wet cooling. Boric acid and graphite were comparable, and with the passage of cutting time, the performance of Boric acid fond good [151]. Graphite was used in grinding, and it was found that it had numerous effects on the process. The major difference was in the surface finish of the workpiece as in other conventional cutting oils, which were very much improved [152].

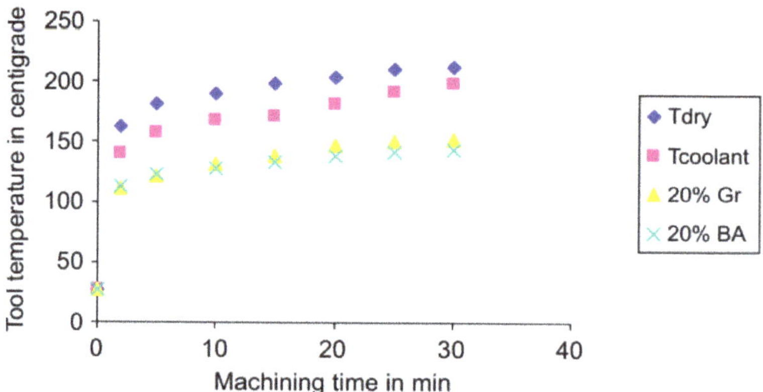

Figure 29. Tool temperature with time in different cutting techniques, reprinted with permission from ref. [151]. Copyright 2008 Elsevier Ltd.

In the machining of AISI 1040, the effect of nanoparticles in cutting fluid was noted, and it was found that thermal conductivity increased, and heat transfer rate increased about 6%, which increased tool life. It was found that about 1% addition of nanoparticles in cutting fluids is optimal [153].

Different types of solid lubricants like MoS_2, CuO, SiO_2, and CaF_2, etc., are useful due to the low strength of bonding between these shears off rapidly. They are also nontoxic and produce a good lubricity effect [154]. In the turning of bearing steel, the effect of Cu nano-fluid with vegetable oil under minimum quantity lubrication was noted. It was found that surface roughness was improved by about 51% due to self-laminated film formation between the tool and workpiece, which reduced the friction. Due to the better thermal conductivity of Cu nanofluid, a reduction in cutting zone temperature was observed, about 21%, compared to vegetable oil machining [155]. Solid lubricant-assisted machining is an ecofriendly technique that contributes to improving the economical aspect of any industry. Improved tool life and higher productivity were observed in the machining of AISI 304 steel. Surface roughness was improved up to 39%, which was improved due to less wear of the tool tip [156]. All lubricants were supplied to the machining area with the help of a special feeding system, as shown in Figure 30 [157]. Solid lubricants have several drawbacks over other sustainable techniques such as (i) high wear rate with a high coefficient of friction, (ii) some lubricants have poor heat dissipation due to low thermal conductivity, like polymers lubricants, (iii) comprises poor self-absorption of heat ability which disturbs the durability of lubricants [158].

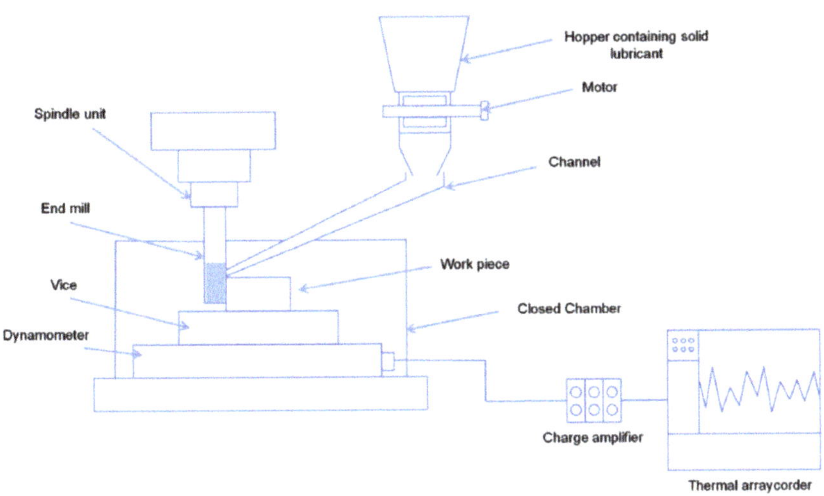

Figure 30. Lubricant feeding system, reprinted with permission from ref. [157]. Copyright 2005 Elsevier Ltd.

4.6. Alternative Cutting Fluids

The use of cutting oils/lubricants causes diseases in employees. To minimize the effect of these oils, some user-friendly oils like vegetable-based oils and other bio-degradable oils can be used. The usage of these oils improved the surface finish and enhanced the tool life. Due to less coefficient of friction than other mineral oils, the machining efficiency improved, and cutting forces were reduced. These are less toxic than other mineral oils, etc. Table 11 shows the positive and negative impacts of vegetable-based oils on energy, cost, and environment. These are efficient in terms of all these parameters like in enhancement of tool life, less requirement of energy due to reduction in forces and eco-friendly. However, there are some negative issues like fume generation and cleaning problems, as chips adhere to oil [159].

Table 11. A critical analysis of environmental aspects in bio-degradable oil aided machining, reprinted with permission from ref. [159]. Copyright 2019 Elsevier Ltd.

Influence	Performance Issues	Energy	Cost	Environment
Positive	Improved Tool Wear profile	Lower specific cutting energy due to reduced force	Improve performance of bio-based oils reduce overall cost	Eco-friendly cooling lubricating agent
	Increased Tool life	Reduced temperature	Cost of recycling	Recycling of oils can be done
	improved surface finish and less friction	Energy consumption during production of bio-oils	Cost of bio-oil coolant sometimes higher than conventional coolant	Fluids from chips need to be separated before chip processing
Negative	Conventional application mode lacks penetration	Mode of oil application determines the additional energy consumption	Cost of additives if used	MQL spray cause inhalation problem
	Adhere with chips-separation of oils from chips are required	-	-	Fumes can cause problems to human

In experimental machining of AISI 304, two types of vegetable-based cutting oils were used. One was sunflower oil, and the other was canola oil. The comparison was made with the semi-synthetic mineral oil. It was noted that the above two oils performed well in the context of being environmentally friendly and in cost reduction. The surface finish was also improved. It was noted that the performance of canola oil with the additive was best in the overall scenario [160].

Soybean and sunflower oils were tested as metalworking fluids, and it was noted that these had a good impact on the environment and were suitable for cutting and forming operations. These are the best alternative to cutting oils [161]. Table 12 shows the different advantages and disadvantages of vegetable oils. They are cost-efficient and less toxic than mineral oils. The low rate of environmental pollution and high biodegradability make these safer for use. One drawback is low thermal stability [162–170].

Table 12. Different Advantages and disadvantages of vegetable oils used as lubricants.

Advantages	Disadvantages
High Biodegradability	Poor Corrosion Protection
Less environmental pollution	Low Thermal Stability
Low Volatility	High Freezing Points
Lesser production cost	Oxidative Stability
High Flash Points	
Low Toxicity	
High Viscosity indices	
Wide Production Possibilities	
Compatibility with other additives	

About 95% usage of vegetable-based oil in Brazil was reported in contrast to petroleum-based oils due to their biodegradable properties and ability to be extracted from natural resources. Soybean oil is the most commonly used in industrial applications [171]. In addition to their positive impact on the environment, it was reported that surface roughness was improved by 31.6% when vegetable oils were used in MQL [100].

In the turning of alloy steel ASIS 9310, vegetable oil performed excellently in terms of the material removal rate, which lead to high productivity. Machining performance was increased by using these oils by 117% in terms of tool life, and thrust forces were also reduced. Also, it has a less negative environmental impact [7]. Vegetable oils have a high boiling point and molecular weight, due to which the chances of vaporization are less than other neat oils. Less smoke is produced, so it is less hazardous for the working environment as well as for people. The product quality is improved by the effect of the lubricating film. Friction and heat generation were lower [118]. In the turning of AISI 4340 stainless steel, three different oils, palm oil, sunflower oil, and coconut oil, were used. Sunflower oil performed well in terms of surface finish and chip compression ratio. One drawback of vegetable oils is the generation of smoke due to a lower flash point [172].

4.7. Air/Gas/Vapor Cooling

The use of cutting fluids causes environmental damage and health-related issues. In order to avoid these issues, a green cutting environment is being created. In this environment, the use of water vapor plays a major role because there is no need for recycling or disposal, and it is non-toxic and environmentally friendly. The setup diagram is below in Figure 31.

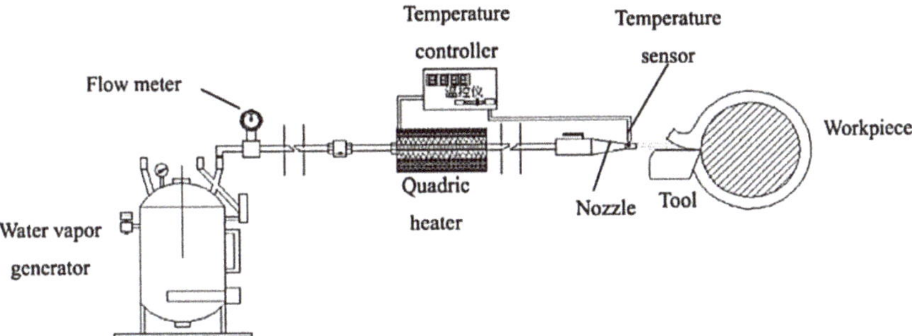

Figure 31. Vapor generator device and feeding system, reprinted with permission from ref. [173]. Copyright 2004 Elsevier Ltd.

Temperature reduction, cutting force reduction and improvement in the surface finish is a positive impact of this technology. Below, Figure 32 shows the temperature comparison between different modes of cutting lubrication techniques in which the use of water vapor is the best technique compared to dry cutting, compressed air, and oil-water emulsion [173].

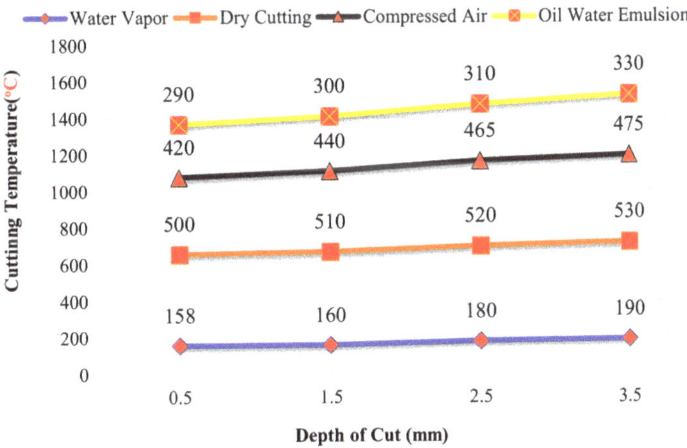

Figure 32. Cutting Temperature Variation on different Depth of cut, reprinted with permission from ref. [173]. Copyright 2004 Elsevier Ltd.

Cold air cooling is best during machining as it mitigates the environmental and health issues caused due to use of coolants. Energy consumption increased by 20%, but coolant cost was reduced by 80%, which is an economically good impact [174]. In the literature, different gases have been exploited as a coolant for sustainable machining of steel, i.e., carbon dioxide (CO_2), argon, water vapor, oxygen, and nitrogen, as depicted by Kim et al. [175] and Yamazaki et al. [176] in their investigations. Contrarily, it comprises some drawbacks; for example, rough turning is not appropriate for the gas/air cooling method. It also acquires an additional setup for the supplement of gas particles to the machining area. As a coolant, compressed air is not suitable for machining a superalloy like Inconel alloy. From an environmental perspective, CO_2 as a gas is not compatible for greenhouse effect.

The past studies warrant the use of air or gas as a coolant to sustain the process environmentally. Liu et al. [177] performed machining on ANSI 1045 steel against a P10 carbide tool under different concentrations of gases and oils. For instance, water vapors (WV), a mixture of CO_2 and O_2, a combination of WV and CO_2, a grouping of WV and O_2, dry machining, and wet machining under oil-H_2O emulsion were prepared for processing. They deduced that cutting forces improved significantly with increased tool life up to 4 to 5 times and 2 to 3 times with CO_2 state and WV, respectively. Junyan et al. [178] collated the two different machining contexts; process under WV and state of dry machining. They evaluated the impact of the K20 carbide insert on the performance of ANSI 304 stainless steel in the aforementioned two machining situations. They extrapolated that better results were obtained with WV followed by dry machining in terms of improvement in tool life, a reduction in cutting forces of 25 to 30%, and modification in surface integrity.

In the machining of AISI 1040, the comparison was carried out between gases applications, wet and dry machining. Three gases were taken, oxygen, nitrogen, and carbon dioxide. It was found that gas application had better result in surface quality, cutting zone temperature, and cutting forces, etc. CO_2 had a better cooling effect than other gases used, and the cutting forces and thrust forces were less using this gas compared to other gases. At lower feed, good surface quality was achieved with gas compared to wet machining, in which surface quality improved at a high feed. Figure 33 shows the relation of mean cutting force with feed in dry, wet, and different gases. CO_2 was best in all other techniques [179].

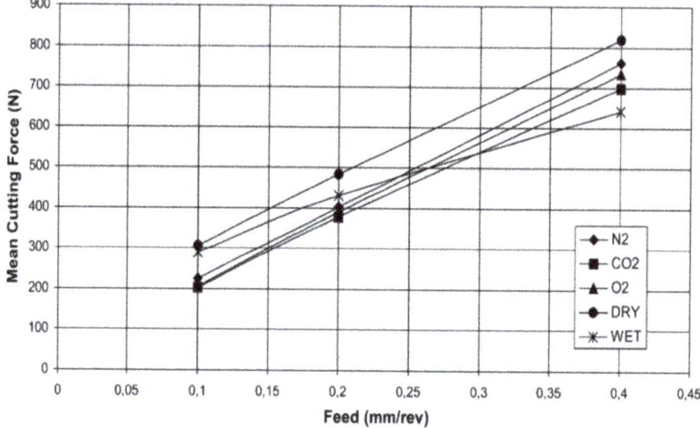

Figure 33. Variation in cutting force by using different machining techniques, reprinted with permission from ref. [179]. Copyright 2004 Elsevier B.V.

4.8. High-Pressure Coolant (HPC)

This is another widely accepted technique in the manufacturing industry. Conventional machining mostly uses one mechanical mechanism, but HPC is usually comprised of three systems: mechanical, thermal, and tribological controls, which makes it impressive and valuable in high-speed machining [180]. High-speed machining is useful in the following conditions, (i) difficult to machine materials, (ii) high speed and feed, (iii) deep-hole drilling, (iv) continuous chips production [181]. HPC generally provides high pressure to the coolant, which allows the deep flow of the fluid between the work-electrode spaces or contact regions of tools and chips, as specified in Figure 34 [182]. The effect of the above phenomenon improves tool life, decreases the consumption of cutting fluid, and maintains the temperature of the work part [183]. It has been found from the literature that HPC not only offers less TWR but also gives superior cooling properties, which results in lessened contact distance as the force of coolant pressure lifts the chip away from tool faces [184]. Ezugwu et al. [185] investigated that boron nitride (BN) and ceramic tools are not fit for high-speed processing of Ti-alloys with HPC supply because it begins the nose rupture and generates discontinuous chips which damage the cutting edges.

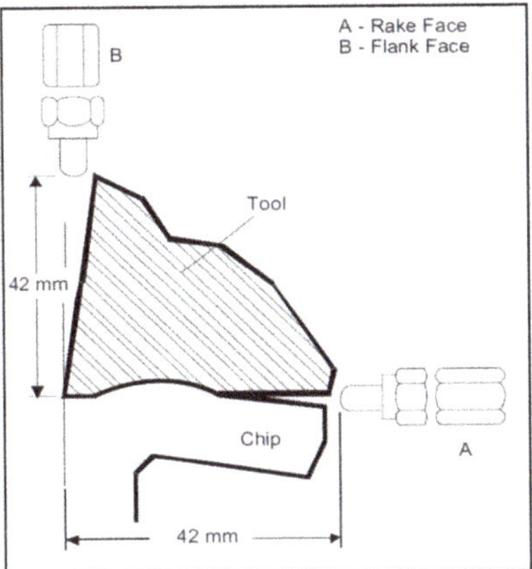

Figure 34. Position of the tool with respect to the workpiece, reprinted with permission from ref. [182]. Copyright 2006 Elsevier Ltd.

It was mentioned earlier that an increase in coolant supply with greater pressure increases the tool life. A study confirmed that tool life is raised by 740% when pressure and coolant speed are set at 203 bar and 50 m/min, respectively. In addition, chip formation is also affected by varying the cutting conditions and coolant pressure at acceptable levels [186]. Kumar et al. [187] have evaluated the effect of HPC on the machining performance of ASSAB 718 steel. The improvement in tool wear, flank wear, chip shape and thickness, and cutting forces are governed by HPC. Dhar et al. [188] assessed the consequence of HPC on chips, tool life, and roundness deviation while drilling of AISI 4340 steel. They investigated the outcomes under HPC drilling with the dry drilling process. The results summarized that small chip thickness, less roundness, and minimal tool wear were observed via HPC drilling. Thus, researchers called it a more beneficial process than drilling under conventional coolant. Naves et al. [189] presented the machinability of AISI 316 austenitic steel by employing HPC. They used 5% and 10% vegetable oil with coolant at different ranges of pressure (100, 150, 200 bar) against carbide tool inserts. They concluded that flank wear was significantly lower when pressure up to 100 bars with a 10% concentration of fluid was applied. The above literature successfully showed that HPC is a highly effective method for achieving long tool life, minimum chip size, and, most importantly, the consumption of fluid is decreased by 50%.

5. Discussion

This section broadly investigates the sustainability aspect of steel governed by different machining techniques, such as expressed in Figure 10. To build a state-of-the-art review, comprehensive literature has been studied regarding the sustainability point of view for the manufacture of steel. For this reason, the key letters and strings were treated to reveal the different studies relevant to the above-mentioned case for the literature survey. The literature survey comprises published work obtained from various sources of Journals, including Science Direct, Emerald, Springer, and other publishers. The last 25–30 years articles, 43% from the past five years, were cited in this study as presented in the graph shown in Figure 35.

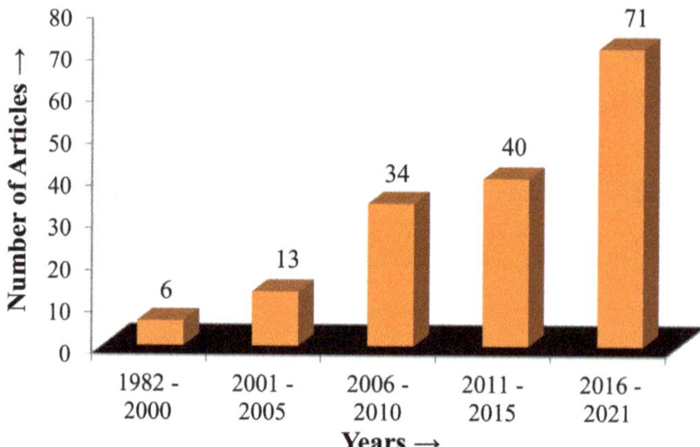

Figure 35. Graphical representation of the number of cited articles in different year bands.

It has been regarded that steel is the most popular material and is preferred widely in diverse application sectors, i.e., automotive, aerospace, and manufacturing industries, etc. However, advancement in manufacturing areas, together with the requirement of cleaner production, demands sustainable machining techniques. Therefore, this review article is contributing towards the sustainable methods needed for the machining of steel. In addition to the above discussion, this study is also presented as a guide for the selection of the best technique that gives net zero-emission in their manufacturing products. The important published work corresponds to sustainable techniques employed for the machining of steel is described below.

Numerous researchers claimed that MQL is the most suitable technique for achieving a sustainable machining environment followed by conventional processes such as drilling, milling, and grinding, etc. [190–193]. Najiha et al. [194] said that MQL is considered a cleaner production process due to its cost-effectiveness and ensuring the safety of both workers and the environment. This statement is also validated by other researchers; for instance, Boswell et al. [111] and Eltaggaz et al. [195] predicted that MQL consumed a minimum quantity of cutting liquid which directly reduces the emission of hazardous fumes, and thus the performance of MQL process was upgraded. Moreover, vegetable oil and non-natural esters are the most commonly used fluids in MQL owing to superb biodegradability and non-toxicity, as stated by Boswell et al. in their study [111]. Dhar and Khan [196] explained that some benefits of the aforementioned fluids over conventional metalworking lubricants are:

i. They provide high MRR with a small cutting time;
ii. The small electrode erosion rate;
iii. They are good absorbers at high pressure;
iv. Minimal vaporization and evaporation lead to being environmentally sustainable.

Synthetic ester, sometimes also known as vegetable oil, is also a promising fluid in order to sustain the machining process due to its high boiling point, excellent flashpoint, and low viscosity, as implied by Dixit et al. [117] in their investigation. Hence, both stated fluids extensively used in the MQL process are the best alternatives in terms of suitability than other conventional liquids.

Cryogenic is another fundamental sustainable approach used for the cutting of steel by manipulating cryogenic fluid at optimum temperature. To keep the cutting temperature low, a coolant like nitrogen gas is used because of its non-corrosive and non-combustible nature. Pereira et al. [67] carried out turning operation on AISI 304 material and inferred that a 50% improvement in tool life with a 30% reduction in cutting speed was commemorated under cryogenic machining conditions. Pusavec et al. [197] evaluated the impact of cryogenic machining on Inconel 718 alloy by taking surface integrity as a responses parameter. They have used various mixtures of cryogenic liquids and found good surface asperities over the machined region with a cryogenic cutting procedure. Another study by Pusavec et al. [198] was written about the effect of various machining applications (such as cryogenic cooling, MQL, dry machining, cryo-lubricant machining) on the same alloy of Ni (Inconel 718) by constructing a response surface methodology (RSM) model. They validated the model with the support of an ANOVA study. The results were explicated that cryogenic cutting fluid or lubrication significantly improved the performance of machining while treating. Machai and Biermann [199] tested the tool life (TL) during machining of Ti-1023 at specific cutting conditions (Cutting rate = 50–150 m/min, Feed = 0.1 mm, machined depth = 0.3 mm, and stroke length = 50–250 m) under wet and CO_2 blend. They summarized that TL is raised approximately by two times with the cryogenic machining as compared to wet operating conditions. Machining under emulsion state also generated large size craters than that of cryogenic condition.

Cutting lubricants improve the design attributes of the machining, but they are strictly avoided by some researchers due to the production of health-hazardous fumes and gases during processing which causes serious diseases for the workers. Researchers have practiced the machining operation without any fluid referred to as "dry machining (DM)". Many manufacturing companies, especially those which produce metallic products, are adopting this technology owing to the freedom from environmental impacts. However, DM has certain drawbacks, which prevent its usage at a level as high as the rest of the techniques which used cutting fluids. Gyanendra and Prabir [129] enlisted some disadvantages of DM, which are mentioned below:

i.   Excessively raise the temperature of the cutting zone, which results in poor TL;
ii. Heat affected zones are enlarged, which tends to decrease the strength of the specimen;
iii. Surface finish compromised at such elevated conditions;
iv. Geometric accuracy and dimensional accuracy of the work part is significantly altered;
v.   The DM has a challenge to machine the difficult-to-cut materials;
vi. In comparison to other sustainable machining techniques, DM has high costs and less productivity.

There are numerous studies based on DM. For example, Servaraj et al. [200] assessed the effect of cutting speed (80–120 m/min) and feed rate (0.04–0.12 mm/rev) on cutting forces at a DOC of 0.5 mm under DM state. They have tested all experiments on the stainless steel (SS) material. They proposed that magnitude of cutting forces is significantly remodeled with a small alteration in feed rate. Fernandez-Abia et al. [201] also performed experimentation under DM environment at a cutting speed of 37–870 m/min, feed rate of 0.2 mm/rev, and 0.1 mm DOC while turning AISI 303 SS. They have taken two responses viz cutting capabilities and chips geometry. They deduced that a cutting rate greater than 450 m/min yielded minimal cutting forces along with satisfactorily decreased chip thickness. Salem and Ahmad [202] optimized the design parameters, such that surface integrity and power consumption, of 316 SS under DM situation using Boron Nitride (BN) electrode. An RSM methodology was developed to validate the machining parameters. The results pointed toward the reduction in power consumption up to 6.8%, with an increase in the surface finish of about 13.9%.

Another prime technique used to ascertain the sustainability perspective of steel can be accomplished with the help of using various cutting fluids. Many pieces of research claimed the amplification of machining performance in terms of output responses like cutting rate, MRR, TWR, and SR when different cutting fluids in the dielectric are exercised. For instance, Kashif et al. [203] comprehensively examined the dispersion of graphene nano-powder mixed in the dielectric medium onto the output parameters (i.e., MRR & TWR) of electric discharge machining (EDM) using three electrodes (copper, brass, and aluminum). They proposed that graphene particles in the dielectric disperse the sparking by raising the plasma channel, which leads to raise the MRR and reduce the TWR. Although there are numerous cutting fluids, vegetable oil-based dielectric combinations are mostly preferred owing to being environmentally friendly, renewable, non-toxic, non-hazardous, and high biodegradability [125]. Vegetable oil also possesses a high freezing point, good corrosion resistance, and is thermally considered stable [204]. Cetin et al. [205] conducted a study to illustrate the impact of canola oil, sunflower oil, and mineral oils on the machining performance of an AISI 304L work part by noticing the SR, cutting forces, and feed forces. They suggested that both vegetable oils provide better surface asperities over the machined surface. Other than this, vegetable oil is outperformed in terms of less feed and cutting forces, followed by mineral oil. Hence, different powder mixed dielectric influences the machining processes and improved the results.

Table 13 shows the comparison in terms of some response parameters like surface finish and MRR, cutting temperature, tool life, and cutting forces. Tool life was best in cryogenic cooling and Cryo treatment of tools. Dry cutting was not found comparable with other techniques in terms of all these parameters, but from an environmental point of view, it was good. The surface finish was best in cryogenic cooling and solid lubricants technique. The material removal rate was observed to be greater in the solid lubricants technique.

Table 13. Comparison between different sustainable Techniques. (× Bad ×× Good ××× Better ×××× Best).

Technique	Tool Life	Surface Finish	MRR	Cutting Temperature	Cutting Forces
Cryogenic Cooling	××××	××××	×××	××××	×××
MQL	×××	×××	××	×××	×××
Dry Cutting	×	××	××	×	××
Solid Lubricants	×××	××××	××××	××××	×××
Air/Vapor/Gas	××	××	×××	×××	×××
Cryogenic treatment	××××	×××	××	×××	××××
Alternative cutting fluids	×××	×××	×××	×××	×××

After careful review of the presented literature, it has been inferred that sustainable techniques are environmental-friendly and manifest as non-hazardous for human beings. If the demand of manufacturers is to achieve a high MRR with a good surface finish, then the use of solid lubricants will be preferred. While, if the need is limited to high tool life with excellent surface asperities on the machined part, then cryogenic cooling will be favored. Similarly, sometimes studies are only restricted to minimum cutting forces, then dry machining would be used irrespective of the other machining attributes. The different challenges possessed by the sustainable techniques while their implementation is presented in the subsequent section. Then the paper is summarized in the conclusion section with the discussion of future directions.

6. Challenges in Sustainable Machining

Different challenges faced for the implementation of sustainable techniques are following.

6.1. Lack of Awareness Regarding These Techniques

Most industries are not aware of these latest techniques, which can contribute considerably in terms of productivity improvement and a green and safe environment for workers and surroundings. They only trust conventional methods and consider these a technical requirement. Top management should be equipped with the latest knowledge of the world's reforms in the field of machining of parts.

6.2. Lack of Management Commitment

For the implementation of these techniques in any manufacturing industry, a change mindset is most important. Management should be willing to provide all needed resources for effectiveness, but in most industries, this commitment is not present.

6.3. High Equipment Cost

No doubt these techniques are very helpful and of paramount importance to the industries for today and tomorrow, but some techniques like cryogenic machining need much attention because right now, their setup cost is high. Indeed, most industries do not know the payback of these, which makes them reluctant to implement.

6.4. Usage of Old Technology

In most industries, old and conventional machines are being used for the manufacturing of parts. There is very little or absence of provision for installation of equipment for working of these techniques. For installation, a huge cost is required for the replacement of the existing system.

6.5. Fear of Losing Business during Adoption of These Techniques

It is a myth in conventional industries that whenever they try a new system, it will lead them towards loss of production. It seems that there is not enough time for the trial of any new technology, which is the wrong concept. Without taking a risk, no improvement will occur.

7. Post-Processing Challenges of Additive Manufactured Steel

Additive manufactured steel parts are widely used in different engineering applications. However, geometric inaccuracy and poor surface integrity disallow the use of steel components after their manufacturing through additive manufacturing techniques. From this perspective, post-processing operations are performed to mitigate the above issues. Those post-processes include drilling, milling, grinding, blasting, and tapping, etc. The challenges inherent with the aforesaid traditional processes have been explained by the National Institute of Standards and Technology (NIST), as far as additive manufactured (AM) steel is concerned.

The residual, tensile (outer surface), and compressive (inner surface) stresses still remain in the steel component after its production from AM, which hinder the use of such parts of steel via milling, drilling, and other conventional processes because it can be generated trust forces, vibrations, and high frequency that may decrease the tool life rapidly, as discussed by Brandon and Eric in their study [206]. They also reported that residual stress also altered the chips' formation owing to high tensile forces at the outer surface when plastic deformation is reached.

The fabrication of steel parts via direct energy deposition (DED) has inhomogeneities because of inappropriate cooling and porosity that tend to affect the surface finish of the desired steel parts when turning operation is carried out at certain input variables. Teo et al. [207] also studied the effect of the post-processing technique (sandblasting) on the AM 316L stainless steel part. The surface quality issue governed by the DED process has been successfully eliminated by the sandblasting process, but the introduction of surface damage and peel-off layer takes place. They reported that the peel-off layer could be removed either by electro-polishing technique or by adding abrasive particles; however, it leads to another problem of corrosion. Therefore, steel parts comprising limitations towards machining after their fabrication through AM process.

8. Conclusions

Sustainable machining techniques became the need of the hour to fulfill environmental regulations and to improve operator's safety. These techniques play an important role in any industry in terms of economic, social, and environmental benefits. To maximize productivity and to make products market competitive, sustainable machining techniques should be adopted. The following main points are concluded for understanding and implementation:

- Cryogenic machining is a very popular technique in terms of its excellent cooling impact, which leads to longer tool life and good surface integrity. It does not require

residual cleaning as in conventional cutting, but some drawbacks are still there like, chips cleaning problem and frostbite hazard associated with the operator's health.
- Dry machining is good in terms of an environmental point of view but leads to low surface integrity of the product and higher tooling cost.
- MQL technique is an intermittent solution between dry and cryogenic in terms of machining cost and product quality. However, problem is chip evacuation is a problem.
- In cryogenic treatment of cutting tools, machinability increases, and due to good thermal conductivity, the cutting temperature decreases. This type of tool is not suitable for interrupted cutting due to breakage problems.
- Solid lubricants are found more effective in terms of surface integrity of parts, and tool life increases as the heat transfer rate increased.
- Vegetable oils are found to be good from an environmental point of view as there is no need to dispose of these as compared to mineral oils. Due to low flash points, smoke is produced during the turning of steel. Sometimes cleaning may be problematic as chips engaged with oil which is difficult to separate.
- Air-gas cooling is a good technique for the environment compared to conventional coolants, but energy costs increased about 20%.

Numerous studies have been conducted on the sustainability of steel in various application conditions. As steel is emerging mostly in automotive sectors, along with other large setups like aerospace, nuclear power plant, marine areas, and biomedical equipment, etc. Organizations are conscious of environmental problems, which ultimately put the life of humans at risk. Therefore, it is much more necessary to examine the sustainability point of discussion on steel material. This review was compiled to investigate the sustainable machining techniques for the steel material. Further, the details presented in this review can act as a guide in selecting the best solution to be integrated towards achieving net zero-emission in their manufacturing products.

9. Future Implications

This review highlights the major limitations like frostbite hazard in cryogenic machining and initial setup cost, which is difficult to afford by any local industry. A comprehensive investigation is required to mitigate the aforesaid issue of cryogenic machining by ensuring a controlled temperature environment. The mathematical modeling of the sustainable cutting mechanisms with respect to the cutting of steel is still an area that needs special attention.

Author Contributions: All authors contributed equally to the data-curation, investigation, writing, and reviewing process. All authors have read and agreed to the published version of the manuscript.

Funding: This study has not received any funding.

Data Availability Statement: The raw/processed data required to reproduce these findings cannot be shared at this time as the data also forms part of an ongoing study.

Conflicts of Interest: There is not any conflict of interest between parts.

Compliance with Ethical Standards: There is not any potential conflict of interest.

Consent to Publish: The authors provide their consent to publish this work.

Nomenclature

This section describes the nomenclature of various abbreviations used in this study:

MQL	Minimum Quantity Lubrication
CEO	Chief executive officer
OEM	Original Equipment Manufacturers

LAS	Low Alloy Steel
HAS	High Alloy Steel
LNG	Liquefied Natural Gas
AISC	American Institute of Steel Construction
HSLA	High Strength Low Alloy
MPa	Mega Pascal
GPa	Giga Pascal
SS	Structural Steel
HRS	Heat Resistant Steels
AISI	American Iron and Steel Institute
PRISMA	Preferred Reporting Items for Systematic Reviews and Meta-Analysis
SR	Surface Roughness
MRR	Material Removal Rate
TWR	Tool Wear Rate
CBN	Cubic Boron Nitride
HTMF	Hard Turning with Minimal Fluid
SEM	Scanning Electron Microscope
HRC	Hardness Rockwell C Scale
CVD	Chemical Vapour Deposition
MWF	Metalworking Fluid
MoS_2	Molybdenum disulfide
DOC	Depth of Cut
DM	Dry Machining
TL	Tool Life
EDM	Electric Discharge Machining
LOF	Lack-of-fusion

References

1. Rao, P.N. Sustainable Manufacturing: Principles, Applications and Directions. *Effic. Manuf.* **2013**, *28*, 16.
2. Siva Rama Krishna, L.; Srikanth, P.J. Evaluation of environmental impact of additive and subtractive manufacturing processes for sustainable manufacturing. *Mater. Today Proc.* **2021**, *45*, 3054–3060. [CrossRef]
3. Rajput, S.P.S.; Datta, S. Sustainable and green manufacturing—A narrative literature review. *Mater. Today Proc.* **2020**, *26*, 2515–2520. [CrossRef]
4. Lee, H.-T.; Song, J.-H.; Min, S.-H.; Lee, H.-S.; Song, K.Y.; Chu, C.N.; Ahn, S.-H. Research Trends in Sustainable Manufacturing: A Review and Future Perspective based on Research Databases. *Int. J. Precis. Eng. Manuf.-Green Technol.* **2019**, *6*, 809–819. [CrossRef]
5. Pathak, P.; Singh, M.P. Sustainable Manufacturing Concepts: A Literature Review. *Int. J. Eng. Technol. Manag. Res.* **2020**, *4*, 1–13. [CrossRef]
6. Abdalla, H.S.; Baines, W.; McIntyre, G.; Slade, C. Development of novel sustainable neat-oil metal working fluids for stainless steel and titanium alloy machining. Part 1. Formulation development. *Int. J. Adv. Manuf. Technol.* **2007**, *34*, 21–33. [CrossRef]
7. Lawal, S.A.; Choudhury, I.A.; Nukman, Y. Application of vegetable oil-based metalworking fluids in machining ferrous metals—A review. *Int. J. Mach. Tools Manuf.* **2012**, *52*, 1–12. [CrossRef]
8. Chetan; Ghosh, S.; Venkateswara Rao, P. Application of sustainable techniques in metal cutting for enhanced machinability: A review. *J. Clean. Prod.* **2015**, *100*, 17–34. [CrossRef]
9. Dahmus, J.B.; Gutowski, T.G. An Environmental Analysis of Machining. In Proceedings of the ASME 2004 International Mechanical Engineering Congress and Exposition, Anaheim, CA, USA, 13–19 November 2004; pp. 643–652.
10. Sartal, A.; Bellas, R.; Mejías, A.M.; García-Collado, A. The sustainable manufacturing concept, evolution and opportunities within Industry 4.0: A literature review. *Adv. Mech. Eng.* **2020**, *12*, 168781402092523. [CrossRef]
11. Gaurav Aggarwal, M.W. The Automotive Industry in the Era of Sustainability. In *European Vehicle Market Statistics*; Diaz, S., Mock, P., Bernard, Y., Bieker, G., Pniewska, I., Ragon, P.-L., Rodriguez, F., Tietge, U., Wappelhorst, S., Eds.; Capgemini Research Institute: Paris, France, 2020.
12. Luo, Z.; Dubey, R.; Gunasekaran, A.; Childe, S.J.; Papadopoulos, T.; Hazen, B.; Roubaud, D. Sustainable production framework for cement manufacturing firms: A behavioural perspective. *Renew. Sustain. Energy Rev.* **2017**, *78*, 495–502. [CrossRef]
13. Cai, W.; Lai, K. Sustainability assessment of mechanical manufacturing systems in the industrial sector. *Renew. Sustain. Energy Rev.* **2021**, *135*, 110169. [CrossRef]
14. Malek, J.; Desai, T.N. A systematic literature review to map literature focus of sustainable manufacturing. *J. Clean. Prod.* **2020**, *256*, 120345. [CrossRef]
15. Hami, N.; Muhamad, M.R.; Ebrahim, Z. The Impact of Sustainable Manufacturing Practices and Innovation Performance on Economic Sustainability. *Procedia CIRP* **2015**, *26*, 190–195. [CrossRef]

16. Murali Krishna, I.V.; Manickam, V. *Environmental Management—Science and Engineering for Industry*; Butterworth-Heinemann: Oxford, UK, 2017; p. 664. ISBN 978-0-12-811990-7.
17. Jayal, A.D.; Badurdeen, F.; Dillon, O.W.; Jawahir, I.S. Sustainable manufacturing: Modeling and optimization challenges at the product, process and system levels. *CIRP J. Manuf. Sci. Technol.* **2010**, *2*, 144–152. [CrossRef]
18. Agan, Y.; Acar, M.F.; Borodin, A. Drivers of environmental processes and their impact on performance: A study of Turkish SMEs. *J. Clean. Prod.* **2013**, *51*, 23–33. [CrossRef]
19. Zein, A. Towards Implementation. In *Transition Towards Energy Efficient Machine Tools*; Zein, A., Ed.; Springer: Berlin, Germany, 2012; pp. 111–130. ISBN 978-3-642-32246-4.
20. Sarkis, J. Manufacturing's role in corporate environmental sustainability—Concerns for the new millennium. *Int. J. Oper. Prod. Manag.* **2001**, *21*, 666–686. [CrossRef]
21. Lu, T.; Gupta, A.; Jayal, A.D.; Badurdeen, F.; Feng, S.C.; Dillon, O.W.; Jawahir, I.S. A Framework of Product and Process Metrics for Sustainable Manufacturing. In *Advances in Sustainable Manufacturing*; Seliger, G., Khraisheh, M.M.K., Jawahir, I.S., Eds.; Springer: Berlin, Germany, 2011; pp. 333–338. ISBN 978-3-642-20182-0.
22. Jawahir, I.S.; Dillon, O.W., Jr. Sustainable Manufacturing Processes: New Challenges for Developing Predictive Models and Optimization Techniques. In Proceedings of the First International Conference on Sustainable Manufacturing, Montreal, QC, Canada, 18–19 October 2007; pp. 1–19.
23. Hegab, H.A.; Darras, B.; Kishawy, H.A. Towards sustainability assessment of machining processes. *J. Clean. Prod.* **2018**, *170*, 694–703. [CrossRef]
24. Waas, T.; Hugé, J.; Block, T.; Wright, T.; Benitez-Capistros, F.; Verbruggen, A. Sustainability Assessment and Indicators: Tools in a Decision-Making Strategy for Sustainable Development. *Sustainability* **2014**, *6*, 5512–5534. [CrossRef]
25. Lovins, A.B.; Lovins, H.; Hawken, P. A Road Map for Natural Capitalism. *Underst. Bus. Environ.* **1999**, *16*, 250–263.
26. Diaz-Elsayed, N.; Jondral, A.; Greinacher, S.; Dornfeld, D.; Lanza, G. Assessment of lean and green strategies by simulation of manufacturing systems in discrete production environments. *CIRP Ann.* **2013**, *62*, 475–478. [CrossRef]
27. Abdul Rashid, S.H.; Evans, S.; Longhurst, P. A comparison of four sustainable manufacturing strategies. *Int. J. Sustain. Eng.* **2008**, *1*, 214–229. [CrossRef]
28. Rusinko, C. Green Manufacturing: An Evaluation of Environmentally Sustainable Manufacturing Practices and Their Impact on Competitive Outcomes. *IEEE Trans. Eng. Manag.* **2007**, *54*, 445–454. [CrossRef]
29. Gimenez, C.; Sierra, V.; Rodon, J. Sustainable operations: Their impact on the triple bottom line. *Int. J. Prod. Econ.* **2012**, *140*, 149–159. [CrossRef]
30. Chen, J.; Yin, X.; Mei, L. Holistic Innovation: An Emerging Innovation Paradigm. *Int. J. Innov. Stud.* **2018**, *2*, 1–13. [CrossRef]
31. Sanchez, J.A.; Pombo, I.; Alberdi, R.; Izquierdo, B.; Ortega, N.; Plaza, S.; Martinez-Toledano, J. Machining evaluation of a hybrid MQL-CO_2 grinding technology. *J. Clean. Prod.* **2010**, *18*, 1840–1849. [CrossRef]
32. Jawahir, I.S. *Sustainable Manufacturing: The Driving Force for Innovative Products, Processes and Systems for Next Generation Manufacturing*; Institute for Sustainable Manufacturing: Lexington, KY, USA, 2010; Volume 19.
33. Zheng, H.; Liu, K. Machinability of Engineering Materials. In *Handbook of Manufacturing Engineering and Technology*; Nee, A., Ed.; Springer: London, UK, 2013; pp. 1–34. ISBN 978-1-4471-4976-7.
34. Conejo, A.N.; Birat, J.-P.; Dutta, A. A review of the current environmental challenges of the steel industry and its value chain. *J. Environ. Manag.* **2020**, *259*, 109782. [CrossRef] [PubMed]
35. Diva Metal Ltd. Company. What Is Stainless Steel? Available online: http://www.divametal.com/EN/PaslanmazCelik.html (accessed on 1 January 2014).
36. Desu, R.K.; Nitin Krishnamurthy, H.; Balu, A.; Gupta, A.K.; Singh, S.K. Mechanical properties of Austenitic Stainless Steel 304L and 316L at elevated temperatures. *J. Mater. Res. Technol.* **2016**, *5*, 13–20. [CrossRef]
37. Sanjana, E.; Vishnu, A.V.; Naidu, G.G.; Kumar, K.S. Machinability of Alloy Steels—A Review. *Int. J. Sci. Res. Dev.* **2016**, *10*, 50–53.
38. Srikanth, M.; Asmatulu, R. Nanotechnology Safety in the Construction and Infrastructure Industries. In *Nanotechnology Safety*; Asmatulu, R., Ed.; Elsevier: Amsterdam, The Netherlands, 2013; pp. 99–113. ISBN 978-0-444-59438-9.
39. De J. Jorge, L.; Cândido, V.S.; da Silva, A.C.R.; da C. Garcia Filho, F.; Pereira, A.C.; da Luz, F.S.; Monteiro, S.N. Mechanical properties and microstructure of SMAW welded and thermically treated HSLA-80 steel. *J. Mater. Res. Technol.* **2018**, *7*, 598–605. [CrossRef]
40. Rasul, M.G. *Thermal Power Plants*; InTech: Rijeka, Croatia, 2011; ISBN 978-953-307-952-3.
41. Song, S.-H.; Zhuang, H.; Wu, J.; Weng, L.-Q.; Yuan, Z.-X.; Xi, T.-H. Dependence of ductile-to-brittle transition temperature on phosphorus grain boundary segregation for a 2.25Cr1Mo steel. *Mater. Sci. Eng. A* **2008**, *486*, 433–438. [CrossRef]
42. Pham, A.H.; Ohba, T.; Morito, S.; Hayashi, T. Effect of Chemical Composition on Average γ/α' Orientation Relationship in Carbon and Low Alloy Steels. *Mater. Today Proc.* **2015**, *2*, S663–S666. [CrossRef]
43. Basirat, M.; Shrestha, T.; Barannyk, L.; Potirniche, G.; Charit, I. A Creep Damage Model for High-Temperature Deformation and Failure of 9Cr-1Mo Steel Weldments. *Metals* **2015**, *5*, 1487–1506. [CrossRef]
44. Klueh, R.L.; Vitek, J.M. Elevated-temperature tensile properties of irradiated 9 Cr-1 MoVNb steel. *J. Nucl. Mater.* **1985**, *132*, 27–31. [CrossRef]
45. Klueh, R.L.; Vitek, J.M. The resistance of 9 Cr-1 MoVNb and 12 Cr-1 MoVW steels to helium embrittlement. *J. Nucl. Mater.* **1983**, *117*, 295–302. [CrossRef]

46. Chuaiphan, W.; Srijaroenpramong, L. Optimization of gas tungsten arc welding parameters for the dissimilar welding between AISI 304 and AISI 201 stainless steels. *Def. Technol.* **2019**, *15*, 170–178. [CrossRef]

47. Nayak, M.; Sehgal, R.; Sharma, R.K. Mechanical Characterization and Machinability Behavior of Annealed AISI D6 Cold Working Steel. *Indian J. Mater. Sci.* **2015**, *2015*, 196178. [CrossRef]

48. Baek, G.Y.; Shin, G.Y.; Lee, K.Y.; Shim, D.S. Mechanical Properties of Tool Steels with High Wear Resistance via Directed Energy Deposition. *Metals* **2019**, *9*, 282. [CrossRef]

49. Nagy, A.I.; Fábián, E.R.; Horváth, R.; Terek, P. Difficulties in the Machining Super Duplex Stainless Steels. *Műszaki Tudományos Közlemények* **2019**, *11*, 141–144. [CrossRef]

50. Magadum, S.; Kumar, A.; Srinivasa, C. Cryogenic Machining of SS304 Steel. In Proceedings of the 5th International & 26th All India Manufacturing Technology, Guwahati, India, 12–14 December 2014; p. 5.

51. Ingle Sushil, V.; Patil Amit, S.; Patel Rohit, J. Machining Challenges in Stainless Steel—A Review. *Int. J. Adv. Res. Ideas Innov. Technol.* **2017**, *3*, 8.

52. Laleh, M.; Hughes, A.E.; Xu, W.; Gibson, I.; Tan, M.Y. Unexpected erosion-corrosion behaviour of 316L stainless steel produced by selective laser melting. *Corros. Sci.* **2019**, *155*, 67–74. [CrossRef]

53. Thompson, S.W. Interrelationships between yield strength, low-temperature impact toughness, and microstructure in low-carbon, copper-precipitation-strengthened, high-strength low-alloy plate steels. *Mater. Sci. Eng. A* **2018**, *711*, 424–433. [CrossRef]

54. Durmusoglu, Ş.; Türker, M.; Tosun, M. Effect of welding parameters on the mechanical properties of GMA-welded HY-80 steels. *Mater. Test.* **2015**, *57*, 866–871. [CrossRef]

55. Rajbongshi, S.K.; Annebushan Singh, M.; Kumar Sarma, D. A comparative study in machining of AISI D2 steel using textured and non-textured coated carbide tool at the flank face. *J. Manuf. Process.* **2018**, *36*, 360–372. [CrossRef]

56. Rath, D.; Panda, S.; Pal, K. Dry turning of AISI D3 steel using a mixed ceramic insert: A study. *J. Mech. Eng. Sci.* **2019**, *233*, 6698–6712. [CrossRef]

57. Kajendirakumar, S.V.; Sathish Kumar, G.; Parameswaran, P.; SureshKumar, B. Optimization of process parameters in electric discharge Machining D3 steel using Grey Relational Analysis. *Mater. Today Proc.* **2021**, *37*, 1137–1139. [CrossRef]

58. Guo, C.; Liu, S.; Hu, R.; Liu, C.; Chen, F. Microstructure and Properties of a 2.25Cr1Mo0.25V Heat-Resistant Steel Produced by Wire Arc Additive Manufacturing. *Adv. Mater. Sci. Eng.* **2020**, *2020*, 8470738. [CrossRef]

59. Baddoo, N.R. Stainless steel in construction: A review of research, applications, challenges and opportunities. *J. Constr. Steel Res.* **2008**, *64*, 1199–1206. [CrossRef]

60. Ramana, P.V.; Kharub, M.; Singh, J.; Singh, J. On material removal and tool wear rate in powder contained electric discharge machining of die steels. *Mater. Today Proc.* **2021**, *38*, 2411–2416. [CrossRef]

61. Moher, D.; Liberati, A.; Tetzlaff, J.; Altman, D.G. For the PRISMA Group Preferred reporting items for systematic reviews and meta-analyses: The PRISMA statement. *BMJ* **2009**, *339*, b2535. [CrossRef] [PubMed]

62. Gunjal, S.U.; Patil, N.G. Experimental Investigations into Turning of Hardened AISI 4340 Steel using Vegetable based Cutting Fluids under Minimum Quantity Lubrication. *Procedia Manuf.* **2018**, *20*, 18–23. [CrossRef]

63. Ravi, S.; Gurusamy, P. Role of cryogenic machining: A sustainable manufacturing process. *Mater. Today Proc.* **2021**, *37*, 316–320. [CrossRef]

64. Dhananchezian, M.; Pradeep Kumar, M. Cryogenic turning of the Ti–6Al–4V alloy with modified cutting tool inserts. *Cryogenics* **2011**, *51*, 34–40. [CrossRef]

65. Deshpande, S.; Deshpande, Y. A Review on Cooling Systems Used In Machining Processes. *Mater. Today Proc.* **2019**, *18*, 5019–5031. [CrossRef]

66. Yildiz, Y.; Nalbant, M. A review of cryogenic cooling in machining processes. *Int. J. Mach. Tools Manuf.* **2008**, *48*, 947–964. [CrossRef]

67. Pereira, O.; Rodríguez, A.; Fernández-Abia, A.I.; Barreiro, J.; López de Lacalle, L.N. Cryogenic and minimum quantity lubrication for an eco-efficiency turning of AISI 304. *J. Clean. Prod.* **2016**, *139*, 440–449. [CrossRef]

68. Ravi, S.; Gurusamy, P. Experimental investigation of cryogenic cooling on cutting force, surface roughness and tool wear in end milling of hardened AISI D3 steel using uncoated tool. *Mater. Today Proc.* **2020**, *33*, 3314–3318. [CrossRef]

69. Darwin, J.D.; Mohan Lal, D.; Nagarajan, G. Optimization of cryogenic treatment to maximize the wear resistance of 18% Cr martensitic stainless steel by Taguchi method. *J. Mater. Process. Technol.* **2008**, *195*, 241–247. [CrossRef]

70. Gill, S.S.; Singh, H.; Singh, R.; Singh, J. Cryoprocessing of cutting tool materials—A review. *Int. J. Adv. Manuf. Technol.* **2010**, *48*, 175–192. [CrossRef]

71. Stratton, P.F. Optimising nano-carbide precipitation in tool steels. *Mater. Sci. Eng. A* **2007**, *449–451*, 809–812. [CrossRef]

72. Molinari, A.; Pellizzari, M.; Gialanella, S.; Straffelini, G.; Stiasny, K.H. Effect of deep cryogenic treatment on the mechanical properties of tool steels. *J. Mater. Process. Technol.* **2001**, *118*, 350–355. [CrossRef]

73. Barron, R.F. Cryogenic treatment of metals to improve wear resistance. *Cryogenics* **1982**, *22*, 409–413. [CrossRef]

74. Da Silva, F.J.; Franco, S.D.; Machado, Á.R.; Ezugwu, E.O.; Souza, A.M. Performance of cryogenically treated HSS tools. *Wear* **2006**, *261*, 674–685. [CrossRef]

75. Meng, F.; Tagashira, K.; Azuma, R.; Sohma, H. Role of Eta-carbide Precipitations in the Wear Resistance Improvements of Fe-12Cr-Mo-V-1.4C Tool Steel by Cryogenic Treatment. *ISIJ Int.* **1994**, *34*, 205–210. [CrossRef]

76. Dhar, N.R.; Kamruzzaman, M. Cutting temperature, tool wear, surface roughness and dimensional deviation in turning AISI-4037 steel under cryogenic condition. *Int. J. Mach. Tools Manuf.* **2007**, *47*, 754–759. [CrossRef]

77. Rotella, G.; Dillon, O.W.; Umbrello, D.; Settineri, L.; Jawahir, I.S. The effects of cooling conditions on surface integrity in machining of Ti6Al4V alloy. *Int. J. Adv. Manuf. Technol.* **2014**, *71*, 47–55. [CrossRef]

78. Sivaiah, P.; Chakradhar, D. Effect of cryogenic coolant on turning performance characteristics during machining of 17-4 PH stainless steel: A comparison with MQL, wet, dry machining. *CIRP J. Manuf. Sci. Technol.* **2018**, *21*, 86–96. [CrossRef]

79. Umbrello, D.; Pu, Z.; Caruso, S.; Outeiro, J.C.; Jayal, A.D.; Dillon, O.W.; Jawahir, I.S. The effects of Cryogenic Cooling on Surface Integrity in Hard Machining. *Procedia Eng.* **2011**, *19*, 371–376. [CrossRef]

80. Sivaiah, P.; Chakradhar, D. Performance improvement of cryogenic turning process during machining of 17-4 PH stainless steel using multi objective optimization techniques. *Measurement* **2019**, *136*, 326–336. [CrossRef]

81. Liew, P.J.; Shaaroni, A.; Sidik, N.A.C.; Yan, J. An overview of current status of cutting fluids and cooling techniques of turning hard steel. *Int. J. Heat Mass Transf.* **2017**, *114*, 380–394. [CrossRef]

82. Mayer, P.; Kirsch, B.; Müller, R.; Becker, S.; Harbou, E.V.; Aurich, J.C. Influence of Cutting Edge Geometry on Deformation Induced Hardening when Cryogenic Turning of Metastable Austenitic Stainless Steel AISI 347. *Procedia CIRP* **2016**, *45*, 59–62. [CrossRef]

83. Pereira, O.; Rodríguez, A.; Fernández-Valdivielso, A.; Barreiro, J.; Fernández-Abia, A.I.; López-de-Lacalle, L.N. Cryogenic Hard Turning of ASP23 Steel Using Carbon Dioxide. *Procedia Eng.* **2015**, *132*, 486–491. [CrossRef]

84. Yıldırım, Ç.V. Investigation of hard turning performance of eco-friendly cooling strategies: Cryogenic cooling and nanofluid based MQL. *Tribol. Int.* **2020**, *144*, 106127. [CrossRef]

85. Khare, S.K.; Sharma, K.; Phull, G.S.; Pandey, V.P.; Agarwal, S. Conventional and cryogenic machining: Comparison from sustainability perspective. *Mater. Today Proc.* **2020**, *27*, 1743–1748. [CrossRef]

86. Fratila, D. Environmentally friendly Manufacturing Processes in the Context of Transition to Sustainable Production. In *Comprehensive Materials Processing*; Hashmi, S., Ed.; Elsevier: Amsterdam, The Netherlands, 2014; Volume 8, pp. 163–175. ISBN 978-0-08-096533-8.

87. Arun Kumar, S.; Yoganath, V.G.; Krishna, P. Machinability of Hardened Alloy Steel using Cryogenic Machining. *Mater. Today Proc.* **2018**, *5*, 8159–8167. [CrossRef]

88. Vila, C.; Abellán-Nebot, J.V.; Siller-Carrillo, H.R. Study of Different Cutting Strategies for Sustainable Machining of Hardened Steels. *Procedia Eng.* **2015**, *132*, 1120–1127. [CrossRef]

89. Khanna, N.; Shah, P. Chetan Comparative analysis of dry, flood, MQL and cryogenic CO_2 techniques during the machining of 15-5-PH SS alloy. *Tribol. Int.* **2020**, *146*, 106196. [CrossRef]

90. Cica, D.; Sredanovic, B.; Tesic, S.; Kramar, D. Predictive modeling of turning operations under different cooling/lubricating conditions for sustainable manufacturing with machine learning techniques. *Appl. Comput. Inform.* **2020**, ahead-of-print. [CrossRef]

91. Ravi, S.; Gurusamy, P. Cryogenic machining of AISI p20 steel under liquid nitrogen cooling. *Mater. Today Proc.* **2021**, *37*, 806–809. [CrossRef]

92. Fernandes, M.E.P.; de Melo, A.C.A.; de Oliveira, A.J.; Chesman, C. Hard turning of AISI D6 tool steel under dry, wet and cryogenic conditions: An economic investigation aimed at achieving a sustainable machining approach. *Clean. Eng. Technol.* **2020**, *1*, 100022. [CrossRef]

93. Ravi, S.; Gurusamy, P. Studies on the effect of cryogenic machining of AISI D2 steel. *Mater. Today Proc.* **2021**, *37*, 2391–2395. [CrossRef]

94. Biček, M.; Dumont, F.; Courbon, C.; Pušavec, F.; Rech, J.; Kopač, J. Cryogenic machining as an alternative turning process of normalized and hardened AISI 52100 bearing steel. *J. Mater. Process. Technol.* **2012**, *212*, 2609–2618. [CrossRef]

95. Dhananchezian, M.; Priyan, M.R.; Rajashekar, G.; Narayanan, S.S. Study The Effect Of Cryogenic Cooling On Machinability Characteristics During Turning Duplex Stainless Steel 2205. *Mater. Today Proc.* **2018**, *5*, 12062–12070. [CrossRef]

96. Çetindağ, H.A.; Çiçek, A.; Uçak, N. The effects of CryoMQL conditions on tool wear and surface integrity in hard turning of AISI 52100 bearing steel. *J. Manuf. Process.* **2020**, *56*, 463–473. [CrossRef]

97. Lu, T.; Kudaravalli, R.; Georgiou, G. Cryogenic Machining through the Spindle and Tool for Improved Machining Process Performance and Sustainability: Pt. I, System Design. *Procedia Manuf.* **2018**, *21*, 266–272. [CrossRef]

98. Gholap, T.A.; Mohod, S.A. Review on Experimental Analysis of Cryogenic Cooling on Various Machining Processes. *Int. J. Adv. Res. Sci. Eng.* **2015**, *4*, 10.

99. Dhar, N.R.; Kamruzzaman, M.; Ahmed, M. Effect of minimum quantity lubrication (MQL) on tool wear and surface roughness in turning AISI-4340 steel. *J. Mater. Process. Technol.* **2006**, *172*, 299–304. [CrossRef]

100. Lawal, S.A.; Choudhury, I.A.; Nukman, Y. A critical assessment of lubrication techniques in machining processes: A case for minimum quantity lubrication using vegetable oil-based lubricant. *J. Clean. Prod.* **2013**, *41*, 210–221. [CrossRef]

101. Padmini, R.; Krishna, P.V.; Mahith, S.; Kumar, S. Influence of Green Nanocutting Fluids on Machining Performance Using Minimum Quantity Lubrication Technique. *Mater. Today Proc.* **2019**, *18*, 1435–1449. [CrossRef]

102. Sam Paul, P.; Varadarajan, A.S.; Robinson Gnanadurai, R. Study on the influence of fluid application parameters on tool vibration and cutting performance during turning of hardened steel. *Eng. Sci. Technol. Int. J.* **2016**, *19*, 241–253. [CrossRef]

103. Sharma, A.K.; Tiwari, A.K.; Dixit, A.R. Effects of Minimum Quantity Lubrication (MQL) in machining processes using conventional and nanofluid based cutting fluids: A comprehensive review. *J. Clean. Prod.* **2016**, *127*, 1–18. [CrossRef]

104. Beatrice, B.A.; Kirubakaran, E.; Thangaiah, P.R.J.; Wins, K.L.D. Surface Roughness Prediction using Artificial Neural Network in Hard Turning of AISI H13 Steel with Minimal Cutting Fluid Application. *Procedia Eng.* **2014**, *97*, 205–211. [CrossRef]

105. Rahim, E.A.; Ibrahim, M.R.; Rahim, A.A.; Aziz, S.; Mohid, Z. Experimental Investigation of Minimum Quantity Lubrication (MQL) as a Sustainable Cooling Technique. *Procedia CIRP* **2015**, *26*, 351–354. [CrossRef]

106. Özbek, O.; Saruhan, H. The effect of vibration and cutting zone temperature on surface roughness and tool wear in eco-friendly MQL turning of AISI D2. *J. Mater. Res. Technol.* **2020**, *9*, 2762–2772. [CrossRef]

107. Hegab, H.; Kishawy, H.A.; Darras, B. Sustainable Cooling and Lubrication Strategies in Machining Processes: A Comparative Study. *Procedia Manuf.* **2019**, *33*, 786–793. [CrossRef]

108. Hamran, N.N.N.; Ghani, J.A.; Ramli, R.; Haron, C.H.C. A review on recent development of minimum quantity lubrication for sustainable machining. *J. Clean. Prod.* **2020**, *268*, 122165. [CrossRef]

109. Benedicto, E.; Carou, D.; Rubio, E.M. Technical, Economic and Environmental Review of the Lubrication/Cooling Systems Used in Machining Processes. *Procedia Eng.* **2017**, *184*, 99–116. [CrossRef]

110. Tai, B.L.; Stephenson, D.A.; Furness, R.J.; Shih, A.J. Minimum Quantity Lubrication (MQL) in Automotive Powertrain Machining. *Procedia CIRP* **2014**, *14*, 523–528. [CrossRef]

111. Boswell, B.; Islam, M.N.; Davies, I.J.; Ginting, Y.R.; Ong, A.K. A review identifying the effectiveness of minimum quantity lubrication (MQL) during conventional machining. *Int. J. Adv. Manuf. Technol.* **2017**, *92*, 321–340. [CrossRef]

112. Singh, G.; Aggarwal, V.; Singh, S. Critical review on ecological, economical and technological aspects of minimum quantity lubrication towards sustainable machining. *J. Clean. Prod.* **2020**, *271*, 122185. [CrossRef]

113. Singh, G.; Aggarwal, V.; Singh, S. An outlook on the sustainable machining aspects of minimum quantity lubrication during processing of difficult to cut materials. *Mater. Today Proc.* **2020**, *33*, 1592–1598. [CrossRef]

114. Abellan-Nebot, J.V.; Rogero, M.O. Sustainable machining of molds for tile industry by minimum quantity lubrication. *J. Clean. Prod.* **2019**, *240*, 118082. [CrossRef]

115. Mia, M.; Gupta, M.K.; Singh, G.; Królczyk, G.; Pimenov, D.Y. An approach to cleaner production for machining hardened steel using different cooling-lubrication conditions. *J. Clean. Prod.* **2018**, *187*, 1069–1081. [CrossRef]

116. Sharma, V.S.; Dogra, M.; Suri, N.M. Cooling techniques for improved productivity in turning. *Int. J. Mach. Tools Manuf.* **2009**, *49*, 435–453. [CrossRef]

117. Dixit, U.S.; Sarma, D.K.; Davim, J.P. Machining with Minimal Cutting Fluid. In *Environmentally Friendly Machining*; Dixit, U.S., Sarma, D.K., Davim, J.P., Eds.; Springer: Boston, MA, USA, 2012; pp. 9–17, ISBN 978-1-4614-2307-2.

118. Khan, M.M.A.; Mithu, M.A.H.; Dhar, N.R. Effects of minimum quantity lubrication on turning AISI 9310 alloy steel using vegetable oil-based cutting fluid. *J. Mater. Process. Technol.* **2009**, *209*, 5573–5583. [CrossRef]

119. Braga, D.U.; Diniz, A.E.; Miranda, G.W.A.; Coppini, N.L. Using a minimum quantity of lubricant (MQL) and a diamond coated tool in the drilling of aluminum–silicon alloys. *J. Mater. Process. Technol.* **2002**, *122*, 127–138. [CrossRef]

120. Kishawy, H.A.; Dumitrescu, M.; Ng, E.-G.; Elbestawi, M.A. Effect of coolant strategy on tool performance, chip morphology and surface quality during high-speed machining of A356 aluminum alloy. *Int. J. Mach. Tools Manuf.* **2005**, *45*, 219–227. [CrossRef]

121. Fluid Application—MQL (Minimum Quantity Lubrication). Available online: http://pdocs.masterchemical.com/mcc/docs/db-docs/tb_us-english/Fluid_Application_MQL.pdf (accessed on 6 May 2012).

122. Schultheiss, F.; Zhou, J.; Gröntoft, E.; Ståhl, J.-E. Sustainable machining through increasing the cutting tool utilization. *J. Clean. Prod.* **2013**, *59*, 298–307. [CrossRef]

123. Khanna, N.; Shah, P.; Maruda, R.W.; Krolczyk, G.M.; Hegab, H. Experimental investigation and sustainability assessment to evaluate environmentally clean machining of 15-5 PH stainless steel. *J. Manuf. Process.* **2020**, *56*, 1027–1038. [CrossRef]

124. Bagaber, S.A.; Yusoff, A.R. Sustainable Optimization of Dry Turning of Stainless Steel based on Energy Consumption and Machining Cost. *Procedia CIRP* **2018**, *77*, 397–400. [CrossRef]

125. Debnath, S.; Reddy, M.M.; Yi, Q.S. Environmental friendly cutting fluids and cooling techniques in machining: A review. *J. Clean. Prod.* **2014**, *83*, 33–47. [CrossRef]

126. Diaz, N.; Choi, S.; Helu, M.; Chen, Y.; Jayanathan, S.; Yasui, Y.; Kong, D.; Pavanaskar, S.; Dornfeld, D. Machine Tool Design and Operation Strategies for Green Manufacturing. In Proceedings of the 4th CIRP International Conference on High Performance Cutting, Gifu, Japan, 24–26 October 2010; p. 6.

127. Pusavec, F.; Krajnik, P.; Kopac, J. Transitioning to sustainable production—Part I: Application on machining technologies. *J. Clean. Prod.* **2010**, *18*, 174–184. [CrossRef]

128. Bart, J.C.J.; Gucciardi, E.; Cavallaro, S. Environmental life-cycle assessment (LCA) of lubricants. In *Biolubricants*; Bart, J.C.J., Gucciardi, E., Cavallaro, S., Eds.; Elsevier: Amsterdam, The Netherlands, 2013; pp. 527–564, ISBN 978-0-85709-263-2.

129. Goindi, G.S.; Sarkar, P. Dry machining: A step towards sustainable machining—Challenges and future directions. *J. Clean. Prod.* **2017**, *165*, 1557–1571. [CrossRef]

130. Chetan; Narasimhulu, A.; Ghosh, S.; Rao, P.V. Study of Tool Wear Mechanisms and Mathematical Modeling of Flank Wear During Machining of Ti Alloy (Ti6Al4V). *J. Inst. Eng. India Ser. C* **2015**, *96*, 279–285. [CrossRef]

131. Kumar, S.; Khedkar, N.K.; Jagtap, B.; Singh, T.P. The Effects of Cryogenic Treatment on Cutting Tools. *IOP Conf. Ser. Mater. Sci. Eng.* **2017**, *225*, 012104. [CrossRef]

132. Podgornik, B.; Leskovšek, V.; Vižintin, J. Influence of Deep-Cryogenic Treatment on Tribological Properties of P/M High-Speed Steel. *Mater. Manuf. Process.* **2009**, *24*, 734–738. [CrossRef]

133. Yong, A.Y.L.; Seah, K.H.W.; Rahman, M. Performance of cryogenically treated tungsten carbide tools in milling operations. *Int. J. Adv. Manuf. Technol.* **2007**, *32*, 638–643. [CrossRef]

134. Chang, Y.-P.; Wang, G.; Horng, J.-H.; Chu, L.-M.; Hwang, Y.-C. Effects of Deep Cryogenic Treatment on Wear Mechanisms and Microthermal Expansion for the Material of Drive Elements. *Adv. Mater. Sci. Eng.* **2013**, *2013*, 1–7. [CrossRef]

135. Ramji, B.R.; Narasimba Murthy, H.N.; Krishna, M. Performance Analysis of Cryogenically Treated Carbide Drills in Drilling White Cast Iron. *Int. J. Appl. Eng. Res.* **2010**, *1*, 553–560.

136. Gill, S.S.; Singh, J.; Singh, R.; Singh, H. Effect of Cryogenic Treatment on AISI M2 High Speed Steel: Metallurgical and Mechanical Characterization. *J. Mater. Eng. Perform.* **2012**, *21*, 1320–1326. [CrossRef]

137. Thakur, D.; Ramamoorthy, B.; Vijayaraghavan, L. Influence of Different Post Treatments on Tungsten Carbide–Cobalt Inserts. *Mater. Lett.* **2008**, *62*, 4403–4406. [CrossRef]

138. Priyadarshini, A. *A Study of the Effect of Cryogenic Treatment on the Performance of High Speed Steel Tools and Carbide Inserts*; National Institute of Technology: Rourkela, India, 2007.

139. Thamizhmanii, S.; Nagib, M.; Sulaiman, H. Performance of deep cryogenically treated and non-treated PVD inserts in milling. *J. Achiev. Mater. Manuf. Eng.* **2011**, *49*, 7.

140. Gill, S.S.; Singh, J.; Singh, R.; Singh, H. Metallurgical principles of cryogenically treated tool steels—A review on the current state of science. *Int. J. Adv. Manuf. Technol.* **2011**, *54*, 59–82. [CrossRef]

141. Palanisamy, D.; Devaraju, A.; Manikandan, N.; Arulkirubakaran, D. Performance evaluation of cryo-treated tungsten carbide inserts in machining PH stainless steel. *Mater. Today Proc.* **2020**, *22*, 487–491. [CrossRef]

142. Palanisamy, D.; Senthil, P. Development of ANFIS model and machinability study on dry turning of cryo-treated PH stainless steel with various inserts. *Mater. Manuf. Process.* **2017**, *32*, 654–669. [CrossRef]

143. Candane, D. Effect of Cryogenic Treatment on Microstructure and Wear Characteristics of AISI M35 HSS. *Int. J. Mater. Sci. Appl.* **2013**, *2*, 56. [CrossRef]

144. Çiçek, A.; Kıvak, T.; Uygur, I.; Ekici, E.; Turgut, Y. Performance of cryogenically treated M35 HSS drills in drilling of austenitic stainless steels. *Int. J. Adv. Manuf. Technol.* **2012**, *60*, 65–73. [CrossRef]

145. Kalsi, N.S.; Sehgal, R.; Sharma, V.S. Effect of tempering after cryogenic treatment of tungsten carbide–cobalt bounded inserts. *Bull. Mater. Sci.* **2014**, *37*, 327–335. [CrossRef]

146. Kara, F.; Karabatak, M.; Ayyıldız, M.; Nas, E. Effect of machinability, microstructure and hardness of deep cryogenic treatment in hard turning of AISI D2 steel with ceramic cutting. *J. Mater. Res. Technol.* **2020**, *9*, 969–983. [CrossRef]

147. Deshpande, R.G. Machining C-45 Steel with Cryogenically Treated and Microwave Irradiated Tungsten Carbide Cutting Tool Inserts. *Int. J. Innov. Res. Sci. Eng. Technol.* **2015**, *4*, 7.

148. Gunda, R.K.; Reddy, N.S.K.; Kishawy, H.A. A Novel Technique to Achieve Sustainable Machining System. *Procedia CIRP* **2016**, *40*, 30–34. [CrossRef]

149. Dilbag, S.; Rao, P.V. Performance improvement of hard turning with solid lubricants. *Int. J. Adv. Manuf. Technol.* **2008**, *38*, 529–535. [CrossRef]

150. Mukhopadhyay, D.; Banerjee, S.; Reddy, N.S.K. Investigation to Study the Applicability of Solid Lubricant in Turning AISI 1040 steel. *J. Manuf. Sci. Eng.* **2007**, *129*, 520–526. [CrossRef]

151. Vamsi Krishna, P.; Nageswara Rao, D. Performance evaluation of solid lubricants in terms of machining parameters in turning. *Int. J. Mach. Tools Manuf.* **2008**, *48*, 1131–1137. [CrossRef]

152. Shaji, S.; Radhakrishnan, V. Analysis of process parameters in surface grinding with graphite as lubricant based on the Taguchi method. *J. Mater. Process. Technol.* **2003**, *141*, 51–59. [CrossRef]

153. Srikant, R.R.; Rao, D.N.; Subrahmanyam, M.S.; Krishna, V.P. Applicability of cutting fluids with nanoparticle inclusion as coolants in machining. *Proc. Inst. Mech. Eng. Part B J. Eng. Tribol.* **2009**, *223*, 221–225. [CrossRef]

154. Venkatesan, K.; Devendiran, S.; Sachin, D.; Swaraj, J. Investigation of machinability characteristics and comparative analysis under different machining conditions for sustainable manufacturing. *Measurement* **2020**, *154*, 107425. [CrossRef]

155. Junankar, A.A.; Yashpal; Purohit, J.K.; Gohane, G.M.; Pachbhai, J.S.; Gupta, P.M.; Sayed, A.R. Performance evaluation of Cu nanofluid in bearing steel MQL based turning operation. *Mater. Today Proc.* **2021**, *44*, 4309–4314. [CrossRef]

156. Agarwal, V.; Agarwal, S. Performance profiling of solid lubricant for eco-friendly sustainable manufacturing. *J. Manuf. Process.* **2021**, *64*, 294–305. [CrossRef]

157. Suresh Kumar Reddy, N.; Venkateswara Rao, P. Experimental investigation to study the effect of solid lubricants on cutting forces and surface quality in end milling. *Int. J. Mach. Tools Manuf.* **2006**, *46*, 189–198. [CrossRef]

158. Sunil, T.; Sandeep, M.; Kumaraswami, R.; Shravan, A. A Critical Review on Solid Lubricants. *Int. J. Mech. Eng. Technol.* **2016**, *7*, 193–199.

159. Krolczyk, G.M.; Maruda, R.W.; Krolczyk, J.B.; Wojciechowski, S.; Mia, M.; Nieslony, P.; Budzik, G. Ecological trends in machining as a key factor in sustainable production—A review. *J. Clean. Prod.* **2019**, *218*, 601–615. [CrossRef]

160. Kuram, E.; Ozcelik, B.; Bayramoglu, M.; Demirbas, E.; Simsek, B.T. Optimization of cutting fluids and cutting parameters during end milling by using D-optimal design of experiments. *J. Clean. Prod.* **2013**, *42*, 159–166. [CrossRef]

161. Shashidhara, Y.M.; Jayaram, S.R. Vegetable oils as a potential cutting fluid—An evolution. *Tribol. Int.* **2010**, *43*, 1073–1081. [CrossRef]
162. Adhvaryu, A.; Erhan, S.Z. Epoxidized soybean oil as a potential source of high-temperature lubricants. *Ind. Crops Prod.* **2002**, *15*, 247–254. [CrossRef]
163. Mercurio, P.; Burns, K.A.; Negri, A. Testing the ecotoxicology of vegetable versus mineral based lubricating oils: 1. Degradation rates using tropical marine microbes. *Environ. Pollut.* **2004**, *129*, 165–173. [CrossRef] [PubMed]
164. Pop, L.; Puşcaş, C.; Bandur, G.; Vlase, G.; Nuţiu, R. Basestock Oils for Lubricants from Mixtures of Corn Oil and Synthetic Diesters. *J. Am. Oil Chem. Soc.* **2008**, *85*, 71–76. [CrossRef]
165. Cermak, S.C.; Isbell, T.A. Synthesis and physical properties of estolide-based functional fluids. *Ind. Crops Prod.* **2003**, *18*, 183–196. [CrossRef]
166. Ohkawa, S.; Konishi, A.; Hatano, H.; Ishihama, K.; Tanaka, K.; Iwamura, M. Oxidation and Corrosion Characteristics of Vegetable-Base Biodegradable Hydraulic Oils. *SAE Trans.* **1995**, *104*, 737–745.
167. Gapinski, R.E.; Joseph, I.E.; Layzell, B.D. *A Vegetable Oil Based Tractor Lubricant*; SAE Technical Paper 941758; SAE International: Warrendale, PA, USA, 1994.
168. Becker, R.; Knorr, A. An evaluation of antioxidants for vegetable oils at elevated temperatures. *Lubr. Sci.* **1996**, *8*, 95–117. [CrossRef]
169. Honary, A.T. An investigation of the use of soybean oil in hydraulic systems. *Bioresour. Technol.* **1996**, *56*, 41–47. [CrossRef]
170. Zeman, A.; Sprengel, A.; Niedermeier, D.; Späth, M. Biodegradable lubricants—Studies on thermo-oxidation of metal-working and hydraulic fluids by differential scanning calorimetry (DSC). *Thermochim. Acta* **1995**, *268*, 9–15. [CrossRef]
171. Said, D.; Belinato, G.; Sarmiento, G.S.; Otero, R.L.S.; Totten, G.E.; Gastón, A.; Canale, L.C.F. Comparison of Oxidation Stability and Quenchant Cooling Curve Performance of Soybean Oil and Palm Oil. *J. Mater. Eng. Perform.* **2013**, *22*, 1929–1936. [CrossRef]
172. Majak, D.; Olugu, E.U.; Lawal, S.A. Analysis of the effect of sustainable lubricants in the turning of AISI 304 stainless steel. *Procedia Manuf.* **2020**, *43*, 495–502. [CrossRef]
173. Liu, J.; Han, R.; Sun, Y. Research on experiments and action mechanism with water vapor as coolant and lubricant in Green cutting. *Int. J. Mach. Tools Manuf.* **2005**, *45*, 687–694. [CrossRef]
174. Ginting, Y.R.; Boswell, B.; Biswas, W.K.; Islam, M.N. Environmental Generation of Cold Air for Machining. *Procedia CIRP* **2016**, *40*, 648–652. [CrossRef]
175. Kim, S.W.; Lee, D.W.; Kang, M.C.; Kim, J.S. Evaluation of machinability by cutting environments in high-speed milling of difficult-to-cut materials. *J. Mater. Process. Technol.* **2001**, *111*, 256–260. [CrossRef]
176. Yamazaki, T.; Miki, K.; Sato, U.; Sato, M. Cooling air cutting of Ti-6Al-4V alloy. *J. Jpn. Inst. Light Met.* **2003**, *53*, 416–420. [CrossRef]
177. Liu, J.; Han, R.; Zhang, L.; Guo, H. Study on lubricating characteristic and tool wear with water vapor as coolant and lubricant in green cutting. *Wear* **2007**, *262*, 442–452. [CrossRef]
178. Junyan, L.; Huanpeng, L.; Rongdi, H.; Yang, W. The study on lubrication action with water vapor as coolant and lubricant in cutting ANSI 304 stainless steel. *Int. J. Mach. Tools Manuf.* **2010**, *50*, 260–269. [CrossRef]
179. Çakır, O.; Kıyak, M.; Altan, E. Comparison of gases applications to wet and dry cuttings in turning. *J. Mater. Process. Technol.* **2004**, *153–154*, 35–41. [CrossRef]
180. M'Saoubi, R.; Axinte, D.; Soo, S.L.; Nobel, C.; Attia, H.; Kappmeyer, G.; Engin, S.; Sim, W.-M. High performance cutting of advanced aerospace alloys and composite materials. *CIRP Ann.* **2015**, *64*, 557–580. [CrossRef]
181. Shete, H.; Sohani, M.; Todkar, A. Machining in high pressure coolant environment—A strategy to improve machining performance: A review. *Int. J. Manuf. Technol. Manag.* **2019**, *33*, 25. [CrossRef]
182. Diniz, A.E.; Micaroni, R. Influence of the direction and flow rate of the cutting fluid on tool life in turning process of AISI 1045 steel. *Int. J. Mach. Tools Manuf.* **2007**, *47*, 247–254. [CrossRef]
183. Wertheim, R.; Rotberg, J.; Ber, A. Influence of High-pressure Flushing through the Rake Face of the Cutting Tool. *CIRP Ann.* **1992**, *41*, 101–106. [CrossRef]
184. Ezugwu, E.O. Key improvements in the machining of difficult-to-cut aerospace superalloys. *Int. J. Mach. Tools Manuf.* **2005**, *45*, 1353–1367. [CrossRef]
185. Ezugwu, E.O.; Bonney, J.; Da Silva, R.B.; Çakir, O. Surface integrity of finished turned Ti–6Al–4V alloy with PCD tools using conventional and high pressure coolant supplies. *Int. J. Mach. Tools Manuf.* **2007**, *47*, 884–891. [CrossRef]
186. Ezugwu, E.O.; Bonney, J. Effect of high-pressure coolant supply when machining nickel-base, Inconel 718, alloy with coated carbide tools. *J. Mater. Process. Technol.* **2004**, *153–154*, 1045–1050. [CrossRef]
187. Senthil Kumar, A.; Rahman, M.; Ng, S.L. Effect of High-Pressure Coolant on Machining Performance. *Int. J. Adv. Manuf. Technol.* **2002**, *20*, 83–91. [CrossRef]
188. Dhar, N.R.; Rashid, M.H.; Siddiqui, A.T. Effect of high pressure coolant on chip, roundness deviation and tool wear in drilling AISI-4340 steel. *ARPN J. Eng. Appl. Sci.* **2006**, *1*, 8.
189. Naves, V.T.G.; Da Silva, M.B.; Da Silva, F.J. Evaluation of the effect of application of cutting fluid at high pressure on tool wear during turning operation of AISI 316 austenitic stainless steel. *Wear* **2013**, *302*, 1201–1208. [CrossRef]
190. Marques, A.; Paipa Suarez, M.; Falco Sales, W.; Rocha Machado, Á. Turning of Inconel 718 with whisker-reinforced ceramic tools applying vegetable-based cutting fluid mixed with solid lubricants by MQL. *J. Mater. Process. Technol.* **2019**, *266*, 530–543. [CrossRef]

191. Osman, K.A.; Ünver, H.Ö.; Şeker, U. Application of minimum quantity lubrication techniques in machining process of titanium alloy for sustainability: A review. *Int. J. Adv. Manuf. Technol.* **2019**, *100*, 2311–2332. [CrossRef]
192. Paturi, U.M.R.; Maddu, Y.R.; Maruri, R.R.; Narala, S.K.R. Measurement and Analysis of Surface Roughness in WS2 Solid Lubricant Assisted Minimum Quantity Lubrication (MQL) Turning of Inconel 718. *Procedia CIRP* **2016**, *40*, 138–143. [CrossRef]
193. Sharif, M.N.; Pervaiz, S.; Deiab, I. Potential of alternative lubrication strategies for metal cutting processes: A review. *Int. J. Adv. Manuf. Technol.* **2017**, *89*, 2447–2479. [CrossRef]
194. Najiha, M.S.; Rahman, M.M.; Yusoff, A.R. Environmental impacts and hazards associated with metal working fluids and recent advances in the sustainable systems: A review. *Renew. Sustain. Energy Rev.* **2016**, *60*, 1008–1031. [CrossRef]
195. Eltaggaz, A.; Hegab, H.; Deiab, I.; Kishawy, H.A. Hybrid nano-fluid-minimum quantity lubrication strategy for machining austempered ductile iron (ADI). *Int. J. Interact. Des. Manuf. IJIDeM* **2018**, *12*, 1273–1281. [CrossRef]
196. Khan, M.M.A.; Dhar, N.R. Performance evaluation of minimum quantity lubrication by vegetable oil in terms of cutting force, cutting zone temperature, tool wear, job dimension and surface finish in turning AISI-1060 steel. *J. Zhejiang Univ.-Sci. A* **2006**, *7*, 1790–1799. [CrossRef]
197. Pusavec, F.; Hamdi, H.; Kopac, J.; Jawahir, I.S. Surface integrity in cryogenic machining of nickel based alloy—Inconel 718. *J. Mater. Process. Technol.* **2011**, *211*, 773–783. [CrossRef]
198. Pusavec, F.; Deshpande, A.; Yang, S.; M'Saoubi, R.; Kopac, J.; Dillon, O.W.; Jawahir, I.S. Sustainable machining of high temperature Nickel alloy—Inconel 718: Part 2—Chip breakability and optimization. *J. Clean. Prod.* **2015**, *87*, 941–952. [CrossRef]
199. Machai, C.; Biermann, D. Machining of β-titanium-alloy Ti–10V–2Fe–3Al under cryogenic conditions: Cooling with carbon dioxide snow. *J. Mater. Process. Technol.* **2011**, *211*, 1175–1183. [CrossRef]
200. Philip Selvaraj, D.; Chandramohan, P.; Mohanraj, M. Optimization of surface roughness, cutting force and tool wear of nitrogen alloyed duplex stainless steel in a dry turning process using Taguchi method. *Measurement* **2014**, *49*, 205–215. [CrossRef]
201. Fernández-Abia, A.I.; Barreiro, J.; Lacalle, L.N.L.; de Martínez, S. Effect of very high cutting speeds on shearing, cutting forces and roughness in dry turning of austenitic stainless steels. *Int. J. Adv. Manuf. Technol.* **2011**, *57*, 61–71. [CrossRef]
202. Bagaber, S.A.; Yusoff, A.R. Multi-responses optimization in dry turning of a stainless steel as a key factor in minimum energy. *Int. J. Adv. Manuf. Technol.* **2018**, *96*, 1109–1122. [CrossRef]
203. Ishfaq, K.; Asad, M.; Anwar, S.; Pruncu, C.I.; Saleh, M.; Ahmad, S. A Comprehensive Analysis of the Effect of Graphene-Based Dielectric for Sustainable Electric Discharge Machining of Ti-6Al-4V. *Materials* **2021**, *14*, 23. [CrossRef]
204. Siniawski, M.T.; Saniei, N.; Adhikari, B.; Doezema, L.A. Influence of fatty acid composition on the tribological performance of two vegetable-based lubricants. *J. Synth. Lubr.* **2007**, *24*, 101–110. [CrossRef]
205. Cetin, M.H.; Ozcelik, B.; Kuram, E.; Demirbas, E. Evaluation of vegetable based cutting fluids with extreme pressure and cutting parameters in turning of AISI 304L by Taguchi method. *J. Clean. Prod.* **2011**, *19*, 2049–2056. [CrossRef]
206. Hung, W. Post-Processing of Additively Manufactured Metal Parts. *J. Mater. Eng. Perform.* **2021**, 1–22. [CrossRef]
207. Teo, A.Q.A.; Yan, L.; Chaudhari, A.; O'Neill, G.K. Post-Processing and Surface Characterization of Additively Manufactured Stainless Steel 316L Lattice: Implications for BioMedical Use. *Materials* **2021**, *14*, 1376. [CrossRef]

MDPI

St. Alban-Anlage 66

4052 Basel

Switzerland

www.mdpi.com

Materials Editorial Office

E-mail: materials@mdpi.com

www.mdpi.com/journal/materials